Oxidative Stress in Applied Basic Research and Clinical Practice

Editor-in-Chief
Donald Armstrong

More information about this series at
http://www.springer.com/series/8145

Note from the Editor-in-Chief

All books in this series illustrate point-of-care testing and critically evaluate the potential of antioxidant supplementation in various medical disorders associated with oxidative stress. Future volumes will be updated as warranted by emerging new technology, or from studies reporting clinical trials.

Donald Armstrong
Editor-in-Chief

Phyllis A. Dennery • Giuseppe Buonocore
Ola Didrik Saugstad
Editors

Perinatal and Prenatal Disorders

Humana Press

Editors
Phyllis A. Dennery
Division of Neonatology
Children's Hospital of Philadelphia
University of Pennsylvania School of Medicine
Philadelphia, PA, USA

Giuseppe Buonocore
Department of Pediatrics
Obstetrics and Reproductive Medicine
University of Siena
Siena, Italy

Ola Didrik Saugstad
Department of Pediatrics
Faculty of Medicine
Rikshospitalit Medical Center
University of Oslo
Oslo, Norway

ISSN 2197-7224 ISSN 2197-7232 (electronic)
ISBN 978-1-4939-1404-3 ISBN 978-1-4939-1405-0 (eBook)
DOI 10.1007/978-1-4939-1405-0
Springer New York Heidelberg Dordrecht London

Library of Congress Control Number: 2014950663

Printed on acid-free paper

Humana Press is a brand of Springer
Springer is part of Springer Science+Business Media (www.springer.com)

Preface

The embryonic and fetal period is wrought with oxidative challenges, which can impact development (Chap. 1). These lead to long-lasting consequences in neonatal life and beyond. This book will evaluate the continuum of perinatal to postnatal oxidative stress and address possible therapeutic strategies to mitigate the deleterious effects.

In the prenatal period, maternal constitutive or acquired exposure to oxidative injury can result in pathology that impacts both the mother and the fetus. The examples of diabetes (Chap. 3) and maternal smoking (Chap. 4) will be discussed in the book. In the instance of Down syndrome (Chap. 7), the adverse oxidant environment resulting from excess manganese superoxide dismutase leads to abnormal organogenesis and development.

Maternal infection (Chap. 2) and poor placentation can result in adverse neonatal effects related to oxidative stress. In terms of poor placentation, a fetus may present with intrauterine growth restriction and have oxidative stress-mediated maladaptation of organs and tissues (Chap. 6). This is well illustrated in adults born to malnourished mothers during the Dutch famine. It is thought that the resultant intrauterine growth restriction and "thrifty" substrate utilization predisposed these infants to develop diabetes, hypertension, and obesity, or syndrome X, as adults at a higher proportion than in normally grown infants. These observations raise the question of whether this predisposition could also occur in preterm infants who have abnormal growth and poor nutrition during a critical period of development. Onset of preterm labor may also be determined by oxidative stress. The evidence for this is discussed in Chap. 5.

Postnatally, there are many consequences of the extrauterine environment on a premature or sick host. A baby born prematurely will have decreased oxidative defenses and, therefore, will be more susceptible to oxidative stress as tissues and organs are still developing. The eyes are particularly vulnerable to changes in hyperoxia, leading to abnormal retinal blood vessel proliferation or retinopathy of prematurity (Chap. 8). The relationship between oxidative stress and the development of this condition was found only after physicians had instituted the use of liberal oxygen delivery to prevent equally adverse consequences related to hypoxemia, illustrating the double-edged sword of oxygen. Necrotizing enterocolitis is another condition that may result from enhanced oxidative stress in the gut of an immature infant (Chap. 9).

Enhanced oxidative burden is not restricted to preterm infants; several conditions lead to adverse oxidative stress in term infants including the prenatal hypoxia. This has caused a reevaluation of how we resuscitate infants in the delivery room (Chap. 11). The dire consequences of oxidative stress can lead to brain injury (Chap. 14), pain (Chap. 16), and seizures (Chap. 20). Oxidative stress can also lead to aggravation of pulmonary vasoreactivity as in persistent pulmonary hypertension of the neonate (PPHN), as discussed in Chap. 10. One of the key discoveries of the last century was nitric oxide (NO). Its use was shown to prevent or ameliorate pulmonary hypertension in the neonate. In fact, use of inhaled NO has improved the survival of infants with pulmonary hypertension and may promote alveolar development. Nevertheless, it is well known that NO in the presence of oxygen can generate the peroxynitrite radical. In animal models, this combination results in pulmonary toxicity, perhaps explaining why inhaled NO has been fraught with challenges in preterm infants since these vulnerable hosts may not have sufficient antioxidant defenses. Although NO is not recommended for use in prematures, surfactant that is used to alleviate respiratory distress syndrome also has significant antioxidative properties as discussed (Chap. 12). Another common concern in neonates is hyperbilirubinemia (Chap. 15). Interestingly, this provides antioxidant defenses to the neonate in the transition to a relatively hyperoxic environment but carries a risk of significant toxicity.

Once we understand the impact of oxidative stress in the uterus, placenta, fetus, and neonates, we need to develop strategies to mitigate these effects. General mechanisms of how oxidative stress affects the neonate and how glutathione is involved are explored in Chap. 13. Also, we gain a better understanding of how endogenous antioxidant defenses develop (Chap. 17). This sets the stage for the development of antioxidant therapies (Chap. 18) and for understanding the importance of nutrition to antioxidant protection (Chap. 19). The development of new therapeutic interventions to prevent and obviate oxidative stress is also examined (Chap. 22).

Overall, the examples in this book illustrate that the perinatal and neonatal periods are an important time of vulnerability to oxidative stress and that the effects of excess oxidative stress can lead to lifelong adverse events (Chap. 21). It is the hope that, with a better understanding of the mechanisms by which reactive oxygen species impact the fetal and neonatal host, therapeutic interventions can be developed to obviate such effects.

We are extremely grateful to John Bretz for his diligence and outstanding commitment to the execution of this book.

Philadelphia, PA, USA — Phyllis A. Dennery
Siena, Italy — GIusseppe Buonocore
Oslo, Norway — Ola Didrik Saugstad

Contents

Contributors

Richard L. Auten Jr Neonatal-Perinatal Research Institute, Duke University Medical Center, Durham, NC, USA

Ali Aycicek Pediatric Hematology/Oncology Clinic, Eskisehir State Hospital, Eskisehir, Turkey

John Anthony Bauer Division of Neonatology, Department of Pediatrics, Kentucky Children's Hospital, University of Kentucky College of Medicine, Lexington, KY, USA

Carlo V. Bellieni Department of Pediatrics, Obstetrics and Reproductive Medicine, University of Siena, Siena, Italy

Elizabeth Bonney Department of Obstetrics, Gynecology, and Reproductive Sciences, University of Vermont College of Medicine, Burlington, VT, USA

Giuseppe Buonocore Departments of Pediatrics, Obstetrics and Reproductive Medicine and Molecular and Developmental Medicine, University of Siena, Siena, Italy

D. Allan Butterfield Department of Chemistry, Sanders-Brown Center on Aging, and Center of Membrane Sciences, University of Kentucky, Lexington, KY, USA

Carlo Dani Department of Critical Care Medicine and Surgery, Section of Neonatology, Careggi University Hospital, Florence, Italy

Jonathan M. Davis Division of Newborn Medicine, Department of Pediatrics, The Floating Hospital for Children at Tufts Medical Center, Tufts University School of Medicine, Boston, MA, USA

Jennifer Dawson Division of Neonatology, The Royal Women's Hospital, Melbourne, Australia

Javier Escobar Neonatal Research Group Laboratory, Health Research Institute La Fe, Valencia, Spain

Ursula Felderhoff-Müser Department of Pediatrics, Division of Neonatology, University Hospital Essen, University Duisburg-Essen, Essen, Germany

James Friel Richardson Center for Functional Foods, University of Manitoba, Winnipeg, MB, Canada

Ursula Geronzi Pediatric Neurology-Immunology and Endocrinology Unit, Department of Pediatrics, Gynecology, Obstetrics, and Reproductive Medicine, University of Siena, Siena, Italy

Peter J. Giannone Division of Neonatology, Department of Pediatrics, Kentucky Children's Hospital, University of Kentucky College of Medicine, Lexington, KY, USA

Salvatore Grosso Pediatric Neurology-Immunology and Endocrinology Unit, Department of Pediatrics, Gynecology, Obstetrics, and Reproductive Medicine, University of Siena, Siena, Italy

Anna-Lena Hård Section of Pediatric Ophthalmology, The Queen Silvia Children's Hospital, The Sahlgrenska Academy at University of Gothenburg, Göteborg, Sweden

Ann Hellström Section of Pediatric Ophthalmology, The Queen Silvia Children's Hospital, The Sahlgrenska Academy at University of Gothenburg, Göteborg, Sweden

Helmut Hummler Division of Neonatology, Universitätsklinikum Ulm, Ulm, Germany

Alicia Jawerbaum Laboratory of Reproduction and Metabolism, CEFYBO-CONICET School of Medicine, University of Buenos Aires, Buenos Aires, Argentina

Alan H. Jobe Cincinnati Children's Hospital Medical Center, University of Cincinnati, Cincinnati, OH, USA

Suhas G. Kallapur Cincinnati Children's Hospital Medical Center, University of Cincinnati, Cincinnati, OH, USA

Boris W. Kramer Department of Pediatrics, School of Oncology and Developmental Biology, School of Mental Health and Neuroscience, Maastricht University Medical Center, Maastricht, The Netherlands

Julia Kuligowski Neonatal Research Group Laboratory, Health Research Institute La Fe, Valencia, Spain

Martha Lappas Obstetrics, Nutrition and Endocrinology Group, Department of Obstetrics and Gynaecology, University of Melbourne, Heidelberg, VIC, Australia

Mercy Perinatal Research Centre, Mercy Hospital for Women, Heidelberg, VIC, Australia

Jennifer W. Lee Tufts University School of Medicine, Boston, MA, USA

Andrea Leiva Cellular and Molecular Physiology Laboratory (CMPL), Division of Obstetrics and Gynecology, Faculty of Medicine, School of Medicine, Pontificia Universidad Católica de Chile, Santiago, Chile

Ramkumar Menon Department of Obstetrics and Gynecology, The University of Texas Medical Branch at Galveston, Galveston, TX, USA

Lutfiya Miller-Pinsler Department of Pharmacology and Toxicology, Faculty of Medicine, University of Toronto, Toronto, ON, Canada

Craig A. Nankervis Department of Pediatrics, The Ohio State University College of Medicine and Nationwide Children's Hospital and Research Institute, Columbus, OH, USA

Fabián Pardo Cellular and Molecular Physiology Laboratory (CMPL), Division of Obstetrics and Gynecology, Faculty of Medicine, School of Medicine, Pontificia Universidad Católica de Chile, Santiago, Chile

Marzia Perluigi Department of Biochemical Sciences, Sapienza University of Rome, Rome, Italy

Serafina Perrone Department of Molecular and Developmental Medicine, University of Siena, Siena, Italy

Chiara Poggi Department of Critical Care Medicine and Surgery, Section of Neonatology, Careggi University Hospital, Florence, Italy

Denise Rook Department of Pediatrics, Erasmus Medical Centre - Sophia Children's Hospital, Rotterdam, The Netherlands

Antonino Santacroce Department of Molecular and Developmental Medicine, University of Siena, Siena, Italy

Brandon Schanbacher Division of Neonatology, Department of Pediatrics, Kentucky Children's Hospital, University of Kentucky College of Medicine, Lexington, KY, USA

Aaron M. Shapiro Division of Biomolecular Sciences, Faculty of Pharmacy, University of Toronto, Toronto, ON, Canada

Rebecca A. Simmons Department of Pediatrics Children's Hospital Philadelphia, University of Pennsylvania, Philadelphia, PA, USA

Lois Smith Harvard Medical School, Boston Children's Hospital, Boston, MA, USA

Luis Sobrevía Cellular and Molecular Physiology Laboratory (CMPL), Division of Obstetrics and Gynecology, Faculty of Medicine, School of Medicine, Pontificia Universidad Católica de Chile, Santiago, Chile

The University of Queensland Centre for Clinical Research, Herston, QLD, Australia

Robin H. Steinhorn Department of Pediatrics, Children's Hospital, University of California Davis Medical Center, Sacramento, CA, USA

Michael R. Stenger Department of Pediatrics, The Ohio State University College of Medicine and Nationwide Children's Hospital and Research Institute, Columbus, OH, USA

Roland Stocker Vascular Biology Division, Victor Chang Cardiac Research Institute, Liverpool, NSW, Australia

School of Medical Sciences, Faculty of Medicine, University of New South Wales, Kensington, Australia

Maria Luisa Tataranno Department of Molecular and Developmental Medicine, University of Siena, Siena, Italy

Johannes B. van Goudoever Department of Pediatrics, AMC-Emma Children's Hospital, Amsterdam, The Netherlands

Máximo Vento Division of Neonatology and Health Research Institute La Fe, University and Polytechnic Hospital La Fe, Valencia, Spain

Stephen Wedgwood Department of Pediatrics, University of California Davis Medical Center, Sacramento, CA, USA

Peter G. Wells Division of Biomolecular Sciences, Faculty of Pharmacy, and Department of Pharmacology and Toxicology, Faculty of Medicine, University of Toronto, Toronto, ON, Canada

About the Editors

Phyllis A. Dennery, M.D., FAAP is a Professor of Pediatrics at the University of Pennsylvania School of Medicine in Philadelphia, Pennsylvania. She holds the Werner and Gertrude Henle endowed chair in pediatrics at The Children's Hospital of Philadelphia (CHOP). She is Chief of the Division of Neonatology and Newborn Services at CHOP and the University of Pennsylvania Health System. Dr. Dennery obtained her medical degree from Howard University in Washington, D.C., and completed a residency in pediatrics at Children's Hospital National Medical Center in Washington, D.C., and a fellowship in neonatology at Case Western Reserve University (Rainbow Babies and Children's Hospital) in Cleveland, Ohio. Dr. Dennery's research interests are in neonatal oxidative stress, lung chronobiology, and circadian gene regulation; her clinical interests are in neonatal jaundice, bronchopulmonary dysplasia, and the long-term consequences of prematurity.

Professor Giuseppe Buonocore is Professor and Chairman of the School of Specialization in Pediatrics and Neonatology in the Department of Molecular and Developmental Medicine, University of Siena, Tuscany. He is internationally recognized for his research on hyperbilirubinemia, biomarkers of oxidative stress, free radical–related diseases of the newborn, and the early identification of the fetus and newborn at high risk of brain damage. He is the President of the Union of European Neonatal Perinatal Societies (UENPS), President of the international research foundation Europe Against Infant Brain Injury (EURAIBI), and President of the Tuscan Section of the Italian Society of Neonatology.

Dr. Ola Didrik Saugstad is Professor of Pediatrics, and Director of Research in the Department of Pediatrics at the University of Oslo in Norway. He is a consultant of neonatology at Oslo University Hospital, Rikshospitalet. Dr. Saugstad obtained his M.D. and Ph.D. from University of Oslo. He completed research fellowships in neonatology from the University of Oslo and the University of Uppsala in Sweden, and he carried out a postdoctoral fellowship from the University of California, San Diego. His main research interests are asphyxia, oxidative stress in premature infants, reoxygenation, and resuscitation of newborn infants. He holds an Honorary Doctorate from the University of Athens in Greece and is a member of several editorial boards of pediatric and perinatal journals.

Chapter 1
Impact of Oxidative Stress on Development

Peter G. Wells, Lutfiya Miller-Pinsler, and Aaron M. Shapiro

Introduction

The developing embryo and fetus, collectively termed the conceptus, are highly susceptible to the adverse effects of oxidative stress initiated by endogenous processes and by xenobiotics that enhance the formation of reactive oxygen species (ROS). This susceptibility is due in part to the high conceptal rates of cellular division and differentiation and the complex processes involved in the formation of organ structures and development of functional systems, including brain activity. Alterations in these processes can result in in utero or perinatal death, or structural and functional birth defects, termed "teratogenesis." The instability of ROS, and particularly hydroxyl radicals, means that proximate formation of ROS within the conceptus, rather than distal maternal formation, plays a critical role in teratogenesis. Susceptibility is compounded by relatively high levels of embryonic and fetal enzymes involved in ROS formation and xenobiotic bioactivation to free radical

---

These authors contributed equally.

P.G. Wells, Pharm.D. (✉)
Division of Biomolecular Sciences, Faculty of Pharmacy, and Department of Pharmacology and Toxicology, Faculty of Medicine, University of Toronto, Toronto, ON, Canada
e-mail: pg.wells@utoronto.ca

L. Miller-Pinsler
Department of Pharmacology and Toxicology, Faculty of Medicine, University of Toronto, 144 College Street, Toronto, ON M5S 3M2, Canada

A.M. Shapiro
Division of Biomolecular Sciences, Faculty of Pharmacy, University of Toronto, 144 College Street, Toronto, ON M5S 3M2, Canada

P.A. Dennery et al. (eds.), *Perinatal and Prenatal Disorders*,
Oxidative Stress in Applied Basic Research and Clinical Practice,
DOI 10.1007/978-1-4939-1405-0_1

Human Development and "Critical Periods" for Drug Exposure

GESTATIONAL TIME (weeks)	1 - 2	3	4	5	6	7	8	12	16	20 - 36	38
STAGE	From fertilization to blastocyst; Implantation	**Embryonic** Period					**Fetal** Period				
DEVELOPMENTAL PROCESS	Cellular division	Cellular differentiation and organogenesis					Histological differentiation and functional development				
TERATOLOGICAL CONSEQUENCE	Prenatal death	Major morphological abnormalities					Functional defects and minor morphological abnormalities				

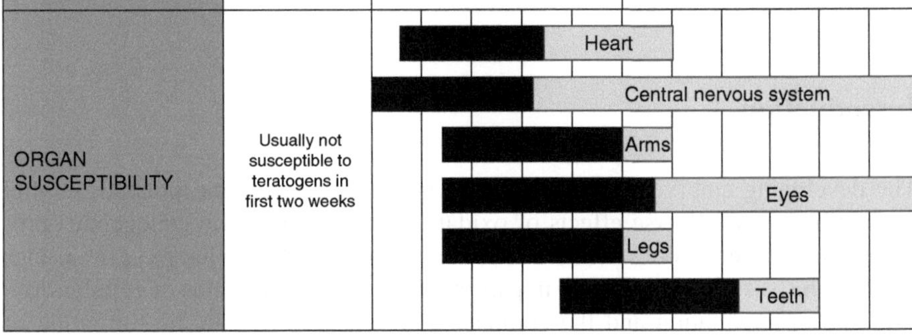

Fig. 1.1 Human development and critical periods for exposure to developmentally toxic agents (Modified from Wells et al. [98]). Exposure to toxic agents during the first 2 weeks following fertilization can result in either the death of the conceptus or a morphologically normal child, or possibly birth defects resulting from genetic or epigenetic changes. Exposure during the embryonic period, which encompasses the period of cellular differentiation and organogenesis, can result in either *in utero* death or major gross morphological birth defects. Exposure during the fetal period, which encompasses histological differentiation and functional development, can result in altered functions of organs and biochemical pathways and systems

intermediates and conversely low levels of protective antioxidative enzymes. Enhanced ROS levels can adversely affect development by altering signal transduction and/or by oxidatively damaging conceptal cellular macromolecules such as lipids, proteins, and DNA, the latter of which may be mitigated by DNA repair enzymes. The risk of teratogenesis is therefore largely determined at the conceptal level, at which in mouse models those littermates with an unfavorable imbalance among pathways of ROS formation and detoxification, and DNA repair, exhibit greater structural and/or functional birth defects than littermates with a favorable balance.

Embryonic and Fetal Development

Following fertilization, development in the human begins with a 2-week period of conceptal cell division, followed by an embryonic period of organ development and a fetal period of functional development (Fig. 1.1). Human exposure to developmentally toxic agents during the initial 2 weeks following fertilization traditionally has been believed to result in either the death of the conceptus or a normal child, since cells have not yet begun to differentiate and a sufficient residual of unaffected cells can compensate for the lost cells. However, this concept was largely based upon teratogens that caused structural birth defects by receptor-mediated processes, and it is likely that exposure to agents that damage DNA via either genetic or epigenetic mechanisms may alter later developmental processes. Later effects also could arise from early exposure to teratogens with long half-lives of elimination. Exposure during the *embryonic* period, which encompasses the period of cellular differentiation and organogenesis, largely results in in utero death or major structural birth defects (shortened or missing limbs, cleft palate, spinal bifida, etc.), although postnatal functional deficits can be initiated by exposure to some xenobiotics during this period. The risk of developmental toxicities exhibits a remarkable pattern not found in other areas of toxicology, namely, critical periods of susceptibility, exemplified in Fig. 1.1 by the filled bars representing periods in which specific organs are at maximal risk for alterations leading to major structural defects [61, 104]. Exposure to teratogens outside of the critical period for a given organ typically will not affect the structural development of that organ, although prior exposure to a xenobiotic with slow elimination may result in toxic concentrations being sustained into the critical period. Exposures in the later component of the critical period, exemplified by the open bars, which for some organs extend into the fetal period, result in only minor structural defects, and exposures beyond the critical period have no teratogenic effect. Exposures during the *fetal* period, during which histological differentiation and functional development occur, typically result in altered functions of organs and biochemical pathways and systems like the immune system. Susceptibility to the initiation of postnatal cancer by transplacental carcinogens at least in mouse models is highest during this period. Fetal exposures can be particularly important for organs like the brain, in which functional development continues throughout gestation and postnatally until around the age of 20 years in humans. As the risk of functional teratogenesis from fetal exposures has become more evident, the traditional focus on the first trimester of pregnancy, encompassing the embryonic period, has shifted to a more balanced view of developmental risk encompassing the entire duration of pregnancy and even into the postnatal period.

The broad spectrum of developmental outcomes is shown in Fig. 1.2 and can vary by the nature of the developmental toxin, the exposure level, and gestational timing of the exposure.

Conceptal Reactive Oxygen Species Formation

Developmental toxicity can be initiated via a spectrum of mechanisms that are not necessarily mutually exclusive, including the receptor-mediated effects of xenobiotics and macromolecular damage caused by the bioactivation of xenobiotics to electrophilic and free radical intermediates. Teratological mechanisms involving

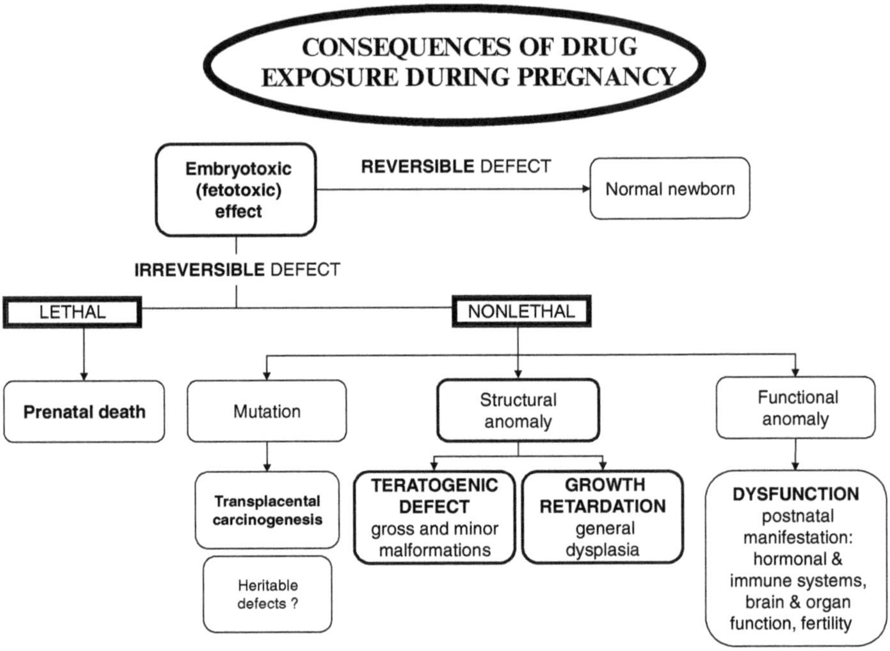

Fig. 1.2 Consequences of exposure to developmentally toxic agents during pregnancy. A reversible defect will result in a normal newborn. An irreversible defect, if lethal, can lead to prenatal death. If the irreversible defect is nonlethal, three general consequences are possible: (1) mutations leading to transplacental carcinogenesis or heritable defects; (2) structural or gross morphological anomalies, either teratogenic defects or growth retardation; or (3) functional anomalies, including a spectrum of postnatal metabolic dysfunctions (Modified from Neubert et al. [65])

receptor-mediated processes and the covalent binding of xenobiotic electrophilic reactive intermediates to cellular macromolecules have been discussed elsewhere ([97–99, 120]). Reactive oxygen species (ROS) and oxidative stress can be initiated proximally in the embryo and fetus by xenobiotics and agents like ionizing radiation via several mechanisms (Fig. 1.3). These include redox cycling of quinone metabolites of xenobiotics, cardiac suppression by xenobiotics followed by reperfusion (e.g., phenytoin), activation and/or induction of ROS-producing NADPH oxidases (NOXs) (e.g., ethanol and methanol), and interruption of the mitochondrial electron transport chain. A number of endogenous pathways can lead to ROS formation, including redox cycling (e.g., quinone metabolites of dopamine in the brain), as well as enhanced ROS formation in diseases like diabetes [25, 99, 116]. In diabetes, high sugar levels and possibly additional factors can dysregulate the mitochondrial electron transport chain [114] as well as induce NOXs [28], both of which result in enhanced ROS formation. Xenobiotics can also be converted to ROS-initiating free radical intermediates by enzymes like prostaglandin H synthases (PHSs) and lipoxygenases (LPOs) (Fig. 1.4) as well as by cytochromes P450 (CYPs). Unlike CYPs, PHSs and LPOs are highly expressed in the

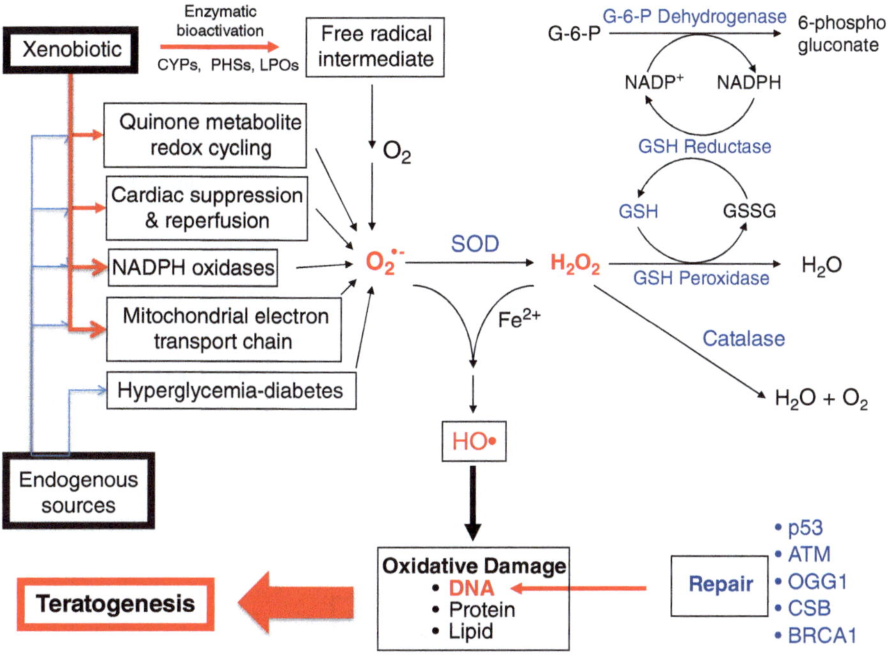

Fig. 1.3 Biochemical pathways for endogenous and xenobiotic-enhanced formation and detoxification of reactive oxygen species (ROS) and repair of oxidatively damaged cellular macromolecules. Teratogenesis is postulated to result from ROS-mediated alterations in signal transduction (not shown) and/or embryonic or fetal macromolecular damage. If embryonic or fetal ROS formation exceeds the proximal capacity for ROS detoxification and/or repair of cellular macromolecules, this imbalance can result in enhanced teratogenesis, even at a therapeutic drug concentration or generally "safe" exposure level for an environmental chemical. *Blue arrows* indicate pathways through which ROS are formed endogenously. Many of the same ROS-forming pathways that are enhanced by xenobiotics are also responsible for endogenous ROS formation. Abbreviations: *ATM* ataxia telangiectasia mutated, *BRCA1* breast cancer 1, *CSB* Cockayne syndrome B, *CYPs* cytochromes P450, *Fe* iron, *G-6-P* glucose-6-phosphate, *GSH* glutathione, *GSSG* glutathione disulfide, *LPOs* lipoxygenases, *NADP*$^+$ nicotinamide adenine dinucleotide phosphate, *OGG1* oxoguanine glycosylase 1, *PHSs* prostaglandin H synthases, *SOD* superoxide dismutase (Modified from Wells et al. [100])

embryo and fetus, and these enzymes have been shown in animal models to contribute to the bioactivation and teratogenicity of several drugs including benzo[a]pyrene, thalidomide, phenytoin, and structurally related antiepileptic drugs.

The nature of ROS is discussed elsewhere in this book. Briefly, ROS include hydrogen peroxide (H_2O_2), superoxide anion ($O_2^{\bullet-}$), and hydroxyl radical (HO•). Unlike $O_2^{\bullet-}$ and H_2O_2, which are relatively stable and diffusible forms of ROS, HO• is highly unstable and nondiffusible, reacting rapidly with nearby molecular targets within the subcellular compartment in which it is formed. ROS are an important component of many physiological signal transduction pathways; however, if levels of ROS exceed the capacity for detoxification, this results in "oxidative stress," including the oxidation of

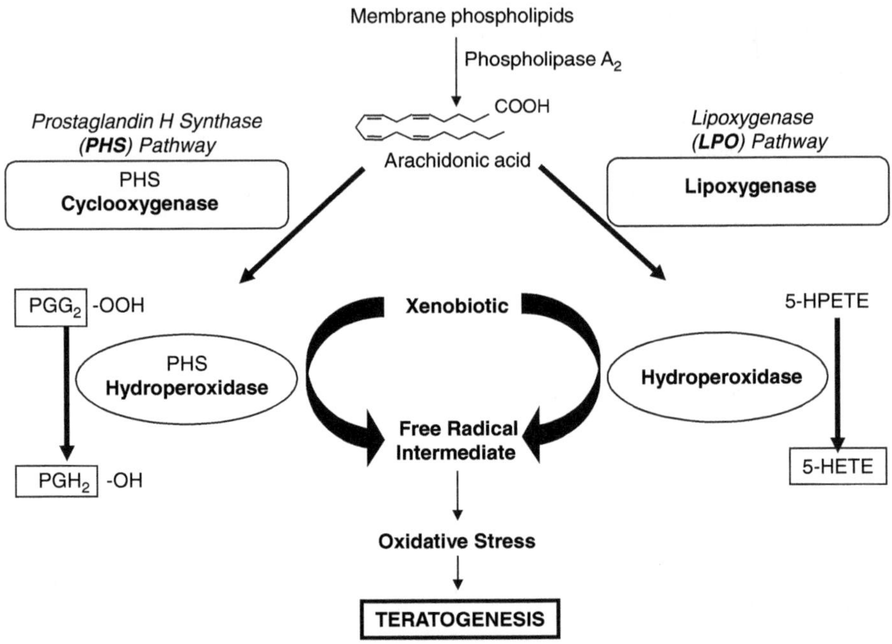

Fig. 1.4 Bioactivation of xenobiotics catalyzed by prostaglandin H synthases (PHSs) and lipoxygenases (LPOs). The hydroperoxidase component of PHSs, and hydroperoxidases associated with LPOs, can oxidize xenobiotics to free radical intermediates that can initiate ROS formation, oxidative stress, and teratogenesis. Arachidonic acid is released from membrane phospholipids by the action of phospholipase A2 and can then serve as a substrate in both the cyclooxygenase- and lipooxygenase-dependent eicosanoid pathways generating the corresponding hydroperoxides, which are then reduced by hydroperoxidases to the corresponding alcohols. In this pathway, xenobiotic may serve as a reducing co-substrate, itself being oxidized to a reactive free radical intermediate. If not detoxified, this free radical can initiate oxidative stress resulting in ROS-mediated oxidative macromolecular damage and teratogenesis (Modified from Yu and Wells [115])

proteins, lipids, DNA, and other cellular macromolecules and/or alterations in signal transduction pathways, which can play a role in disease, aging, and xenobiotic toxicity, including developmental abnormalities (Fig. 1.5) [24, 25, 99].

The instability of ROS, and particularly hydroxyl radicals, means that proximate formation of ROS within the conceptus, rather than distal maternal formation, plays a critical role in teratogenesis. Embryo culture studies have shown that the embryo has the necessary enzymes for bioactivating "proteratogens" to free radical reactive intermediates, as well as ROS-producing NOXs, balanced by variable but critically important activities of protective antioxidative and DNA repair enzymes. As discussed in the next section on mechanisms, this has important implications for the

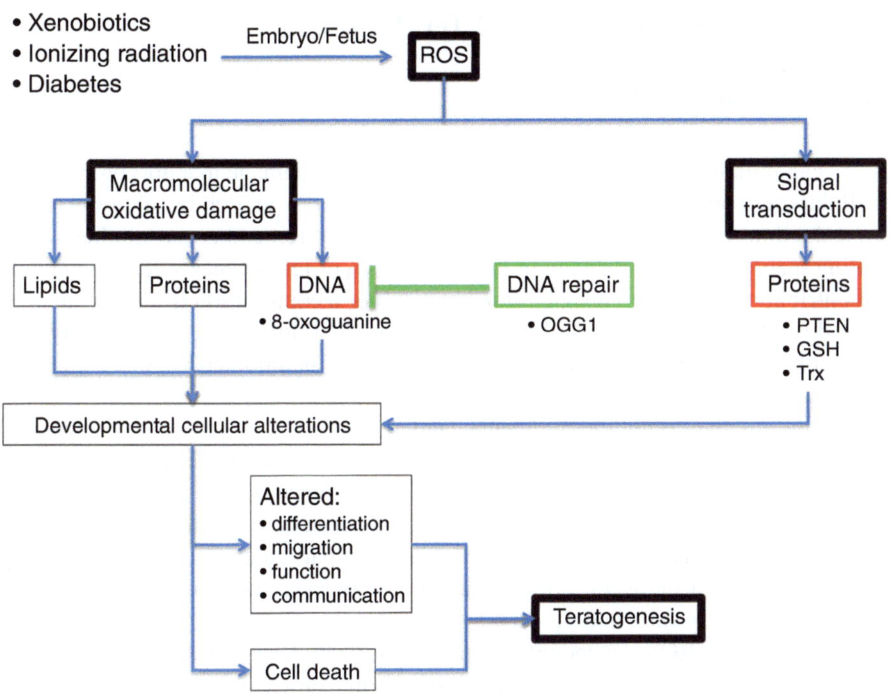

Fig. 1.5 Alternative biochemical effects of reactive oxygen species (ROS): oxidative damage to cellular macromolecules and altered signal transduction. Physiological or environmentally enhanced levels of embryonic and fetal ROS can adversely affect development via one or both of the following two mechanisms: (1) Irreversible oxidative damage to cellular macromolecules including DNA, proteins, peptides, and lipids. 8-Oxoguanine is the most prevalent of over 20 DNA lesions initiated by ROS, and this lesion is repaired by oxoguanine glycosylase 1 (OGG1), one of several DNA repair proteins known to protect the developing embryo and fetus (see Fig. 1.3). (2) Reversible oxidative modification of signaling molecules, exemplified by phosphatase and tensin homolog (PTEN), glutathione (GSH), and thioredoxin (Trx). The degree to which either or both of these two alternative pathways contribute to the pathogenic mechanism may vary with the particular nature and timing of ROS initiation and the conceptal target tissue, among other factors. These biochemical changes can lead to non-apoptotic alterations in cellular function including differentiation, migration, function, and communication or may result in cell death. If these cellular alterations occur during critical windows of development, they may cause gross morphological birth defects or postnatal functional abnormalities such as neurodevelopmental deficits.

determinants of teratological risk as well as for the analysis of data when teratogenesis is initiated by the formation of ROS, as distinct from receptor-mediated mechanisms. Evidence for a role of these pathways in embryonic and fetal ROS formation, as well as the involvement of ROS in mechanisms of teratogenesis, is derived primarily if not exclusively from animal models, with no human studies to provide corroboration.

Mechanisms of ROS-Initiated Teratogenesis

ROS-Dependent Mechanism

ROS can initiate developmental toxicity via oxidative damage to embryonic or fetal cellular macromolecules (lipids, proteins, DNA, etc.) and/or by altering signal transduction (Fig. 1.5) [14, 25, 99]. It is important to remember that studies implicating macromolecular damage in the mechanism of teratogenesis do not preclude a contribution from altered signal transduction nor a contribution from mechanisms that do not involve ROS, as discussed later for thalidomide. It seems likely that particular types of developmental abnormalities can be initiated via more than one mechanism.

Representative xenobiotics, agents, and diseases for which a ROS-mediated mechanism of teratogenesis has been implicated are listed in Table 1.1, although other mechanisms including receptor-mediated actions may also contribute [98]. It is also possible that different mechanisms may lead to a similar developmental abnormality and that the relative contribution of a particular mechanism may vary with the nature of the ROS-initiating agent, the exposure concentration, the gestational time of exposure, and the target tissue or cell type.

Evidence of a ROS-dependent mechanism of teratogenesis has been obtained by a variety of approaches involving the use of pharmacological probes and genetically modified animal models (Table 1.2). Pharmacological probes all have actions in addition to their intended activity, so the degree of certainty is enhanced by the use of multiple probes that modulate complementary pathways and by corroborating studies using animal models in which the pathway in question has been genetically modified. Even results from studies using genetically modified animals such as knockout mice can be confounded by unappreciated biochemical changes in these mice in response to the loss of an important gene, so corroborating approaches improve the confidence of an interpretation. Studies supporting ROS-dependent mechanisms of developmental toxicity include: (1) Blocking xenobiotic bioactivation by enzymes like prostaglandin H synthases (PHSs) to a free radical intermediate, using PHS inhibitors such as acetylsalicylic acid (ASA, aspirin) or eicosatetraynoic acid (ETYA), which reduce the teratogenicity of drugs like phenytoin, benzo[a]pyrene, and thalidomide in vivo and/or in embryo culture [47, 99]. These studies have been corroborated by the use of knockout mice lacking a PHS isozyme, which exhibit reduced embryonic bioactivation of xenobiotics like phenytoin and benzo[a]pyrene, and are protected from the resulting increase in oxidatively damaged cellular macromolecules and developmental toxicity in vivo and/or in embryo culture [99]. In the case of drugs like ethanol and methanol, which induce the expression of ROS-producing NADPH oxidases (NOXs), pretreatment with a NOX inhibitor reduces embryonic DNA oxidation and teratogenesis [18, 100]. (2) Altering antioxidative pathways, as discussed below in section "Antioxidative Pathways," where increases in antioxidants and antioxidative pathways protect the embryo and fetus from xenobiotic-initiated oxidative damage to cellular macromolecules and from the associated

Table 1.1 Representative drugs and environmental chemicals (xenobiotics) and other agents and diseases for which ROS have been implicated in the mechanism of teratogenesis

Alcohol (ethanol)*
Cocaine*
Cyclophosphamide*
Benzo[a]pyrene
Diethylstilbestrol*
Hydroxyurea
Methamphetamine
Methanol
Methylmercury*
Phenytoin and structurally related antiepileptic drugs*
Thalidomide*
Tobacco smoke*
Valproic acid*
Ionizing radiation*
Diabetes*

From studies in animal models. Asterisks indicate known human teratogens [100], although xenobiotics like benzo [a]pyrene, a component of tobacco smoke, are presumed human teratogens despite the absence of human studies evaluating them in isolation. Evidence for ROS involvement does not preclude a contribution from other mechanisms, including receptor-mediated initiation, which has been implicated for many of the above xenobiotics. Teratological outcomes include structural birth defects (e.g., ethanol, thalidomide) and postnatal functional abnormalities including neurodevelopmental deficits (e.g., ethanol, valproic acid, and methamphetamine) and cancer (e.g., diethylstilbestrol) (Based upon data from papers and recent reviews [15, 25, 85, 94, 99])

developmental abnormalities in vivo and/or in embryo culture. A converse increase in conceptal macromolecular damage and developmental toxicity is observed when antioxidants and antioxidative pathways are reduced. (3) Pretreatment with free radical spin trapping agents like phenylbutylnitrone (PBN), which block oxidative damage to cellular macromolecules and the associated developmental toxicity of drugs like phenytoin, thalidomide, methanol and ethanol in vivo and in embryo culture [6, 47, 99, 100], and lipopolysaccharide-initiated toxicity in vivo [113]. In addition to implicating ROS in the teratological mechanism, such approaches provide insights into the relative contribution of ROS, as distinct from other mechanisms, such as the covalent binding of xenobiotic electrophilic reactive intermediates to embryonic and fetal cellular macromolecules or receptor-mediated effects of the parent compound or a stable metabolite.

Table 1.2 Representative approaches implicating reactive oxygen species (ROS) in the mechanism of developmental toxicity

Process	Drug	Probe/outcome	Citation
Xenobiotic free radical formation	Phenytoin and structurally related drugs	ESR/EPR spectrometry	[43, 73]
	Ethanol (EtOH)	ESR/EPR spectrometry	[4, 40]
	Methanol	ESR/EPR spectrometry	[32, 91]
	Methamphetamine	ESR/EPR spectrometry	[30]
ROS production			
Fluorescent probes	Valproic acid	DCF fluorescence	[94]
Hydroxyl radical formation	Phenytoin	Salicylate hydroxylation	[36]
Oxidatively damaged macromolecules			
Lipids	See Table 1.3	MDA formation	
		HNE formation	
		Isoprostane formation	
		TBARS formation	
Proteins and peptides	See Table 1.3	Carbonyl formation	
		Protein thiol and GSH oxidation	
DNA			
Oxidation	EtOH	8-oxoguanine formation	[18, 53–56]
	Thalidomide and two hydrolysis products	8-oxoguanine formation	[47, 75]
Strand brakes	Ionizing irradiation	Comet assay	[50]
Double-strand breaks	Methylmercury	H2AX phosphorylation	[70]
	Valproic acid	Gamma-H2AX foci	[87]
Reduced xenobiotic bioactivation to a free radical intermediate			
PHS, LPOs	Phenytoin	Enzyme inhibitors (ASA, ETYA, and others)	[43, 58, 115]
	Thalidomide	Enzyme inhibitor (ASA, ETYA)	[5, 47]
	Benzo[a]pyrene	PHS knockout mice	[74]

Reduced activity of ROS-producing NOX enzymes	EtOH	Enzyme inhibitor (DPI)	[18]
Reduced antioxidative protection	See Table 1.5	Genetically deficient mice	
		Dietary depletion	
		GSH depletion	
Enhanced antioxidative protection	See Table 1.5	Antioxidants	
		GSH precursors	
		Antioxidative enzymes	
		Genetically enhanced mice	
Free radical spin trapping agents	EtOH	PBN	[54]
	Phenytoin	PBN	[43, 101]
	Thalidomide	PBN	[47, 75]
	Cocaine	PBN	[119]
	Lipopolysaccharide	PBN	[117]

Abbreviations: *ASA* acetylsalicylic acid, *DCF* dichlorofluorescein, *DPI* diphenylene iodonium, *EPR* electron paramagnetic resonance, *ESR* electron spin resonance, *ETYA* 5,8,11,14-eicosatetraynoic acid, *EtOH* ethanol, *gamma-H2AX* phosphorylated histone 2AX protein, *GSH* glutathione, *HNE* 4-hydroxy-2-nonenal, *LPO* lipoxygenase, *MDA* malondialdehyde, *NOX* NADPH oxidase, *PBN* phenylbutylnitrone, *PHS* prostaglandin H synthase, *ROS* reactive oxygen species, *TBARS* thiobarbituric acid reactive substances

The high reactivity of ROS, and particularly hydroxyl radicals, means that ROS cannot readily travel from the mother to the embryo. Therefore, proximal embryonic and fetal pathways for ROS formation, as distinct from distal maternal pathways, are major determinants of risk. An illustration of the importance of proximate pathways can be seen for ROS formation in a mouse knockout model for PHS2, in which progeny from the same dam and litter may be wild type, or heterozygous or homozygous deficient for the *phs2* gene. In this example, among all littermates, the PHS2-deficient progeny, lacking bioactivating activity, are protected from benzo[a]pyrene teratogenicity, at least at lower doses, while the wild-type littermates are susceptible, although all are littermates from the same dam and are exposed to the same teratogen concentration [74]. Even strains of mice that are not genetically modified exhibit substantial littermate variability in embryonic determinants of risk, as exemplified with CD-1 mice, in which embryonic catalase activity can vary up to fourfold among embryos of the same litter [3]. Accordingly, for mechanisms involving the formation of ROS, the conceptal genotype is the fundamental determinant of risk rather than the litter or maternal genotype. Traditionally, teratological data are analyzed by litter, grouping all fetuses in the litter as equivalent, largely to compensate for the occasionally confounding occurrence of exceptional developmental anomalies in an isolated litter, often termed the "litter effect." This approach usually works well for teratogens acting through a receptor-mediated mechanism, as long as there is no genetic modification or substantial spontaneous littermate variability in the receptor gene. However, in the case of a ROS-mediated mechanism, although it is important to evaluate a sufficient number of litters to dilute the impact of an anomalous litter response, the individual conceptus rather than the litter is the essential parameter. As discussed below in sections "Antioxidative Pathways" and "DNA Repair," the conceptal activity for antioxidative and DNA repair pathways is similarly critical. Accordingly, the balance among the conceptal pathways of ROS formation and detoxification, and DNA repair, within a single embryo or fetus is the fundamental determinant of developmental risk for that conceptus.

ROS-Initiated Macromolecular Damage

ROS oxidize or oxidatively damage all embryonic and fetal cellular macromolecules, including lipids, proteins, peptides (e.g., glutathione [GSH]), RNA, and DNA, as well as alter signal transduction (Fig. 1.5), so it is difficult to know which molecular events are contributing to developmental toxicity and to what degree. Proteins and lipids in particular are oxidatively damaged by enhanced conceptal ROS formation (Table 1.3) and would be expected to contribute to mechanisms of ROS-mediated teratogenesis, in which case the repair or removal of such macromolecular lesions would likely modulate the developmental toxicity of a ROS-initiating agent. However, approaches to definitively test this hypothesis have yet to be developed, and studies providing evidence of a causal role for oxidatively damaged lipids and proteins in the pathogenic mechanism are lacking. On the other hand, a number of genetically modified animal models have been developed that lack key proteins and

Table 1.3 Representative studies of reactive oxygen species (ROS) formation, lipid peroxidation, and protein and peptide oxidation in embryonic and fetal tissues exposed to ROS-initiating teratogens

Species	Drug	Exposure duration	Probe	Evidence for oxidative stress and damage	Reference
Lipid peroxidation					
Cell culture					
Fetal rat astrocytes (GD 21)	25, 50 mM EtOH	7–10 days	L-cysteine, L-cystine, anti-CYP2E1 antibody	↑ ROS (DCF), lipid peroxidation (MDA and 4-HNE), ↓ GSH; protection with GSH addition	[60]
Sprague-Dawley fetal rat hepatocytes (GD 20)	2 mg/ml EtOH	24 h	None	↑ MDA (TBARS)	[16]
Whole embryo culture					
C57BL/6 mouse	5 mg/ml EtOH in ECM	6 h drug, 30 h medium	SOD co-treatment	EtOH ↑ superoxide anion production (NBT) lipid peroxidation (MDA), ↓ by SOD co-exposure	[42]
Rat (strain not specified)	*WEC*: 17–171 mM EtOH; *Midbrain cell culture*: 6–856 mM EtOH	WEC: 48 h Cell culture: 96 h	Catalase co-treatment, vitamin E (no protection)	*WEC*: ↑ 8-OHdG (ELISA), ROS (DCF); *Cell culture*: ↓ cytotoxicity with probes	[48]
In vivo					
C57BL/6 mouse	5–20 % EtOH in drinking water (p.o.)	2 weeks prior to gestation + throughout gestation to GD 18	Transgenic SOD mice	↓ Fetal hepatic 8-isoprostanes in mice with enhanced SOD expression	[102]
Wistar rat	2 g/kg EtOH (i.p.)	15 days prior to mating + entire gestation period	None	↑ Lipid peroxidation (TBARS) in fetal liver, brain, kidneys, testes	[64]
CD-1 mouse	65 mg/kg phenytoin (i.p.)	GD 12	Pretreatment with PBN or ASA	Phenytoin ↑ TBARS; ↓ by PBN and ASA	[49]
Wistar rat	1.5, 3 g/kg methanol (i.p.)	N/A	Acute	↑ Lipid peroxidation (TBARS), ↓ GSH-Px, SOD, GSSG reductase activities, ↓ ascorbate levels	[92]
CD-1 mouse	400, 600 mg/kg hydroxyurea (i.p.)	GD 9	None	Hydroxyurea ↑ HNE-protein adduct formation (western blot)	[85]

(continued)

Table 1.3 (continued)

Species	Drug	Exposure duration	Probe	Evidence for oxidative stress and damage	Reference
Protein and peptide oxidation					
Whole embryo culture					
CD-1 mouse	80 uM phenytoin; 10 uM benzo[a]pyrene	4 h (biochemistry); 24 h embryopathies	PBN	Phenytoin and benzo[a]pyrene ↑ protein oxidation (western) and phenytoin ↑ the oxidation of GSH to GSSG (HPLC)	[105]
Sprague-Dawley rat	50–500 uM diamide	2 h	BCNU, BSO	Diamide ↑ embryotoxicity and the oxidation of GSH to GSSG; potentiated by BCNU and BSO	[27]
In vivo					
CD-1 mouse	65 mg/kg phenytoin (i.p.)	GD 12	Pretreatment with PBN or ASA	Phenytoin ↑ protein oxidation (carbonyl formation) and degradation (release of primary amines); blocked by PBN and ASA	[49]
CFT Wistar rat	400, 800 uM/kg t-butyl hydroperoxide (i.p.)	GD 5–7 or 8–10; sacrificed GD 13	None	t-Butyl hydroperoxide ↑ embryonic protein carbonyls (hydrazone derivatives), ROS formation (DCF), and depletion of GSH and protein thiols (fluorometry)	[89]
Wistar rat	1.5, 3 g/kg methanol (i.p.)	N/A	None	↓ Soluble and protein sulfhydryl groups	[92]

Abbreviations: *8-OHdG* 8-hydroxy-2′-deoxyguanosine, *ASA* acetylsalicylic acid, *CYP* cytochrome P450, *DCF* dichlorofluorescein, *ELISA* enzyme-linked immunosorbent assay, *EtOH* ethanol, *GD* gestational day, *GSH* reduced glutathione, *GSH-Px* glutathione peroxidase, *GSSG* glutathione disulfide, *HNE* hydroxynonenal, *HPLC* high-performance liquid chromatography, *MDA* malondialdehyde, *NBT* nitroblue tetrazolium, *PBN* phenylbutylnitrone, *ROS* reactive oxygen species, *SOD* superoxide dismutase, *TBARS* thiobarbituric acid reactive substances, *WEC* whole embryo culture

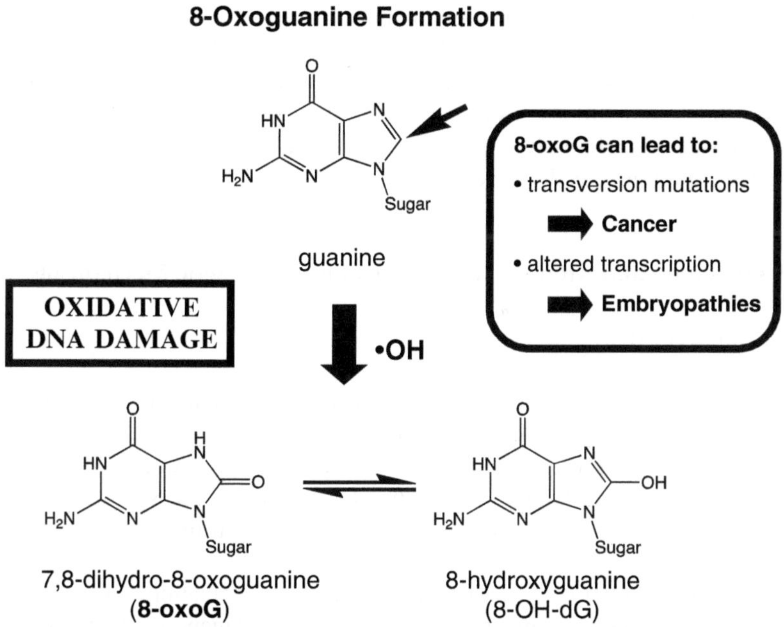

Fig. 1.6 Reaction of hydroxyl radicals (HO•) with guanine residues of DNA to form the molecular lesion 7,8-dihydro-8-oxoguanine (8-oxoguanine, 8-oxoG). If not repaired, this DNA lesion can cause mutagenic changes in gene sequences, and/or altered gene transcription, which respectively may lead to the development of postnatal cancer in the progeny, and/or structural or functional teratogenesis (From Wells et al. [99])

enzymes involved in the recognition of oxidatively damaged DNA and its repair, as discussed below in section "DNA Repair." Knockout or conditional knockout mice lacking these DNA repair components exhibit an increase in oxidatively damaged DNA in embryos and in fetal brain, and in the associated embryopathies and postnatal neurodevelopmental deficits, caused by agents like benzo[a]pyrene, methamphetamine, ethanol, and ionizing radiation [80, 99, 100]. These studies show that DNA oxidation, as distinct from oxidative damage to other types of cellular macromolecules, is a pathogenic molecular event. They also provide insight into the relative teratogenic contribution of oxidatively damaged DNA as distinct from ROS-mediated alterations in signal transduction. The next two subsections accordingly focus upon DNA as a macromolecular target of ROS, which can cause both mutagenic changes in gene sequence and direct and indirect epigenetic changes with no change in gene sequence.

Mutagenic Mechanisms

ROS initiate over 20 different lesions in DNA, with 8-oxoguanine (Fig. 1.6) being the most prevalent [31]. This lesion is mutagenic, leading to a change in the DNA structure, and likely is involved in the mechanism of transplacental carcinogenesis, perhaps including postnatal cancer in the children of mothers treated with the synthetic estrogen diethylstilbestrol [82]. However, aside from cancers, other forms of developmental toxicity, including structural birth defects and postnatal neurodevelopmental deficits, likely result not from mutagenesis but rather from the direct effects of 8-oxoG and possibly other ROS-initiated DNA lesions on embryonic and fetal gene expression, as discussed below.

DNA Oxidation and Altered Gene Expression: Epigenetic Mechanisms

Direct Effects of ROS

Unlike the ROS-initiated genetic damage leading to mutagenesis, ROS-initiated lesions in DNA, and particularly the oxidation of DNA by hydroxyl radicals, may directly alter gene expression without changes to the DNA sequence, by interfering with transcriptional machinery [35, 37–39, 76]. At least for the 8-oxoG lesion, some of its developmental effects may occur via a mechanism that could be viewed as epigenetic.

Epigenetic modulation of transcription involves the regulation of gene expression in a sequence-independent manner. One key form of epigenetic regulation is DNA methylation of cytosine residues within CpG islands, which either renders DNA available as euchromatin to promote transcription or sequesters DNA into heterochromatin to hinder transcription. The periods during gametogenesis and immediately following fertilization are marked by the global demethylation of genomic DNA with subsequent remethylation, catalyzed by DNA methyltransferases (DNMTs), in a process known as epigenetic reprogramming [77]. This comprehensive reprogramming provides an enhanced window of sensitivity to modulation by epigenetic regulation [10]. Experimentally, this has been shown in mice using the DNMT inhibitor 5-azacytidine shortly after fertilization. Inhibition of the remethylation of DNA by 5-azacytidine at this critical stage in development, in the days prior to the start of organ development, resulted in neural tube and ocular defects [83], suggesting that loss of epigenetic control during reprogramming can lead to teratogenesis.

DNA silencing by methylation involves the binding of methyl-CpG-binding domain proteins (MBDs), which leads to the incorporation of histones and subsequent conversion to heterochromatin [52]. Methylation of DNA can be reversed either enzymatically or by ROS and particularly the 8-oxoG lesion (Fig. 1.7). Enzymatic hydroxylation of the 5-methyl group of cytosine by Tet methylcytosine dioxygenase 1 (Tet1) forms 5-hydroxymethylcytosine (5-hmC) [93], which prevents MBD binding [52].

Similar to the inhibition of MBD binding by 5-hmC formation, oxidized guanine, in the form of 8-oxoguanine, when adjacent to 5-mC can block MBD binding, thereby maintaining DNA as euchromatin [95] (Fig. 1.7). Oxidative stress during development may therefore initiate teratogenesis via the interruption of epigenetic sequestering of DNA to heterochromatin.

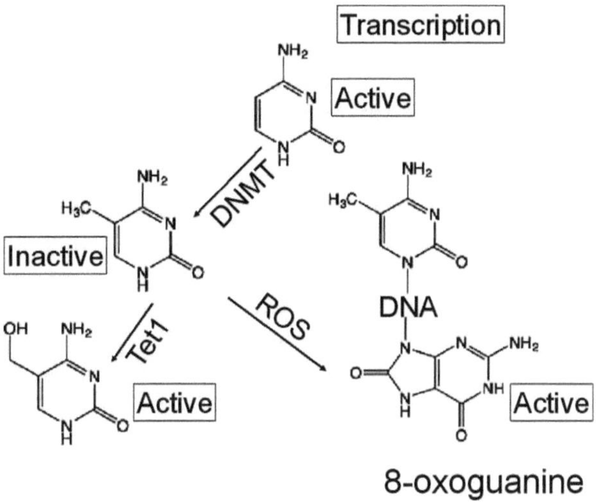

Fig. 1.7 Oxidatively "damaged" DNA in the activation of gene transcription. Hydroxylation of guanine residues by hydroxyl radicals forms 8-oxoguanine, which when adjacent to 5-methylcytosine in CpG islands can block the binding of methyl-CpG binding domain proteins (MBDs) to DNA, thereby maintaining DNA as euchromatin and genes in their active state. This could be viewed as an epigenetic effect of DNA oxidation

Indirect Epigenetic Effects

Xenobiotics also may inhibit or induce DNMTs, resulting in gene activation or silencing. Similarly, xenobiotic-initiated alterations in the acetylation, methylation, or phosphorylation of histone proteins may alter chromatin structure and the binding of transcription factors to DNA. Finally, xenobiotics may alter the expression of related components like MBDs that are necessary gene transcription. It is not yet clear the extent to which, if any, ROS may contribute to the modulation of these epigenetic regulators, but if so this contribution would likely vary with the nature of the developmental toxicant and possibly the stage of gestation and conceptal target tissue. For example, the ROS-initiating teratogen ethanol in rodent embryos or fetuses alters the methylation status of several genes in more severely affected embryos, as well as gene expression of several DNMTs and at least one MBD, and DNMT activity in fetal brains; however, the involvement of ROS was not investigated ([121], [122]).

Macromolecular Oxidative Damage Versus Signal Transduction

ROS-initiated changes in signal transduction have been suggested by some investigators to play a primary role in developmental toxicity, with ROS-initiated macromolecular damage contributing only at higher xenobiotic exposures. There appears to be little evidence to support this hierarchy, and oxidative damage to cellular macromolecules including DNA is typically evident at minimally developmentally toxic doses and concentrations of numerous ROS-initiating teratogens. For example, DNA oxidation in fetal brain is measurable in the absence of ROS-enhancing conditions and is elevated by a dose of ethanol that does not cause structural birth defects when administered during embryogenesis [55]. Both oxidative damage and altered signal transduction are likely to play important roles in teratogenesis (Fig. 1.5). While macromolecular damage to lipids, proteins, and DNA are all potentially important mechanisms of teratogenesis, approaches for proving the selective mechanistic contributions of lipid and protein damage, as distinct from DNA damage, to teratogenesis have yet to be developed. An increase in conceptal lipid peroxidation and/or protein oxidation has been reported for embryopathies or birth defects caused by numerous teratogens (Table 1.4); however, while such results are biomarkers for oxidative stress, this association is not evidence for a causal role of these macromolecular lesions in the pathogenic mechanism. Accordingly, it is unclear in which cases the oxidative damage to lipids and proteins is causally involved in or merely associated with increased teratogenesis. In the case of oxidatively damaged DNA, on the other hand, the developmentally pathogenic role of 8-oxoG and possibly other DNA lesions is particularly demonstrable in genetically modified rodent models with deficient DNA repair. Indeed, physiological oxidative stress alone in the developing embryo and fetus, in the absence of xenobiotic exposure, can result in postnatal neurodevelopmental deficits in progeny with genetic deficiencies in DNA repair enzymes like ataxia telangiectasia mutated (ATM) [9] or oxoguanine glycosylase 1 (OGG1) [100]. However, ROS-initiated mechanisms of developmental toxicity involving macromolecular damage and alterations in signal transduction are not mutually exclusive, and the relative teratological contributions of each mechanism could vary with the teratogen, time of gestation, and conceptal target tissue.

There also is some cross-talk between ROS-mediated macromolecular damage and altered signal transduction, which further complicates the potential mechanism of teratogenesis, as exemplified by phosphatase and tensin homolog (PTEN). PTEN is a protein tyrosine phosphatase that controls cellular proliferation through the phosphatidylinositide 3-kinases (PI3K)/protein kinase B (PKB/Akt)/mammalian target of rapamycin (mTOR) (PI3K/Akt/mTOR) pathway [46]. PTEN regulates the PI3K/Akt/mTOR pathway by dephosphorylating phosphatidylinositol (3,4,5)-triphosphate (PIP3) to phosphatidylinositol (4,5)-diphosphate (PIP2), which leads to the inhibition of cell growth and promotion of apoptosis [46].

In addition to the dephosphorylation of PIP3, a number of additional signaling pathways have been identified as PTEN dependent, including DNA damage repair. PTEN is important for chromosomal stability through its association with centromere-specific binding protein C (CENPB) and induction of Rad51, which is involved in homologous recombination repair (HRR) [78]. The loss of PTEN results in an inability to repair DNA as indicated by the presence of p53 binding protein 1 (53BP1) foci through a mechanism that is independent of nuclear phosphatase signaling. PTEN contains a phosphorylation site for ATM kinase activity, which promotes the SUMOylation of PTEN. This SUMOylation is thought to promote nuclear translocation and is necessary for double-strand break repair by HRR [7].

In the presence of ROS, PTEN can become reversibly inactivated through the generation of a disulfide bridge at Cys^{71} and Cys^{124} [44]. PTEN inhibition has been extensively studied in the context of neurodevelopment. Clinically, a number of nonsense mutations in the PTEN gene have been identified and are associated with Bannayan–Riley–Ruvalcaba syndrome, which is characterized by macrocephaly,

Table 1.4 Redox-regulated transcription factors involved in development

Transcription factor	Regulation and developmental role
HIF-1	Activated in hypoxia
	Induced by reducing agents
	Controls:
	Vascular genes
	Iron regulatory elements
	Erythropoietic genes
NF-κB	Activated in oxidative stress
	Developmental expression
	Controls:
	Apoptosis
	Proliferation
AP-1	Activated in oxidative stress
	Developmental expression
	Controls:
	DNA repair (Ref-1)
	Apoptosis
	Antioxidative genes
Nrf1	Activated in oxidative stress
	Developmental expression
	Controls:
	Apoptosis
	Redox balance
	Antioxidative genes
Nrf2	See section "Regulation of Embryonic and Fetal ROS-Protective Pathways by Nuclear Factor Erythroid 2-Related Factor 2 (Nrf2) in Teratogenesis" and Fig. 1.12
Wnt	Activated by hydrogen peroxide
	Developmental expression
	Controls
	Body patterning
	Organogenesis

Abbreviations: *AP-1* activator protein-1, *HIF-1* hypoxia-inducible factor-1, *NF-κB* nuclear factor kappa B, *Nrf1* nuclear factor erythroid 2-related factor 1, *Nrf2* nuclear factor erythroid 2-related factor 1, *Wnt* wingless and integration site for mouse mammary tumor virus (From Dennery [14])

developmental delay, and a number of neoplastic lesions [86]. In PTEN knockout mice, the inhibition of mTOR complex 1 (mTORC1), a downstream effector of the PI3K/Akt/mTOR pathway, prevented macrocephaly, the incidence of seizures, and behavioral issues including anxiety and social interaction [118]. Mice deficient in the DNA repair protein ATM, which promotes nuclear translocation of PTEN, are at greater risk of xenobiotic-initiated embryopathies [8]. Taken together, these results suggest that the inhibition of PTEN can lead to adverse developmental outcomes by both alterations in signal transduction and loss of DNA repair.

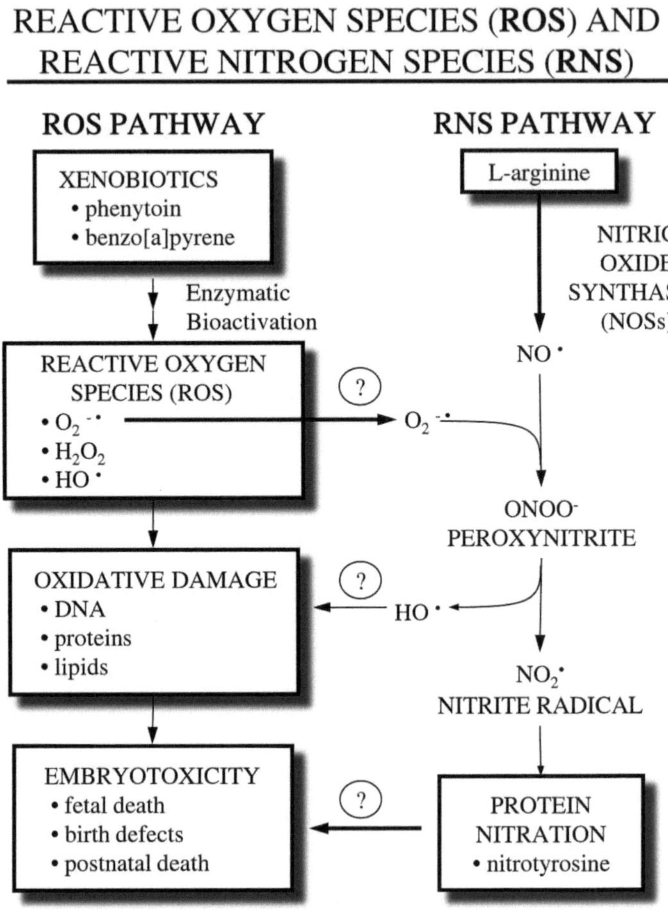

Fig. 1.8 Molecular damage by reactive nitrogen species (RNS) and potential interactions between the RNS and ROS pathways leading to enhanced damage to embryonic and fetal cellular macromolecules (From Kasapinovic et al. [33])

Reactive Nitrogen Species (RNS) in Teratogenesis

Little is known about the role of RNS, produced via embryonic and fetal nitric oxide synthases (NOSs), in errors of development. However, it seems likely that RNS alone and/or in combination with ROS can adversely affect development (Fig. 1.8) via altered signal transduction and/or macromolecular damage. RNS have been implicated in the mechanism of teratogenesis for several xenobiotics [20], including phenytoin [33], although in the latter case using NOS knockout mice, RNS cannot fully account for the teratogenic effects observed. However, the contribution of RNS in phenytoin embryopathies revealed in NOS knockout mice was carried out in embryo culture, demonstrating that proximate embryonic NOS, as distinct from maternal activity, was the source of RNS.

ROS-Mediated Signal Transduction

The potential for dual ROS mechanisms contributing to ROS-mediated teratogenesis is exemplified by the antiepileptic drug phenytoin. Although ROS-initiated macromolecular damage has been implicated in the mechanism of phenytoin teratogenesis, this drug also increases the embryonic levels of signal transduction proteins Ras and NF-κB (Fig. 1.9) [34, 107], as is typically observed during oxidative stress. The embryopathic effects of phenytoin were blocked by pretreatment with either a farnesyltransferase that prevents posttranslational Ras activation or an antisense oligonucleotide inhibitor of NF-κB, implicating this signal transduction pathway in the mechanism of teratogenesis. Thalidomide is another example of a drug for which multiple mechanisms of teratogenesis have been implicated, including some potentially unrelated to oxidative stress [47]. In regard to oxidative stress, thalidomide and at least two of its metabolites can enhance embryonic ROS formation, at least in part via bioactivation by embryonic prostaglandin H synthases (PHSs) to a reactive intermediate, which can alter embryonic signal transduction pathways (see below) and/or oxidatively damage embryonic DNA [25, 47, 99].

Mechanisms of ROS-initiated signal transduction in development have been reviewed in detail elsewhere [14, 25, 99]. Signal transduction pathways involving ROS and RNS are tightly and temporally regulated within cell types, subcellular organelles, and microenvironments within individual proteins and lipids [29], which likely accounts at least in part for the distinctive patterns of abnormal structural and/or functional development caused by different teratogens. Among the various forms of ROS, H_2O_2 is relatively stable and diffusible and hence is commonly implicated in signal transduction via the selective oxidation of sulfhydryl groups of specific cysteine residues, which is reversible at lower H_2O_2 concentrations, although higher concentrations can result in irreversible changes [24]. This usually reversible oxidation is determined by the redox state of the cell, which is regulated by the ratios of several redox "couples" (Fig. 1.10), including cysteine/cystine, GSH/glutathione disulfide, and, for both forms of thioredoxin (TRX1, TRX2), $TRX_{reduced}/TRX_{oxidized}$, all of which are localized in different cellular compartments. These redox couples are believed to regulate developmental pathways for cellular proliferation, differentiation, apoptosis, and necrosis in a fashion that is selective for the species, strain, time of gestation, target tissue and cell, and cellular compartment, thereby accounting at least in part for the distinctive teratological profiles that are characteristic of different teratogens [25]. As the cellular redox state of the developing embryo moves from an anaerobic reducing environment, which favors cellular proliferation, to an oxidizing environment, which favors differentiation, the contributions of these redox couples change according to the distinctive redox potentials of their components, contributing to the variably distinctive consequences of in utero exposure to different ROS-initiating teratogens [25]. ROS-mediated oxidation of selective cysteine residues is implicated in signaling by several redox-sensitive endogenous ligands (Table 1.5) [14] and by tumor necrosis factor alpha and epidermal growth factor, both of which stimulate ROS formation [25], and for teratogens like thalidomide, which is postulated to disrupt NF-κB signaling leading to changes in multiple genes involved in limb growth (Fig. 1.11) [25].

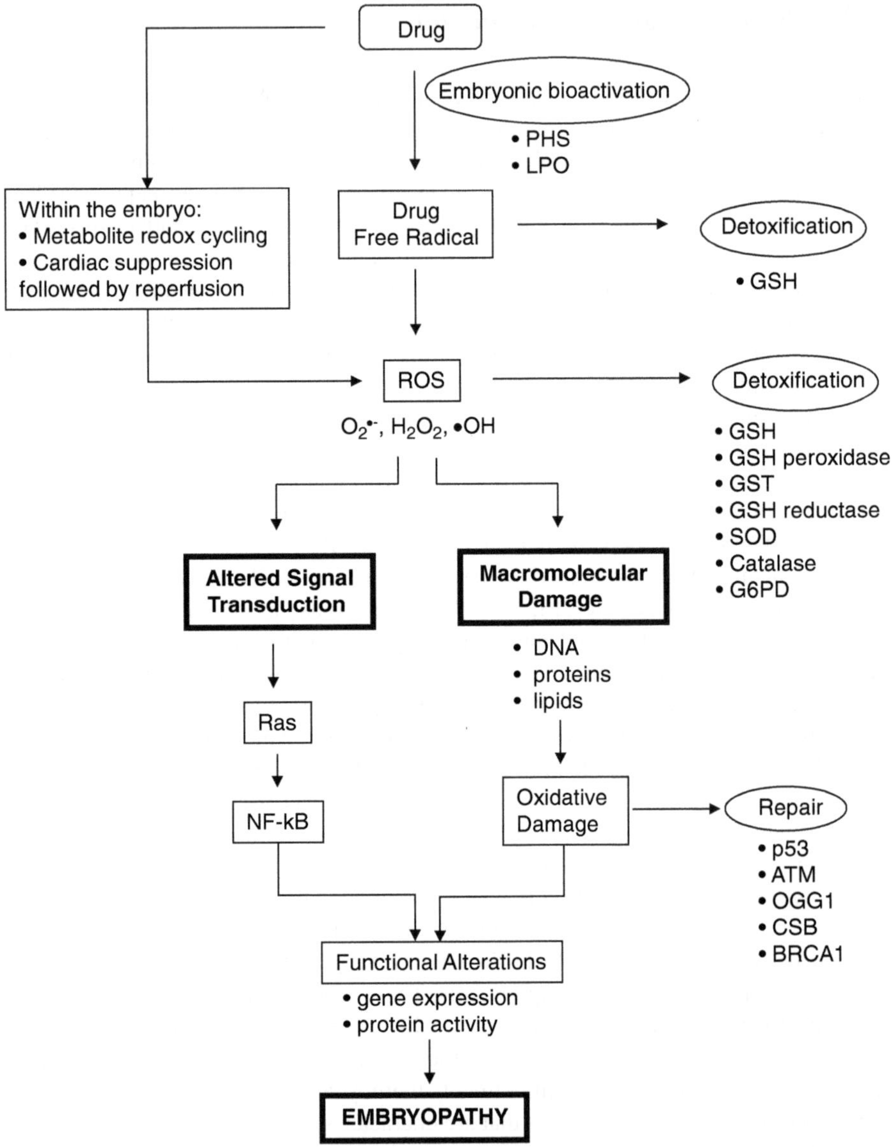

Fig. 1.9 Dual effects of ROS in initiating signal transduction and oxidatively damaging structural macromolecules, which are not mutually exclusive. Potential contribution of Ras and NF-κB proteins in signal transduction pathways initiated by drug-enhanced formation of ROS (Modified from Kennedy et al. [34])

Oxidative stress can affect signaling in other ways, including, as examples, (1) reducing DNA and histone methylation reactions indirectly via the diversion of homocysteine to GSH synthesis, (2) altered gene transcription inactivation via reduced stability of hypoxia-inducible factor (HIF), and (3) altered gene transcription via an increase in sirtuin-dependent deacetylation of some transcription factors (increased transcription) or histone proteins (decreased transcription) [99].

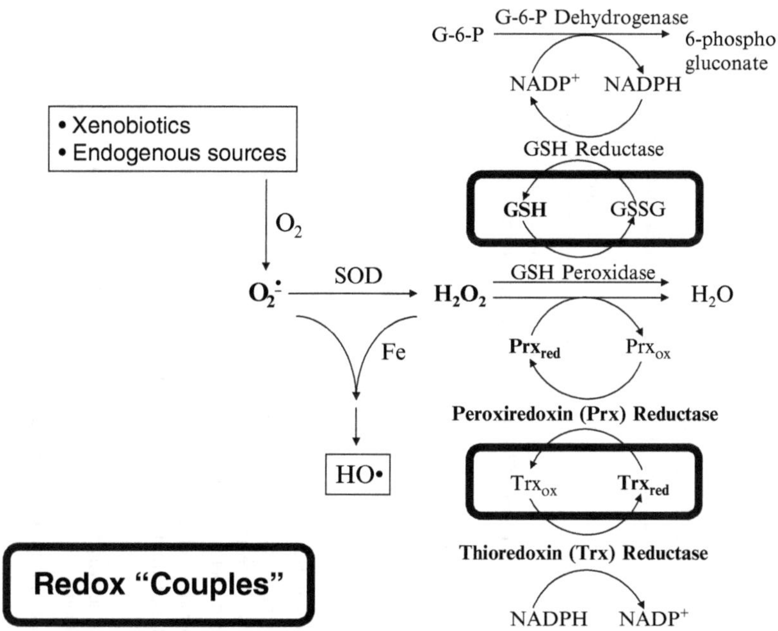

Fig. 1.10 Redox couples modulating the cellular effects of endogenous or xenobiotic-enhanced oxidative stress. Redox couples in addition to those circled include NADPH/$NADP^+$ and Prx_{red}/Prx_{ox} shown above and cysteine/cystine (not shown). Abbreviations: *ox* oxidized, *red* reduced (From Wells et al. [99])

Antioxidative Pathways

In the absence of adequate antioxidative protection within the embryo and fetus, even physiological levels of conceptal ROS production can be developmentally toxic (Table 1.6). Untreated progeny with genetic deficiencies in glucose-6-phosphate dehydrogenase (G6PD) [67] or catalase [2, 3] exhibit increased embryopathies compared to wild-type littermates with normal activities of these antioxidative enzymes, even though embryonic and fetal activities of enzymes like catalase are less than 10 % of maternal activity. A role for endogenous ROS has been similarly implicated in the in utero initiation of postnatal carcinogenesis in cancer-prone p53 knockout mice: dietary supplementation with low-dose vitamin E reduced conceptal DNA oxidation and postnatal tumorigenicity [12], whereas high-dose vitamin had an opposite, pro-oxidant effect, enhancing conceptal DNA oxidation and postnatal tumorigenicity [13].

As discussed earlier for ROS formation and the predominant importance of individual embryonic or fetal activity over maternal activity or the average activity for a litter, due to the instability of ROS, proximate antioxidative protection within the conceptus, as distinct from maternal activities, is critical. The same will be reiterated for conceptal DNA repair in the following section. This observation is evident in genetically altered mice with either deficient or enhanced activities of catalase, in

Table 1.5 Modulation of representative antioxidants and antioxidative enzymes implicating endogenous and/or xenobiotic-enhanced reactive oxygen species (ROS) in the mechanism of developmental toxicity

Component	Drug/model	Modulation and outcome	Citation
Antioxidants			
Vitamin E	Phenytoin	Maternal pretreatment with lower doses of vitamin E reduces phenytoin teratogenicity	[12, 97]
	Phenytoin	Paradoxically, higher vitamin E doses have a pro-oxidant effect, enhancing phenytoin teratogenicity	[13, 96]
	p53 knockout mice	Low-dose dietary supplementation with vitamin E reduces oxidatively damaged DNA in fetal tissues and reduces spontaneous postnatal tumorigenicity in cancer-prone p53 mice	[12, 97]
	p53 knockout mice	High-dose dietary supplementation with vitamin E has a paradoxical pro-oxidant effect, enhancing DNA oxidation in fetal tissues of p53 knockout mice and increasing spontaneous tumorigenicity	[13]
	Ethanol (EtOH)	Maternal dietary supplementation with vitamin E decreased fetal malformations and fetal hepatic 8-iso-PGF(2alpha) levels	[103]
Caffeic acid	Phenytoin	Maternal pretreatment with caffeic acid reduces phenytoin teratogenicity	[101]
Propyl gallate	Hydroxyurea	Maternal treatment with propyl gallate reduces hydroxyurea teratogenicity	[15]
Glutathione (GSH)	DEM	Depletion with diethyl maleate (DEM) enhances the teratogenicity of thalidomide, 5-fluorouracil, cadmium hydrochloride, and phenytoin	[41, 63, 109]
	BSO	Inhibition of GSH synthesis with buthionine sulfoximine (BSO) decreases baseline growth parameters in embryo culture and enhances the teratogenic effects of methanol, phenytoin, 5-fluorouracil, and diabetes	[23, 26, 58, 62, 84, 109]
	GSH precursors	Pretreatment with GSH precursors including N-acetylcysteine and GSH esters reduces the teratogenic effects of cyclophosphamide, thalidomide, phenytoin, and diabetes	[22, 41, 84, 110]
Antioxidative enzymes			
GSH peroxidase	Phenytoin	Reduction in GSH peroxidase activity by dietary restriction of selenium cofactor enhances phenytoin teratogenicity	[72]
	High glucose	Addition of GSH peroxidase to embryo culture medium reduces glucose-initiated embryopathies	[19]
GSH reductase	Phenytoin	Inhibition of GSH reductase by a low, non-teratogenic dose of bis-chloroethyl-nitrosourea (BCNU) enhances phenytoin teratogenicity	[111]

Superoxide dismutase (SOD)	Phenytoin, benzo[a]pyrene	Addition of polyethylene glycol-conjugated superoxide dismutase (PEG-SOD) to embryo culture medium reduces oxidative macromolecular damage and embryopathies caused by phenytoin and benzo[a]pyrene (this protection is not observed in vivo, where maternal pretreatment with PEG-SOD paradoxically enhances phenytoin teratogenicity)	[105, 106]
	High glucose	Addition to SOD to embryo culture medium reduces glucose-initiated embryopathies	[19]
	Transgenic SOD expression with EtOH, diabetes	Transgenic mice expressing increased SOD were protected against malformations caused by EtOH and streptozotocin-induced diabetes, while SOD wild-type mice exhibited malformations and fetal loss	[21, 102]
	SOD knockout mice with EtOH	SOD knockout mice were more susceptible to EtOH-induced teratogenesis and fetal hepatic isoprostane formation	[102]
Catalase	Phenytoin, benzo[a]pyrene	Addition of PEG-catalase to embryo culture medium reduces oxidative macromolecular damage and embryopathies caused by phenytoin and benzo[a]pyrene	[2, 105]
	Phenytoin, EtOH	In vivo maternal pretreatment with PEG-catalase reduces phenytoin and EtOH teratogenicity	[3, 56]
	Genetic catalase-deficiency untreated and with phenytoin, EtOH, methanol	Untreated mutant mice deficient in catalase exhibit greater embryopathies than wild-type controls and are more susceptible to macromolecular damage and embryopathies caused phenytoin, EtOH, and methanol in embryo culture and/or in vivo	[2, 56, 57]
	Transgenic catalase expression untreated and with phenytoin, EtOH, methanol	Untreated transgenic mice with enhanced catalase activity exhibit fewer embryopathies than wild-type controls and are protected from macromolecular damage and embryopathies caused by phenytoin, EtOH, and methanol in embryo culture and/or in vivo	[2, 56, 57]
	High glucose	Addition of catalase to embryo culture medium reduces glucose-initiated embryopathies	[19]
Glucose-6-phosphate dehydrogenase (G6PD)	Genetic G6PD-deficiency	Untreated mutant mice deficient in G6PD exhibit enhanced in utero and postnatal death compared to wild-type littermates	[67]
	Genetic G6PD-deficiency with phenytoin	G6PD-deficient progeny are more susceptible to phenytoin-initiated DNA oxidation and teratogenicity	[67]
Thioredoxin (Trx2)	*trx2* knockout mice	Untreated homozygous *trx2* knockout mice exhibit open neural tubes and apoptosis at embryonic day (E) 10.5 and die by E 12.5	[68]

Abbreviations: *BCNU* bis-chloroethyl-nitrosourea, *BSO* buthionine sulfoximine, *DEM* diethyl maleate, *E* embryonic day, *EtOH* ethanol, *G6PD* glucose-6-phosphate dehydrogenase, *GSH* glutathione, *p53* protein 53, *PEG* polyethylene glycol-conjugated, *ROS* reactive oxygen species, *Trx2* thioredoxin 2

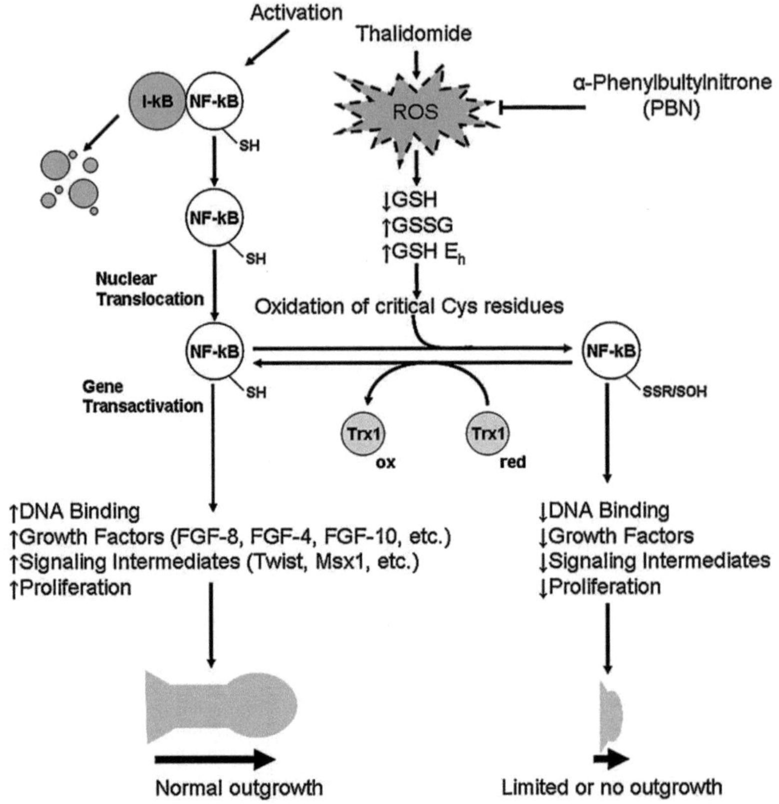

Fig. 1.11 A postulated role for ROS signaling in thalidomide teratogenicity. Normal limb outgrowth requires both the activation of NF-κB in the cytosol via oxidative and non-oxidative mechanisms, followed by translocation of NF-κB to the nucleus and binding to DNA. Thalidomide enhances ROS, which disrupt NF-κB activation and/or DNA binding, causing a loss of gene expression, with limited or absent limb growth. This effect is blocked by the free radical spin trapping agent phenylbutylnitrone (PBN). Abbreviations: *cys* cysteine, *FGF* fibroblast growth factor, *GSH* reduced glutathione, *GSSG* oxidized glutathione, *I-κB* inhibitor of NF-κB, *NF-κB* nuclear factor kappa-light-chain-enhancer of activated B cells, *ox* oxidized, *PBN* phenylbutylnitrone, *red* reduced, *ROS* reactive oxygen species, *SH* sulfhydryl group, *SOH* oxidized sulfhydryl group, *SSR* reduced sulfhydryl group, *Trx* thioredoxin, *Eh* redox potential (From Hansen and Harris [25])

which embryopathic susceptibility to phenytoin [2, 3], methanol [57], or ethanol [56] is dependent upon the embryonic genotype, with increasing effect of the genetic alteration going from wild-type to heterozygous to homozygous gene modification (knockout deficiency or transgenic enhancement), even though these embryos are littermates and exposed to the same drug concentrations and uterine environment. A similar embryonic gene determinant of susceptibility, as distinct from a maternal or litter determinant, has been observed for G6PD in the developmental toxicity of phenytoin.

Table 1.6 Modulation of DNA damage response and repair proteins and enzymes implicating oxidatively damaged DNA, and particularly the 8-oxoguanine (8-oxoG) lesion, in the pathogenic mechanism of physiological and xenobiotic-enhanced oxidative stress and developmental toxicity

Protein	Modulation and outcome	Reference
ATM	Untreated *atm* knockout mice exhibit enhanced embryopathies compared to wild-type controls	[9]
	atm knockout mice are more susceptible than wild-type littermates to developmental toxicity caused by ionizing radiation and phenytoin in vivo and/or in embryo culture	[8, 45]
p53	*p53* knockout mice are more susceptible to benzo[a]pyrene teratogenicity	[66]
	Cultured embryonic limb buds from *p53* knockout mice exhibited increased malformations caused by 4-hydroperoxycyclophosphamide, the activated metabolite of cyclophosphamide	[59]
	p53 knockout mice are more susceptible to malformations caused by ionizing radiation	[69]
OGG1	In vitro studies using *ogg1* knockout cells, which cannot repair the 8-oxoG lesion, show a higher level of DNA double-strand breaks (gamma-H2AX) and cytotoxicity when exposed to methylmercury	[70]
	In vitro studies using transgenic cells overexpressing OGG1 show a lower level of 8-oxoG, gamma-H2AX, and cytotoxicity compared to wild-type control cells when exposed to H_2O_2, menadione, cisplatin, oxaliplatin, phenytoin, and/or methylmercury	[71, 79]
	In vivo, untreated *ogg1* knockout mice exhibit postnatal neurodevelopmental deficits compared to wild-type controls	[53]
	ogg1 knockout mice exhibit higher levels of oxidatively damaged DNA than wild-type controls and are more susceptible to birth defects and postnatal neurodevelopmental deficits caused by *in utero* exposure to methamphetamine and ethanol	[53, 108]
FPG	In vitro studies using transgenic cells expressing high levels of FPG, the bacterial homolog of OGG1, show a lower level of cytotoxicity compared to wild-type control cells when exposed to H_2O_2, menadione, cisplatin, oxaliplatin, and/or phenytoin	[79]
CSB	*csb* knockout mice exhibit higher levels of oxidatively damaged DNA than wild-type controls and are more susceptible to birth defects and postnatal neurodevelopmental deficits caused by in utero exposure to methamphetamine	[51]
BRCA1	*brca1* conditional knockout embryos exposed in culture to methamphetamine and alcohol exhibit higher levels of 8-oxoG, gamma H2AX, and embryopathies compared to wild-type controls	[88]

Abbreviations: *8-oxoG* 8-oxoguanine, *ATM* ataxia telangiectasia mutated, *BRCA1* breast cancer susceptibility protein 1, *CSB* Cockayne syndrome B, *FPG* formamidopyrimidine glycosylase, *gamma-H2AX* phosphorylated histone 2 AX protein, OGG1 oxoguanine glycosylase 1, *p53* protein 53

Pretreatment with lipid-soluble vitamin E or water-soluble caffeic acid reduces the teratogenicity of drugs like phenytoin (Table 1.6) [99]. Similarly, preincubation with exogenous polyethylene glycol (PEG)-conjugated (stabilized) forms of superoxide dismutase (SOD) and catalase reduces the oxidative macromolecular damage and embryopathies caused by xenobiotics like phenytoin and benzo[a]pyrene in embryo culture, as is also observed in genetically modified mice overexpressing endogenous catalase activity when exposed to phenytoin, ethanol, and methanol [1–3, 56, 57, 99, 100]. In vivo, maternal pretreatment with PEG-catalase increases embryonic catalase activity and reduces phenytoin teratogenicity, although in vivo pretreatment with PEG-SOD conversely enhances phenytoin teratogenicity, possibly due to the accumulation of H_2O_2 [106]. Catalase overexpressing mice are similarly resistant to structural birth defects and/or postnatal neurodevelopmental deficits caused by in utero exposure to phenytoin and ethanol [1, 2, 56].

Conversely, depletion of cellular antioxidants like GSH, inhibition of antioxidative enzymes like GSH peroxidase and GSH reductase, or the use of genetically modified mice with a deficiency in catalase or G6PD enhance the macromolecular damage and teratogenicity caused by in utero exposure to xenobiotics like phenytoin, ethanol, and methanol in embryo culture and/or in vivo [1–3, 56, 57, 99]. While results from embryo culture and in vivo studies are similar for most teratogens, one exception is methanol, for which evidence of a ROS-dependent mechanism of developmental toxicity in embryo culture is comprehensive [57, 100], while in vivo results suggest an alternative mechanism [90].

DNA Repair

The genetic engineering of cells and mouse models with altered DNA repair activity has provided the opportunity for evaluating the pathogenic role of oxidatively damaged DNA in developmental toxicity (Table 1.6), as distinct from the contributions of oxidative damage to other cellular macromolecules like proteins and lipids or receptor-mediated mechanisms. In addition, these models allow a selective evaluation of the relative contribution of ROS-initiated DNA oxidation as distinct from ROS-initiated alterations in signal transduction. The complementary insight provided by these models is the importance of embryonic and fetal DNA repair in protecting the conceptus from the developmental toxicity of endogenous and xenobiotic-enhanced conceptal ROS formation.

As with antioxidative pathways, studies in untreated genetically modified mice with deficient DNA repair proteins have revealed a potentially pathogenic role for oxidatively damaged DNA resulting from physiological levels of developmental ROS formation. For example, the ataxia telangiectasia-mutated (ATM) protein and p53 are critical components of the pathways for the detection of oxidatively damaged DNA and the transduction of the signal for DNA repair or apoptotic cell death

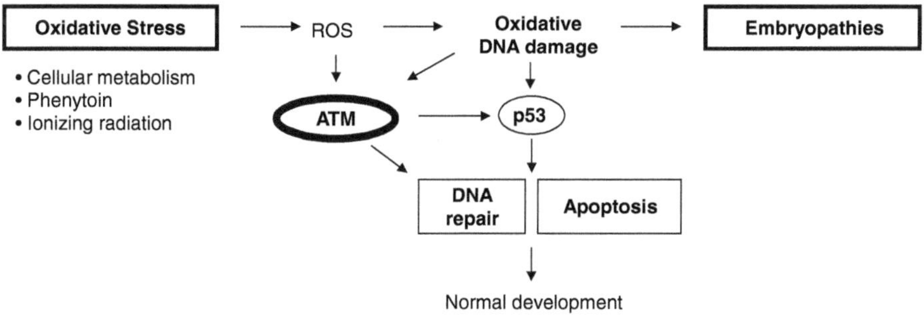

Fig. 1.12 The relation of the ATM and p53 proteins in the cellular DNA repair process or apoptosis following DNA damage initiated by oxidative stress (From Bhuller and Wells [9])

(Fig. 1.12). The DNA repair-deficient progeny of untreated ATM knockout mice exhibit enhanced embryopathies in culture compared to wild-type littermates [9]. Similarly, untreated knockout mice lacking oxoguanine glycosylase 1 (OGG1), the major enzyme for repair of the 8-oxoG lesion in DNA, exhibit enhanced postnatal neurodevelopmental deficits compared to their wild-type littermates [100].

In genetically modified cells and/or mouse models with a deficiency in ATM, p53, OGG1, the DNA repair protein Cockayne syndrome B (CSB) or the DNA repair-coordinating protein breast cancer 1 (BRCA1), DNA oxidation and the cytotoxicity and/or embryopathic effects of xenobiotics like phenytoin, benzo[a]pyrene, methamphetamine, and ethanol, as well as ionizing radiation, in vivo and/or in embryo culture, are enhanced [51, 80, 88, 100]. Similarly in vitro, *ogg1* knockout cells are more sensitive to DNA damage and cytotoxicity caused by methylmercury [70, 71]. In some cases, the picture can be less straightforward. In the case of p53, which in response to DNA damage can transduce a signal either for DNA repair or apoptosis, the modulatory effect of this protein may vary with the drug and target tissue. Accordingly, while p53 knockout mice are more susceptible to the full spectrum of teratogenic effects of benzo[a]pyrene [66], they are protected from the ocular teratogenic effect of 2-chloro-2′-deoxyadenosine [112], in the latter case presumably due to a reduced signal for apoptosis.

Conversely, cells genetically engineered to express enhanced levels of OGG1 or its bacterial homolog, formamidopyrimidine glycosylase (FPG), exhibit reduced levels of oxidatively damaged DNA, DNA double-strand breaks, identified by the phosphorylation of histone protein 2AX (gamma H2AX), and cytotoxicity caused by H_2O_2, menadione, cisplatin, oxaliplatin, phenytoin, and/or methylmercury [71, 80].

The protective effects of the DNA damage detection and response proteins ATM, p53, and BRCA1 suggest that oxidatively damaged DNA plays an important pathogenic role in the developmental toxicity of endogenous ROS or xenobiotic-enhanced ROS formation, while the protective effects of OGG1, CSB, and FPG suggest that 8-oxoG specifically is a pathogenic lesion.

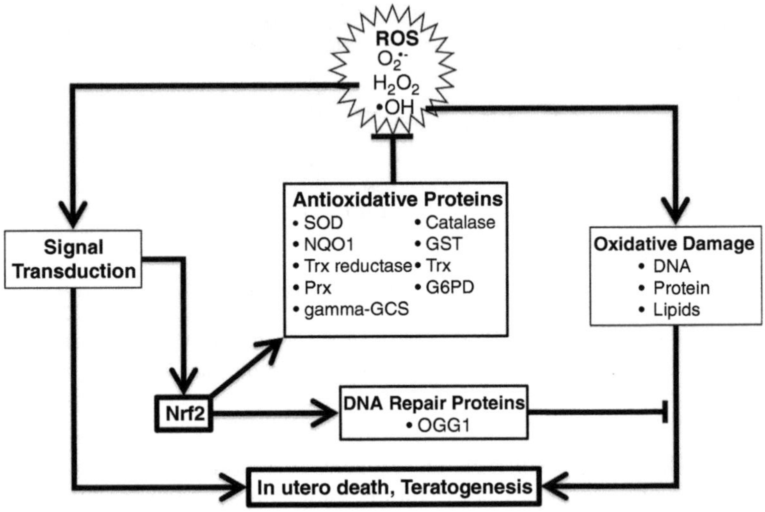

Fig. 1.13 Central role of embryonic and fetal nuclear factor erythroid 2-related factor 2 (Nrf2) in protecting progeny from endogenous and xenobiotic-enhanced oxidative stress. ROS can enhance both the activation of conceptal Nrf2 in the cytoplasm and its binding to DNA in the nucleus, stimulate the comprehensive expression of a battery of proteins and enzymes that blocks redox cycling, detoxify reactive oxygen species, and repair oxidatively damaged DNA. Abbreviations: *gamma-GCS* gamma-glutamyl-cysteine synthase, *G6PD* glucose-6-phosphate dehydrogenase, *GST* glutathione S-transferase, *NQO1* NAD(P)H quinone oxidoreductase, *Prx* peroxiredoxin, *SOD* superoxide dismutase, *Trx* thioredoxin

Regulation of Embryonic and Fetal ROS-Protective Pathways by Nuclear Factor Erythroid 2-Related Factor 2 (Nrf2) in Teratogenesis

Nrf2 is a transcription factor that regulates the transcription of numerous proteins, and particularly several important proteins and enzymes that protect against oxidative stress, including (1) NAD(P)H:quinone oxidoreductase (NQO1), which blocks the redox cycling of catechol metabolites; (2) antioxidative peptides and enzymes, including the GSH synthetic enzyme gamma-glutamyl-cysteine synthase, the glutamate/cystine transporter, TRX1, TRX reductase-1, peroxiredoxins, SOD, catalase, and G6PD; and (3) the DNA repair enzyme OGG1 (Fig. 1.13) [25, 81]. Nrf2 in the cytoplasm is activated by the ROS-initiated oxidation of sulfhydryl groups in cysteine residues of its repressor protein, Kelch-like ECH-associated protein 1 (Keap1), which releases Nrf2 from Keap1 and allows it to translocate to the nucleus, where it heterodimerizes with other proteins and binds to the antioxidant response element (ARE), thereby activating gene transcription. The DNA binding of Nrf2 also is redox-sensitive. Cytosolic activation of Nrf2 is primarily controlled by the GSH redox couple, while Nrf2 DNA binding is controlled by the nuclear Trx1 couple. Little is known about the role of Nrf2 in development. Although the viability and apparently normal development of Nrf2 knockout mice suggests a negligible role, Nrf2 mRNA and protein are expressed during organogenesis [11]. Moreover, pretreatment of pregnant mice with 3H-1,2-dithiole-3-thione (D3T), a Nrf2 activator, protects their progeny from apoptosis and malformations caused by in utero exposure to the ROS-initiating teratogen ethanol [17], suggesting that Nrf2 serves an important embryoprotective role. More recently, the central role of Nrf2 in protecting the fetus from oxidative stress was corroborated in Nrf2 knockout mice, the Nrf2-deficient progeny of which when exposed in utero to the ROS-initiating teratogen methamphetamine exhibited increased oxidatively damaged DNA in fetal brain, and more severe postnatal neurodevelopmental deficits in activity and olfaction, compared to wild-type littermates [81]. Accordingly, as in adult tissues, Nrf2 appears to serve in the embryo and fetus as a master switch that, in response to either endogenous or xenobiotic-enhanced oxidative stress, activates the transcription of a comprehensive battery of enzymes and proteins that interrupt ROS formation, detoxify ROS, and repair DNA that has been oxidatively damaged by ROS (Fig. 1.12).

Conclusions and Future Considerations

Oxidative stress in the developing embryo and fetus can adversely affect both gross morphological and functional development. This stress can occur with both physiological levels of ROS when conceptal ROS detoxification or macromolecular repair is deficient and with elevated conceptal ROS levels initiated by xenobiotics, ionizing radiation, and other ROS-initiating environmental factors. Mechanisms of teratogenesis include both alterations in ROS-mediated signal transduction pathways and ROS-initiated macromolecular damage, which are not necessarily mutually exclusive and in some cases may also involve ROS-independent mechanisms. At least in animal models (primarily rodents and rabbits), susceptibility is substantially determined by the conceptal balance among pathways for ROS formation, ROS detoxification, and repair of oxidative macromolecular damage, any of which can differ markedly among littermates.

Our understanding of the full role of ROS-mediated macromolecular damage will benefit from the development of more definitive approaches to establish a causal or pathogenic role for specific conceptal cases of lipid peroxidation and protein oxidation, similar to the use of knockout mice deficient in particular DNA repair proteins for revealing the embryopathic role of DNA oxidation and repair. This includes the role of lipid peroxidation products in subsequent reactions with proteins and DNA. More definitive studies of the cross-talk between ROS-mediated signal transduction and DNA oxidation, the latter of which could be considered a form of epigenetic signaling, will likely provide important new insights into the mechanisms of ROS-mediated teratogenesis.

Acknowledgment Research from the authors' laboratory was supported by grants from the Canadian Institutes of Health Research (CIHR). AMS was supported in part by a CIHR Frederick Banting and Charles Best Canada Graduate Scholarship.

References

1. Abramov JP, Tran A, Shapiro AM, Wells PG. Protective role of endogenous catalase in baseline and phenytoin-enhanced neurodevelopmental and behavioral deficits initiated in utero and in aged mice. Reprod Toxicol. 2012;33:361–73.
2. Abramov JP, Wells PG. Embryonic catalase protects against endogenous and phenytoin-enhanced DNA oxidation and embryopathies in acatalasemic and human catalase-expressing mice. FASEB J. 2011;25:2188–200.
3. Abramov JP, Wells PG. Embryoprotective role of endogenous catalase in acatalasemic and human catalase-expressing mouse embryos exposed in culture to developmental and phenytoin-enhanced oxidative stress. Toxicol Sci. 2011;120:428–38.
4. Albano E, Tomasi A, Goria-Gatti L, Dianzani MU. Spin trapping of free radical species produced during the microsomal metabolism of ethanol. Chem Biol Interact. 1988;65:223–34.
5. Arlen RR, Wells PG. Inhibition of thalidomide teratogenicity by acetylsalicylic acid: evidence for prostaglandin H synthase-catalyzed bioactivation of thalidomide to a teratogenic reactive intermediate. J Pharmacol Exp Ther. 1996;277:1649–58.
6. Azarbayjani F, Borg LAH, Danielsson BR. Increased susceptibility to phenytoin teratogenicity: excessive generation of reactive oxygen species or impaired antioxidant defense? Basic Clin Pharmacol Toxicol. 2006;99:305–11.
7. Bassi C, Ho J, Srikumar T, Dowling RJO, Gorrini C, Miller SJ, Mak TW, Neel BG, Raught B, Stambolic V. Nuclear PTEN controls DNA repair and sensitivity to genotoxic stress. Science. 2013;341:395–9.
8. Bhuller Y, Jeng W, Wells PG. Variable in vivo embryoprotective role for ataxia-telangiectasia-mutated against constitutive and phenytoin-enhanced oxidative stress in atm knockout mice. Toxicol Sci. 2006;93:146–55.
9. Bhuller Y, Wells PG. A developmental role for ataxia-telangiectasia mutated in protecting the embryo from spontaneous and phenytoin-enhanced embryopathies in culture. Toxicol Sci. 2006;93:156–63.
10. Cantone I, Fisher AG. Epigenetic programming and reprogramming during development. Nat Struct Mol Biol. 2013;20:282–9.
11. Chan K, Lu R, Chang JC, Kan YW. Proc Natl Acad Sci U S A. 1996;93:13943–8.
12. Chen CS, Squire JA, Wells PG. Reduced tumorigenesis in p53 knockout mice exposed in utero to low-dose vitamin E. Cancer. 2009;115:1563–75.
13. Chen CS, Wells PG. Enhanced tumorigenesis in p53 knockout mice exposed in utero to high-dose vitamin E. Carcinogenesis. 2006;27:1358–68.
14. Dennery PA. Effects of oxidative stress on embryonic development. Birth Defects Res C Embryo Today. 2002;81:155–62.
15. DeSesso JM. Amelioration of teratogenesis. I. Modification of hydroxyurea-induced teratogenesis by the antioxidant propyl gallate. Teratology. 1981;24:19–35.
16. Devi BG, Schenker S, Mazloum B, Henderson GI. Ethanol-induced oxidative stress and enzymatic defenses in cultured fetal rat hepatocytes. Alcohol. 1996;13:327–32.
17. Dong J, Sulik KK, S-y C. Nrf2-mediated transcriptional induction of antioxidant response in mouse embryos exposed to ethanol in vivo: implications for the prevention of fetal alcohol spectrum disorders. Antioxid Redox Signal. 2008;10:2023–33.
18. Dong J, Sulik KK, S-y C. The role of NOX enzymes in ethanol-induced oxidative stress and apoptosis in mouse embryos. Toxicol Lett. 2010;193:94–100.
19. Eriksson UJ, Borg LA. Protection by free oxygen radical scavenging enzymes against glucose-induced embryonic malformations in vitro. Diabetologia. 1991;34:325–31.
20. Fantel AG, Person RE. Further evidence for the role of free radicals in the limb teratogenicity of L-NAME. Teratology. 2002;66:24–32.
21. Hagay ZJ, Weiss Y, Zusman I, Peled-Kamar M, Reece EA, Eriksson UJ, Groner Y. Prevention of diabetes-associated embryopathy by overexpression of the free radical scavenger copper zinc superoxide dismutase in transgenic mouse embryos. Am J Obstet Gynecol. 1995;173:1036–41.

22. Hales BF. Modification of the teratogenicity and mutagenicity of cyclophosphamide with thiol compounds. Teratology. 1981;23:373–81.
23. Hales BF, Brown H. The effect of in vivo glutathione depletion with buthionine sulfoximine on rat embryo development. Teratology. 1991;44:251–7.
24. Halliwell B, Gutteridge JMC. Free radicals in biology and medicine. 4th ed. New York: Oxford University Press; 2007.
25. Hansen JM, Harris C. Redox control of teratogenesis. Reprod Toxicol. 2013;35:165–79.
26. Harris C, Dixon M, Hansen JM. Glutathione depletion modulates methanol, formaldehyde and formate toxicity in cultured rat conceptuses. Cell Biol Toxicol. 2004;20:133–45.
27. Hiranruengchok R, Harris C. Glutathione oxidation and embryotoxicity elicited by diamide in the developing rat conceptus in vitro. Toxicol Appl Pharmacol. 1993;120:62–71.
28. Inoguchi T, Li P, Umeda F, Yu HY, Kakimoto M, Imamura M, Aoki T, Etoh T, Hashimoto T, Naruse M, Sano H, Utsumi H, Nawata H. High glucose level and free fatty acid stimulate reactive oxygen species production through protein kinase C – dependent activation of NAD(P)H oxidase in cultured vascular cells. Diabetes. 2000;49:1939–45.
29. Janssen-Heininger YMW, Mossman BT, Heintz NH, Forman HJ, Kalyanaraman B, Finkel T, Stamler JS, Rhee SG, van der Vliet A. Redox-based regulation of signal transduction: principles, pitfalls, and promises. Free Radic Biol Med. 2008;45:1–17.
30. Jeng W, Ramkissoon A, Parman T, Wells PG. Prostaglandin H synthase-catalyzed bioactivation of amphetamines to free radical intermediates that cause CNS regional DNA oxidation and nerve terminal degeneration. FASEB J. 2006;20:638–50.
31. Jovanovic SV, Simic MG. One-electron redox potentials of purines and pyrimidines. J Phys Chem. 1986;90:974–8.
32. Kadiiska MB, Mason RP. Acute methanol intoxication generates free radicals in rats: an ESR spin trapping investigation. Free Radic Biol Med. 2000;28:1106–14.
33. Kasapinovic S, McCallum GP, Wiley MJ, Wells PG. The peroxynitrite pathway in development: phenytoin and benzo[a]pyrene embryopathies in inducible nitric oxide synthase (iNOS) knockout mice. Free Radic Biol Med. 2004;37:1703–11.
34. Kennedy JC, Memet S, Wells PG. Antisense evidence for nuclear factor kB-dependent embryopathies initiated by phenytoin-enhanced oxidative stress. Mol Pharmacol. 2004;66:404–12.
35. Khobta A, Anderhub S, Kitsera N, Epe B. Gene silencing induced by oxidative DNA base damage: association with local decrease of histone H4 acetylation in the promoter region. Nucleic Acids Res. 2010;38:4285–95.
36. Kim PM, Wells PG. Phenytoin-initiated hydroxyl radical formation: characterization by enhanced salicylate hydroxylation. Mol Pharmacol. 1996;49:172–81.
37. Kitsera N, Stathis D, Luhnsdorf B, Muller H, Carell T, Epe B, Khobta A. 8-Oxo-7,8-dihydroguanine in DNA does not constitute a barrier to transcription, but is converted into transcription-blocking damage by OGG1. Nucleic Acids Res. 2011;39:5926–34.
38. Klungland A, Bjelland S. Oxidative damage to purines in DNA: role of mammalian Ogg1. DNA Repair. 2007;6:481–8.
39. Klungland A, Rosewell I, Hollenbach S, Larsen E, Daly G, Epe B, Seeberg E, Lindahl T, Barnes DE. Accumulation of premutagenic DNA lesions in mice defective in removal of oxidative base damage. Proc Natl Acad Sci U S A. 1999;96:13300–5.
40. Knecht KT, Bradford BU, Mason RP, Thurman RG. In vivo formation of a free radical metabolite of ethanol. Mol Pharmacol. 1990;38:26–30.
41. Knobloch J, Reimann K, Klotz LO, Ruther U. Thalidomide resistance is based on the capacity of the glutathione-dependent antioxidant defense. Mol Pharm. 2008;5:1138–44.
42. Kotch LE, Chen SY, Sulik KK. Ethanol-induced teratogenesis: free radical damage as a possible mechanism. Teratology. 1995;52:128–36.
43. Kubow S, Wells PG. In vitro bioactivation of phenytoin to a reactive free radical intermediate by prostaglandin synthetase, horseradish peroxidase, and thyroid peroxidase. Mol Pharmacol. 1989;35:504–11.
44. Kwon J, Lee S-R, Yang K-S, Ahn Y, Kim YJ, Stadtman ER, Rhee SG. Reversible oxidation and inactivation of the tumor suppressor PTEN in cells stimulated with peptide growth factors. Proc Natl Acad Sci U S A. 2004;101:16419–24.

45. Laposa RR, Henderson JT, Xu E, Wells PG. Atm-null mice exhibit enhanced radiation-induced birth defects and a hybrid form of embryonic cell death indicating a teratological suppressor function for ATM. FASEB J. 2004;18:896–8.
46. Le Belle JE, Orozco NM, Paucar AA, Saxe JP, Mottahedeh J, Pyle AD, Wu H, Kornblum HI. Proliferative neural stem cells have high endogenous ROS levels that regulate self-renewal and neurogenesis in a PI3K/Akt-dependant manner. Cell Stem Cell. 2011;8:59–71.
47. Lee CJJ, Goncalves LL, Wells PG. Embryopathic effects of thalidomide and its hydrolysis products in rabbit embryo culture: evidence for a prostaglandin H synthase (PHS)-dependent, reactive oxygen species (ROS)-mediated mechanism. FASEB J. 2011;25:2468–83.
48. Lee RD, An SM, Kim SS, Rhee GS, Kwack SJ, Seok JH, Chae SY, Park CH, Choi YW, Kim HS, Cho HY, Lee BM, Park KL. Neurotoxic effects of alcohol and acetaldehyde during embryonic development. J Toxicol Environ Health. 2005;68:2147–62.
49. Liu L, Wells PG. In vivo phenytoin-initiated oxidative damage to proteins and lipids in murine maternal hepatic and embryonic tissue organelles: potential molecular targets of chemical teratogenesis. Toxicol Appl Pharmacol. 1994;125:247–55.
50. Maruyama A, Kusama T. Measurement of DNA damage in ICR mouse embryos at preimplantation stage using a comet assay. Congenit Anom. 1998;38:375–83.
51. McCallum GP, Wong AW, Wells PG. Cockayne syndrome B (CSB) protects against methamphetamine-enhanced oxidative DNA damage in murine fetal brain and postnatal neurodevelopmental deficits. Antioxid Redox Signal. 2011;14:747–56.
52. Mellén M, Ayata P, Dewell S, Kriaucionis S, Heintz N. MeCP2 binds to 5hmC enriched within active genes and accessible chromatin in the nervous system. Cell. 2012;151: 1417–30.
53. Miller L, Pinto D, Wells PG. Embryonic DNA repair and ethanol-initiated behavioural deficits in oxoguanine glycosylase 1 (OGG1) knockout mice: a role for oxidatively damaged DNA and protection by a free radical spin trapping agent. Toxicol Sci: (Supplement: The Toxicologist) 2013;132(131):218 (Abstract No. 1022).
54. Miller L, Pinto DJ, Wells PG. Oxidative DNA damage in the in utero initiation of postnatal neurodevelopmental deficits by normal fetal and ethanol-enhanced oxidative stress in oxoguanine glycosylase 1 (ogg1) knockout mice (2014); (submitted).
55. Miller L, Shapiro AM, Cheng J, Wells PG. The free radical spin trapping agent phenylbutylnitrone reduces fetal brain DNA oxidation and postnatal cognitive deficits caused by in utero exposure to a non-structurally teratogenic dose of ethanol: a role for oxidative stress. Free Radic Biol Med. 2013;60:223–32.
56. Miller L, Shapiro AM, Wells PG. Embryonic catalase protects against ethanol-initiated DNA oxidation and teratogenesis in acatalasemic and human catalase-expressing mice. Toxicol Sci. 2013;134:400–11.
57. Miller L, Wells PG. Altered methanol embryopathies in embryo culture with mutant catalase-deficient mice and transgenic mice expressing human catalase. Toxicol Appl Pharmacol. 2011;252:55–61.
58. Miranda AF, Wiley MJ, Wells PG. Evidence for embryonic peroxidase-catalyzed bioactivation and glutathione-dependent cytoprotection in phenytoin teratogenicity: modulation by eicosatetraynoic acid and buthionine sulfoximine in murine embryo culture. Toxicol Appl Pharmacol. 1994;124:230–41.
59. Moallem SA, Hales BF. The role of p53 and cell death by apoptosis and necrosis in 4-hydroperoxycyclophosphamide-induced limb malformations. Development. 1998;125: 3225–34.
60. Montoliu C, Sancho-Tello M, Azorin I, Burgal M, Valles S, Renau-Piqueras J, Guerri C. Ethanol increases cytochrome P4502E1 and induces oxidative stress in astrocytes. J Neurochem. 1995;65:2561–70.
61. Moore K, Persaud TVN. The developing human: clinically oriented embryology. 8th ed. Philadelphia: W.B. Saunders; 2007.
62. Naya M, Mataki Y, Takahira H, Deguchi T, Yasuda M. Effects of phorone and/or buthionine sulfoximine on teratogenicity of 5-fluorouracil in mice. Teratology. 1990;41:275–80.

63. Naya M, Yasuda M. Effects of glutathione and related compounds on teratogenicity of 5-fluorouracil or cadmium hydrochloride in mice*. Congenit Anom. 1997;37:337–44.
64. Nayanatara AK, Nagaraja HS, Ramaswamy C, Bhagyalakshmi K, Ramesh Bhat M, Harini N. Estimation of tissue lipid peroxidation level and organ weight in litters of wistar rats exposed to prenatal alcohol ingestion. J Physiol Biomed Sci. 2009;22:44–7.
65. Neubert D, Barrach HJ, Merker HJ. Drug-induced damage in the embryo or fetus. New York: Springer; 1980.
66. Nicol CJ, Harrison ML, Laposa RR, Gimelshtein IL, Wells PG. A teratologic suppressor role for p53 in benzo[a]pyrene-treated transgenic p53-deficient mice. Nat Genet. 1995;10: 181–7.
67. Nicol CJ, Zielenski J, Tsui L-C, Wells PG. An embryoprotective role for glucose-6-phosphate dehydrogenase in developmental oxidative stress and chemical teratogenesis. FASEB J. 2000;14:111–27.
68. Nonn L, Williams RR, Erickson RP, Powis G. The absence of mitochondrial thioredoxin 2 causes massive apoptosis, exencephaly, and early embryonic lethality in homozygous mice. Mol Cell Biol. 2003;23:916–22.
69. Norimura T, Nomoto S, Katsuki M, Gondo Y, Kondo S. p53-dependent apoptosis suppresses radiation-induced teratogenesis. Nat Med. 1996;2:577–80.
70. Ondovcik SL, Tamblyn L, McPherson JP, Wells PG. Oxoguanine glycosylase 1 (OGG1) protects cells from DNA double-strand break damage following methylmercury exposure. Toxicol Sci. 2012;128:272–83.
71. Ondovcik SL, Tamblyn L, McPherson JP, Wells PG. Sensitivity to methylmercury toxicity is enhanced in oxoguanine glycosylase 1 knockout murine embryonic fibroblasts and is dependent on cellular proliferation capacity. Toxicol Appl Pharmacol. 2013;270:23–30.
72. Ozolins TR, Siksay DL, Wells PG. Modulation of embryonic glutathione peroxidase activity and phenytoin teratogenicity by dietary deprivation of selenium in CD-1 mice. J Pharmacol Exp Ther. 1996;277:945–53.
73. Parman T, Chen G, Wells PG. Free radical intermediates of phenytoin and related teratogens. Prostaglandin H synthase-catalyzed bioactivation, electron paramagnetic resonance spectrometry, and photochemical product analysis. J Biol Chem. 1998;273:25079–88.
74. Parman T, Wells PG. Embryonic prostaglandin H synthase-2 (PHS-2) expression and benzo[a]pyrene teratogenicity in PHS-2 knockout mice. FASEB J. 2002;16:1001–9.
75. Parman T, Wiley MJ, Wells PG. Free radical-mediated oxidative DNA damage in the mechanism of thalidomide teratogenicity. Nat Med. 1999;5:582–5.
76. Pastoriza-Gallego M, Armier J, Sarasin A. Transcription through 8-oxoguanine in DNA repair-proficient and Csb(−)/Ogg1(−) DNA repair-deficient mouse embryonic fibroblasts is dependent upon promoter strength and sequence context. Mutagenesis. 2007;22: 343–51.
77. Perera F, Herbstman J. Prenatal environmental exposures, epigenetics, and disease. Reprod Toxicol. 2011;31:363–73.
78. Planchon SM, Waite KA, Eng C. The nuclear affairs of PTEN. J Cell Sci. 2008;121: 249–53.
79. Preston TJ, Henderson JT, McCallum GP, Wells PG. Base excision repair of reactive oxygen species-initiated 7,8-dihydro-8-oxo-2′-deoxyguanosine inhibits the cytotoxicity of platinum anticancer drugs. Mol Cancer Ther. 2009;8:2015–26.
80. Preston TJ, Henderson JT, McCallum GP, Wells PG. Base excision repair of reactive oxygen species-initiated 7,8-dihydro-8-oxo-2′-deoxyguanosine inhibits the cytotoxicity of platinum anticancer drugs. Mol Cancer Ther. 2010;8:2015–26.
81. Ramkissoon A, Wells PG. Developmental role of nuclear factor-E2-related factor 2 (Nrf2) in protecting against methamphetamine fetal toxicity and postnatal neurodevelopmental deficits. Free Radic Biol Med. 2013;65:620–31.
82. Roy D, Liehr JG. Estrogen, DNA damage and mutations. Mutat Res. 1999;424:107–15.
83. Rutledge JC. Developmental toxicity induced during early stages of mammalian embryogenesis. Mutat Res. 1997;396:113–27.

84. Sakamaki H, Akazawa S, Ishibashi M, Izumino K, Takino H, Yamasaki H, Yamaguchi Y, Goto S, Urata Y, Kondo T, Nagataki S. Significance of glutathione-dependent antioxidant system in diabetes-induced embryonic malformations. Diabetes. 1999;48:1138–44.
85. Schlisser AE, Yan J, Hales BF. Teratogen-induced oxidative stress targets glyceraldehyde-3-phosphate dehydrogenase in the organogenesis stage mouse embryo. Toxicol Sci. 2010;118:686–95.
86. Schwab JG, Pena L, Waggoner D, Pytel P. Two Children with macrocephaly, developmental delay, and PTEN mutation. Clin Pediatr (Phila). 2009;48:89–92.
87. Sha K, Winn LM. Characterization of valproic acid-initiated homologous recombination. Birth Defects Res B Dev Reprod Toxicol. 2010;89:124–32.
88. Shapiro AM, Miller L, Wells PG. Breast cancer 1 (BRCA1)-deficient mice develop normally but are more susceptible to ethanol- and methamphetamine-initiated embryopathies (Abstract). Toxicol Sci (supplement: The Toxicologist) 2013;132:213.
89. Shivananjappa MM, Muralidhara. Differential oxidative stress induction and lethality of rat embryos after maternal exposure to t-butyl hydroperoxide during postimplantation period. Drug Chem Toxicol. 2013;36:209–16.
90. Siu M, Wiley MJ, Wells PG. Methanol teratogenicity in mutant mice with deficient catalase activity and transgenic mice expressing human catalase. Reprod Toxicol. 2013;36:33–9.
91. Skrzydlewska E, Elas M, Farbiszewski R, Roszkowska A. Effect of methanol intoxication on free-radical induced protein oxidation. J Appl Toxicol. 2000;20:239–43.
92. Skrzydlewska E, Farbiszewski R. Liver and serum antioxidant status after methanol intoxication in rats. Acta Biochim Pol. 1997;44:139–45.
93. Tahiliani M, Koh KP, Shen Y, Pastor WA, Bandukwala H, Brudno Y, Agarwal S, Iyer LM, Liu DR, Aravind L, Rao A. Conversion of 5-methylcytosine to 5-hydroxymethylcytosine in mammalian DNA by MLL partner TET1. Science. 2009;324:930–5.
94. Tung EWY, Winn LM. Valproic acid increases formation of reactive oxygen species and induces apoptosis in postimplantation embryos: a role for oxidative stress in valproic acid-induced neural tube defects. Mol Pharmacol. 2011;80:979–87.
95. Valinluck V, Tsai H-H, Rogstad DK, Burdzy A, Bird A, Sowers LC. Oxidative damage to methyl-CpG sequences inhibits the binding of the methyl-CpG binding domain (MBD) of methyl-CpG binding protein 2 (MeCP2). Nucleic Acids Res. 2004;32:4100–8.
96. Wells PG, Bhuller Y, Chen CS, Jeng W, Kasapinovic S, Kennedy JC, Kim PM, Laposa RR, McCallum GP, Nicol CJ, Parman T, Wiley MJ, Wong AW. Molecular and biochemical mechanisms in teratogenesis involving reactive oxygen species. Toxicol Appl Pharmacol. 2005;207:354–66.
97. Wells PG, Kim PM, Laposa RR, Nicol CJ, Parmana T, Winn LM. Oxidative damage in chemical teratogenesis. Mutat Res. 1997;396:65–78.
98. Wells PG, Lee CJJ, McCallum GP, Perstin J, Harper PA. Receptor- and reactive intermediate-mediated mechanisms of teratogenesis. In: Adverse Drug Reactions, Uetrecht JP, editor. Heidelberg: Springer; 2009. p. 131–62.
99. Wells PG, McCallum GP, Chen CS, Henderson JT, Lee CJ, Perstin J, Preston TJ, Wiley MJ, Wong AW. Oxidative stress in developmental origins of disease: teratogenesis, neurodevelopmental deficits, and cancer. Toxicol Sci. 2009;108:4–18.
100. Wells PG, McCallum GP, Miller L, Siu M, Sweeting JN. Oxidative stress and species differences in the metabolism, developmental toxicity and carcinogenic potential of methanol and ethanol. In: The Toxicology of Methanol, Clary JJ, editor. Hoboken: Wiley; 2013. p. 169–253.
101. Wells PG, Zubovits JT, Wong ST, Molinari LM, Ali S. Modulation of phenytoin teratogenicity and embryonic covalent binding by acetylsalicylic acid, caffeic acid, and alpha-phenyl-N-t-butylnitrone: implications for bioactivation by prostaglandin synthetase. Toxicol Appl Pharmacol. 1989;97:192–202.
102. Wentzel P, Eriksson UJ. Ethanol-induced fetal dysmorphogenesis in the mouse is diminished by high antioxidative capacity of the mother. Toxicol Sci. 2006;92:416–22.

103. Wentzel P, Rydberg U, Eriksson UJ. Antioxidative treatment diminishes ethanol-induced congenital malformations in the rat. Alcohol Clin Exp Res. 2006;30:1752–60.
104. Wilson JG. Experimental studies on congenital malformations. J Chronic Dis. 1959;10: 111–30.
105. Winn LM, Wells PG. Evidence for embryonic prostaglandin H synthase-catalyzed bioactivation and reactive oxygen species-mediated oxidation of cellular macromolecules in phenytoin and benzo[a]pyrene teratogenesis. Free Radic Biol Med. 1997;22:607–21.
106. Winn LM, Wells PG. Maternal administration of superoxide dismutase and catalase in phenytoin teratogenicity. Free Radic Biol Med. 1999;26:266–74.
107. Winn LM, Wells PG. Evidence for Ras-dependent signal transduction in phenytoin teratogenicity. Toxicol Appl Pharmacol. 2002;184:144–52.
108. Wong AW, McCallum GP, Jeng W, Wells PG. Oxoguanine glycosylase 1 protects against methamphetamine-enhanced fetal brain oxidative DNA damage and neurodevelopmental deficits. J Neurosci. 2008;28:9047–54.
109. Wong M, Helston LM, Wells PG. Enhancement of murine phenytoin teratogenicity by the gamma-glutamylcysteine synthetase inhibitor L-buthionine-(S, R)-sulfoximine and by the glutathione depletor diethyl maleate. Teratology. 1989;40:127–41.
110. Wong M, Wells PG. Effects of N-acetylcysteine on fetal development and on phenytoin teratogenicity in mice. Teratog Carcinog Mutagen. 1988;8:65–79.
111. Wong M, Wells PG. Modulation of embryonic glutathione reductase and phenytoin teratogenicity by 1,3-bis(2-chloroethyl)-1-nitrosourea (BCNU). J Pharmacol Exp Ther. 1989;250: 336–42.
112. Wubah JA, Ibrahim MM, Gao X, Nguyen D, Pisano MM, Knudsen TB. Teratogen-induced eye defects mediated by p53-dependent apoptosis. Curr Biol. 1996;6:60–9.
113. Xu DX, Chen YH, Zhao L, Wang H, Wei W. Reactive oxygen species are involved in lipopolysaccharide-induced intrauterine growth restriction and skeletal development retardation in mice. Am J Obstet Gynecol. 2006;195:1707–14.
114. Yu T, Jhun BS, Yoon Y. High-glucose stimulation increases reactive oxygen species production through the calcium and mitogen-activated protein kinase-mediated activation of mitochondrial fission. Antioxid Redox Signal. 2011;14:425–37.
115. Yu WK, Wells PG. Evidence for lipoxygenase-catalyzed bioactivation of phenytoin to a teratogenic reactive intermediate: in vitro studies using linoleic acid-dependent soybean lipoxygenase, and in vivo studies using pregnant CD-1 mice. Toxicol Appl Pharmacol. 1995;131:1–12.
116. Zabihi S, Loeken MR. Understanding diabetic teratogenesis: where are we now and where are we going? Birth Defects Res A Clin Mol Teratol. 2010;88:779–90.
117. Zhao L, Chen YH, Wang H, Ji YL, Ning H, Wang SF, Zhang C, Lu JW, Duan ZH, Xu DX. Reactive oxygen species contribute to lipopolysaccharide-induced teratogenesis in mice. Toxicol Sci. 2008;103:149–57.
118. Zhou J, Blundell J, Ogawa S, Kwon CH, Zhang W, Sinton C, Powell CM, Parada LF. Pharmacological inhibition of mTORC1 suppresses anatomical, cellular, and behavioral abnormalities in neural-specific Pten knock-out mice. J Neurosci. 2009;29:1773–83.
119. Wells, P. G., McCallum, G. P., Lam, K. C. H., Henderson, J. T. and Ondovcik, S. L. (2010). Oxidative DNA damage and repair in teratogenesis and neurodevelopmental deficits. Birth Defects Research Part C: Embryo Today: Reviews 90, 103–109.
120. Liu, Y., Balaraman, Y., Wang, G., Nephew, K. P. and Zhou, F. C. (2009). Alcohol exposure alters DNA methylation profiles in mouse embryos at early neurulation. Epigenetics 4, 500–511.
121. Perkins, A., Lehmann, C., Lawrence, R. C. and Kelly, S. J. (2013). Alcohol exposure during development: Impact on the epigenome. International Journal of Developmental Neuroscience 31, 391–397.
122. Zimmerman EF, Potturi RB, Resnick E, Fisher JE. Role of oxygen free radicals in cocaine-induced vascular disruption in mice. Teratology. 1994;49:192–201.

Chapter 2
Chorioamnionitis and Oxidative Stress: New Ideas from Experimental Models

Boris W. Kramer, Suhas G. Kallapur, and Alan H. Jobe

Introduction

Chorioamnionitis is commonly associated with preterm labor and preterm delivery [1]. Although chorioamnionitis can be diagnosed by clinical criteria [2] or by microbiology [3], the most accepted definition is histologic inflammation of the amnion/chorion/decidua [4]. Fetal inflammation is diagnosed by the presence of funisitis or chorionic vasculitis [4]. In addition to triggering preterm labor, chorioamnionitis is also associated with an increased risk for inflammatory diseases in the preterm infant including bronchopulmonary dysplasia, necrotizing enterocolitis [5], and injury to the white matter of the brain. Infection in chorioamnionitis is largely restricted to the amniotic compartment and is not systemically disseminated since only 5 % of preterm infants with histologic chorioamnionitis have early onset sepsis by organisms recovered in routine cultures [6]. However, a higher percentage is blood culture positive for genital mycoplasma and ureaplasma when selective culture methods are used [7, 8]. Therefore, inflammatory diseases of different fetal organs are thought to be mediated by fetal inflammatory response syndrome (FIRS) induced by chorioamnionitis with colonization of the fetus by bacteria in the amniotic fluid. FIRS is the systemic response of the fetus to chorioamnionitis, generally diagnosed clinically as inflammatory cells in the umbilical cord or increased cytokines/acute phase reactants in the cord blood [9, 10].

B.W. Kramer, M.D., Ph.D. (✉)
Department of Pediatrics, School of Oncology and Developmental Biology, School of Mental Health and Neuroscience, Maastricht University Medical Center, Maastricht, The Netherlands
e-mail: b.kramer@maastrichtuniversity.nl

S.G. Kallapur • A.H. Jobe
Cincinnati Children's Hospital Medical Center, University of Cincinnati, Cincinnati, OH, USA

P.A. Dennery et al. (eds.), *Perinatal and Prenatal Disorders*,
Oxidative Stress in Applied Basic Research and Clinical Practice,
DOI 10.1007/978-1-4939-1405-0_2

Oxidative stress can both cause and be induced by inflammation. Therefore, it is logical to ask “what is the contribution of oxidative stress to chorioamnionitis-induced fetal organ injury?” However, chorioamnionitis is a unique infection and inflammatory disorder: (1) the organisms causing chorioamnionitis have low virulence, e.g., ureaplasma/mycoplasma species [11]. (2) The FIRS from chorioamnionitis is compartmentalized in experimental models with the most affected organs being those that are in contact with the amniotic fluid – the lung, amnion/chorion, gut, and the skin [12–15]. Unlike the cytokine storm associated with adult systemic inflammatory response [16], FIRS is a low-grade inflammatory disease [17]. (3) Although chorioamnionitis can occur in term infants, it is most prevalent in preterm infants and therefore largely a disease of a developmentally immature host, and (4) since the host is a fetus with low ambient intrauterine oxygen tension, the oxidant load is much lower than that in post natal life. Not much is known about oxidant stress in human fetuses, and therefore this review will focus on data from animal models of chorioamnionitis.

Animal Models of Chorioamnionitis

Most chorioamnionitis in humans is an ascending infection, where the organisms in the lower genital area ascend in to the chorio-decidual space or the chorioamnion space through the cervix [18]. Organisms are thought to spread diffusely through the chorio-decidual or the chorioamnion plane and then invade the amniotic cavity. However, a recent study using molecular microbiologic techniques in human placentae demonstrated that the initial event is a localized chorio-decidual infection, which then invades locally into the amniotic cavity and thereby infecting amniotic fluid and the fetus prior to diffuse chorio-decidual inflammation [19]. This sequence is consistent with experiments in the Rhesus macaque demonstrating that localized chorio-decidual infection with live *Group B streptococci* did not trigger preterm labor until the amniotic fluid was colonized [20]. However, a transient chorio-decidual infection induced cytokine production in the amniotic fluid, which resulted in fetal lung inflammation without overt infection of amniotic fluid or preterm labor [21]. Therefore, animal models of chorioamnionitis resulting from injection of inflammatory agents or organisms into the amniotic fluid may reproduce the pathology of most cases of chorioamnionitis associated with preterm delivery. Models of chorioamnionitis have been described with intrauterine injection of proinflammatory agonists or live bacteria in the mouse [22, 23] and the rabbit [24–26]. Chorioamnionitis can also be induced by intra-amniotic injection in the sheep using proinflammatory agonists that include IL-1ß [27], IL-1α [27], LPS (ligand for TLR4) [28], and live *Ureaplasma parvum* [29]. In the Rhesus macaque, intra-amniotic injections of *Group B streptococci* [30], *Ureaplasma parvum* [31], IL-1ß [32, 33], or TNF [32] cause chorioamnionitis. Since most of the data on oxidants in chorioamnionitis are described in the preterm sheep model, we will focus on this experimental system.

Inflammation and Oxidants

The major endogenous sources of the reactive oxygen species are the reduced nicotinamide adenine dinucleotide phosphate (NADPH) oxidases, mitochondrial respiratory chain, enzymatic activation of cytochrome p450, and others [34]. Although the general belief is that reactive oxygen species are invariably harmful, they also mediate important biological functions that are adaptive in nature [35]. NADPH oxidases are activated by insulin, platelet-derived growth factors and other growth factors, and proinflammatory molecules such as TNF, complement 5a, leukotriene B4, and others [34]. ROS can regulate the activity of NF-kB, prolyl hydroxylases which can in turn modulate the activity of hypoxia-inducible factor (HIF)-1α. Thus, ROS can regulate cell proliferation, survival, differentiation, and redox homeostasis [36]. The classic role of ROS in innate immunity is killing of microorganisms by phagocytes [37]. Accordingly, patients with chronic granulomatous disease, a disorder characterized by increased susceptibility to infections, have mutations in NOX_2, a subunit of NADPH oxidase, the main source of ROS in neutrophils and monocytes [38]. Further, ROS can also activate the inflammasome NLRP3 leading to production of IL-1ß [39]. ROS also regulate adaptive immune responses via their action in B-cells. Ca^{2+} and reactive oxygen intermediates generated upon B cell receptor activation rapidly engage in a cooperative interaction that acts in a feedback manner to amplify the early signal generated and thereby influence downstream pathways [40]. Therefore, ROS effects range from adaptive to maladaptive with tissue injury depending on the concentration of these molecules and host defense against the ROS. The ROS effects are regulated by suppression of pathways leading to ROS production, catabolism of ROS, and repair of ROS-mediated damage [35]. To what extent these mechanisms operate in the preterm are not understood but constitute an important area of future study.

Developmental Aspects of Oxidants/Antioxidants

Endogenously produced reactive oxygen species are part of the normal development and the homeostasis. The highly toxic superoxide radical O_2^- is, for example, produced by NAD(P)H oxidases. The balance between oxidants and antioxidants is however essential for the normal development, especially after preterm birth [41].

The antioxidant system in the fetus is not upregulated until very late in gestation. During the last 15 % of gestation, the antioxidant enzyme system is increasingly expressed [42, 43]. Nonenzymatic antioxidants also cross the placenta at this late gestation. The enzyme activity of superoxide dismutase (SOD), catalase, and glutathione peroxidase (GP) increased with the development of the lungs in rabbit fetuses during late gestation. The expression of these enzymes increased in parallel with the surfactant synthesis in late gestation [44]. SOD and catalase are secreted into pulmonary surfactant probably to mediate protection of the lipid layer in the alveolus for gas exchange.

The function of SOD is to transform the highly toxic superoxide radical O_2^- into hydrogen peroxide (H_2O_2) and water. Hydrogen peroxide is then further converted to water by catalase, GP, and glutathione reductase. The enzyme activity increases by more than 150 % in the last 15 % of the gestation [44].

Gas exchange in the placenta is also causing oxidative stress for which SOD, catalase, and thioredoxin are locally expressed in gestation-dependent pattern. In human placenta, there was a positive association between the detection of mitochondrial manganese SOD mRNA and chorioamnionitis [45]. The functional implications are unclear.

Apparently, evolution does not prepare the fetus for the abrupt transition from the low oxygen environment in utero to air breathing with higher oxygen exposure and oxidative stress until very late in gestation. The treatment with antenatal maternal glucocorticoids matures the antioxidant enzymes in the preterm fetus which may be part of the beneficial effects of antenatal glucocorticoids [46, 47].

Chorioamnionitis and Oxidative Stress

There are limited data on oxidative stress to the fetus and the counter-regulatory antioxidant enzymes after exposure to chorioamnionitis in the preterm fetus. Intra-amniotic (IA) injection of *E. coli* LPS in preterm sheep at about 80 % gestation causes chorioamnionitis (Table 2.1). The predominant inflammation is in the fetal lung, although modest inflammation, immune modulation, and injury response is also seen in the ileum [48], skin [13], brain [49, 50], thymus [51], and liver [52]. The lung inflammation is maximally induced 1–2 days after IA LPS exposure with large increases in IL-1ß, IL-8, and IL-6 expression and a robust recruitment of neutrophils and monocytes to the lung [53, 54]. Compared to controls, protein carbonyls, markers of oxidative stress, increased by threefold in the airways of fetuses 7 days after exposure to IA LPS but not at 2 days after IA LPS [55]. Similar increases were also noted in myeloperoxidase levels in the airways 7 days after IA LPS [55]. Mirroring the increased oxidants in the lungs, protein carbonyls also increased in the plasma of fetuses 7 days after IA LPS but not at 2 days post LPS [55]. However, the increase in plasma protein carbonyls was more modest compared to the lung response. Since the maximal inflammation preceded increases in oxidants, the oxidant stress response appears to be a consequence rather than a cause of inflammation in this model. It should also be noted that these oxidant stress responses are modest compared to increases after postnatal hyperoxia.

A major mechanism to reduce oxidant damage is the antioxidant enzyme system. The major enzymes include superoxide dismutase, catalase, and glutathione peroxidase. In preterm sheep delivered at 125 days gestational age (80 % gestation) after 7 days exposure to IA LPS, the antioxidant enzyme activity of glutathione peroxidase, catalase, and superoxide dismutase increased in the fetal lung in a dose-dependent fashion [56]. Pulmonary glutathione peroxidase activity increased at 2 days post IA LPS, whereas superoxide dismutase increased by 4 days, and catalase

Table 2.1 Summary of results in sheep models of chorioamnionitis

Chorioamnionitis induced by	Organ	Readout	Reference
LPS intra-amniotically	Lung	Hydrogen peroxide production in cells from fetal airways	[53]
	Lung	Glutathione peroxidase activity increased within 2 days	[56]
		Catalase activity increased within 4 days	
		Superoxide dismutase activity increased within 7 days	
		No sustained effects after 15 days	
	Airways	Protein carbonyls increased after 7 days	[55]
		Myeloperoxidase increased after 7 days	
	Lung	No difference in protein carbonyls, superoxide dismutase, and peroxiredoxin 1 after 7 days	[55]
	Plasma	Protein carbonyl was increased after 7 days	[55]
Interleukin-1 alpha intra-amniotically	Lung	Superoxide dismutase activity, catalase activity, and glutathione peroxidase activity did not increase	[57]

activity increased by 7 days after IA LPS. The induction of antioxidant enzyme systems in the lung was short-lived since the values declined to near control levels 15 days after IA LPS [56]. However, antioxidant enzyme activity did not increase in the fetal sheep lung after exposure to IA IL-1 alpha, although the lung inflammatory response is roughly equivalent to LPS responses [57]. These experiments suggest a time-dependent and ligand-dependent antioxidant enzyme induction in the preterm sheep in different chorioamnionitis models.

Interactive Phenomenon Between Antenatal Inflammation and Postnatal Oxidative Stress

Postnatal hyperoxia inhibits alveolar development in the developing neonatal lungs of mice [58]. Antenatal inflammation in the preterm sheep also inhibits alveolar development, although the effects are milder [59]. Similar changes of inhibited alveolar development were also demonstrated in the rats exposed to IA LPS [60]. Interestingly, IA-induced aberrant lung development was reversed after moderate but not severe postnatal hyperoxia [60]. These results demonstrate the complex interactions between chorioamnionitis or antenatal inflammation and injury resulting from postnatal hyperoxia. In these experiments, the oxidant stress response was not measured, and therefore it is not clear whether the effects were attributable to oxidative lung damage or other mechanisms leading to lung injury. These interactive phenomena are reminiscent of other experiments showing adaptive effects of oxygen exposure.

Adult rats exposed to 95 % oxygen for 7 days had 100 % mortality [61]. However, if the rats were exposed to 95 % oxygen for 2 days followed by 1 day of either exposure to room air or 50–75 % oxygen followed by exposure to 95 % oxygen for continued periods, the mortality was prevented [61]. This tolerance to hyperoxia after exposure to a sublethal oxygen exposure was attributed to increased pulmonary antioxidant enzyme activity [61]. These experiments demonstrate that inflammation and oxidative stress responses can either potentiate injury responses or can cause tolerance leading to reduced organ injury response. Preterm infants are exposed to multiple injurious insults both antenatally and postnatally, and the net result on organ injury can be unpredictable based on the animal studies. The magnitude of the effects of oxidants from chorioamnionitis to fetal organ injury seems to be much less than that following preterm birth and oxygen exposure.

References

1. Goldenberg RL, et al. Infection-related stillbirths. Lancet. 2010;375(9724):1482–90.
2. Tita AT, Andrews WW. Diagnosis and management of clinical chorioamnionitis. Clin Perinatol. 2010;37(2):339–54.
3. DiGiulio DB, et al. Microbial prevalence, diversity and abundance in amniotic fluid during preterm labor: a molecular and culture-based investigation. PLoS One. 2008;3(8):e3056.
4. Andrews WW, et al. The Alabama Preterm Birth study: polymorphonuclear and mononuclear cell placental infiltrations, other markers of inflammation, and outcomes in 23- to 32-week preterm newborn infants. Am J Obstet Gynecol. 2006;195(3):803–8.
5. Been JV, et al. Chorioamnionitis as a risk factor for necrotizing enterocolitis: a systematic review and meta-analysis. J Pediatr. 2013;162(2):236–42 e2.
6. Strunk T, et al. Histologic chorioamnionitis is associated with reduced risk of late-onset sepsis in preterm infants. Pediatrics. 2012;129(1):e134–41.
7. Viscardi RM. Ureaplasma species: role in diseases of prematurity. Clin Perinatol. 2010; 37(2):393–409.
8. Goldenberg RL, et al. The Alabama Preterm Birth Study: umbilical cord blood Ureaplasma urealyticum and Mycoplasma hominis cultures in very preterm newborn infants. Am J Obstet Gynecol. 2008;198(1):43 e1–5.
9. Redline RW, et al. Amniotic infection syndrome: nosology and reproducibility of placental reaction patterns. Pediatr Dev Pathol. 2003;6(5):435–48.
10. Gomez R, et al. The fetal inflammatory response syndrome. Am J Obstet Gynecol. 1998; 179:194–202.
11. Onderdonk AB, et al. Detection of bacteria in placental tissues obtained from extremely low gestational age neonates. Am J Obstet Gynecol. 2008;198(1):110 e1–7.
12. Kemp MW, et al. Inflammation of the fetal ovine skin following in utero exposure to Ureaplasma parvum. Reprod Sci. 2011;18(11):1128–37.
13. Kemp MW, et al. Exposure to in utero lipopolysaccharide induces inflammation in the fetal ovine skin. Reprod Sci. 2011;18(1):88–98.
14. Kramer BW, et al. Modulation of fetal inflammatory response on exposure to lipopolysaccharide by chorioamnion, lung, or gut in sheep. Am J Obstet Gynecol. 2010;202(1):77 e1–9.
15. Newnham JP, et al. The fetal maturational and inflammatory responses to different routes of endotoxin infusion in sheep. Am J Obstet Gynecol. 2002;186(5):1062–8.
16. Matsuda N, Hattori Y. Systemic inflammatory response syndrome (SIRS): molecular pathophysiology and gene therapy. J Pharmacol Sci. 2006;101(3):189–98.

17. Gotsch F, et al. The fetal inflammatory response syndrome. Clin Obstet Gynecol. 2007;50(3):652–83.
18. Goldenberg RL, Hauth JC, Andrews WW. Intrauterine infection and preterm delivery. N Engl J Med. 2000;342(20):1500–7.
19. Kim MJ, et al. Widespread microbial invasion of the chorioamniotic membranes is a consequence and not a cause of intra-amniotic infection. Lab Invest. 2009;89(8):924–36.
20. Grigsby PL, et al. Choriodecidual inflammation: a harbinger of the preterm labor syndrome. Reprod Sci. 2010;17(1):85–94.
21. Adams Waldorf KM, et al. Choriodecidual group B streptococcal inoculation induces fetal lung injury without intra-amniotic infection and preterm labor in Macaca nemestrina. PLoS One. 2011;6(12):e28972.
22. Hirsch E, Saotome I, Hirsh D. A model of intrauterine infection and preterm delivery in mice. Am J Obstet Gynecol. 1995;172(5):1598–603.
23. Elovitz MA, et al. Intrauterine inflammation, insufficient to induce parturition, still evokes fetal and neonatal brain injury. Int J Dev Neurosci. 2011;29(6):663–71.
24. Davies JK, et al. Histologic inflammation in the maternal and fetal compartments in a rabbit model of acute intra-amniotic infection. Am J Obstet Gynecol. 2000;183(5):1088–93.
25. Kannan S, et al. Microglial activation in perinatal rabbit brain induced by intrauterine inflammation: detection with 11C-(R)-PK11195 and small-animal PET. J Nucl Med. 2007;48(6): 946–54.
26. Yoon BH, et al. Experimentally induced intrauterine infection causes fetal brain white matter lesions in rabbits. Am J Obstet Gynecol. 1997;177(4):797–802.
27. Willet KE, et al. Intra-amniotic injection of IL-1 induces inflammation and maturation in fetal sheep lung. Am J Physiol Lung Cell Mol Physiol. 2002;282:L411–20.
28. Jobe AH, et al. Effects of antenatal endotoxin and glucocorticoids on the lungs of preterm lambs. Am J Obstet Gynecol. 2000;182:401–8.
29. Moss TJ, et al. Intrauterine Ureaplasma infection accelerates fetal lung maturation and causes growth restriction in sheep. Am J Obstet Gynecol. 2005;192:1179–86.
30. Gravett MG, et al. Fetal and maternal endocrine responses to experimental intrauterine infection in rhesus monkeys. Am J Obstet Gynecol. 1996;174(6):1725–31; discussion 1731–3.
31. Novy MJ, et al. Ureaplasma parvum or Mycoplasma hominis as sole pathogens cause chorioamnionitis, preterm delivery, and fetal pneumonia in rhesus macaques. Reprod Sci. 2009; 16(1):56–70.
32. Sadowsky DW, et al. Preterm labor is induced by intraamniotic infusions of interleukin-1beta and tumor necrosis factor-alpha but not by interleukin-6 or interleukin-8 in a nonhuman primate model. Am J Obstet Gynecol. 2006;195(6):1578–89.
33. Kallapur SG, et al. Intra-amniotic IL-1ß induces fetal inflammation in rhesus monkeys and alters the regulatory T cell/IL-17 balance. J Immunol. 2013;191:1102–9.
34. Nathan C, Cunningham-Bussel A. Beyond oxidative stress: an immunologist's guide to reactive oxygen species. Nat Rev Immunol. 2013;13(5):349–61.
35. Buonocore G, Perrone S, Tataranno ML. Oxygen toxicity: chemistry and biology of reactive oxygen species. Semin Fetal Neonatal Med. 2010;15(4):186–90.
36. Bae YS, et al. Regulation of reactive oxygen species generation in cell signaling. Mol Cells. 2011;32(6):491–509.
37. Sbarra AJ, Karnovsky ML. The biochemical basis of phagocytosis. I. Metabolic changes during the ingestion of particles by polymorphonuclear leukocytes. J Biol Chem. 1959;234(6): 1355–62.
38. Roos D, et al. Hematologically important mutations: X-linked chronic granulomatous disease (third update). Blood Cells Mol Dis. 2010;45(3):246–65.
39. Zhou R, et al. A role for mitochondria in NLRP3 inflammasome activation. Nature. 2011;469(7329):221–5.
40. Singh DK, et al. The strength of receptor signaling is centrally controlled through a cooperative loop between Ca2+ and an oxidant signal. Cell. 2005;121(2):281–93.

41. Davis JM, Auten RL. Maturation of the antioxidant system and the effects on preterm birth. Semin Fetal Neonatal Med. 2010;15(4):191–5.
42. Sosenko IR, Nielsen HC, Frank L. Lack of sex differences in antioxidant enzyme development in the fetal rabbit lung. Pediatr Res. 1989;26(1):16–9.
43. Frank L, Groseclose EE. Preparation for birth into an O2-rich environment: the antioxidant enzymes in the developing rabbit lung. Pediatr Res. 1984;18(3):240–4.
44. Matalon S, et al. Characterization of antioxidant activities of pulmonary surfactant mixtures. Biochim Biophys Acta. 1990;1035(2):121–7.
45. Than NG, et al. Mitochondrial manganese superoxide dismutase mRNA expression in human chorioamniotic membranes and its association with labor, inflammation, and infection. J Matern Fetal Neonatal Med. 2009;22(11):1000–13.
46. Walther FJ, Jobe AH, Ikegami M. Repetitive prenatal glucocorticoid therapy reduces oxidative stress in the lungs of preterm lambs. J Appl Physiol. 1998;85(1):273–8.
47. Been JV, et al. Antenatal steroids and neonatal outcome after chorioamnionitis: a meta-analysis. BJOG. 2011;118(2):113–22.
48. Wolfs TG, et al. Endotoxin induced chorioamnionitis prevents intestinal development during gestation in fetal sheep. PLoS One. 2009;4(6):e5837.
49. Nitsos I, et al. Chronic exposure to intra-amniotic lipopolysaccharide affects the ovine fetal brain. J Soc Gynecol Investig. 2006;13(4):239–47.
50. Gavilanes AW, et al. Chorioamnionitis induced by intraamniotic lipopolysaccharide resulted in an interval-dependent increase in central nervous system injury in the fetal sheep. Am J Obstet Gynecol. 2009;200(4):437 e1–8.
51. Kuypers E, et al. Ovine fetal thymus response to lipopolysaccharide-induced chorioamnionitis and antenatal corticosteroids. PLoS One. 2012;7(5):e38257.
52. Wilson TC, et al. Pulmonary and systemic induction of SAA3 after ventilation and endotoxin in preterm lambs. Pediatr Res. 2005;58(6):1204–9.
53. Kramer BW, et al. Dose and time response for inflammation and lung maturation after intra-amniotic endotoxin in preterm lambs. Am J Respir Crit Care Med. 2001;164:982–8.
54. Kallapur SG, et al. Intra-amniotic endotoxin: chorioamnionitis precedes lung maturation in preterm lambs. Am J Physiol Lung Cell Mol Physiol. 2001;280(3):L527–36.
55. Cheah FC, et al. Oxidative stress in fetal lambs exposed to intra-amniotic endotoxin in a chorioamnionitis model. Pediatr Res. 2008;63(3):274–9.
56. Sosenko IR, Jobe AH. Intraamniotic endotoxin increases lung antioxidant enzyme activity in preterm lambs. Pediatr Res. 2003;53(4):679–83.
57. Sosenko IR, et al. IL-1alpha causes lung inflammation and maturation by direct effects on preterm fetal lamb lungs. Pediatr Res. 2006;60(3):294–8.
58. Warner B, et al. Functional and pathological effects of prolonged hyperoxia in neonatal mice. Am J Physiol. 1998;275:L110–7.
59. Willet KE, et al. Antenatal endotoxin and glucocorticoid effects on lung morphometry in preterm lambs. Pediatr Res. 2000;48:782–8.
60. Tang JR, et al. Moderate postnatal hyperoxia accelerates lung growth and attenuates pulmonary hypertension in infant rats after exposure to intra-amniotic endotoxin. Am J Physiol Lung Cell Mol Physiol. 2010;299(6):L735–48.
61. Frank L, et al. New "rest period" protocol for inducing tolerance to high O2 exposure in adult rats. Am J Physiol. 1989;257(4 Pt 1):L226–31.

Chapter 3
Oxidative Stress in Pregnancies Complicated by Diabetes

Martha Lappas, Andrea Leiva, Fabián Pardo, Luis Sobrevía, and Alicia Jawerbaum

Introduction

GDM is defined as any degree of glucose intolerance with onset or first recognition during pregnancy [99]. GDM and its complications impose significant economic constraints on health systems and in some countries, during pregnancy, labour and the puerperium. In addition, there is also a personal cost to individuals and families of monitoring and potential adverse outcomes. Unless preventative measures are implemented, diabetes-related expenditure will increasingly dominate the health budgets of many countries [173].

M. Lappas (✉)
Obstetrics, Nutrition and Endocrinology Group, Department of Obstetrics and Gynaecology, University of Melbourne, Level 4/163 Studley Road, Heidelberg 3084, VIC, Australia

Mercy Perinatal Research Centre, Mercy Hospital for Women, Level 4/163 Studley Road, Heidelberg 3084, VIC, Australia
e-mail: mlappas@unimelb.edu.au

A. Leiva • F. Pardo
Cellular and Molecular Physiology Laboratory (CMPL), Division of Obstetrics and Gynaecology, Faculty of Medicine, School of Medicine, Pontificia Universidad Católica de Chile, Santiago, Chile

L. Sobrevía
Cellular and Molecular Physiology Laboratory (CMPL), Division of Obstetrics and Gynaecology, Faculty of Medicine, School of Medicine, Pontificia Universidad Católica de Chile, Santiago, Chile

The University of Queensland Centre for Clinical Research, Herston, QLD, Australia

A. Jawerbaum
Laboratory of Reproduction and Metabolism, CEFYBO-CONICET School of Medicine, University of Buenos Aires, Buenos Aires, Argentina

P.A. Dennery et al. (eds.), *Perinatal and Prenatal Disorders*, Oxidative Stress in Applied Basic Research and Clinical Practice, DOI 10.1007/978-1-4939-1405-0_3

Although many women who develop GDM during pregnancy revert back to normal glucose tolerance (NGT) postdelivery, a portion of women go on to develop diabetes later in life [35, 92]. One of the largest postpartum study performed in Boston, USA, demonstrated that incidence of DM (type 1 and type 2) amongst women who previously acquired GDM was 49.9 % in comparison to 7.0 % in controls [109]. GDM women are at a higher risk of developing cardiovascular disease [139] particularly in women with a family history of type 2 DM [19]. Of recent, studies have also suggested a relationship with GDM and the development of certain cancers, including pancreatic cancer [119] and breast cancer postmenopause [118].

Diabetic women are at higher risk of pregnancy-induced hypertension and, in severe circumstances, preeclampsia [98, 176]. This can impact on the pregnancy occasionally leading to iatrogenic delivery, caesarean delivery and intrauterine growth-restricted (IUGR) offspring [98, 140, 176].

Complications include, but are not limited to, macrosomia, neonatal hypoglycaemia, congenital malformation, polycythemia, hyperbilirubinemia, hypocalcaemia, respiratory distress syndrome and unexplained stillbirths [21, 77, 137, 142]. Macrosomia occurs in 20–30 % of infants whose mothers have GDM [77] and can also impede on the likelihood of a vaginal birth.

Increased risk of small for gestational age (SGA) babies has also been shown in women with GDM. Studies have shown that women receiving intensive insulin treatment have increased risk for SGA babies in comparison to NGT controls [44]. Furthermore, population studies have demonstrated a negative association between maternal BMI and SGA only in males suggesting sexual dimorphisms may play a role in the risk of macrosomia or SGA offspring [129].

Intrauterine exposure to maternal diabetes leaves offspring with an increased risk of obesity and subsequently altered glucose metabolism and increased risk of type 2 DM and cardiovascular disease in adulthood [98, 162].

Pathophysiology of Diabetes in Pregnancy

Chronic Insulin Resistance

During normal pregnancy, insulin resistance develops to shift the balance of maternal energy utilisation and mobilisation towards supporting the developing foetus. Hence, she utilises less glucose (thus sparing it as a constitutive substrate for her foetus) and stops suppressing lipolysis at the adipocyte, leading to increased plasma triglycerides providing alternative substrates for the mother and allowing for lipid delivery to the placenta and ultimate transfer of fatty acids across the placenta to the foetus.

Pregnancy-derived hormones, including human placental lactogen (hPL) and human placental growth hormone (hPGH), are believed to be contributing factor in reprogramming maternal physiology to an insulin-resistant state to allow for optimum maternal-foetal nutrient transfer [9]. Cytokines secreted from adipose tissues, or adipokines, have also been implicated in insulin resistance. For example, in pregnancy, circulating levels of TNF-α correlate with whole body insulin sensitivity [76].

In women with GDM, peripheral insulin resistance is even more pronounced (Catalano et al. 2002; Colomiere et al. 2010), which results in greater substrate availability for foetal growth and development (Lain and Catalano 2007). Approximately 90 % of women with GDM present with chronic insulin resistance [16]. Previous studies have demonstrated increased whole body insulin resistance in women prior to GDM [22]. Deteriorated β-cell function is believed to be the underlying cause; hyperbolic relationships between insulin sensitivity and insulin secretion revealed progressively diminished pancreatic β-cell function in women with GDM [70] both before and during pregnancy [15, 73, 132]. These studies have altered the perception of GDM, demonstrating both insulin resistance and decreased β-cell function may define the pathophysiology of GDM rather than just insulin resistance [132].

Inflammation

Low grade chronic inflammation is a central feature of GDM. Although the initial stimulus triggering inflammation in pregnancy with GDM is not currently known, potential candidates include dietary and environmental factors such as caloric overload or changes in microbiota in the pregnant women [17, 86].

Maternal systemic inflammation with increased concentration of adipokines and inflammatory cytokines is increased by the accumulation of activated macrophages in the interstitial stroma of the placenta of women with GDM [11, 31, 78, 124]. Cross-sectional studies have shown that circulating concentrations of TNF-α are higher in pregnant women with GDM when compared to the normal pregnant women [43, 174]. Importantly, TNF-α is a significant predictor of insulin resistance during pregnancy [76]. An overproduction of placental TNF-α has been demonstrated in patients with type I diabetes and GDM [90] and is associated with increased foetal adiposity [124]. Other circulating proinflammatory cytokines like monocyte chemoattractant protein-1 (MCP-1) [78], IL-1β [161] and IL-6 [7] were also found to be positively correlated with GDM. Further to this, there is significantly greater release of the proinflammatory cytokine TNF-α under conditions of high glucose concentrations in the GDM placenta [29]. Given that proinflammatory cytokines have been implicated in the regulation of glucose and lipid metabolism, and in insulin resistance, these data are consistent with the hypothesis that proinflammatory cytokines may be involved in the pathogenesis and/or progression of GDM.

Current dogma suggests that maternal adipose tissue as well as the placenta contributes to inflammation in GDM pregnancies by synthesising and secreting a number of cytokine hormones. Indeed, our own recent studies have shown that the secretion of the proinflammatory cytokine IL-1β is higher in omental adipose tissue from women with GDM when compared to BMI-matched controls [83].

Apoptosis

Apoptosis is crucial to the maintenance of placental homeostasis and any imbalance in this process may compromise placenta function and, consequently, pregnancy success. Hyperglycaemia modulates the expression of apoptosis regulatory genes in the preimplantation blastocyst stage [101]. In placental trophoblast, there is a higher apoptosis index in diabetic placentas [138]. In women with GDM, maternal leukocyte activity of the proapoptotic protein poly(ADP-ribose) polymerase (PARP) is elevated as early as the middle of pregnancy [55]. Furthermore, there is also a positive linear correlation that was observed between the severity of carbohydrate intolerance and PARP activity of circulating maternal leukocytes [55]. Additionally, hyperglycaemia upregulates p53, triggering the mitochondrial death cascade pathway in the mouse placenta [100], and increases the rate of apoptosis in cultured trophoblast cell lines [167]. Placental apoptosis may be initiated by a variety of stimuli, including hypoxia and oxidative stress [141].

Endothelial and Vascular Dysfunction

Human fetoplacental endothelial dysfunction is a feature in GDM [150] with metabolic changes that include elevated synthesis of NO paralleled by increased expression of endothelial NO synthase (eNOS), high membrane transport of L-arginine and reduced membrane transport of adenosine in primary cultures of human umbilical vein (HUVEC, macrovascular) and placenta microvillus (hPMEC, microvascular) endothelial cells [50, 133, 171]. The alterations in HUVEC and hPMEC are associated with increased eNOS and inducible NOS (iNOS) expression and activity, respectively. In addition, HUVEC and hPMEC exhibit increased expression and activity of the human cationic amino acid transporter 1 (hCAT-1) and reduced expression and activity of the equilibrative nucleoside transporters 1 (hENT1) and hENT2, as well as for the corresponding *SLC7A1* (coding for hCAT-1) *SLC29A1* (coding for hENT1) and *SLC29A2* (coding for hENT2) gene promoter activities. Interestingly, the above-mentioned alterations in the expression of these genes result from increased NO level, as well as increased activity of protein kinase C (PKC) as well as p44 and p42 mitogen-activated protein kinases (p44/42^{mapk}), and activation of the transcription factor complex hCHOP-C/EBPα [38]. The latter is evidenced as an increased hCHOP-C/EBPα binding to the corresponding consensus sequences in these genes. In addition, a key role is played by the specific protein 1 (Sp1) transcription factor as repressor of *SLC29A1* and activator of *SLC7A1* expression via a NO/p44/42^{mapk}/PKC signalling axis in foetal endothelial cells from GDM. Whether an increase in the bioactive NO level is always beneficial is arguable since abnormally elevated levels of this gas could lead to formation of nitrogen-derived reactive species including the highly active peroxynitrate (ONOO-). Thus, a physiological increase of endothelial-derived NO level is crucial to maintain the correct vascular responsiveness to vasoactive molecules; however, in pathological conditions, a supraphysiological increase of NO synthesis and bioavailability could result to be detrimental [122, 135].

Markers of endothelial dysfunction have been reported in GDM [85]. These include increased levels of endothelium-derived reactive oxygen and nitrative-derived species (a phenomenon where NADPH oxidase and xanthine oxidase, amongst other enzymes, play key roles in the foetal and maternal circulation), increased levels of symmetric dimethylarginine (ADMA) and C-reactive protein maternal plasma levels and altered expression of junctional adhesion molecules. All these alterations could result from endothelial dysfunction in this syndrome. Several recent evidences show that a reduced ENTs-mediated adenosine transport in GDM is associated with stimulation of L-arginine transport and NO synthesis (the referred ALANO signalling pathway) in GDM [135, 160]. Since adenosine is also a potent antioxidant and because its plasma level is increased in the foetal blood in GDM, it is likely that this nucleoside will be acting a role as such in this syndrome. Indeed, we hypothesise that the ALANO signalling pathway itself could be a signalling mechanism through which adenosine acts as antioxidant in GDM resulting in improvement of endothelial function (Fig. 3.1). ALANO pathway is also associated to a reduced expression and activity of hENT1, thus accumulating adenosine in the extracellular space to facilitate its several biological actions. However, even being a relatively new point of view, a potential involvement of hENT2 expression and activity in restoring (at least partially) the GDM-associated reduced adenosine transport and fetoplacental vascular endothelial dysfunction in this syndrome is also likely [150]. Insulin also exerts a differential modulation of endothelial cells from macrocirculation compared with microcirculation, possibly due to expression of different insulin receptor isoforms in this type of tissues [133, 171]. The latter is essentially proposed on the bases that endothelial cells from the macrocirculation of the human placenta a predominant expression of insulin receptor A (IR-A) subtype compared with IR-B subtype associated with activity (i.e. phosphorylated over total) $p42/44^{mapk}$/Akt ratio >1, suggestive of a preferential mitogenic rather than a metabolic insulin effect in these cell types ([153], and F Westermeier, L Sobrevia, *unpublished results*). However, a similar expression of IR-A and IR-B subtypes, with activity $p42/44^{mapk}$/Akt ratio ~1, has been reported in cells from the microcirculation of the human placenta [120]. Insulin action on its receptors is also modulated by activation of other types of receptors, such as adenosine receptors in the human fetoplacental endothelium [48, 50]. Thus, a differential biological action of insulin as modulator of endothelial function subjected to co-regulation by other (parallel) signalling molecules is proposed. However, potentially common functional characteristic leading to changes in the bioavailability of adenosine and metabolism of L-arginine is evidenced by human foetal micro- and macrovascular endothelium accounting for endothelial dysfunction in GDM.

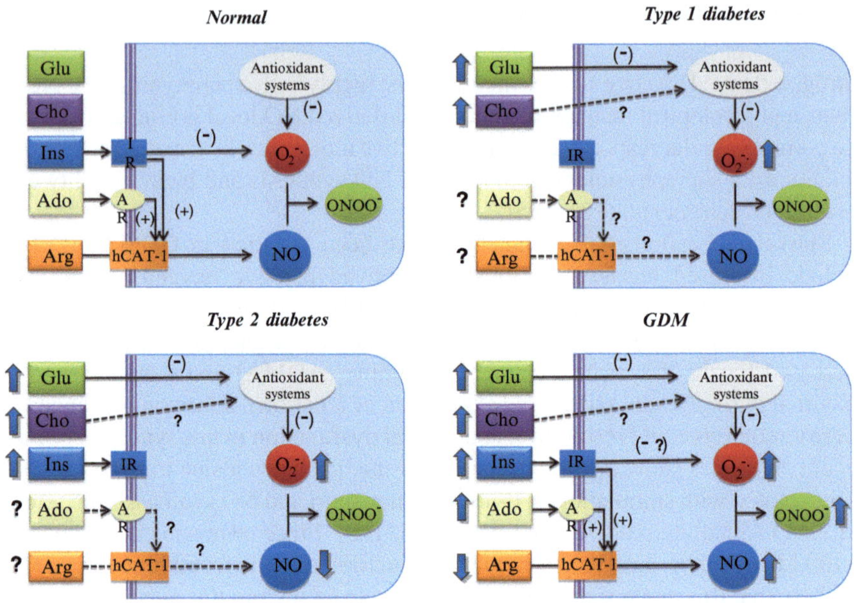

Fig. 3.1 Placenta endothelial dysfunction in normal and diabetic pregnancies. In cells from normal pregnancies (*Normal*), the production of reactive oxygen species is controlled by antioxidant systems whose consequence is an efficient inhibition (–) of the synthesis of superoxide anion ($O_2^{-\cdot}$). Normal circulating levels of D-glucose (Glu) and cholesterol (Cho) seem to be in equilibrium with the activity of the antioxidant systems; however, circulating insulin (Ins) plays a key role diminishing (–) the production of $O_2^{-\cdot}$ via activation of insulin receptors (IR) in these cells. In addition, adenosine (Ado) activates adenosine receptors (AR) leading to stimulation (+) of L-arginine (Arg) transport via the human cationic amino acid transporter 1 (hCAT-1), a phenomenon apparently required for the synthesis of nitric oxide (NO). Insulin also stimulates the latter phenomenon, keeping a basal synthesis of NO in placental endothelial cells. The $O_2^{-\cdot}$ and NO will then react to form a minimal level of peroxynitrite ($ONOO^-$). In endothelial cells from type 1 diabetes, the production of $O_2^{-\cdot}$ is increased due to a reduced activity of antioxidant systems caused by the increased levels of circulating D-glucose and perhaps (*dotted lines*, ?) in response to the seen elevated plasma levels of total cholesterol. This phenomenon is worsened since in this syndrome circulating insulin is negligible or absent, thus limiting this hormone's antioxidant capacity in reducing $O_2^{-\cdot}$ overproduction. It is also unknown (?) whether circulating levels of adenosine and/or L-arginine, as well as AR activation by adenosine and associated signalling modulating hCAT-1 activity and/or expression, are altered in these patients, thus limiting the biological actions of adenosine on the L-arginine/NO signalling pathway. Since $O_2^{-\cdot}$ is likely increased in this syndrome, a larger production of $ONOO^-$ is expected. However, this phenomenon could be limited by a normal or reduced synthesis of NO. In type 2 diabetes, insulin levels are increased, but insulin receptors are not responsive to insulin in this syndrome, thus since not signalling associated to this hormone is expected, its antioxidant capacity in reducing $O_2^{-\cdot}$ overproduction is limited or absent. In this syndrome, a reduced activity of antioxidant systems is likely due to increased plasma D-glucose and cholesterol overproduction of $O_2^{-\cdot}$. However, since a reduced synthesis of NO is also feasible in this syndrome, formation of $ONOO^-$ could be negligible or reaching values with an evident consequence in the endothelial function. The role of adenosine and L-arginine is again unclear in this syndrome. In GDM, there is an increased plasma D-glucose, cholesterol, insulin and adenosine levels. Elevated extracellular D-glucose reduces the activity of antioxidant systems leading to an increase in the $O_2^{-\cdot}$ synthesis, a phenomenon that could be minimally regulated (?) by insulin. However, insulin in concordance with adenosine increases hCAT-1 expression and activity leading to overproduction of NO, thus facilitating the formation of pathophysiological levels of $ONOO^-$ in this cell type. The increased L-arginine uptake via hCAT-1 results in reduced circulating levels of L-arginine in the human fetoplacental circulation. It is also proposed that elevated maternal cholesterol in pregnancy could result in reduced activity on the antioxidant systems in women whose pregnancies course with GDM and maternal supraphysiological hypercholesterolemia

Oxidative Stress

It has been postulated that oxidative stress is a causative agent in GDM-related placental disorders such as preeclampsia [2, 105] with ROS and antioxidant enzyme systems playing crucial roles in pregnancy. Although measurements of markers of oxidative stress in maternal blood and urine show that pregnancy per se is a state of oxidative stress, this condition is higher in pregnancies complicated with diabetes [85, 105].

ROS include highly chemically reactive molecules such as $O_2^{\bullet-}$, hydroxyl radical ($^{\bullet}OH$), peroxide radicals ($ROO^{\bullet}$) and carbon monoxide (CO) and non-radicals that are oxidising agents easily converted into radicals, such as hypochlorous acid (HOCl), ozone (O_3), singlet oxygen ($^1O_2$) and hydrogen peroxide (H_2O_2). The $O_2^{\bullet-}$ can be synthesised by NADPH oxidase (NOX), xanthine oxidoreductase (XOR), complexes I and III of the electron transport chain, uncoupled NOS, heme-oxygenase (HO) and the P_{450} enzymes family and enzymes of the arachidonic acid metabolism. In addition, reactive nitrative species (RNS) relate mainly to nitric oxide (NO), which is normally synthesised by the nitric oxide synthase (NOS); however, it can be converted into species such as nitrosonium cation (NO^+), nitroxyl anion (NO^-) and peroxynitrite ($ONOO^-$). During pregnancy there is an elevation of nitrative stress due to increased RNS [105]. A prolonged nitrative stress is seen in women with GDM, which may be involved in the development of carbohydrate intolerance later in life or in the development of late cardiovascular complications as a sign or evidence of foetal programming of adult diseases due to posttranslational modifications under a nitrative stress as well [55, 85].

Prolonged nitrative stress is evident in women with GDM and type 2 diabetes [18, 55, 85], which may be involved in the development of carbohydrate intolerance later in life or in the development of late cardiovascular complications. Indeed, nitrative stress can induce damage to proteins, lipids and nucleic acids, disrupts multiple signalling pathways during organ development and impairs NO availability affecting fetoplacental vascular function. These alterations can either persist or generate adaptations in the offspring and lead to increases in the risk of metabolic and cardiovascular diseases [64, 120, 122, 131].

Mitochondria are considered to be the major generators of ROS. There is very limited data on the role of the electron transport chain in diabetic pregnancies. However, placental mitochondrial mass increases with gestational age, and placental mitochondria is a source of oxidative stress in other placental pathologies such as preeclampsia, suggesting increased contribution of the electron transport chain to ROS generation [166]. It is possible that excessive ROS production by placental mitochondria may be released in the foetal circulation and may be critical in foetal programming of atherosclerosis [88].

There are several antioxidant systems controlling the damage produced by oxidative stress. Nonenzymatic antioxidant mechanisms include ascorbic acid (vitamin C), α-tocopherol (vitamin E) and glutathione (GSH). The latter is the major cellular redox buffer. In addition, cells express enzymatic systems such as superoxide dismutase (SOD), which converts $O_2^{\bullet-}$ into H_2O_2 and O_2; glutathione peroxidase (GPx) that catalyses the reduction of H_2O_2 oxidising GSH to oxidised glutathione (GSSG), which oxidises cysteine residues of proteins (i.e. *S*-glutathiolation); and catalase that metabolises H_2O_2 to form O_2 and H_2O. In patients with GDM, there is a decrease in the antioxidant defence paralleled by increased levels of protein oxidation markers associated with oxidative stress [85].

Oxidative Stress Mediates Cellular and Metabolic Damage in the Intrauterine Compartment

The maternal circulating levels of free radicals and antioxidants are altered in diabetic pregnancies with the pro-oxidant/antioxidant imbalance towards prooxidation. In addition, in the placenta, there is an overproduction of free radicals, oxidation reactions are accelerated and the radical scavenger function mechanisms are impaired. Given that the placenta provides the interface of the maternal and foetal circulations, it may play a crucial role in protecting the foetus from adverse effects of the maternal diabetic milieu, while disturbances in placental function may exacerbate this state.

In the placenta, diabetes-induced oxidative stress may affect a number of critical pathways, including implantation and decidualisation, embryo organogenesis and placental nutrient transport (Fig. 3.2). These pathways play an important role in many aspects of pregnancy, including implantation, decidualisation, embryogenesis and placentation.

Implantation and Decidualisation

The human embryo undergoes interstitial implantation by invading the maternal decidua at the blastocyst stage. It has been suggested that diabetes induces oxidative stress which can affect fertilisation and induce apoptosis, resulting in embryo fragmentation, implantation failure or abortion. It is well established that high glucose concentrations can impair preimplantation embryo development or induce embryo degeneration and cause a reduction in total cell number of blastocysts [12, 101, 116]. Mild diabetes is sufficient to trigger alterations in maternal organs, leading to impaired decidua development contributing to failure in embryonic implantation and early embryonic losses [147]. Diabetic metabolic factors may induce embryotoxicity in preimplantation embryos through derangement of the antioxidant defence mechanism [112] and impaired NO synthesis [63, 108].

Adequate uteroplacental blood flow is critical for the survival and growth of the developing foetus. In early pregnancy, a subpopulation of foetal trophoblast cells, the extravillous trophoblast, invades the decidualised (differentiated) endometrial stromal compartment (interstitial invasion) and its spiral arteries (endovascular invasion). The ability of the trophoblast to invade the uterus is related to NO production during implantation and during the remodelling of uteroplacental arteries [47, 51, 72, 156].

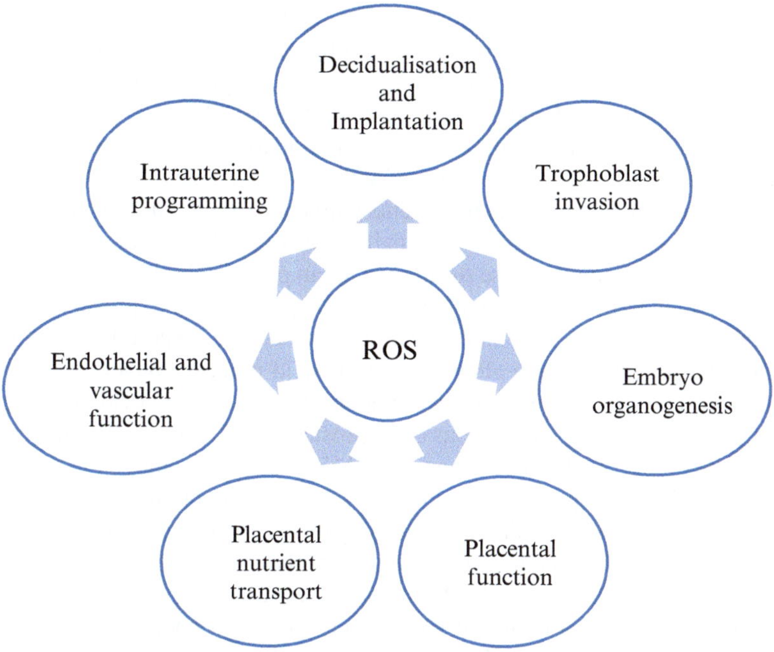

Fig. 3.2 In normal pregnancy, ROS plays a number of important roles throughout pregnancy, including embryo development, implantation, angiogenesis, placental development and function and thus foetal development

Organogenesis

Evidence of increased ROS has been found as early as the oocyte and early embryo stage in hyperglycaemic and insulin-resistant conditions and related to mitochondrial defects, increased apoptotic events and early loss in embryos from diabetic rats [68, 114].

Spontaneous abortions and structural inborn anomalies are the main complications of diabetic pregnancy [121]. Studies performed in the early organogenesis stage have clearly established the relevance of the relationship between oxidative stress and the induction of congenital malformations [93]. Indeed, in vitro and in vivo studies in animal models of diabetes and pregnancy have shown that maternal diabetes and hyperglycaemia generate ROS and impair antioxidant capacity in parallel to the induction of congenital malformations [36].

The increases in isoprostanes and protein carbonyls evidence the oxidative stress in embryos from diabetic rodents during early organogenesis, the period during which most malformations are induced. The antioxidant capacity plays a fundamental role in counteracting oxidative stress and is affected by maternal diabetes [36, 66]. Indeed, studies performed in diabetic experimental models have shown a decreased expression of Cu/Zn SOD, Mn SOD and glutathione peroxidase in embryos from diabetic rats when compared to controls [23, 148, 168]. Additionally, the importance of antioxidant enzymes is highlighted by experiments that show that animal strains resistant to the induction of congenital malformations in streptozotocin-induced diabetes have increased MnSOD concentrations. In this context, the different susceptibility to congenital malformations in genetic models of diabetes is related to the expression of antioxidant enzymes [34, 111].

As expected, diabetic animal treatments with antioxidants such as vitamin E and C prevent the induction of congenital malformations, although there are still difficulties in the translation of these results to human pregnancies [85]. It is interesting that folic acid in the dose used in diabetic human pregnancies to prevent congenital malformations seems to have antioxidant effects. Indeed, experimental models of diabetes and pregnancy have shown that folic acid reduces congenital malformations, at least in part, by exerting antioxidant effects. This is evidenced by their capacity to reduce the concentrations of isoprostanes and regulate the balance of the activity of matrix metalloproteinases (MMPs) and its endogenous inhibitors (TIMPs), markers of a proinflammatory state during early organogenesis [52].

Oxidative stress-induced damage during embryo organogenesis has been related to the disruption of signalling pathways that change the balance between cell growth, differentiation and death. During organogenesis, apoptosis is required in an appropriate location and temporal pattern, and increased apoptosis, related to the increased oxidative stress, has been evidenced in experimental models of diabetes and pregnancy [178]. In fact, when embryos from diabetic rats are compared to non-malformed ones, decreased expression of glutathione peroxidase and increased apoptosis have been found in the developing heart [169].

The paired box protein PAX-3 is a transcription factor needed for the correct processes of neural tube closure and neural crest cell migration. In maternal diabetes, an impaired induction of PAX-3 occurs due to increased oxidative stress [104]. This leads to an impaired process of closure of the neural tube and to apoptosis of neural crest cells, which result in cephalic and cardiac malformations. Moreover, prevention of apoptosis and induction of hyperglycaemia-induced neural tube and heart malformations, together with increases in PAX-3 expression, can be achieved by treatments with antioxidants such as α-tocopherol and glutathione ethyl ester [178]. Treatments with resveratrol have also been shown to reduce oxidative stress and malformation rates in experimental models of diabetes [146].

Peroxynitrite is another potent oxidant formed in the embryo during organogenesis in diabetic experimental models [63]. Peroxynitrites, which are the result of excessive production of ROS and nitric oxide (NO), have the capacity to induce oxidative damage to DNA, proteins and lipids. Peroxynitrite-induced protein modifications, which can be evidenced by nitrotyrosine immunostaining, interfere with the function of relevant enzymes during development and are involved in endoplasmic reticulum stress and apoptosis [65, 183]. The involvement of ROS and NO interaction in embryo dysmorphogenesis is highlighted by the reduced ROS- and diabetes-induced teratogenesis observed in inducible NO synthase (iNOS) knockout mice [152].

Lipids are also affected by oxidative stress, and increased lipoperoxidation in the embryos from diabetic rats not only is a marker of increased ROS but also indicates the loss of function of bioactive lipids, especially those of polyunsaturated nature like essential fatty acids (EFAs), which cannot be synthesised by mammals and should thus be provided through the diet. The EFA linoleic acid (n-6) is the precursor of arachidonic acid, an n-6 long chain polyunsaturated fatty acid (LC- PUFA) highly required during embryo organogenesis, whereas α-linolenic acid (n-3) is the precursor of docosahexaenoic acid (DHA), an n-3 LC-PUFA highly required for foetal visual, cognitive and immune system development [80]. Arachidonic acid is the precursor of prostaglandins (PGs) and leukotrienes (LT), bioactive lipids involved in developmental processes. Indeed, arachidonic acid deficiencies are clearly related to the induction of congenital malformations in maternal diabetes [63]. At least in part, this is because the embryo requires PGE2 during organogenesis to sustain the rapid growth of the neural folds during the process of closure of the neural tube. Besides, PGI2 is a positive regulator of PGE2 and phospholipid synthesis and is an endogenous agonist of the nuclear receptor PPARδ [62].

Peroxisome proliferator-activated receptors (PPARs) are ligand-activated transcription factors that heterodimerise with the RXR receptor for retinoic acid and, after ligand binding, lead to the regulation of the transcription of multiple genes involved in metabolic homeostasis and anti-inflammatory processes both through transactivation and transrepression pathways [62]. Indeed, antioxidant genes such as catalase are positively regulated by PPAR activation, while enzymes involved in the proinflammatory response such as iNOS and MMPs are negatively regulated through PPAR activation. Besides, PPARs have relevant roles in cell differentiation and development. Amongst the three known PPAR isoforms (PPARα, PPARγ and PPARδ), PPARδ is the only one that has been found to be expressed during embryo early organogenesis [14]. Interestingly, PPARδ has an important role in embryo organogenesis and decidualisation and is decreased in embryos from diabetic rats [62]. Since increased lipoperoxidation, deficiency in EFAs and decreased PPARs impair embryo development in maternal diabetes, maternal diets enriched in unsaturated fatty acids capable of activating PPARs have been studied as possible regulators of embryo impairments [53]. These dietary treatments have been shown to positively regulate PGE_2 and PGI_2 concentrations and negatively regulate ROS and NO production and MMPs activity, regulations that occur in parallel to a reduction in malformation and resorption rates [62].

It has been hypothesised that activation of stress-activated protein kinases may be a primary mechanism for the induction of diabetic embryopathy that is mediated by oxidative stress-induced excessive apoptosis. In rats, streptozotocin-induced diabetes is associated with c-Jun N-terminal kinase (JNK)1/2 activation in the embryos and yolk sacs [127]. Of note, vitamin E supplementation reduced the activation of JNK in the embryos and yolk sacs from diabetic mice, which results in the prevention of or rescue from diabetic embryopathy [128].

Therefore, in maternal diabetes, embryos are exposed to increased oxidative and nitrative stress, which affects the availability and function of bioactive lipids, induces post-transductional changes that affect protein function and damages the

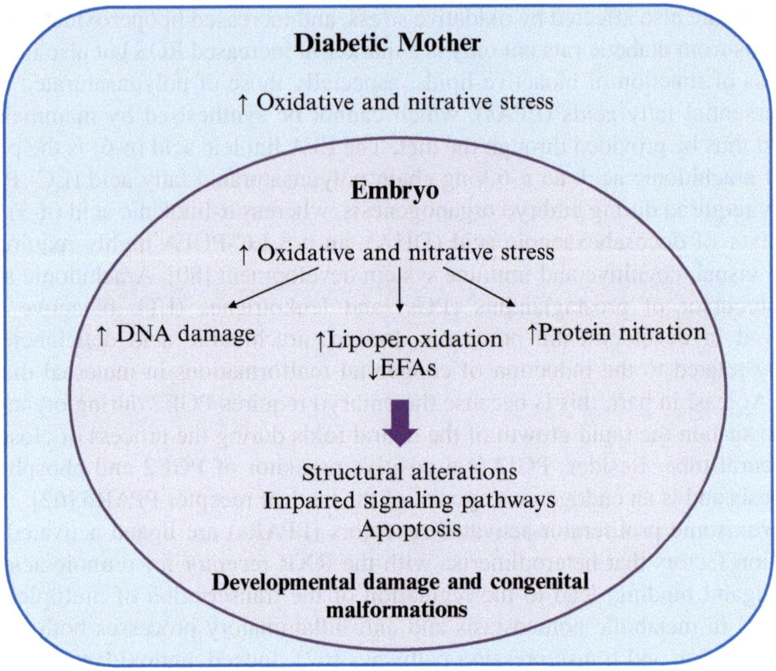

Fig. 3.3 Schematic representation of maternal diabetes-induced oxidative damage during embryo organogenesis

DNA structure, leading to aberrant apoptotic pathways, causing congenital malformations and inducing developmental alterations that can further affect the foetal and postnatal stages (Fig. 3.3).

It should be noted that congenital malformations constitute the most severe postnatal consequences of diabetes and pregnancy and are related to oxidative and nitrative stress during early organogenesis. As congenital malformations are mostly induced before the 8th week of pregnancy, efforts in conception planning, in maintaining glucose homeostasis and in assuring folate intake in diabetic women at reproductive age are needed to prevent their induction.

Fetoplacental Development

Experimental studies have shown that oxidative stress increases with gestational age in healthy pregnancies. This is also evident in diabetic pregnancies, in which ROS are increased when compared to healthy pregnancies at different gestational ages [64]. In cord blood from both GDM and type 1 mothers with diabetes, there is evidence of increase ROS and decreased antioxidant enzyme activities [74].

Oxidative stress can induce oxidative damage to proteins, lipids and nucleic acids, which can result in disruption of normal structure and function. Damage to chromatin can cause both genetic and epigenetic changes and can influence gene expression and differentiation during fetoplacental development [158].

Studies in experimental models of diabetes have shown increased ROS in foetuses and placentas, possibly related to the excessive transfer of metabolic substrates from maternal circulation. Foetal oxidative stress is evidenced even in mild diabetic experimental models, implying that the antioxidant capacity can be exceeded at mild hyperglycaemia. Moreover, according to the severity and genetic background of the experimental models of diabetes and pregnancy, the activity of antioxidant enzymes is increased or decreased in placentas and foetuses, being the damage induced by ROS to proteins, lipids and nucleic acids greater when the antioxidant defences are impaired [66, 75, 113].

Increased ROS generation, together with increased NO production and the consequent formation of the potent oxidant peroxynitrite, is involved in the induction of fetoplacental damage during development in experimental models of diabetes [64]. Indeed, peroxynitrite-induced protein nitration affects the activity of multiple enzymes, including Cu/Zn superoxide dismutase (whose activity and thus antioxidant capacity are reduced), MMPs (whose activity is increased, which leads to the generation of a proinflammatory environment) and PGI_2 synthase (whose activity is reduced, which impairs vasodilation and anti-inflammatory pathways). Reductions in the concentrations of PGI_2 (or prostacyclin) affect signalling pathways both through the membrane prostaglandin I_2 receptor (resulting in impaired vasodilation) and the nuclear receptors PPARδ. In the placentas from diabetic rats, PPARδ activation has been shown to negatively regulate NO production, lipid mass and lipid peroxidation [81]. On the other hand, 15-deoxy-$\Delta^{12,14}$-PGJ_2 (15d-PGJ_2), an endogenous ligand of PPARγ, has been also found reduced in the placenta and foetuses from diabetic rats, and its activation has potent anti-inflammatory effects and leads to the reduction of NO and MMPs activity in placentas and foetuses [123]. MMPs are proteolytic enzymes involved in remodelling processes of the extracellular matrix. MMPs overactivity, evident in both foetuses and placentas during development, constitutes a marker of a proinflammatory state, which can be observed even in sera from pregnant diabetic rats. Both ROS and NO have been shown to positively regulate MMPs activity in foetuses and placentas from diabetic rats. In experimental studies with diabetic animals that were fed with diets enriched in PPAR ligands during pregnancy, increased concentrations of 15d-PGJ2 in foetuses and placentas and a reduction in placental MMPs can be observed, together with a reduction in serum MMPs concentrations [94]. Besides, nitrative stress is increased and leads to specific nitration of MMPs in placentas from type 2 diabetic patients, where MMPs are overactivated [18].

Experimental models have shown that increases in NO and MMPs are also evident in the initial phases of implantation in maternal diabetes [63], the increased pro-oxidant and proinflammatory environment affects the formation, the development and the function of the placenta.

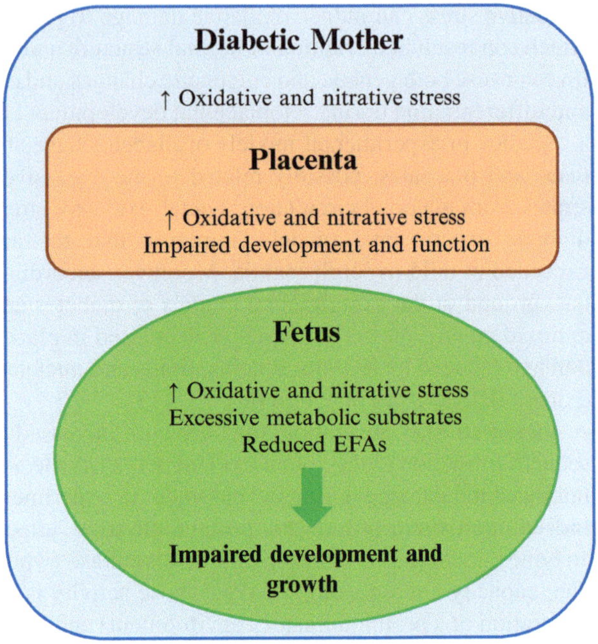

Fig. 3.4 Schematic representation of maternal diabetes-induced oxidative damage in the placenta and the foetus

Overall, both as a consequence of placental alterations and as a direct response to the abnormal metabolic milieu that reaches the foetal compartment, the foetuses are exposed to oxidative stress, challenging its development and growth (Fig. 3.4).

Placental Transport Mechanisms

Foetal growth and development is dependent on placental uptake and transfer of nutrients, including lipids, from the maternal circulation. Thus, pathological placental development and/or function may contribute to changes in the rate of foetal growth, via alterations in the transport of nutrients across the placental barrier into the foetal circulation. Foetal overgrowth in pregnancies complicated by type 1 diabetes [61] but not by gestational diabetes mellitus (GDM) [59] is associated with increased placental glucose transporter (GLUT) activity and protein expression. The capacity of the placenta to transport the essential amino acid L-leucine is increased in GDM complicated by foetal overgrowth [58], and placental system A amino acid transporter activity is increased in both type 1 diabetes and GDM [58].

Leptin [57] and adiponectin [67] have been shown to regulate placental system A activity. There are, however, very few studies that have explored the role of oxidative stress in regulating placental transport mechanisms. Recent studies have shown that antioxidants, including dehydroascorbic acid, N-acetyl cysteine (NAC) and catalase and SOD, can regulate placental system A amino acid transporter activity [8]. We have recently also shown that in placental explants, increased oxidative stress resulted in reduced glucose uptake and lowered GLUT1 expression [84]. In accordance with these studies, in mice overexpressing thioredoxin-1, placental oxidative stress is reduced, which was shown to be associated with increased protein expression of placental GLUT1 [159]. Furthermore, chronic hypoxia in vivo with high altitude pregnancies also reduces expression of GLUT1 [179].

Perinatal Period

At term pregnancy, evidence of increased oxidative stress is observed in different foetal organs, in foetal serum and in the placenta from experimental models of diabetes and pregnancy [64]. In foetuses and neonates from diabetic rats, the heart, brain, liver, kidney and lungs are organs that show increased oxidative stress, alterations that occur in parallel to reduced antioxidant defences [75, 82, 95, 126]. Treatments with vitamins E and C have been shown to reduce oxidative stress in foetal organs such as the brain and the heart in foetuses from diabetic rats [75, 113, 144].

Natural medicinal herbs and plants have been shown to also be protective against pregestational and gestational diabetes-induced fetopathy and thus may serve as a therapeutic supplement under diabetic pregnancy. *Tinospora cordifolia* is an herbaceous vine of the family Menispermaceae, found in tropical and subtropical regions of India. It protects diabetes-induced oxidative stress in foetal brain and liver tissues [143]. Extract of leaves of *Morus nigra* (usually known as black mulberry) also shows capacity to reduce oxidative stress and reduced malformations in foetuses from diabetic rats, a beneficial effect observed despite the lack of regulation of maternal glycaemia [163]. There are no clinical studies that have evaluated the effect of natural medicinal products on the foetus.

Alterations in the signalling pathways that regulate antioxidant pathways are, at least in part, related to the increased reactive oxygen and nitrogen species observed at term in experimental models of diabetes in pregnancy. This includes PPARs signalling, which is profoundly impaired in foetal organs at term diabetic pregnancies. In the foetal lung, PPARα is decreased in male foetuses from diabetic rats, and the activation of this nuclear receptor is able to prevent the overproduction of NO in lungs from both male and female foetuses [82]. In the foetal liver, there is increased production of NO, increased lipid mass and increased lipoperoxidation, alterations that are prevented by the activation of the nuclear receptor PPARα [95]. Animal dietary treatments with PUFAs, providing PPARs ligands and preventing the deficit in EFAs, have been found to regulate ROS and NO production in different foetal organs and to reduce lipoperoxidation in neonatal serum in offspring from diabetic rats [82, 95, 177].

Therefore, there is a growing evidence of increasing knowledge of increased ROS in the intrauterine compartment affecting the function of foetal organs in the perinatal period, and various animal studies that show that the impaired intrauterine pro-oxidant/proinflammatory state can be reduced through treatments compatible with the pregnant state, which deserve to be evaluated in human pregnancies.

Postnatal Consequences of Oxidative Stress

The described placental impairments related to oxidative stress in maternal diabetes are related to alterations in foetal development with consequences in postnatal development. Indeed, the placenta has been proposed to have a relevant role in the intrauterine programming of metabolic and cardiovascular diseases [60]. Besides, oxidative and nitrative stress and the consequent induction of a pro-oxidant and proinflammatory intrauterine environment play a relevant role in the programming of metabolic and cardiovascular diseases in the offspring that is common to various pathologies such as obesity, diabetes and intrauterine growth restriction [85, 145, 157, 158]. Maternal diabetes programmes intrauterine metabolic and cardiovascular diseases [40, 54], and this may be in part due to the impaired generation of reactive oxygen and nitrogen species during pregnancy [89]. Indeed, animal models of diabetes and pregnancy show that vascular, cardiac and renal impairments in adult offspring are related to impairments in NO availability, peroxynitrite formation and antioxidant enzyme activities [27, 71, 130, 154].

The pancreas is highly susceptible to both oxidative stress and changes in glucose concentrations, leading to changes in β cell mass, which are related to the neonatal macrosomic and microsomic phenotypes and to the impaired glucose tolerance in the adult offspring [40, 54]. On the other hand, in neonates from diabetic rats, there is impaired generation of orexigenic and anorexigenic hypothalamic neurons that regulate food intake, an alteration related to ROS formation, PPARs function and leptin resistance, as well as to the programming of metabolic diseases [32, 42].

Overall, as schematised in Fig. 3.5, the programming of metabolic and cardiovascular diseases is highly related to intrauterine and perinatal oxidative and nitrative stress, which affects the function of bioactive molecules, cells, tissues and organs during development. This leads to impaired signalling pathways that affect the complex and interrelated mechanisms that regulate insulin sensitivity, metabolic pathways, appetite and endothelial dysfunction in the adult offspring.

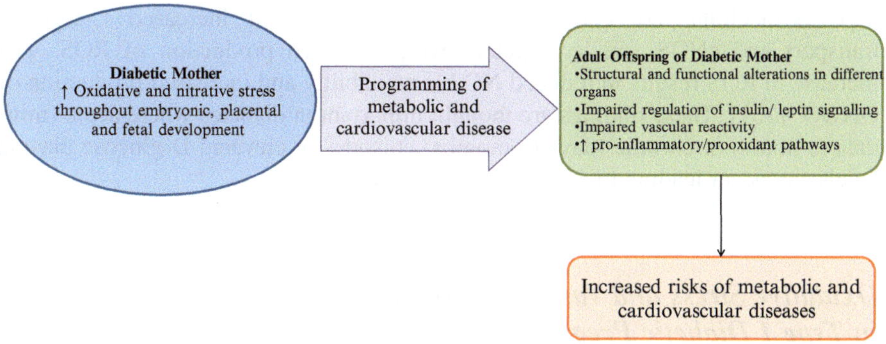

Fig. 3.5 Schematic representation of maternal diabetes-induced oxidative damage in the offspring

Oxidative Stress Mediates Endothelial and Vascular Dysfunction in the Placenta

An abnormal function of the placenta (placental dysfunction) is seen in diabetic pregnancies, which is associated with vascular and endothelial dysfunction [20, 37, 105, 135, 150, 170]. It is thought that an imbalance in the antioxidant and pro-oxidant mechanisms is determinant of a normal or altered vascular and endothelial dysfunction in humans [151].

Oxidative Stress and Vascular Alterations in Pregnancy

Diabetes mellitus is the most common endocrine disorder leading to the development of macrovascular and microvascular dysfunction that contribute to patient morbidity and mortality. In pregnancy, these conditions affect not only the mother but also the developing foetus [45]. Numerous studies demonstrate impaired endothelial function (defined as an altered capacity of the endothelium to take up and metabolise the cationic amino acid L-arginine (the substrate for NO synthesis via NOS)) both in pregestational and GDM where the oxidative stress and ROS and RNS contribute to the progression of the disease [6, 45]. The development of cardiovascular disease in diabetic patients is associated with increased oxidative stress caused by higher activity of the enzymes NOX and xanthine oxidase (XO) together with uncoupling of NOS [13]. These phenomena induce cell dysfunction through oxidation of lipoproteins, nucleic acids, carbohydrates and proteins [85].

The development of endothelial dysfunction in diabetes mellitus is related with the degree of hyperglycaemia. Free radicals, such as superoxide anion, reduce the bioavailability of NO by limiting NOS expression and activity. In fact, human umbilical vein endothelial cells (HUVEC) from normal pregnancies exposed to high extracellular concentrations of D-glucose exhibit increased L-arginine transport and eNOS activity together with increased production of ROS. The increase in ROS results in reduced NO bioavailability and increased production of $ONOO^-$ [151]. These findings are thought important in diabetes where acute (minutes to hours) or chronic (days to months) episodes of elevated D-glucose plasma levels can occur leading to endothelial dysfunction.

Oxidative Stress and Vascular Alterations in Type 1 Diabetic Pregnancies

Endothelial dysfunction in the maternal vasculature is well described in pregnant women with type 1 diabetes and is associated with the development of pregnancy complications such as preeclampsia and restriction of foetal growth [28, 56]. Vascular alterations due to type 1 diabetes include microangiopathies such as retinopathy, which progresses during pregnancy and up to the first-year postpartum [1], and nephropathy, which is the main risk factor to develop preeclampsia in women with type 1 diabetes (40–60 % of the cases). This condition is directly related with an increase in the level of maternal glycosilated Hb_{A1c} and proteinuria [69]. In pregnancies with type 1 diabetes, foetal macrosomia is expected. However, when pregnancy courses with maternal vascular alterations, foetal growth restriction occurs mainly associated with the manifestation of preeclampsia [56]. In this regard, the maternal endothelial dysfunction in type 1 diabetic pregnancies evaluated as elevated blood levels of soluble E-selectin correlates with increased foetal growth; however, in these patients an elevated plasma level of the vascular cell adhesion molecule-1 (sVCAM-1) correlates with reduced foetal growth [180].

The endothelial function evaluated in the skin microvasculature using isolated microvessels or in situ perfused vessels from women with type 1 diabetes is reduced compared with nondiabetic control subjects [79, 125]. However, microvascular endothelial dysfunction was not different when it was adjusted by the level of glycosylated Hb_{A1c} in these patients [125]. Additionally, pregnant women with well-controlled type 1 diabetes have both normal endothelial and smooth muscle function in abdominal arteries [5]. Thus, glycaemic control is relevant regarding the endothelial function. In fact, incubation of HUVEC from normal pregnancies with plasma from type 1 diabetic pregnant women exhibited increased eNOS activity [134]. Interestingly, the resulting increase in NO level was rapidly reduced by its inactivation due to increased circulating ROS in these patients. This phenomenon has also been described in HUVEC primary cultures from pregnancies cursing with GDM [48].

Offspring of mothers with diabetes mellitus have increased risk of developing type 1 diabetes, and a reduced percentage of these patients develop type 2 diabetes or obesity in adulthood [107]. Children followed for up to 30 years old that suffered of exposure in uterus to diabetes (type 1, type 2 and GDM) are predisposed to develop diseases of the circulatory system in the adulthood, including atherosclerosis, hypertension and hard attack amongst others [175]. The impact of type 1 diabetes in the placenta includes enhanced placental angiogenesis, mainly described as longitudinal growth more than increased placental capillary diameter [96], which associates with increased expression of vascular endothelial growth factor (VEGF) and its receptors 1 (VEGFR-1), 2 (VEGFR-2) and 3 (VEGFR-3) [33]. Endothelial cells from type 1 diabetic women exhibit functional and structural alterations including enhanced expression of placental vascular endothelial cadherin and β-catenin, molecules that play key roles in the process of barrier formation and angiogenesis in the placenta from patients with this syndrome in pregnancy [87]. In addition, a reduced resistance to shear stress, reduced uptake of D-glucose [136] and altered plasma membrane structure together with increased mitochondrial area [26] have been described in patients with type 1 diabetes. Thus, a wide range of adaptations and mis-adaptations occur in the maternal and foetal endothelium in pregnancies coursing with type 1 diabetes.

Oxidative Stress and Vascular Alterations in Type 2 Diabetic Pregnancies

Type 2 diabetes is mainly associated with maternal macroangiopathies such as coronary artery disease [69]. Type 2 diabetes leads to maternal and foetal endothelial dysfunction reported to be a consequence of the deregulation of the glycaemic control [117] associated with overproduction of ROS [115].

Pregnant women with type 2 diabetes exhibit placental endothelial dysfunction leading to serious foetal consequences. In the foetus, the maternal hyperglycaemia leads to higher foetal hyperinsulinemia, which associates with foetal overgrowth [164]. Studies performed in primary cultures of HUVEC from normal pregnancies that were exposed to high extracellular concentrations of D-glucose have demonstrated an increase in the L-arginine transport and eNOS activity together with increased production of ROS. These phenomena lead to a reduced NO bioavailability and increased production of $ONOO^-$ [151]. Interestingly, HUVEC isolated from newborns from mothers with familiar history of type 2 diabetes has a reduced synthesis of NO and an impaired expression of eNOS in response to elevated extracellular D-glucose (25 mM) [3].

Oxidative Stress and Vascular Alterations in GDM Pregnancies

In the maternal vasculature, GDM courses with endothelial dysfunction evaluated as reduced acetylcholine response in small subcutaneous artery rings [79], reduced flow-mediated dilatation in brachial artery [117] and reduced vasodilatory responses in brachial artery in women with euglycaemia with previous GDM [4]. Additionally, women with GDM and their offspring have increased blood concentration of markers of endothelial dysfunction such as the soluble intercellular adhesion molecule-1 (sICAM-1) and sVCAM-1. Both sICAM-1 and sVACM-1 are associated with reduced SOD activity suggesting that oxidative stress plays a key role in the endothelial dysfunction in pregnancies coursing with this syndrome [103].

GDM associates with abnormal regulation of the vascular tone in placental and foetal tissues [46], and in contrast to the increased angiogenesis described in type 1 diabetes, angiogenesis seen in GDM is unperturbed [96]. The effect of GDM in the placental endothelial function includes alteration in the adenosine/L-arginine/NO (ALANO) signalling pathway [135, 160]. HUVECs from GDM pregnancies show increased L-arginine transport and NO synthesis as well as impairment in the action of insulin due to hyperglycaemia [160]. Thus, vascular dysfunction resulting from this syndrome may be a consequence of a functional dissociation between the synthesis of NO and its bioavailability to the vascular endothelium and smooth muscle cells in the human placental circulation. In this regard, GDM is associated with oxidative stress where an overproduction of ROS and other free radicals is characteristic and mainly associated with reduced activity of antioxidant enzymes such as NOX and XO [85].

Interestingly, GDM alterations of foetal endothelial function, including reduced adenosine transport and increased NO synthesis [50, 135, 171], are mimicked by exposure of HUVEC from normal pregnancies to elevated extracellular D-glucose (25 mM, high D-glucose) [41]. The latter is also a condition associated with increased formation of ROS [39]. Adenosine transport occurs via the human equilibrative nucleoside transporters 1 (hENT1) and hENT2 and L-arginine transport via the human cationic amino acid transporter 1 (hCAT-1) in HUVECs. GDM has been shown to alter these membrane transport mechanisms through activation of cell signalling cascades involving PI3K, PKC (most likely PKCα), NO and p44 and p44 kDa mitogen-activated protein kinases ($p44/p42^{mapk}$) [38, 160]. One of the direct consequences of the stimulation of these signalling molecules is the activation of transcription factors inhibiting the promoter activity of *SLC29A1* gene (coding for hENT1) leading to reduced transcript and protein abundance with the subsequent reduced adenosine uptake. This could be a mechanism by which adenosine antioxidant properties could be facilitated since increasing extracellular concentrations of this nucleoside are reported in the umbilical whole blood (vein and arteries) [171] as well as in the umbilical vein, but not in the artery blood [133] in GDM.

Adenosine and Oxidative Stress

Several studies refer to adenosine as an endogenous antioxidative molecule, which contributes to lower ROS production in endothelial cells by regulating NOX activity [155, 182]. However, even when in GDM an increase in the fetoplacental circulation of adenosine concentration is seen [133, 171], its antioxidative capacity has not been evaluated [150, 151]. In addition, as in GDM, this potential property of adenosine has not been reported in pregnant women with type 1 or type 2 diabetes [150, 151].

Insulin and Oxidative Stress

Even when there is good control of glycaemia in type 1 diabetes, the changes from hypo- to hyperglycaemia worsen endothelial function, increasing oxidative stress [24]. Insulin is also considered as an endogenous antioxidative, since decreases in ROS generation by suppression of NADPH oxidase [30]. Moreover, it is suggested that when placental endothelium is exposed to higher concentrations of insulin than what is detected in the fetoplacental blood, an increase in the NO bioavailability and maintenance of vascular haemostasis in GDM and other pathologies is seen [120, 153, 171]. Additionally, in HUVEC, insulin leads to an increase in NO synthesis [171] by a mechanism dependent of phosphatidylinositol 3-kinase (PI3K) and Akt signalling pathway [165, 181]. The increase in NO synthesis leads to vasodilation and to increase glucose uptake in target tissues [153, 171]; nevertheless insulin resistance produce an augment in vasoconstriction, mostly due to the impaired insulin signalling [25].

Several studies describe that oxidative stress decreases insulin sensitivity in patients with insulin resistance [97]. Moreover, in type 2 diabetes, diminished eNOS expression and Akt/PKB phosphorylation in arterial tissue is seen, probably due to insulin resistance [110]. This effect is more prolonged during pregnancy since hyperinsulinemia is present from the actual placental implantation day. Since endothelial cells from human placental macro- [171] and microvasculature [133] express insulin receptor and GDM patient courses with higher levels of insulin [172], mostly in the second and third semesters of pregnancy, this hormone could be playing a role as an antioxidant defence mechanism for the growing foetus and the mother. Interestingly, insulin is in fact playing key roles in the haemostasis of human placental macro- and microvascular endothelial dysfunction in GDM. Since adenosine is also increased in the mother and the foetus blood and because this endogenous nucleoside improves the response to insulin in HUVECs [50], it is feasible that added antioxidant efficiency of insulin and the nucleoside adenosine is key in the improvement of fetoplacental endothelial dysfunction in GDM.

Recent studies show that HUVEC and hPMEC in GDM exhibit differential expression of insulin receptor isoforms A (IR-A) and B (IR-B) [133, 171]. HUVEC preferentially expresses IR-A over IR-B isoforms in GDM, and IR-A mRNA is higher in GDM compared with cells from normal pregnancies. On the contrary, in hPMEC a preferential expression of IR-B in cells from GDM compared with normal pregnancies is seen. However, IR-A expression is lower in GDM compared with normal cells. Since IR-A and IR-B activate mainly the $p42/44^{mapk}$ and Akt signalling pathways, respectively, a differential cell signalling mechanism is evident in human fetoplacental endothelium [120]. Thus, specific mechanisms of regulation of insulin receptor isoforms expression could account in endothelial cells from macro- compared with microvasculature in the human placenta [49]. Eventually, if in fact adenosine is involved in the response of these cell types to insulin via a mechanism controlling the redox state of the cells, a differential role of adenosine as an antioxidant could be feasible for the response to IR-A or IR-B in human foetal endothelium.

Cholesterol and Oxidative Stress

A physiological increase of the maternal total cholesterol occurs during pregnancy to satisfy the increasing metabolic demand of lipids by the growing foetus, a condition referred as maternal physiological hypercholesterolemia (MPH) [10, 102]. However, when a mis-adaptation occurs and the level of cholesterol goes over a physiological range, another condition is recognised, i.e. maternal supraphysiological hypercholesterolemia in pregnancy (MSPH) [89, 102]. Even when a strong correlation between maternal cholesterolaemia in pregnancy and endothelial placental dysfunction as well as the size of atherosclerotic lesions in arteries of the foetus is documented [106], MSPH effect on the foetal endothelial function of pregnancies coursing with GDM is unknown [89].

A potential pathophysiological interaction between maternal levels of cholesterol and the oxidative status of the mother, the placenta and the foetus in the development of foetal atherosclerosis have been proposed [89]. Maternal factors associated with increased levels of total cholesterol in pregnancy correlates with a reduced catalase activity, but increased lipid peroxidation and oxidised LDL in MSPH. This phenomenon leads to alterations in the placenta, including reduced catalase, SOD and GPx activity, and in the foetus, including reduced catalase and GPx activity, but increased lipid peroxidation and oxLDL levels [91]. This pro-oxidant environment has been reported leading to endothelial dysfunction in the maternal vasculature; however, whether foetal vasculature is modified in MSPH is unknown [91]. It is suggested that in pregnancy, the mother, the placenta and the foetus are subjected to oxidative stress in a pathological condition of the mother such as MSPH. This is actually worrying since these phenomena lead to vascular alterations including endothelial dysfunction and atherosclerosis [91, 149]. We have recently enrolled a group of pregnant women with GDM that exhibited MSPH, which corresponded to 30 % of the GDM population. These women exhibited increased total and LDL-cholesterol compared with MSPH, thus suggesting that the latter is a condition worsened in GDM (A Leiva, L Sobrevia, *unpublished*). This could be a relevant pathophysiological mechanism accounting for endothelial dysfunction in GDM (A Leiva, L Sobrevia, *unpublished*). How are the pro- and antioxidant mechanisms been operative under this new scenario, in the case of GDM plus MSPH, is not fully known.

Figure 3.1 is a schematic diagram summarising placenta endothelial dysfunction in normal and diabetic pregnancies.

Conclusions

Oxidative stress induces damage in GDM in the placenta and the vasculature of the mother, the placenta and the umbilical cord, produced as a resulting consequence of exacerbated NO and ROS production. Oxidative stress disturbs placental function leading to perpetrations in foetal growth and development. There is also a clear beneficial effect of antioxidants on diabetes and pregnancy animal models, indicating the relevance of oxidative stress in diabetes-induced pregnancy complications. However, whether or not antioxidant supplementation or eating a diet rich in antioxidants can improve the consequences of oxidative stress in the offspring is yet to be elucidated. More studies are thus required to fully understand the short- and long-term health benefit of reducing oxidative stress during diabetic pregnancies.

Acknowledgements ML is supported by Career Development Fellowships from National Health and Medical Research Council (NHMRC) (#454777 and #1047025). The work from Associate Professor Lappas's laboratory was funded by project grants from the Medical Research Foundation for Women and Babies, NHMRC (grant no. 454310) and Diabetes Australia Research Trust (DART). Work from Dr. Jawerbaum's laboratory was supported by ANPCYT (PICT 2010–00034) and by CONICET (PIP 11220100100002). Work from Prof. Sobrevia's laboratory was supported by Fondo Nacional de Desarrollo Científico y Tecnológico (FONDECYT 1110977, 11110059, 3130583), Programa de Investigación Interdisciplinario (PIA) from Comisión Nacional de Investigación en Ciencia y Tecnología (CONICYT, Anillos ACT-73) (Chile).

References

1. Effect of pregnancy on microvascular complications in the diabetes control and complications trial. The Diabetes Control and Complications Trial Research Group. *Diabetes Care.* 2000;23:1084–91.
2. Al-Gubory KH, Fowler PA, Garrel C. The roles of cellular reactive oxygen species, oxidative stress and antioxidants in pregnancy outcomes. Int J Biochem Cell Biol. 2010;42:1634–50.
3. Alvarado-Vasquez N, Zapata E, Alcazar-Leyva S, Masso F, Montano LF. Reduced NO synthesis and eNOS mRNA expression in endothelial cells from newborns with a strong family history of type 2 diabetes. Diabetes Metab Res Rev. 2007;23:559–66.
4. Anastasiou E, Lekakis JP, Alevizaki M, Papamichael CM, Megas J, Souvatzoglou A, Stamatelopoulos SF. Impaired endothelium-dependent vasodilatation in women with previous gestational diabetes. Diabetes Care. 1998;21:2111–5.
5. Ang C, Hillier C, Johnston F, Cameron A, Greer I, Lumsden MA. Endothelial function is preserved in pregnant women with well-controlled type 1 diabetes. BJOG. 2002;109:699–707.
6. Ang C, Lumsden MA. Diabetes and the maternal resistance vasculature. Clin Sci (Lond). 2001;101:719–29.
7. Ategbo JM, Grissa O, Yessoufou A, Hichami A, Dramane KL, Moutairou K, Miled A, Grissa A, Jerbi M, Tabka Z, Khan NA. Modulation of adipokines and cytokines in gestational diabetes and macrosomia. J Clin Endocrinol Metab. 2006;91:4137–43.
8. Bainbridge SA, von Versen-Hoynck F, Roberts JM. Uric acid inhibits placental system A amino acid uptake. Placenta. 2009;30:195–200.
9. Barbour LA, McCurdy CE, Hernandez TL, Kirwan JP, Catalano PM, Friedman JE. Cellular mechanisms for insulin resistance in normal pregnancy and gestational diabetes. Diabetes Care. 2007;30:S112–9.

10. Basaran A. Pregnancy-induced hyperlipoproteinemia: review of the literature. Reprod Sci. 2009;16:431–7.
11. Basu S, Haghiac M, Surace P, Challier JC, Guerre-Millo M, Singh K, Waters T, Minium J, Presley L, Catalano PM, Hauguel-de Mouzon S. Pregravid obesity associates with increased maternal endotoxemia and metabolic inflammation. Obesity (Silver Spring). 2011;19(3): 476–482. doi:10.1038/oby.2010.215.
12. Beebe LF, Kaye PL. Maternal diabetes and retarded preimplantation development of mice. Diabetes. 1991;40:457–61.
13. Bo S, Menato G, Gallo ML, Bardelli C, Lezo A, Signorile A, Gambino R, Cassader M, Massobrio M, Pagano G. Mild gestational hyperglycemia, the metabolic syndrome and adverse neonatal outcomes. Acta Obstet Gynecol Scand. 2004;83:335–40.
14. Braissant O, Wahli W. Differential expression of peroxisome proliferator-activated receptor-alpha, -beta, and -gamma during rat embryonic development. Endocrinology. 1998;139: 2748–54.
15. Buchanan TA. Pancreatic B-cell defects in gestational diabetes: implications for the pathogenesis and prevention of type 2 diabetes. J Clin Endocrinol Metab. 2001;86:989–93.
16. Buchanan TA, Xiang AH. Gestational diabetes mellitus. J Clin Invest. 2005;115:485–91.
17. Cani PD. Gut microbiota and pregnancy, a matter of inner life. Br J Nutr. 2009;101: 1579–80.
18. Capobianco E, White V, Sosa M, Marco ID, Basualdo MN, Faingold MC, Jawerbaum A. Regulation of matrix metalloproteinases 2 and 9 activities by peroxynitrites in term placentas from type 2 diabetic patients. Reprod Sci. 2012;19:814–22.
19. Carr DB, Utzschneider KM, Hull RL, Tong J, Wallace TM, Kodama K, Shofer JB, Heckbert SR, Boyko EJ, Fujimoto WY, Kahn SE. Gestational diabetes mellitus increases the risk of cardiovascular disease in women with a family history of type 2 diabetes. Diabetes Care. 2006;29:2078–83.
20. Casanello P, Escudero C, Sobrevia L. Equilibrative nucleoside (ENTs) and cationic amino acid (CATs) transporters: implications in foetal endothelial dysfunction in human pregnancy diseases. Curr Vasc Pharmacol. 2007;5:69–84.
21. Casey BM, Lucas MJ, McIntire DD, Leveno KJ. Pregnancy outcomes in women with gestational diabetes compared with the general obstetric population. Obstet Gynecol. 1997;90: 869–73.
22. Catalano PM. Obesity, insulin resistance, and pregnancy outcome. Reproduction. 2010; 140:365–71.
23. Cederberg J, Eriksson UJ. Decreased catalase activity in malformation-prone embryos of diabetic rats. Teratology. 1997;56:350–7.
24. Ceriello A, Novials A, Ortega E, La Sala L, Pujadas G, Testa R, Bonfigli AR, Esposito K, Giugliano D. Evidence that hyperglycemia after recovery from hypoglycemia worsens endothelial function and increases oxidative stress and inflammation in healthy control subjects and subjects with type 1 diabetes. Diabetes. 2012;61:2993–7.
25. Cersosimo E, DeFronzo RA. Insulin resistance and endothelial dysfunction: the road map to cardiovascular diseases. Diabetes Metab Res Rev. 2006;22:423–36.
26. Cester N, Rabini RA, Salvolini E, Staffolani R, Curatola A, Pugnaloni A, Brunelli MA, Biagini G, Mazzanti L. Activation of endothelial cells during insulin-dependent diabetes mellitus: a biochemical and morphological study. Eur J Clin Invest. 1996;26:569–73.
27. Chang SY, Chen YW, Zhao XP, Chenier I, Tran S, Sauve A, Ingelfinger JR, Zhang SL. Catalase prevents maternal diabetes-induced perinatal programming via the Nrf2-HO-1 defense system. Diabetes. 2012;61:2565–74.
28. Clausen P, Ekbom P, Damm P, Feldt-Rasmussen U, Nielsen B, Mathiesen ER, Feldt-Rasmussen B. Signs of maternal vascular dysfunction precede preeclampsia in women with type 1 diabetes. J Diabetes Complications. 2007;21:288–93.
29. Coughlan MT, Oliva K, Georgiou HM, Permezel JM, Rice GE. Glucose-induced release of tumour necrosis factor-alpha from human placental and adipose tissues in gestational diabetes mellitus. Diabet Med. 2001;18:921–7.

30. Dandona P, Ghanim H, Bandyopadhyay A, Korzeniewski K, Ling Sia C, Dhindsa S, Chaudhuri A. Insulin suppresses endotoxin-induced oxidative, nitrosative, and inflammatory stress in humans. Diabetes Care. 2010;33:2416–23.
31. Desoye G, Hauguel-de Mouzon S. The human placenta in gestational diabetes mellitus. The insulin and cytokine network. Diabetes Care. 2007;30 Suppl 2:S120–6.
32. Diano S, Liu ZW, Jeong JK, Dietrich MO, Ruan HB, Kim E, Suyama S, Kelly K, Gyengesi E, Arbiser JL, Belsham DD, Sarruf DA, Schwartz MW, Bennett AM, Shanabrough M, Mobbs CV, Yang X, Gao XB, Horvath TL. Peroxisome proliferation-associated control of reactive oxygen species sets melanocortin tone and feeding in diet-induced obesity. Nat Med. 2011; 17:1121–7.
33. Dubova EA, Pavlov KA, Esayan RM, Degtyareva EI, Shestakova MV, Shchegolev AI, Sukhikh GT. Vascular endothelial growth factor and its receptors in the placenta of women with type 1 diabetes mellitus. Bull Exp Biol Med. 2012;152:367–70.
34. Ejdesjo A, Wentzel P, Eriksson UJ. Influence of maternal metabolism and parental genetics on fetal maldevelopment in diabetic rat pregnancy. Am J Physiol Endocrinol Metab. 2012; 302:E1198–209.
35. Engeland A, Bjorge T, Daltveit AK, Skurtveit S, Vangen S, Vollset SE, Furu K. Risk of diabetes after gestational diabetes and preeclampsia. A registry-based study of 230,000 women in Norway. Eur J Epidemiol. 2011;26:157–63.
36. Eriksson UJ. Congenital anomalies in diabetic pregnancy. Semin Fetal Neonatal Med. 2009; 14:85–93.
37. Escudero C, Sobrevia L. A hypothesis for preeclampsia: adenosine and inducible nitric oxide synthase in human placental microvascular endothelium. Placenta. 2008;29:469–83.
38. Farias M, San Martin R, Puebla C, Pearson JD, Casado JF, Pastor-Anglada M, Casanello P, Sobrevia L. Nitric oxide reduces adenosine transporter ENT1 gene (SLC29A1) promoter activity in human fetal endothelium from gestational diabetes. J Cell Physiol. 2006;208:451–60.
39. Felice F, Lucchesi D, di Stefano R, Barsotti MC, Storti E, Penno G, Balbarini A, Del Prato S, Pucci L. Oxidative stress in response to high glucose levels in endothelial cells and in endothelial progenitor cells: evidence for differential glutathione peroxidase-1 expression. Microvasc Res. 2010;80:332–8.
40. Fetita LS, Sobngwi E, Serradas P, Calvo F, Gautier JF. Consequences of fetal exposure to maternal diabetes in offspring. J Clin Endocrinol Metab. 2006;91:3718–24.
41. Flores C, Rojas S, Aguayo C, Parodi J, Mann G, Pearson JD, Casanello P, Sobrevia L. Rapid stimulation of L-arginine transport by D-glucose involves p42/44(mapk) and nitric oxide in human umbilical vein endothelium. Circ Res. 2003;92:64–72.
42. Franke K, Harder T, Aerts L, Melchior K, Fahrenkrog S, Rodekamp E, Ziska T, Van Assche FA, Dudenhausen JW, Plagemann A. 'Programming' of orexigenic and anorexigenic hypothalamic neurons in offspring of treated and untreated diabetic mother rats. Brain Res. 2005; 1031:276–83.
43. Gao XL, Yang HX, Zhao Y. Variations of tumor necrosis factor-alpha, leptin and adiponectin in mid-trimester of gestational diabetes mellitus. Chin Med J (Engl). 2008;121:701–5.
44. GarcÃa-Patterson A, Corcoy R, Balsells M, Altirriba O, Adelantado JM, Cabero L, de Leiva A. A randomized controlled trial using glycemic plus fetal ultrasound parameters versus glycemic parameters to determine insulin therapy in gestational diabetes with fasting hyperglycemia. Diabetes Care. 2002;25:1261.
45. González M, Muñoz E, Puebla C, Guzmán-Gutiérrez E, Cifuentes F, Nien JK, Abarzúa F, Leiva A, Casanello P, Sobrevia L. Maternal and fetal metabolic dysfunction in pregnancy diseases associated with vascular oxidative and nitrative stress. In: Matata BM, Elahi M, editors. The molecular basis for origin of fetal congenital abnormalities and maternal health: an overview of association with oxidative stress. 2011. p. 98–115.
46. Gonzalez M, Gallardo V, Rodriguez N, Salomon C, Westermeier F, Guzman-Gutierrez E, Abarzua F, Leiva A, Casanello P, Sobrevia L. Insulin-stimulated L-arginine transport requires SLC7A1 gene expression and is associated with human umbilical vein relaxation. J Cell Physiol. 2011;226:2916–24.

47. Groesch KA, Torry RJ, Wilber AC, Abrams R, Bieniarz A, Guilbert LJ, Torry DS. Nitric oxide generation affects pro- and anti-angiogenic growth factor expression in primary human trophoblast. Placenta. 2011;32:926–31.
48. Guzman-Gutierrez E, Abarzua F, Belmar C, Nien JK, Ramirez MA, Arroyo P, Salomon C, Westermeier F, Puebla C, Leiva A, Casanello P, Sobrevia L. Functional link between adenosine and insulin: a hypothesis for fetoplacental vascular endothelial dysfunction in gestational diabetes. Curr Vasc Pharmacol. 2011;9:750–62.
49. Guzmán-Gutiérrez E, Arroyo P, Pardo F, Leiva A, Sobrevia L. The adenosine-insulin signaling axis in the fetoplacental endothelial dysfunction in gestational diabetes. In: Luis Sobrevia, editor. Gestational diabetes – causes, diagnosis and treatment; 2013; in press.
50. Guzman-Gutierrez E, Westermeier F, Salomon C, Gonzalez M, Pardo F, Leiva A, Sobrevia L. Insulin-increased L-arginine transport requires A(2A) adenosine receptors activation in human umbilical vein endothelium. PLoS One. 2012;7:e41705.
51. Harris LK, McCormick J, Cartwright JE, Whitley GS, Dash PR. S-nitrosylation of proteins at the leading edge of migrating trophoblasts by inducible nitric oxide synthase promotes trophoblast invasion. Exp Cell Res. 2008;314:1765–76.
52. Higa R, Kurtz M, Mazzucco MB, Musikant D, White V, Jawerbaum A. Folic acid and safflower oil supplementation interacts and protects embryos from maternal diabetes-induced damage. Mol Hum Reprod. 2012;18:253–64.
53. Higa R, White V, Martinez N, Kurtz M, Capobianco E, Jawerbaum A. Safflower and olive oil dietary treatments rescue aberrant embryonic arachidonic acid and nitric oxide metabolism and prevent diabetic embryopathy in rats. Mol Hum Reprod. 2010;16:286–95.
54. Holemans K, Aerts L, Van Assche FA. Lifetime consequences of abnormal fetal pancreatic development. J Physiol. 2003;547:11–20.
55. Horvath EM, Magenheim R, Kugler E, Vacz G, Szigethy A, Levardi F, Kollai M, Szabo C, Lacza Z. Nitrative stress and poly(ADP-ribose) polymerase activation in healthy and gestational diabetic pregnancies. Diabetologia. 2009;52:1935–43.
56. Howarth C, Gazis A, James D. Associations of Type 1 diabetes mellitus, maternal vascular disease and complications of pregnancy. Diabet Med. 2007;24:1229–34.
57. Jansson N, Greenwood SL, Johansson BR, Powell TL, Jansson T. Leptin stimulates the activity of the system A amino acid transporter in human placental villous fragments. J Clin Endocrinol Metab. 2003;88:1205–11.
58. Jansson T, Ekstrand Y, Björn C, Wennergren M, Powell TL. Alterations in the activity of placental amino acid transporters in pregnancies complicated by diabetes. Diabetes. 2002;51: 2214–9.
59. Jansson T, Ekstrand Y, Wennergren M, Powell TL. Placental glucose transport in gestational diabetes mellitus. Am J Obstet Gynecol. 2001;184:111–6.
60. Jansson T, Powell TL. Role of the placenta in fetal programming: underlying mechanisms and potential interventional approaches. Clin Sci (Lond). 2007;113:1–13.
61. Jansson T, Wennergren M, Powell TL. Placental glucose transport and GLUT 1 expression in insulin-dependent diabetes. Am J Obstet Gynecol. 1999;180:163–8.
62. Jawerbaum A, Capobianco E. Review: effects of PPAR activation in the placenta and the fetus: implications in maternal diabetes. Placenta. 2011;32 Suppl 2:S212–7.
63. Jawerbaum A, Gonzalez E. The role of alterations in arachidonic acid metabolism and nitric oxide homeostasis in rat models of diabetes during early pregnancy. Curr Pharm Des. 2005; 11:1327–42.
64. Jawerbaum A, Gonzalez E. Diabetic pregnancies: the challenge of developing in a pro-inflammatory environment. Curr Med Chem. 2006;13:2127–38.
65. Jawerbaum A, Higa R, White V, Capobianco E, Pustovrh C, Sinner D, Martinez N, Gonzalez E. Peroxynitrites and impaired modulation of nitric oxide concentrations in embryos from diabetic rats during early organogenesis. Reproduction. 2005;130:695–703.
66. Jawerbaum A, White V. Animal models in diabetes and pregnancy. Endocr Rev. 2010; 31:680–701.

67. Jones HN, Jansson T, Powell TL. Full-length adiponectin attenuates insulin signaling and inhibits insulin-stimulated amino Acid transport in human primary trophoblast cells. Diabetes. 2010;59:1161–70.
68. Jungheim ES, Moley KH. The impact of type 1 and type 2 diabetes mellitus on the oocyte and the preimplantation embryo. Semin Reprod Med. 2008;26:186–95.
69. Kaaja R. Vascular complications in diabetic pregnancy. Thromb Res. 2011;127 Suppl 3:S53–5.
70. Kahn SE, Prigeon RL, McCulloch DK, Boyko EJ, Bergman RN, Schwartz MW, Neifing JL, Ward WK, Beard JC, Palmer JP, et al. Quantification of the relationship between insulin sensitivity and beta-cell function in human subjects. Evidence for a hyperbolic function. Diabetes. 1993;42:1663–72.
71. Katkhuda R, Peterson ES, Roghair RD, Norris AW, Scholz TD, Segar JL. Sex-specific programming of hypertension in offspring of late-gestation diabetic rats. Pediatr Res. 2012; 72:352–61.
72. Kaufmann P, Black S, Huppertz B. Endovascular trophoblast invasion: implications for the pathogenesis of intrauterine growth retardation and preeclampsia. Biol Reprod. 2003;69:1–7.
73. Kautzky-Willer A, Prager R, Waldhausl W, Pacini G, Thomaseth K, Wagner OF, Ulm M, Streli C, Ludvik B. Pronounced insulin resistance and inadequate beta-cell secretion characterize lean gestational diabetes during and after pregnancy. Diabetes Care. 1997; 20:1717–23.
74. Kinalski M, Sledziewski A, Telejko B, Kowalska I, Kretowski A, Zarzycki W, Kinalska I. Lipid peroxidation, antioxidant defence and acid-base status in cord blood at birth: the influence of diabetes. Horm Metab Res. 2001;33:227–31.
75. Kinalski M, Sledziewski A, Telejko B, Zarzycki W, Kinalska I. Lipid peroxidation and scavenging enzyme activity in streptozotocin-induced diabetes. Acta Diabetol. 2000;37:179–83.
76. Kirwan JP, Hauguel-De Mouzon S, Lepercq J, Challier JC, Huston-Presley L, Friedman JE, Kalhan SC, Catalano PM. TNF-alpha is a predictor of insulin resistance in human pregnancy. Diabetes. 2002;51:2207–13.
77. Kjos SL, Buchanan TA. Gestational diabetes mellitus. N Engl J Med. 1999;341:1749–56.
78. Klein K, Satler M, Elhenicky M, Brix J, Krzyzanowska K, Schernthaner G, Husslein PW, Schernthaner G-H. Circulating levels of MCP-1 are increased in women with gestational diabetes. Prenat Diagn. 2008;28:845–51.
79. Knock GA, McCarthy AL, Lowy C, Poston L. Association of gestational diabetes with abnormal maternal vascular endothelial function. Br J Obstet Gynaecol. 1997;104:229–34.
80. Koletzko B, Lien E, Agostoni C, Bohles H, Campoy C, Cetin I, Decsi T, Dudenhausen JW, Dupont C, Forsyth S, Hoesli I, Holzgreve W, Lapillonne A, Putet G, Secher NJ, Symonds M, Szajewska H, Willatts P, Uauy R. The roles of long-chain polyunsaturated fatty acids in pregnancy, lactation and infancy: review of current knowledge and consensus recommendations. J Perinat Med. 2008;36:5–14.
81. Kurtz M, Capobianco E, Martinez N, Fernandez J, Higa R, White V, Jawerbaum A. Carbaprostacyclin, a PPARdelta agonist, ameliorates excess lipid accumulation in diabetic rat placentas. Life Sci. 2010;86:781–90.
82. Kurtz M, Martinez N, Capobianco E, Higa R, Fornes D, White V, Jawerbaum A. Increased nitric oxide production and gender-dependent changes in PPARalpha expression and signaling in the fetal lung from diabetic rats. Mol Cell Endocrinol. 2012;362:120–7.
83. Lappas M. Activation of inflammasomes in adipose tissue of women with gestational diabetes. Mol Cell Endocrinol. 2014;382:74–83.
84. Lappas M, Andrikopoulos S, Permezel M. Hypoxanthine-xanthine oxidase down-regulates GLUT1 transcription via SIRT1 resulting in decreased glucose uptake in human placenta. J Endocrinol. 2012;213:49–57.
85. Lappas M, Hiden U, Desoye G, Froehlich J, Mouzon SH, Jawerbaum A. The role of oxidative stress in the pathophysiology of gestational diabetes mellitus. Antioxid Redox Signal. 2011; 15:3061–100.

86. Larsen N, Vogensen FK, van den Berg FW, Nielsen DS, Andreasen AS, Pedersen BK, Al-Soud WA, Sorensen SJ, Hansen LH, Jakobsen M. Gut microbiota in human adults with type 2 diabetes differs from non-diabetic adults. PLoS One. 2010;5:e9085.
87. Leach L, Gray C, Staton S, Babawale MO, Gruchy A, Foster C, Mayhew TM, James DK. Vascular endothelial cadherin and beta-catenin in human fetoplacental vessels of pregnancies complicated by Type 1 diabetes: associations with angiogenesis and perturbed barrier function. Diabetologia. 2004;47:695–709.
88. Leduc L, Levy E, Bouity-Voubou M, Delvin E. Fetal programming of atherosclerosis: possible role of the mitochondria. Eur J Obstet Gynecol Reprod Biol. 2010;149:127–30.
89. Leiva A, Pardo F, Ramirez MA, Farias M, Casanello P, Sobrevia L. Fetoplacental vascular endothelial dysfunction as an early phenomenon in the programming of human adult diseases in subjects born from gestational diabetes mellitus or obesity in pregnancy. Exp Diabetes Res. 2011;2011:349286.
90. Lepercq J, Cauzac M, Lahlou N, Timsit J, Girard J, Auwerx J, Hauguel-de Mouzon S. Overexpression of placental leptin in diabetic pregnancy: a critical role for insulin. Diabetes. 1998;47:847–50.
91. Liguori A, D'Armiento FP, Palagiano A, Balestrieri ML, Williams-Ignarro S, de Nigris F, Lerman LO, D'Amora M, Rienzo M, Fiorito C, Ignarro LJ, Palinski W, Napoli C. Effect of gestational hypercholesterolaemia on omental vasoreactivity, placental enzyme activity and transplacental passage of normal and oxidised fatty acids. BJOG. 2007;114:1547–56.
92. Lobner K, Knopff A, Baumgarten A, Mollenhauer U, Marienfeld S, Garrido-Franco M, Bonifacio E, Ziegler AG. Predictors of postpartum diabetes in women with gestational diabetes mellitus. Diabetes. 2006;55:792–7.
93. Loeken MR. Free radicals and birth defects. J Matern Fetal Neonatal Med. 2004;15:6–14.
94. Martinez N, Sosa M, Higa R, Fornes D, Capobianco E, Jawerbaum A. Dietary treatments enriched in olive and safflower oils regulate seric and placental matrix metalloproteinases in maternal diabetes. Placenta. 2012;33:8–16.
95. Martinez N, White V, Kurtz M, Higa R, Capobianco E, Jawerbaum A. Activation of the nuclear receptor PPARalpha regulates lipid metabolism in foetal liver from diabetic rats: implications in diabetes-induced foetal overgrowth. Diabetes Metab Res Rev. 2011; 27:35–46.
96. Mayhew TM. Enhanced fetoplacental angiogenesis in pre-gestational diabetes mellitus: the extra growth is exclusively longitudinal and not accompanied by microvascular remodelling. Diabetologia. 2002;45:1434–9.
97. Meigs JB, Larson MG, Fox CS, Keaney Jr JF, Vasan RS, Benjamin EJ. Association of oxidative stress, insulin resistance, and diabetes risk phenotypes: the Framingham Offspring Study. Diabetes Care. 2007;30:2529–35.
98. Metzger BE. Long-term outcomes in mothers diagnosed with gestational diabetes mellitus and their offspring. Clin Obstet Gynecol. 2007;50:972–9.
99. Metzger BE, Buchanan TA, Coustan DR, de Leiva A, Dunger DB, Hadden DR, Hod M, Kitzmiller JL, Kjos SL, Oats JN, Pettitt DJ, Sacks DA, Zoupas C. Summary and recommendations of the fifth international workshop-conference on gestational diabetes mellitus. Diabetes Care. 2007;30 Suppl 2:S251–60.
100. Moley KH. Hyperglycemia and apoptosis: mechanisms for congenital malformations and pregnancy loss in diabetic women. Trends Endocrinol Metab. 2001;12:78–82.
101. Moley KH, Chi MMY, Knudson CM, Korsmeyer SJ, Mueckler MM. Hyperglycemia induces apoptosis in pre-implantation embryos through cell death effector pathways. Nat Med. 1998;4:1421–4.
102. Montes A, Walden CE, Knopp RH, Cheung M, Chapman MB, Albers JJ. Physiologic and supraphysiologic increases in lipoprotein lipids and apoproteins in late pregnancy and postpartum. Possible markers for the diagnosis of "prelipemia". Arteriosclerosis. 1984;4:407–17.
103. Mordwinkin NM, Ouzounian JG, Yedigarova L, Montoro MN, Louie SG, Rodgers KE. Alteration of endothelial function markers in women with gestational diabetes and their fetuses. J Matern Fetal Neonatal Med. 2013;26(5):507–512. doi:10.3109/14767058.2012.736564.

104. Morgan SC, Relaix F, Sandell LL, Loeken MR. Oxidative stress during diabetic pregnancy disrupts cardiac neural crest migration and causes outflow tract defects. Birth Defects Res A Clin Mol Teratol. 2008;82:453–63.
105. Myatt L. Review: reactive oxygen and nitrogen species and functional adaptation of the placenta. Placenta. 2010;31(Suppl):S66–9.
106. Napoli C, D'Armiento FP, Mancini FP, Postiglione A, Witztum JL, Palumbo G, Palinski W. Fatty streak formation occurs in human fetal aortas and is greatly enhanced by maternal hypercholesterolemia. Intimal accumulation of low density lipoprotein and its oxidation precede monocyte recruitment into early atherosclerotic lesions. J Clin Invest. 1997;100: 2680–90.
107. Nelson SM, Sattar N, Freeman DJ, Walker JD, Lindsay RS. Inflammation and endothelial activation is evident at birth in offspring of mothers with type 1 diabetes. Diabetes. 2007; 56:2697–704.
108. Norwitz ER, Schust DJ, Fisher SJ. Implantation and the survival of early pregnancy. N Engl J Med. 2001;345:1400–8.
109. O'Suillivan JB. The Boston gestational diabetes studies: review and perspectives. In: Sutherland HW, Stowers JM, Pearsons DWM, editors; 1989. p. 287–94.
110. Okon EB, Chung AW, Rauniyar P, Padilla E, Tejerina T, McManus BM, Luo H, van Breemen C. Compromised arterial function in human type 2 diabetic patients. Diabetes. 2005; 54:2415–23.
111. Ornoy A. Embryonic oxidative stress as a mechanism of teratogenesis with special emphasis on diabetic embryopathy. Reprod Toxicol. 2007;24:31–41.
112. Ornoy A, Kimyagarov D, Yaffee P, Abir R, Raz I, Kohen R. Role of reactive oxygen species in diabetes-induced embryotoxicity: studies on pre-implantation mouse embryos cultured in serum from diabetic pregnant women. Isr J Med Sci. 1996;32:1066–73.
113. Ornoy A, Tsadok MA, Yaffe P, Zangen SW. The Cohen diabetic rat as a model for fetal growth restriction: vitamins C and E reduce fetal oxidative stress but do not restore normal growth. Reprod Toxicol. 2009;28:521–9.
114. Ou XH, Li S, Wang ZB, Li M, Quan S, Xing F, Guo L, Chao SB, Chen Z, Liang XW, Hou Y, Schatten H, Sun QY. Maternal insulin resistance causes oxidative stress and mitochondrial dysfunction in mouse oocytes. Hum Reprod. 2012;27:2130–45.
115. Ozdemir G, Ozden M, Maral H, Kuskay S, Cetinalp P, Tarkun I. Malondialdehyde, glutathione, glutathione peroxidase and homocysteine levels in type 2 diabetic patients with and without microalbuminuria. Ann Clin Biochem. 2005;42:99–104.
116. Pampfer S, Vanderheyden I, McCracken JE, Vesela J, De Hertogh R. Increased cell death in rat blastocysts exposed to maternal diabetes in utero and to high glucose or tumor necrosis factor-alpha in vitro. Development. 1997;124:4827–36.
117. Paradisi G, Biaggi A, Ferrazzani S, De Carolis S, Caruso A. Abnormal carbohydrate metabolism during pregnancy: association with endothelial dysfunction. Diabetes Care. 2002; 25:560–4.
118. Perrin MC, Terry MB, Kleinhaus K, Deutsch L, Yanetz R, Tiram E, Calderon-Margalit R, Friedlander Y, Paltiel O, Harlap S. Gestational diabetes and the risk of breast cancer among women in the Jerusalem Perinatal Study. Breast Cancer Res Treat. 2008;108:129–35.
119. Perrin MC, Terry MB, Kleinhaus K, Deutsch L, Yanetz R, Tiram E, Calderon R, Friedlander Y, Paltiel O, Harlap S. Gestational diabetes as a risk factor for pancreatic cancer: a prospective cohort study. BMC Med. 2007;5:25.
120. Pisaneschi S, Boldrini A, Genazzani AR, Coceani F, Simoncini T. Feto-placental vascular dysfunction as a prenatal determinant of adult cardiovascular disease. Intern Emerg Med. 2013;8 Suppl 1:S41–5.
121. Platt MJ, Stanisstreet M, Casson IF, Howard CV, Walkinshaw S, Pennycook S, McKendrick O. St Vincent's Declaration 10 years on: outcomes of diabetic pregnancies. Diabet Med. 2002;19:216–20.
122. Ponzio BF, Carvalho MH, Fortes ZB, do Carmo Franco M. Implications of maternal nutrient restriction in transgenerational programming of hypertension and endothelial dysfunction across F1-F3 offspring. Life Sci. 2012;90:571–7.

123. Pustovrh MC, Capobianco E, Martinez N, Higa R, White V, Jawerbaum A. MMP/TIMP balance is modulated in vitro by 15dPGJ(2) in fetuses and placentas from diabetic rats. Eur J Clin Invest. 2009;39:1082–90.
124. Radaelli T, Varastehpour A, Catalano P, Hauguel-de Mouzon S. Gestational diabetes induces placental genes for chronic stress and inflammatory pathways. Diabetes. 2003;52:2951–8.
125. Ramsay JE, Simms RJ, Ferrell WR, Crawford L, Greer IA, Lumsden MA, Sattar N. Enhancement of endothelial function by pregnancy: inadequate response in women with type 1 diabetes. Diabetes Care. 2003;26:475–9.
126. Raza H, John A. Glutathione metabolism and oxidative stress in neonatal rat tissues from streptozotocin-induced diabetic mothers. Diabetes Metab Res Rev. 2004;20:72–8.
127. Reece EA, Ma XD, Wu YK, Dhanasekaran D. Aberrant patterns of cellular communication in diabetes-induced embryopathy. I. Membrane signalling. J Matern Fetal Neonatal Med. 2002;11:249–53.
128. Reece EA, Wu YK, Zhao Z, Dhanasekaran D. Dietary vitamin and lipid therapy rescues aberrant signaling and apoptosis and prevents hyperglycemia-induced diabetic embryopathy in rats. Am J Obstet Gynecol. 2006;194:580–5.
129. Ricart W, Lopez J, Mozas J, Pericot A, Sancho MA, Gonzalez N, Balsells M, Luna R, Cortazar A, Navarro P, Ramirez O, Flandez B, Pallardo LF, Hernandez A, Ampudia J, Fernandez-Real JM, Hernandez-Aguado I, Corcoy R. Maternal glucose tolerance status influences the risk of macrosomia in male but not in female fetuses. J Epidemiol Community Health. 2009;63:64–8.
130. Rocha SO, Gomes GN, Forti AL, do Carmo Pinho Franco M, Fortes ZB, de Fatima Cavanal M, Gil FZ. Long-term effects of maternal diabetes on vascular reactivity and renal function in rat male offspring. Pediatr Res. 2005;58:1274–9.
131. Roghair RD, Segar JL, Volk KA, Chapleau MW, Dallas LM, Sorenson AR, Scholz TD, Lamb FS. Vascular nitric oxide and superoxide anion contribute to sex-specific programmed cardiovascular physiology in mice. Am J Physiol Regul Integr Comp Physiol. 2009;296:R651–62.
132. Ryan EA, Imes S, Liu D, McManus R, Finegood DT, Polonsky KS, Sturis J. Defects in insulin secretion and action in women with a history of gestational diabetes. Diabetes. 1995; 44:506–12.
133. Salomon C, Westermeier F, Puebla C, Arroyo P, Guzman-Gutierrez E, Pardo F, Leiva A, Casanello P, Sobrevia L. Gestational diabetes reduces adenosine transport in human placental microvascular endothelium, an effect reversed by insulin. PLoS One. 2012;7:e40578.
134. Salvolini E, Rabini RA, Martarelli D, Moretti N, Cester N, Mazzanti L. A study on human umbilical cord endothelial cells: functional modifications induced by plasma from insulin-dependent diabetes mellitus patients. Metabolism. 1999;48:554–7.
135. San Martin R, Sobrevia L. Gestational diabetes and the adenosine/L-arginine/nitric oxide (ALANO) pathway in human umbilical vein endothelium. Placenta. 2006;27:1–10.
136. Sank A, Wei D, Reid J, Ertl D, Nimni M, Weaver F, Yellin A, Tuan TL. Human endothelial cells are defective in diabetic vascular disease. J Surg Res. 1994;57:647–53.
137. Setji TL, Brown AJ, Feinglos MN. Gestational diabetes mellitus. Clin Diab. 2005; 23:17–24.
138. Sgarbosa F, Barbisan LF, Brasil MA, Costa E, Calderon IM, Goncalves CR, Bevilacqua E, Rudge MV. Changes in apoptosis and Bcl-2 expression in human hyperglycemic, term placental trophoblast. Diabetes Res Clin Pract. 2006;73:143–9.
139. Shah BR, Retnakaran R, Booth GL. Increased risk of cardiovascular disease in young women following gestational diabetes mellitus. Diabetes Care. 2008;31:1668–9.
140. Shand AW, Bell JC, McElduff A, Morris J, Roberts CL. Outcomes of pregnancies in women with pre-gestational diabetes mellitus and gestational diabetes mellitus; a population-based study in New South Wales, Australia, 1998–2002. Diabet Med. 2008;25:708–15.
141. Sharp AN, Heazell AE, Crocker IP, Mor G. Placental apoptosis in health and disease. Am J Reprod Immunol. 2010;64:159–69.

142. Sheffield JS, Butler-Koster EL, Casey BM, McIntire DD, Leveno KJ. Maternal diabetes mellitus and infant malformations. Obstet Gynecol. 2002;100:925–30.
143. Shivananjappa MM, Muralidhara. Abrogation of maternal and fetal oxidative stress in the streptozotocin-induced diabetic rat by dietary supplements of Tinospora cordifolia. Nutrition. 2012;28:581–7.
144. Siman CM, Eriksson UJ. Vitamin E decreases the occurrence of malformations in the offspring of diabetic rats. Diabetes. 1997;46:1054–61.
145. Simmons RA. Role of metabolic programming in the pathogenesis of beta-cell failure in postnatal life. Rev Endocr Metab Disord. 2007;8:95–104.
146. Singh CK, Kumar A, Hitchcock DB, Fan D, Goodwin R, LaVoie HA, Nagarkatti P, DiPette DJ, Singh US. Resveratrol prevents embryonic oxidative stress and apoptosis associated with diabetic embryopathy and improves glucose and lipid profile of diabetic dam. Mol Nutr Food Res. 2011;55:1186–96.
147. Sinzato YK, Volpato GT, Iessi IL, Bueno A, Calderon Ide M, Rudge MV, Damasceno DC. Neonatally induced mild diabetes in rats and its effect on maternal, placental, and fetal parameters. Exp Diabetes Res. 2012;2012:108163.
148. Sivan E, Lee YC, Wu YK, Reece EA. Free radical scavenging enzymes in fetal dysmorphogenesis among offspring of diabetic rats. Teratology. 1997;56:343–9.
149. Skilton MR. Intrauterine risk factors for precocious atherosclerosis. Pediatrics. 2008; 121:570–4.
150. Sobrevia L, Abarzua F, Nien JK, Salomon C, Westermeier F, Puebla C, Cifuentes F, Guzman-Gutierrez E, Leiva A, Casanello P. Review: differential placental macrovascular and microvascular endothelial dysfunction in gestational diabetes. Placenta. 2011;32 Suppl 2:S159–64.
151. Sobrevia L, Gonzalez M. A role for insulin on L-arginine transport in fetal endothelial dysfunction in hyperglycaemia. Curr Vasc Pharmacol. 2009;7:467–74.
152. Sugimura Y, Murase T, Oyama K, Uchida A, Sato N, Hayasaka S, Kano Y, Takagishi Y, Hayashi Y, Oiso Y, Murata Y. Prevention of neural tube defects by loss of function of inducible nitric oxide synthase in fetuses of a mouse model of streptozotocin-induced diabetes. Diabetologia. 2009;52:962–71.
153. Tabit CE, Chung WB, Hamburg NM, Vita JA. Endothelial dysfunction in diabetes mellitus: molecular mechanisms and clinical implications. Rev Endocr Metab Disord. 2010;11: 61–74.
154. Tain YL, Lee WC, Hsu CN, Huang LT, Lee CT, Lin CY. Asymmetric dimethylarginine is associated with developmental programming of adult kidney disease and hypertension in offspring of streptozotocin-treated mothers. PLoS One. 2013;8:e55420.
155. Thakur S, Du J, Hourani S, Ledent C, Li JM. Inactivation of adenosine A2A receptor attenuates basal and angiotensin II-induced ROS production by Nox2 in endothelial cells. J Biol Chem. 2010;285:40104–13.
156. Thaler CD, Epel D. Nitric oxide in oocyte maturation, ovulation, fertilization, cleavage and implantation: a little dab'll do ya. Curr Pharm Des. 2003;9:399–409.
157. Thompson JA, Webb RC. Potential role of Toll-like receptors in programming of vascular dysfunction. Clin Sci (Lond). 2013;125:19–25.
158. Thompson LP, Al-Hasan Y. Impact of oxidative stress in fetal programming. J Pregnancy. 2012;2012:582748.
159. Umekawa T, Sugiyama T, Kihira T, Murabayashi N, Zhang L, Nagao K, Kamimoto Y, Ma N, Yodoi J, Sagawa N. Overexpression of thioredoxin-1 reduces oxidative stress in the placenta of transgenic mice and promotes fetal growth via glucose metabolism. Endocrinology. 2008; 149:3980–8.
160. Vasquez G, Sanhueza F, Vasquez R, Gonzalez M, San Martin R, Casanello P, Sobrevia L. Role of adenosine transport in gestational diabetes-induced L-arginine transport and nitric oxide synthesis in human umbilical vein endothelium. J Physiol. 2004;560:111–22.

161. Vitoratos N, Valsamakis G, Mastorakos G, Boutsiadis A, Salakos N, Kouskouni E, Creatsas G. Pre- and early post-partum adiponectin and interleukin-1beta levels in women with and without gestational diabetes. Hormones (Athens). 2008;7:230–6.
162. Vohr BR, Boney CM. Gestational diabetes: the forerunner for the development of maternal and childhood obesity and metabolic syndrome? J Matern Fetal Neonatal Med. 2008;21:149–57.
163. Volpato GT, Calderon IM, Sinzato S, Campos KE, Rudge MV, Damasceno DC. Effect of Morus nigra aqueous extract treatment on the maternal-fetal outcome, oxidative stress status and lipid profile of streptozotocin-induced diabetic rats. J Ethnopharmacol. 2011;138: 691–6.
164. Wahabi HA, Esmaeil SA, Fayed A, Al-Shaikh G, Alzeidan RA. Pre-existing diabetes mellitus and adverse pregnancy outcomes. BMC Res Notes. 2012;5:496.
165. Wang H, Wang AX, Liu Z, Chai W, Barrett EJ. The trafficking/interaction of eNOS and caveolin-1 induced by insulin modulates endothelial nitric oxide production. Mol Endocrinol. 2009;23:1613–23.
166. Wang Y, Walsh SW. Placental mitochondria as a source of oxidative stress in pre-eclampsia. Placenta. 1998;19:581–6.
167. Weiss U, Cervar M, Puerstner P, Schmut O, Haas J, Mauschitz R, Arikan G, Desoye G. Hyperglycaemia in vitro alters the proliferation and mitochondrial activity of the choriocarcinoma cell lines BeWo, JAR and JEG-3 as models for human first-trimester trophoblast. Diabetologia. 2001;44:209–19.
168. Weksler-Zangen S, Yaffe P, Ornoy A. Reduced SOD activity and increased neural tube defects in embryos of the sensitive but not of the resistant Cohen diabetic rats cultured under diabetic conditions. Birth Defects Res A Clin Mol Teratol. 2003;67:429–37.
169. Wentzel P, Gareskog M, Eriksson UJ. Decreased cardiac glutathione peroxidase levels and enhanced mandibular apoptosis in malformed embryos of diabetic rats. Diabetes. 2008;57: 3344–52.
170. Westermeier F, Puebla C, Vega JL, Farias M, Escudero C, Casanello P, Sobrevia L. Equilibrative nucleoside transporters in fetal endothelial dysfunction in diabetes mellitus and hyperglycaemia. Curr Vasc Pharmacol. 2009;7:435–49.
171. Westermeier F, Salomon C, Gonzalez M, Puebla C, Guzman-Gutierrez E, Cifuentes F, Leiva A, Casanello P, Sobrevia L. Insulin restores gestational diabetes mellitus-reduced adenosine transport involving differential expression of insulin receptor isoforms in human umbilical vein endothelium. Diabetes. 2011;60:1677–87.
172. Westgate JA, Lindsay RS, Beattie J, Pattison NS, Gamble G, Mildenhall LF, Breier BH, Johnstone FD. Hyperinsulinemia in cord blood in mothers with type 2 diabetes and gestational diabetes mellitus in New Zealand. Diabetes Care. 2006;29:1345–50.
173. Williams R. The economic impact of diabetes. In: Allgot B, Gan D, King H et al., eds. Diabetes Atlas. 2nd ed. Brussels: International Diabetes Federation, 2003. p175–192.
174. Winkler G, Cseh K, Baranyi E, Melczer Z, Speer G, Hajos P, Salamon F, Turi Z, Kovacs M, Vargha P, Karadi I. Tumor necrosis factor system in insulin resistance in gestational diabetes. Diabetes Res Clin Pract. 2002;56:93–9.
175. Wu CS, Nohr EA, Bech BH, Vestergaard M, Olsen J. Long-term health outcomes in children born to mothers with diabetes: a population-based cohort study. PLoS One. 2012;7:e36727.
176. Xiong X, Saunders LD, Wang FL, Demianczuk NN. Gestational diabetes mellitus: prevalence, risk factors, maternal and infant outcomes. Int J Gynaecol Obstet. 2001;75:221–8.
177. Yessoufou A, Soulaimann N, Merzouk SA, Moutairou K, Ahissou H, Prost J, Simonin AM, Merzouk H, Hichami A, Khan NA. N-3 fatty acids modulate antioxidant status in diabetic rats and their macrosomic offspring. Int J Obes (Lond). 2006;30:739–50.
178. Zabihi S, Loeken MR. Understanding diabetic teratogenesis: where are we now and where are we going? Birth Defects Res A Clin Mol Teratol. 2010;88:779–90.
179. Zamudio S, Baumann MU, Illsley NP. Effects of chronic hypoxia in vivo on the expression of human placental glucose transporters. Placenta. 2006;27:49–55.

180. Zawiejska A, Wender-Ozegowska E, Pietryga M, Brazert J. Maternal endothelial dysfunction and its association with abnormal fetal growth in diabetic pregnancy. Diabet Med. 2011; 28:692–8.
181. Zeng G, Quon MJ. Insulin-stimulated production of nitric oxide is inhibited by wortmannin. Direct measurement in vascular endothelial cells. J Clin Invest. 1996;98:894–8.
182. Zhang Y, Handy DE, Loscalzo J. Adenosine-dependent induction of glutathione peroxidase 1 in human primary endothelial cells and protection against oxidative stress. Circ Res. 2005;96:831–7.
183. Zhao Z, Eckert RL, Reece EA. Reduction in embryonic malformations and alleviation of endoplasmic reticulum stress by nitric oxide synthase inhibition in diabetic embryopathy. Reprod Sci. 2012;19:823–31.

Chapter 4
Tobacco Smoking and Oxidative Stress in Pregnancy

Ali Aycicek

Abbreviations

CAT	Catalase
GPX	Glutathione peroxidase
LOOH	Lipid hydroperoxide
MDA	Malondialdehyde
OSI	Oxidative stress index
ROS	Reactive oxygen species
SOD	Superoxide dismutase
TAC	Total oxidant capacity
TOS	Total oxidant capacity

Introduction

Cigarette smoking undoubtedly represents one of the greatest current health problems, if not the greatest, facing us. Pregnant women have smoking cessation rates only as high as 15–20 % [1]. Cigarette smoking probably contributes the greatest single share of causality to a variety of lethal and disabling effects on health [2]. It is well known that adverse pregnancy outcomes are related to tobacco smoking during pregnancy [3–6]. Most tobacco toxins have a low molecular weight and high

A. Aycicek, M.D. (✉)
Pediatric Hematology/Oncology Clinic,
Eskisehir State Hospital, 26010 Eskisehir, Turkey
e-mail: ayciceka@hotmail.com

P.A. Dennery et al. (eds.), *Perinatal and Prenatal Disorders*,
Oxidative Stress in Applied Basic Research and Clinical Practice,
DOI 10.1007/978-1-4939-1405-0_4

water solubility and therefore readily cross the placenta [7]. Maternal smoking also impairs placental development and anatomy and functions [8, 9]. Active smoking (AS) or passive smoking (PS) (or secondhand smoking, environmental tobacco smoke) in pregnant women results in intrauterine growth retardation [3] and an increased risk of spontaneous abortion [10]. In addition, prenatal exposure to tobacco smoking may lead to higher risk of sudden infant death syndrome [4], reduction of pulmonary function in healthy neonates [11], and wheezy bronchitis in children [12]. Furthermore, reduced lung function in infancy resulting from prenatal exposure to smoking may lead to abnormal lung function in childhood and track into adulthood [13]. Several mechanisms have been postulated to explain such effects. Cigarette smoke contains an abundance of compounds emitted in gases and condensed tar particles, many of which are oxidants and pro-oxidants capable of producing reactive oxygen species (ROS) [14]. The enhanced production of ROS by smoke is related to increased free radical production, antioxidant depletion, and oxidative stress [15–19], which can result in the oxidation of lipids, induction of DNA single-strand breakage, inactivation of certain proteins, and the disruption of biological membranes [20, 21]. One possibility given credence by several in vitro studies [22–26] is that cigarette smoke, rich in free radicals and oxidizing species, depletes plasma antioxidants [14–19, 27]. Cigarette smoking causes oxidative stress in pregnant women and may have a similar effect in fetuses [15].

Quantifying Tobacco Smoke Exposure

Most studies of the effects of AS or PS during pregnancy and on neonates have used questionnaire reports of exposure, although some have used biochemical measures to validate smoking habits [28, 29]. Apart from the problem of reporting bias, the definition of what a "smoker" is and what characterizes smoking varies between studies. Many early studies have described women simply as smokers or nonsmokers [28]. Because of the variable quality of smoking data obtained from questionnaires, recent studies that quantified smoking have usually assessed maternal serum, urine, or hair samples, obtained during the first half of pregnancy, and umbilical serum, urine, or hair samples, obtained at delivery, by measuring the concentration of the nicotine metabolite cotinine [29, 30]. The lower limit of detection was 0–10 ng/ml for both nicotine and cotinine, with an assay calibration curve of 10–200 ng/ml, and subjects with cotinine levels >2–14 ng/ml were considered smokers [9]. Self-reported abstinence was defined as no smoking during the previous 7–10 days.

Distinguishing between passive smokers and nonsmokers was more difficult, and the assay did not accurately discriminate between these exposure groups in all patient types, especially in pregnant women. There are a number of possible explanations for this [30]. First, there was a real overlap in the samples' cotinine from passive and unexposed subjects. This is largely attributable to the fact that even those reporting no exposure to cigarette smoke can still be exposed to small amounts. Quantifying exposure of nonsmokers has been problematic in most studies. Passive smoking depends on a number of factors, including number of smokers, their proximity to the subject, the number of cigarettes smoked, the size of the space, the ventilation of that space, and the duration of exposure. Moreover, interindividual variability in the conversion of nicotine to cotinine may also make it difficult to discriminate between passive and unexposed nonsmokers. The metabolism of nicotine to cotinine varies with age, ethnicity, and sex [31].

Smoking During Pregnancy

Tobacco use started several centuries ago and increased markedly after the invention of the cigarette-making machine. Behaviors pertaining to tobacco use have changed significantly over the past century. Compared with 1964, smoking prevalence rates have halved from 40 to 20 % in the United States, and as a result there has been a slow but steady decline in the rates of tobacco-induced diseases [32]. While the smoking habit is decreasing in developed countries, tobacco use is increasing in developing countries [33]. Once people start smoking, they find it difficult to quit. This is due to the addictive effect of nicotine in tobacco smoke. Pregnant women have smoking cessation rates only as high as 15–20 % [1]. Various studies showed that maternal nicotine exposure during pregnancy and lactation via tobacco smoke of nicotine replacement therapy program the offspring to develop compromised lung structure later in life with consequent compromised lung function. This implies that nicotine replacement therapy is not an option to assist pregnant or lactating smokers to quit [33]. It is best to quit smoking before getting pregnant.

Thirdhand Smoke

New research shows that thirdhand smoke (THS) is a complex phenomenon resulting from residual tobacco smoke pollutants that cling to the clothing and skin of smokers and to surfaces, couches, and carpets in indoor environments [34]. These pollutants, reemitted into the gas phase or reacting with oxidants or other compounds, persist long after the smoke from a cigarette or cigar has cleared. Thus, THS exposure consists of unintentional intake (mainly through inhalation but also via ingestion and dermal routes) of tobacco smoke and other related chemicals that occurs in the absence of concurrent smoking.

To achieve a better understanding of the health effects attributable to THS, future research should evaluate the risk in pregnant women, fetuses, neonates, and other populations.

Oxidant and Antioxidant Molecules

Pro-oxidant molecules can have free radicals or they can catalyze or initiate reduction/oxidation (redox) reactions that result in the production of free radicals or ROS such as the superoxide anion (O_2^-), hydroxyl radicals (HO.), and hydrogen peroxide (H_2O_2), which initiate oxidative chain reactions resulting in oxidative damage to DNA, proteins, and lipids [14, 20, 24]. These oxidized molecules can be measured in biological fluids, being protein carbonyls, as a marker of protein oxidation and thiobarbituric acid reactive species (TBARS), lipid hydroperoxide (LOOH), malondialdehyde (MDA), and total peroxide as markers of lipid oxidation, the most frequently oxidative stress markers measured in humans [35, 36].

Enzymes that can catalyze free radical-producing redox reactions include certain hydroxylases, oxidases, oxygenases, peroxidases, and synthases. The body, on account of its susceptibility to oxidative insult, is naturally provided with efficient enzymatic and nonenzymatic antioxidant systems. A series of enzymes also act as scavenging systems, including superoxide dismutase (SOD), glutathione peroxidase (GPX), and catalase (CAT). These enzymes are the first line of defense against ROS and are generally referred to as primary antioxidants [15]. The level of CAT, which is a peroxisomal hydrogen peroxide-consuming enzyme, increases in chronic smokers, whereas the levels of several antioxidant enzymes, such as CuZnSOD, glutathione transferase (GST), and GPX, appear to decline in smokers with long smoking histories [20].

Plasma has various nonenzymatic antioxidant molecules. Albumin; uric acid; bilirubin; vitamins C, E, and A; β-carotene; and ceruloplasmin are the major antioxidant components of plasma. Total antioxidant capacity (TAC) represents practically all of them. Albumin has about half of the total antioxidant capacity of plasma [37, 38]. Plasma thiol contents originate from albumin and act as the antioxidant component of plasma.

Analytical methods of oxidant [39–43]/antioxidant [37, 38, 44, 45] parameters were described, from highly complex, time-consuming, and expensive procedures such as chemiluminescence-HPLC assay to more rapid, inexpensive, and sensitive techniques such as automated-colorimetric methods.

Oxidant Status

The values of lipid peroxidation products can be used as an index of oxygen free radical generation. The measurement of LOOH, MDA, and total oxidant status (TOS) provides a sensitive index of lipid peroxidation and oxidative stress [43, 46, 47]. It has been argued that the increased production of ROS associated with smoking may exceed the capacity of the oxidant defense system, resulting in oxidative damage to selected proteins, lipids, and DNA [48–50]. Chelchowska et al. reported that the level of MDA was significantly higher in the cord blood of newborns of

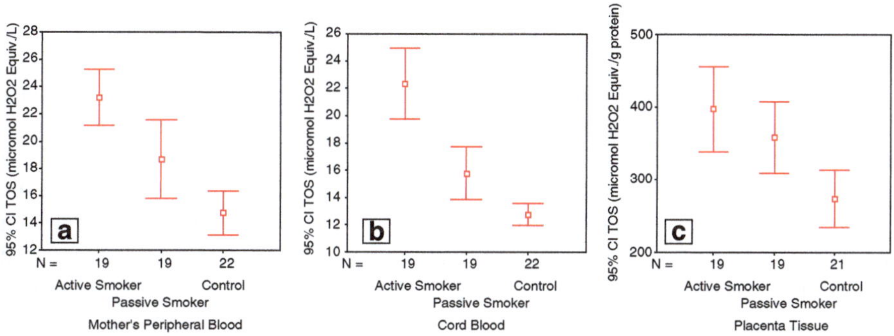

Fig. 4.1 Box plot graphic of mother's peripheral blood (**a**), cord blood (**b**), and placenta tissue (**c**) TOS levels in active smokers, passive smokers, and controls. Differences are significant between the groups ($P < 0.05$)

smoking mothers [36]. Aycicek et al. found that LOOH and TOS levels were significantly higher in active and passive smokers than in the controls. They also reported [15, 27] that the TOS levels of the mother's peripheral blood, cord blood, and placenta tissue were significantly higher in active smokers than in passive smokers and controls. These levels were also significantly higher in the passive smokers than in the controls (Fig. 4.1). Arguelles et al. [51] measured levels of serum lipid peroxides in newborns and their mothers and observed significantly higher peroxidation in the newborns, as well as a positive correlation between the levels measured in the newborns and mothers exposed to tobacco smoke.

Protein carbonyl content levels, a parameter of protein oxidation during pregnancy and in newborns, were increased in pregnant women when compared with nonpregnant subjects [51]. Arguelles et al. reported a significant effect of tobacco smoke exposure on protein carbonyl levels in newborns and mothers [52]. In contrast, Rossner et al. performed a similar study and did not find any relation between tobacco smoke exposure and protein carbonyl levels [51].

Rossner et al. [52] and Daube et al. [53] analyzed oxidative damage to DNA by measuring the levels of 8-hydroxydeoxyguanisine (8-OHdG) in the placenta. They did not find a difference in 8-OHdG levels between the placentas from women exposed and not exposed to tobacco smoke, even though the mean values of plasma cotinine were twofold higher than in subjects from another study [51]. Moreover, in Yin et al. [54], the comparison of 8-OHdG levels in placental DNA from women exposed and not exposed to tobacco smoke based on plasma cotinine levels also did not show a significant difference, although the levels of 8-OHdG in women exposed to tobacco smoke were higher than those in women not exposed. Sbrana et al. [55] demonstrated that the tobacco-mediated metabolic gene pathway perturbations manifest with significant placental accumulation of both 4-hydroxy-2-nonenal (4-HNE, a marker for oxidative lipid damage) and 8-hydroxydeoxyguanisine (8-OHdG, a marker of DNA damage) showed increased levels in the placenta of smokers compared with controls.

Antioxidant Status

The potential damage that can be caused by free radicals is normally minimized by the antioxidant systems. In passive smoking infants, several components of the antioxidant defense system have been reported to be impaired as compared with those of infants not exposed to smoking [40]. Chelchowska et al. [36] reported that the level of TAC was significantly decreased in the cord blood of newborns of smoking mothers. Fayol et al. [18] demonstrated that the TAC level was low in the infant cord blood of passive smoking mothers but that this was not the case in the infants of active smokers. Aycicek and Ipek [15, 27] found that maternal peripheral blood, cord blood, and placental tissue TAC levels were lower in active and passive smokers than in controls (Fig. 4.2), and positive significant correlations were found between placenta tissue TAC and cord blood TAC levels (Fig. 4.3). It is also reported that mothers who smoke, even if they did not smoke during pregnancy, have a higher oxidative stress parameter [51]. Thus, these mothers have a high concentration of serum lipid peroxidation and protein carbonyl content and decreased TAC.

Positive significant correlations were found between maternal cigarette exposure and placenta, cord blood, and maternal peripheral blood TOS and OSI levels, while a negative significant correlation was found between number of cigarettes exposed to and birth weight and head circumference and between placenta and cord blood TAC levels [15, 27].

Uric acid is a well-known low-molecular-weight water-soluble plasma antioxidant [38, 56]. Several clinical studies in humans have demonstrated increased uric acid production as a result of oxidative stress, such as that related to smoking [57]. Fayol et al. reported that the uric acid levels in the cord blood of passive smoking infants were significantly higher than those in the cord blood of controls [18].

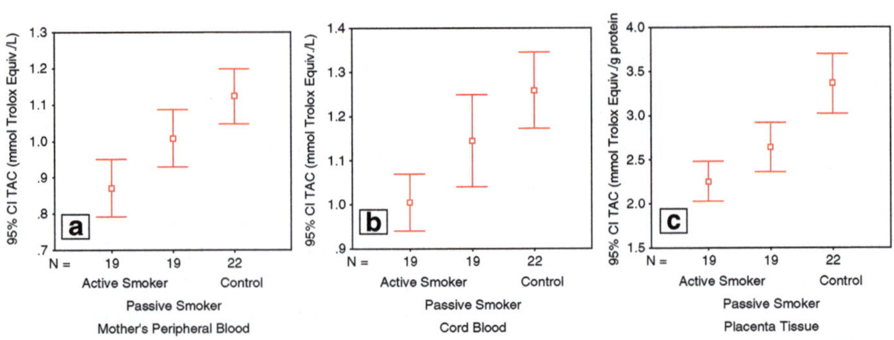

Fig. 4.2 Box plot graphic of maternal peripheral blood (**a**), cord blood (**b**), and placenta tissue (**c**) TAC levels in active smokers, passive smokers, and controls. Differences are more significant between active and passive smokers than controls ($P<0.05$)

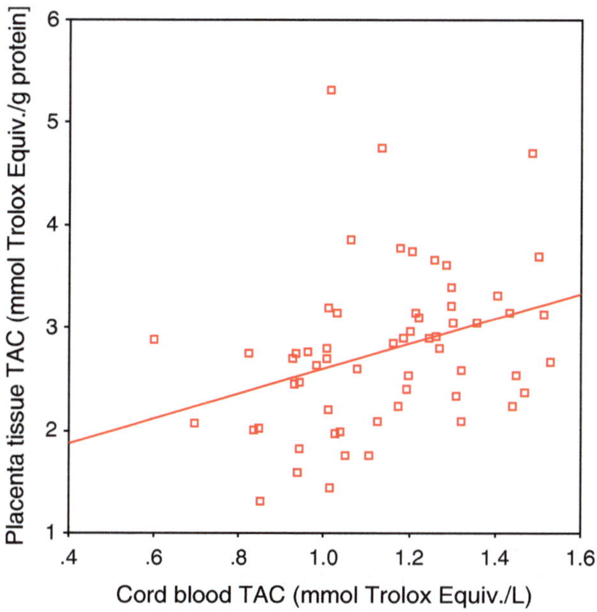

Fig. 4.3 Correlation graphic of TAC levels in placental tissue and cord blood ($r = 0.303$, $P = 0.010$) in smoker mothers

However, Aycicek and Ipek [15] found that plasma uric acid levels were slightly higher in active smokers than in passive smokers and controls, but the difference was not statistically significant.

Oxidative Stress Index

In addition to these markers, OSI, which is a cumulative marker of both oxidative and antioxidative power [15, 17, 40], was significantly increased in the cord blood of active and passive smokers (Fig. 4.4). These results are the first on placental tissue. According to the data obtained in the present study, increased TOS levels, decreased TAC, cumulatively increased OSI, and the other antioxidants together with the routine clinical parameters may implicate the presence of oxidative stress in the placenta subsequently in fetuses. In addition, decreased lipophilic antioxidants may play a role in the pathogenesis of atherosclerosis in the fetuses of mothers who are active or passive smokers through increased susceptibility to lipid peroxidation in utero.

It is reported that highly significant positive significant correlations were found between maternal cigarette exposure and placenta, cord blood, and maternal peripheral blood TOS (Fig. 4.4) and OSI levels, while a negative

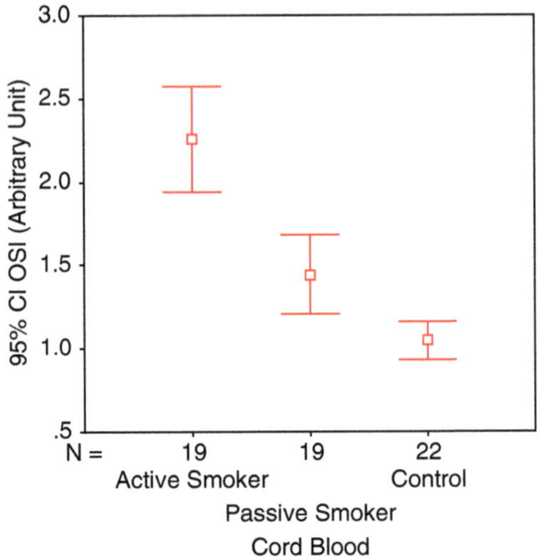

Fig. 4.4 Box plot graphic of placenta tissue OSI levels in active smokers, passive smokers, and controls. Differences are significant between the groups ($P<0.05$)

significant correlation was found between number of cigarettes exposed to and birth weight and head circumference and between placenta and cord blood TAC levels [27].

Antioxidant Enzymes

Cigarette smoking is associated with systemic oxidative stress, leading to an upregulation of antioxidant systems (superoxide dismutase (SOD), catalase (CAT), glutathione peroxidase (GPx), and heme oxygenase (HO)) in some tissues [58]. Smoking throughout pregnancy resulted in elevated expression of the HO enzymes in the placental basal plate region. Smoking did not alter the expression of CAT, SOD, or GPx in any of the placental regions studied. Interestingly, cord blood CAT levels were lower in active and passive smokers than in controls [15].

Ceruloplasmin is a copper-containing glucoprotein with multiple physiological functions, including ferroxidase and oxidase activity. It is induced by inflammatory processes, air pollution, and cigarette smoking [14, 59, 60]. The level of ceruloplasmin was found to be significantly increased in neonates whose mothers were active or passive smokers [61]. However, Aycicek et al. found that ceruloplasmin levels did not differ significantly between each of the three groups (active and passive smokers and controls) and did not find any correlation in their study [15]. Similarly, Fayol et al. reported that cord blood antioxidant parameters strongly correlated with maternal antioxidant status, with the exception of ceruloplasmin [18]. They also stated that an altered neonatal ceruloplasmin concentration may reflect the effect of cigarette exposure on the antioxidant system of the neonate rather than transplacental transfer of maternal ceroloplasmin. Further investigations are needed to address this issue.

Aycicek et al. found that basal/salt-stimulated PON1 activities were significantly lower in patients who were active smokers than in passive smokers and controls, while LOOH and TOS levels were significantly higher [15]. According to the data obtained in the present study, increased LOOH and TOS levels, decreased PON1 activities, and the presence of other antioxidants together with the routine clinical parameters may implicate the presence of oxidative stress in fetuses. In addition, decreased lipophilic antioxidants and PON1 activities may play a role in the pathogenesis of atherosclerosis in infants of active smoking mothers through an increased susceptibility to lipid peroxidation in utero.

Antioxidative Gene Expression

It is reported that smokers showed upregulation of xenobiotic genes (CYP1B1, GSTM1, and CBR3), whose expression was likely directly induced by tobacco smoke exposure. GSTM1 (glutathione S-transferase Mu 1) is implicated in the detoxification of electrophilic compounds, including carcinogens, environmental toxins, and products of oxidative stress [9]. It is stated that cigarette smoke alters the expression of genes involved in oxidative stress, immune and inflammatory responses, xenobiotic metabolism (particularly cytochromes CYP1A1 and CYP1B1), coagulation and fibrinolysis, oncogenesis, DNA repair, structural units of condensed DNA, and extracellular matrix degradation in smokers and their newborns. The majority of genes implicated in the above-mentioned pathways are nonspecifically deregulated and thus represent rather general biomarkers of toxic exposure. In contrast, aldehyde metabolism (e.g., ALDH3) appears to be uniquely modulated by cigarette smoke [62].

Cigarette smoking is associated with altered antioxidant defense enzyme gene expression and regulation in pregnancy. In this context, Vatovova et al. reported increased expression of GPx3 and NXN in placental and fetal cells [9]. GPx3 (glutathione peroxidase 3) is a peroxidase that catalyzes detoxification of hydrogen peroxide and plays a crucial role in protecting proteins and DNA from oxidation caused by smoking [63, 64]. A thioredoxin-related protein, NXM (nucleoredoxin), governs ROS-stimulated Wnt signaling pathway, which is essential for early development and stem cell maintenance [65]. In contrast to expectations, SOD2 (superoxide dismutase 2) expression showed downregulation in peripheral blood cells of smokers, and this finding was confirmed by qRT-PCR [63]. SOD2 encodes a mitochondrial antioxidant enzyme that transforms superoxide into oxygen and hydrogen peroxide, which are less toxic products. A similar decrease in SOD2 levels was found in human neuroblastoma cells exposed to cigarette smoke condensate [66], suggesting that smoking may cause free radical or ROS overcharge, which in turn may be responsible for inhibition of gene expression in antioxidant genes.

Proinflammatory Effect

Passive smoking is increasingly appreciated to have major adverse health effects and to result in a proinflammatory state [67]. It is reported that prenatal and postnatal maternal exposure to environmental tobacco smoke increases neonatal arterial expression of genes that are proinflammatory and induce or contribute to vascular injury while reducing the arterial expression of a gene for angiogenesis [67]. Perinatal and following 1 year of postnatal environmental tobacco smoke exposure, nonhuman primates were found to have increased vascular oxidative stress (protein carbonyls and SOD) and mitochondrial dysfunction/damage (cytochrome oxidase, mitochondrial DNA) that were coupled to reductions in mitochondrial antioxidant capacity and copy number in vascular tissue compared to filtered air-exposed controls [68]. These changes may be responsible for early arterial vascular remodeling that is predisposing to a subsequent vascular disease. Multiple reports have now documented that early and intermediate stage atheroma occurs in children and teenagers [69]. Thus, it is plausible that the potential for increased atherosclerotic disease susceptibility in adulthood following in utero and early life smoke exposure because inflammation and mitochondrial damage and dysfunction in response to tobacco smoke exposure are features of early atherosclerosis.

Summary and Conclusions

Active or passive maternal smoking is associated with important alterations in the oxidant and antioxidant balance in maternal and cord blood and placental tissue and causes potent oxidative stress in all of them. However, future longitudinal studies estimating oxidative markers and antioxidants serially throughout pregnancy may be able to prove the association of elevated oxidant status and diminished antioxidant levels in pregnant women, and the estimation of oxidative stress markers may be predictive of the development health effects of prenatal tobacco smoke.

References

1. French GM, Groner JA, Wewers ME, et al. Staying smoke free: an intervention to prevent postpartum relapse. Nicotine Tob Res. 2007;9:663–70.
2. Breslow L. Cigarette smoking and health. Public Health Rep. 1980;95:451–5.
3. Martinez FD, Wright AL, Taussig LM. The effect of paternal smoking on the birthweight of newborns whose mothers did not smoke. Group Health Medical Associates. Am J Public Health. 1994;84:1489–91.
4. Wisborg K, Kesmodel U, Henriksen TB, et al. Exposure to tobacco smoke in utero and the risk of stillbirth and death in the first year of life. Am J Epidemiol. 2001;154:322–7.
5. Park EY, Hong YC, Lee KH, et al. Maternal exposure to environmental tobacco smoke, GSTM1/T1 polymorphisms and oxidative stress. Reprod Toxicol. 2008;26:197–202.

6. Steyn K, de Wet T, Saloojee Y, et al. The influence of maternal cigarette smoking, snuff use and passive smoking on pregnancy outcomes: the Birth To Ten Study. Paediatr Perinat Epidemiol. 2006;20:90–9.
7. Jauniaux E, Burton GJ. Morphological and biological effects of maternal exposure to tobacco smoke on the feto-placental unit. Early Hum Dev. 2007;83:699–706.
8. Bruchova H, Vasikova A, Merkerova M, et al. Effect of maternal tobacco smoke exposure on the placental transcriptome. Placenta. 2010;31:186–91.
9. Votavova H, Dostalova Merkerova M, Fejglova K, et al. Transcriptome alterations in maternal and fetal cells induced by tobacco smoke. Placenta. 2011;32:763–70.
10. George L, Granath F, Johansson AL, et al. Environmental tobacco smoke and risk of spontaneous abortion. Epidemiology. 2006;17:500–5.
11. Lodrup Carlsen KC, Jaakkola JJ, Nafstad P, et al. In utero exposure to cigarette smoking influences lung function at birth. Eur Respir J. 1997;10:1774–9.
12. Zlotkowska R, Zejda JE. Fetal and postnatal exposure to tobacco smoke and respiratory health in children. Eur J Epidemiol. 2005;20:719–27.
13. Landau LI. Tobacco smoke exposure and tracking of lung function into adult life. Paediatr Respir Rev. 2008;9:39–44.
14. Alberg AJ. The influence of cigarette smoking on circulating concentrations of antioxidant micronutrients. Toxicology. 2002;180:121–37.
15. Aycicek A, Ipek A. Maternal active or passive smoking causes oxidative stress in cord blood. Eur J Pediatr. 2008;167:81–5.
16. Aycicek A, Erel O, Kocyigit A. Decreased total antioxidant capacity and increased oxidative stress in passive smoker infants and their mothers. Pediatr Int. 2005;47:635–9.
17. Aycicek A, Erel O, Kocyigit A. Increased oxidative stress in infants exposed to passive smoking. Eur J Pediatr. 2005;164:775–8.
18. Fayol L, Gulian JM, Dalmasso C, et al. Antioxidant status of neonates exposed in utero to tobacco smoke. Biol Neonate. 2005;87:121–6.
19. Kosecik M, Erel O, Sevinc E, et al. Increased oxidative stress in children exposed to passive smoking. Int J Cardiol. 2005;100:61–4.
20. Durak I, Elgun S, Kemal Bingol N, et al. Effects of cigarette smoking with different tar content on erythrocyte oxidant/antioxidant status. Addict Biol. 2002;7:255–8.
21. Liu X, Lu J, Liu S. Synergistic induction of hydroxyl radical-induced DNA singlestrand breaks by chromium (VI) compound and cigarette smoke solution. Mutat Res. 1999;440:109–17.
22. Frei B, Forte TM, Ames BN, et al. Gas phase oxidants of cigarette smoke induce lipid peroxidation and changes in lipoprotein properties in human blood plasma. Biochem J. 1991;277:133–8.
23. Harats D, Ben-Naim M, Dabach Y, et al. Cigarette smoking renders LDL susceptible to peroxidative modifications and enhanced metabolism by macrophages. Atherosclerosis. 1989;79:245–52.
24. McCall MR, van den Berg JJM, Kuypers FA, et al. Modification of LCAT activity and HDL structure. New links between cigarette smoke and coronary heart disease. Arterioscler Thromb. 1994;14:248–53.
25. Scheffler E, Wiest E, Woehrle J, et al. Smoking influences the atherogenic potential of low-density lipoprotein. Clin Invest. 1992;70:263–8.
26. Yokode M, Kita T, Arai H, et al. Cholesteryl ester accumulation in macrophages incubated with low density lipoprotein pretreated with cigarette smoke extract. Proc Natl Acad Sci U S A. 1988;85:2344–8.
27. Aycicek A, Varma M, Ahmet K, et al. Maternal active or passive smoking causes oxidative stress in placental tissue. Eur J Pediatr. 2011;170:645–51.
28. Bardy AH, Seppälä T, Lillsunde P, et al. Objectively measured tobacco exposure during pregnancy: neonatal effects and relation to maternal smoking. Br J Obstet Gynaecol. 1993;100: 721–6.
29. Peacock JL, Cook DG, Carey IM, et al. Maternal cotinine level during pregnancy and birthweight for gestational age. Int J Epidemiol. 1998;27:647–56.

30. Florescu A, Ferrence R, Einarson TR, et al. Reference values for hair cotinine as a biomarker of active and passive smoking in women of reproductive age, pregnant women, children, and neonates: systematic review and meta-analysis. Ther Drug Monit. 2007;29:437–46.
31. Perez-Stable EJ, Herrera B, Jacob 3rd P, et al. Nicotine metabolism and intake in black and white smokers. JAMA. 1998;280:152–6.
32. Paoletti L, Jardin B, Carpenter MJ, et al. Current status of tobacco policy and control. J Thorac Imaging. 2012;27:213–9.
33. Maritz GS, Mutemwa M. Tobacco smoking: patterns, health consequences for adults, and the long-term health of the offspring. Glob J Health Sci. 2012;4:62–75.
34. Protano C, Vitali M. The new danger of thirdhand smoke: why passive smoking does not stop at secondhand smoke. Environ Health Perspect. 2011;119:A422.
35. Valenzuela A. The biological significance of malondialdehyde determination in the assessment of tissue oxidative stress. Life Sci. 1991;48:301–9.
36. Chelchowska M, Laskowska-Klita T, Leibschang J. The effect of tobacco smoking during pregnancy on concentration of malondialdehyde in blood of mothers and in umbilical cord blood. Ginekol Pol. 2005;76:960–5.
37. Erel O. A novel automated method to measure total antioxidant response against potent free radical reactions. Clin Biochem. 2004;37:112–9.
38. Erel O. A novel automated direct measurement method for total antioxidant capacity using a new generation, more stable ABTS radical cation. Clin Biochem. 2004;37:277–85.
39. Miyazawa T. Determination of phospholipid hydroperoxides in human blood plasma by a chemiluminescence-HPLC assay. Free Radic Biol Med. 1989;7:209–17.
40. Harma M, Harma M, Erel O. Increased oxidative stress in patients with hydatidiform mole. Swiss Med Wkly. 2003;133:563–6.
41. Harma M, Harma M, Erel O. Oxidative stress in women with preeclampsia. Am J Obstet Gynecol. 2005;192:656–7.
42. Kamal A, Gomaa A, Khafif M, et al. Plasma lipid peroxides among workers exposed to silica or asbestos dusts. Environ Res. 1989;49:173–80.
43. Erel O. A new automated colorimetric method for measuring total oxidant status. Clin Biochem. 2005;38:1103–11.
44. Benzie IF, Strain JJ. Ferric reducing/antioxidant power assay: direct measure of total antioxidant activity of biological fluids and modified version for simultaneous measurement of total antioxidant power and ascorbic acid concentration. Methods Enzymol. 1999;299:15–27.
45. Erel O. Automated measurement of serum ferroxidase activity. Clin Chem. 1998;44:2313–9.
46. Arab K, Steghens JP. Plasma lipid hydroperoxides measurement by an automated xylenol orange method. Anal Biochem. 2004;325:158–63.
47. Kim DH, Suh YS, Mun KC. Tissue levels of malondialdehyde after passive smoke exposure of rats for a 24-week period. Nicotine Tob Res. 2004;6:1039–42.
48. Cross CE, O'Neill CA, Reznick AZ, et al. Cigarette smoke oxidation of human plasma constituents. Ann N Y Acad Sci. 1993;686:72–89; discussion 89–90.
49. Schwertner HA. Association of smoking and low serum bilirubin antioxidant. Atherosclerosis. 1998;136:383–7.
50. Yoshie Y, Ohshima H. Synergistic induction of DNA strand breakage by cigarette tar and nitric oxide. Carcinogenesis. 1997;18:1359–63.
51. Argüelles S, Machado MJ, Ayala A, et al. Correlation between circulating biomarkers of oxidative stress of maternal and umbilical cord blood at birth. Free Radic Res. 2006;40:565–70.
52. Rossner Jr P, Milcova A, Libalova H, et al. Biomarkers of exposure to tobacco smoke and environmental pollutants in mothers and their transplacental transfer to the foetus. Part II. Oxidative damage. Mutat Res. 2009;669:20–6.
53. Daube H, Scherer G, Riedel K, et al. DNA adducts in human placenta in relation to tobacco smoke exposure and plasma antioxidant status. J Cancer Res Clin Oncol. 1997;123:141–51.
54. Yin B, Whyatt RM, Perera FP, et al. Determination of 8-hydroxydeoxyguanosine by immunoaffinity chromatography-monoclonal antibody-based ELISA. Free Radic Biol Med. 1995;18:1023–32.

55. Sbrana E, Suter MA, Abramovici AR, et al. Maternal tobacco use is associated with increased markers of oxidative stress in the placenta. Am J Obstet Gynecol. 2011;205:246.e1–7.
56. Polidori MC, Mecocci P, et al. Cigarette smoking cessation increases plasma levels of several antioxidant micronutrients and improves resistance towards oxidative challenge. Br J Nutr. 2003;90:147–50.
57. Pasalic D, Marinkovic N, Feher-Turkovic L. Uric acid as one of the important factors in multifactorial disorders – facts and controversies. Biochem Med (Zagreb). 2012;22:63–75.
58. Sidle EH, Casselman R, Smith GN. Effect of cigarette smoke on placental antioxidant enzyme expression. Am J Physiol Regul Integr Comp Physiol. 2007;293:R754–8.
59. Dalamaga AL, Agroyannis B, Vitoratos N, et al. Effect of smoking on ceruloplasmin and its ferroxidase activity in pregnant women. Gynecol Obstet Invest. 1996;42:13–5.
60. Wisborg K, Kesmodel U, Henriksen TB, et al. Exposure to tobacco smoke in utero and the risk of stillbirth and death in the first year of life. Am J Epidemiol. 2001;154:322–7.
61. Milnerowicz H, Slowińska M. Concentration of metals, ceruloplasmin, metallothionein and the activity of N-acetyl-beta-D-glucosaminidase and gamma-glutamyltransferase in pregnant women who smoke and in those environmentally exposed to tobacco smoke and in their infants. Part I. Int J Occup Med Environ Health. 1997;10:187–202.
62. Sen B, Mahadevan B, Demarini DM. Transcriptional responses to complex mixtures a review. Mutat Res. 2007;636:144–77.
63. Park EM, Park YM, Gwak YS. Oxidative damage in tissues of rats exposed to cigarette smoke. Free Radic Biol Med. 1998;25:79–86.
64. Yildiz L, Kayaoglu N, Aksoy H. The changes of superoxide dismutase, catalase and glutathione peroxidase activities in erythrocytes of active and passive smokers. Clin Chem Lab Med. 2002;40:612.
65. Funato Y, Miki H. Redox regulation of Wnt signalling via nucleoredoxin. Free Radic Res. 2010;44:379–88.
66. Russo M, Cocco S, Secondo A, Adornetto A, Bassi A, Nunziata A, et al. Cigarette smoke condensate causes a decrease of the gene expression of Cu-Zn superoxide dismutase, Mn superoxide dismutase, glutathione peroxidase catalase, and free radical-induced cell Injury in SH-SY5Y human neuroblastoma cells. Neurotox Res. 2009;19:49–54.
67. Villablanca AC, Pinkerton KE, Rutledge JC. Maternal and neonatal exposure to environmental tobacco smoke targets pro-inflammatory genes in neonatal arteries. J Cardiovasc Transl Res. 2010;3:696–703.
68. Westbrook D, Anderson P, Pinkerton K, Ballinger S. Perinatal tobacco smoke exposure increases vascular oxidative stress and mitochondrial damage in non-human primates. Cardiovasc Toxicol. 2010;10:216–26.
69. Berenson GS, Srinivasan SR. Prevention of atherosclerosis in childhood. Lancet. 1999;354: 1223–4.

Chapter 5
Oxidative Stress and Preterm Birth

Ramkumar Menon and Elizabeth Bonney

Preterm Birth (PTB) and Its Risk Factors

Preterm birth (PTB; birth before 37 weeks of gestation) is a major complication of pregnancy [1]. According to a recent report (http://www.who.int/pmnch/media/news/2012/preterm_birth_report/en/index.html), more than 15 million infants are born prematurely each year, accounting for more than 1 in 10 infants born worldwide. Over one million children die each year due to complications of PTB; and many survivors face a lifetime of disability, including learning disabilities and visual and hearing problems. PTB rates are increasing in almost all countries with reliable data, and prematurity due to low birth weight (<2,500 g) is the leading cause of newborn deaths (babies in the first 4 weeks of life) and now the second leading cause of death after pneumonia in children under the age of 5 [1].

PTB is a complex syndrome. A myriad of etiologies (both exogenous and endogenous), like genetic susceptibility, race and ethnic origin, psychosocial and socioeconomic factors, environmental and behavioral factors, and infection, contribute to development of preterm labor [2–4]. Every facet of PTB's understanding is complicated by heterogeneities. It has been suggested that inflammation resulting from these insults comprises a common final pathway leading to PTB [4]. However, a recent biomarker review by Menon and colleagues noted 117 biomarkers have been studied in PTB, but none of them have significant positive or negative predictive

R. Menon, MS, Ph.D. (✉)
Department of Obstetrics and Gynecology, The University of Texas Medical Branch at Galveston, 301 University Blvd, MRB 11.138, Galveston, TX 77555, USA
e-mail: Ram.menon@utmb.edu; ra2menon@utmb.edu

E. Bonney, M.D.
Department of Obstetrics, Gynecology, and Reproductive Sciences, University of Vermont College of Medicine, Burlington, VT, USA

P.A. Dennery et al. (eds.), *Perinatal and Prenatal Disorders*,
Oxidative Stress in Applied Basic Research and Clinical Practice,
DOI 10.1007/978-1-4939-1405-0_5

value as a risk identifier, underlining the pathophysiologic heterogeneities of this complex syndrome [5], the need to take a wider systems-based approach to the problem, and the importance of generation of new hypotheses. Oxidative stress (OS) is thought to underlie several pathological processes [6–15]. The purpose of this review is to further develop the hypothesis that OS may play a significant role in the pathophysiology of this syndrome and to delineate the possible underlying mechanisms [16]. PTB is commonly classified into three categories:

Medically indicated (physician-driven) PTB accounts for approximately 25 % of all PTB (95 % CI = 18.7 − 35.2 %) [17]. There are several reports implicating OS in diseases leading to medically indicated or physician-driven prematurity, and several excellent reviews exist on OS in pregnancy and pregnancy complications like pre-eclampsia, gestational diabetes, and intrauterine growth restriction [18–23]. Moreover, data exists relating OS to poor neonatal outcome in babies born preterm [19–39].

Preterm Premature Rupture of Membranes (pPROM)

Preterm premature rupture of membranes (pPROM) is said to account for 25 % of PTB (95 % CI = 7.1 − 51.2 %); however, 30–40 % of women experience pPROM [17]. This process contributes to physician-driven PTB because of maternal illness or developing fetal compromise [1]. In addition, pPROM is a precursor to spontaneous PTB. Infection is associated with up to 70 % of all pPROM [2, 4] with resulting inflammation that leads to the breakdown of the fetal membranes [40–43]. More than one risk factor is generally associated with pPROM [44]. Few studies have reported a relationship between OS markers and pPROM [41, 45–49].

Spontaneous PTB is occasionally referred to as "idiopathic" and accounts for 50 % of PTB [50]. The term "PTB," as cited in the rest of this chapter, refers to the spontaneous onset of labor followed by preterm delivery. PTB does not arise from a single etiologic factor indicative of etiologic heterogeneity. Inflammation is also thought to contribute significantly to this process. However, the factors that lead to PTB with or without rupture are still unclear, and the factors that specifically lead to PTB in the absence of pPROM are not known. Few studies specifically relate to OS and PTB [51–56] and none of these studies not their reviews have conceptualized a mechanistic role for OS in PTB pathology.

OS

Oxygen-dependent (or cellular) respiration and other energy-building processes generate highly reactive free radicals (molecules with one or more unpaired electrons in their outer orbitals) in mammalian cells. Production of these reactive intermediates is due to incomplete reduction of oxygen and includes superoxide radical (one electron) and hydrogen peroxide (two electrons) [57, 58]. Hydrogen peroxide does not contain unpaired electrons in the valence orbitals, and it is not a free-radical molecule until the O–O bond is lysed by ferrous iron through the Fenton reaction, yielding one of the most powerful oxidants called the "hydroxyl radical (HO•)." These free radicals and oxidants derived from them are called "reactive oxygen species" (ROS). Generation of ROS is an inevitable physiologic process associated with all aerobic processes in mammalian cells. Redox balance is maintained through the production and subsequent elimination of ROS. Cells are able to protect themselves against OS by the finely tuned regulation of redox status through endogenous enzymes, antioxidants, and other cellular mechanisms. At low levels, ROS often generated in biological systems is essential for cell division and survival, cell signaling, inflammation and immune functions, autophagy, and stress response [59–61]. However, an overwhelming redox imbalance compromises a biological system's ability to detoxify these highly reactive molecules or can result in a failure to repair any damage caused by them [62]. Therefore, the former is termed "OS" and the latter "oxidative damage" [63]. Oxidative damage due to ROS generation has been linked to the development of adult diseases including cardiovascular disease, cancer, chronic inflammation, and neurologic disorders.

The two main sources of ROS in human cells are (1) mitochondrial and (2) non-mitochondrial. The mitochondria generate ROS as a by-product of the electron transport chain during respiration, and the rate of ROS production is proportional to the rate of mitochondrial respiration. The ROS production rate is thus obviously higher when metabolic rates are high [64, 65]. A vast majority of ROS in the human body is produced by mitochondria, and mitochondria are the primary sources of superoxide production other than phagocytic cells during innate immune defense.

The Fenton reaction (explained above), where H_2O_2 degradation results in HO•, is an example of non-mitochondrial generation of ROS. One of the first identified non-mitochondrial sources of ROS is phagocytes, where ROS is not the by-product, but rather the product of the primary function of innate immune defense by nicotinamide adenine dinucleotide phosphate (NADPH) oxidases (NOX) [66]. Here, ROS is generated to kill the invading pathogens. However, other cell types and tissues, such as the lung epithelium, also express NOX enzymes, and while these may contribute to innate immunity, they may also contribute to diseases such as asthma [67]. Other pro-oxidant-generating enzymes include nitric oxide synthase [68, 69] and the more recently highlighted dual oxygenases (Duox 1) that are members of NOX family [51, 70, 71].

Oxidative Damage Induced by ROS

OS occurs when the balance between pro-oxidants and antioxidants is disrupted. OS can produce a spectrum of genetic, metabolic, and cellular responses, and pro-oxidants exert their effects on cellular elements—namely, lipids, proteins, and nucleic acids—disrupting their expression, structure, and function [18, 63, 72]. Peroxidation of proteins leads to the loss of the sulfhydryl groups and linking of carbonyl groups with the side chains of other amino acids [73, 74]. Although these proteins are normally proteolytically cleared from the system, under heightened ROS, oxidized proteins do not undergo proteolysis but rather accumulate as long hydrophobic bonds that affect cell function [73]. Cell membrane phospholipids are always a target of ROS activity [75]. The peroxidized cell membrane stiffens, loses its selective permeability, and loses its integrity. Oxidized cell membranes are also susceptible to action by phospholipase enzymes that can activate a series of enzymatic and nonenzymatic breakdowns of oxidized phospholipids. One of the nonenzymatic by-products of this lipid peroxidation is F2-Isoprostane which is considered the biomarker of ROS in a system [76].

The most lethal damage ROS can cause is that exerted on DNA. This damage can result in single- or double-strand breaks, interchanging of sister chromatids, cross-linking of DNA to DNA or protein, and base modifications [77–79]. Although all four bases of DNA are susceptible to ROS-mediated alterations, hydroxylated nucleotide 8-hydroxydeoxyguanosine (8-OHdG) was for the first time referred to as a major product of oxidative DNA damage, and high concentrations of 8-OHdG in biological fluids are considered a biomarker of ROS [80–91]. Damages in DNA are constantly repaired, and simply measuring 8-OHdG levels is not sufficient to measure the extent of ROS; however, higher concentrations of 8-OHdG clearly indicate an underlying pathology that cannot be ignored.

Interaction Between OS and Other Pathways of Tissue Stress and Damage

The idea that ROS are signaling molecules in a variety of cellular processes suggests that an overabundance of ROS in a cell is likely to drive dysregulation at the molecular, cellular, tissue, and organ levels. Parturition comprises uterine contraction, membrane rupture, and dilation of the cervix. The mechanisms underlying these processes include inflammation, apoptosis, senescence, and alterations in collagen and the extracellular matrix. Expanding evidence, briefly mentioned here below, suggests ROS are involved in these mechanisms.

OS and Inflammation

ROS can regulate inflammation on several levels, possibly producing disparate results. For example, ROS can support the function of the NLRP3/NALP3 "inflammasome." This group of interacting proteins cleaves pro-IL-1β and pro-IL-18 to their active forms which then are released and free to bind to their receptors. In contrast, oxidation of caspase 1, another member of the inflammasome, can lead to caspase-1 inactivation, thus suppressing inflammation by this pathway [92]. Toll-like receptors (TLR) are thought to be an important mediator of PTB-induced infection.

While signals through these receptors may themselves generate ROS (Sanlioglu, Williams et al. 2001), these molecules can enhance the TLR signaling pathway to activate more NF-κB and nuclear translocation [93, 94] and subsequent expression of such molecules as cyclooxygenase 2 [95]. Tumor necrosis factor (TNF), a key inflammatory cytokine, binds to TNF membrane receptor and subsequently supports production of ROS which can then mediate cellular necrosis via the MLKL pathway [96].

OS and Apoptosis

OS has been implicated in enhanced apoptosis in degenerative disease, trauma and sepsis, and other disorders. Mitochondrial accumulation of ROS can lead to cytosolic release of proapoptotic proteins and contribute to the formation of the "apoptosome" [97]. Recently, members of the Wnt/b-catenin family have been linked to protection against ROS-induced apoptosis [98]. Both proteins are downregulated in abnormal pregnancy [99]. In addition, ROS can enhance apoptosis via TNF expression [100]; TNF then binds to its receptor and can lead to caspase-mediated apoptosis [96]. Moreover, TNF can participate in a signaling pathway involving RIP1 and or RIP3 leading to regulated necrosis or necroptosis, thus suggesting that ROS may lead to this alternative form of programmed cell death.

OS and Autophagy

Autophagy comprises three processes called "chaperone-mediated autophagy," "microautophagy," and "macroautophagy" and is the regulated process by which abnormal cytoplasmic proteins or organelles are targeted to the lysosome for degradation [101]. Although initially thought to be a response to starvation, growth factor depravation, or energy depletion, autophagy is now thought to play a critical role in immunity and cellular homeostasis. OS enhances expression and function of several key proteins in the generation of the autophagosome [102], and autophagy is thought to help remove proteins damaged by ROS. ROS may also enhance autophagy through suppression of the mTOR pathway [103, 104].

OS and Senescence

It has been hypothesized that placental senescence occurs as a result of OS and is a contributor to stillbirth [105]. A potential underlying mechanism is likely through damage to telomeres, which are rich in residues that are susceptible to oxidative attack [105]. Oxidative damage to telomeres in the context of the normal decrease in telomerase activity that occurs with time during gestation could lead to decreased telomere length and decreased ability for cellular turnover [106].

Within the uterus, decidual senescence due to premature terminal differentiation and restricted growth may also be a cause of PTB [107]. Such senescence has been observed in the context of enhanced mammalian target of rapamycin complex 1 (mTORC1) pathway signaling [107]. ROS regulation of this pathway is complex, as low levels of ROS can enhance, but high levels can inhibit senescence [108]. In addition, activation of this pathway may itself support increased production of ROS [109].

OS and Lysis of Collagen

The extracellular matrix is comprised of a number of polysaccharides and protein fibers. Collagen is an important fiber in the extracellular matrix and undergoes significant alteration in cervical dilation. Matrix metalloproteinases 1, 8, and 13—among others—are important collagenases. There is evidence that ROS may both increase the expression [110] and the activation of MMPs [111].

Antioxidant Mechanisms and the Sources of ROS

Generation of ROS is tightly regulated by an array of enzymatic and nonenzymatic ways to keep redox balance. This balance in humans is maintained by a well-orchestrated antioxidant system that is responsible for homeostatic balance which maintains physiologic functions but prevents oxidative damage. Antioxidants, in general, are substances that decrease the severity of OS by forming less active radicals or by quenching damage created by free-radical chain reactions. Antioxidants can be classified into three major categories and include (1) low molecular antioxidants (glutathione [GSH], vitamins C and E, bilirubin, and urate); (2) enzymes that neutralize free radicals (superoxide dismutase), catalase, glutathione peroxidase, DT diaphorase, and peroxiredoxin; and (3) nonenzymatic proteins (thioredoxin, glutaredoxin, and metallothioneins).

OS and Animal Models of Prematurity

Much of the data from animal models relating OS to poor pregnancy outcomes has occurred in animal models of preeclampsia [37, 112]. Little attention therefore has been paid to the length of gestation in these models. This issue will likely be addressed in future studies. Because of the mechanisms related to both inflammation and infection, it is logical to examine the role of OS in animal models of infection or inflammation-induced prematurity. The only studies specifically addressing this issue in animals utilize the lipopolysaccharide (LPS)-induced preterm labor mouse model [56, 113–115]. 2011). These studies have been used to show an increase in OS in the placenta of mice given LPS during pregnancy [113]. These studies have also shown the protective effect of antioxidant therapy in increasing the time till delivery in mice given LPS [56, 113, 115].

OS in Adverse Pregnancy Outcome and Failure of Antioxidant Clinical Trials

Healthy pregnancy, as with other physiological states, is characterized by a stable redox balance between ROS and antioxidants [20, 21, 116]. Reviews by Myatt et al., Agarwal et al., Dennery et al., and many others have already highlighted ROS-associated damages and their influence on adverse pregnancy outcome [18, 20, 21, 117]. Several clinical trials have been conducted to minimize ROS and improve pregnancy outcome with a marginal success rate [22, 118–123]. Many of these trials were conducted to reduce the risk of preeclampsia, with an underlying ROS pathology with defective placentation. Numerous reasons can be cited for the failure in these trials, but the following primary factors are highlighted:

1. These interventions are based on oxidative damage and not necessarily the type of ROS that leads to oxidative damage. ROS is not likely a causal factor but a pathophysiology secondary to other signals; therefore, it is unlikely that reversing OS would reduce the risk of adverse pregnancy outcome.
2. Clinical trials are conducted based on the generalized theory that OS is an underlying pathology. This is unlikely and may very well be limited to a subset of subjects with specific exposures that can cause ROS. These exposures can be, but are not limited to, cigarette smoking, obesity, and antioxidant nutrient deficiency.
3. There are no biomarkers used for screening subjects for ROS prior to intervention to document the type of ROS and to assess the amount of ROS-associated damage. Even the type of biomarker measures may not be reflective of the type of exposure or extent of ROS as these markers may represent an end-stage indicative of damage and not necessarily risk.
4. Like inflammation, ROS is not a homogeneous phenomenon. Different tissues and cells (and even within the cell itself), cytoplasmic and nuclear membranes, and the contents of different organelles (inner and outer membranes of mitochondria) may react differently to different stressors producing distinct patterns of ROS. The antioxidants used in trials may curtail a particular ROS pathway or production of a specific free radical, but depending on the type of risk in a given individual, the ROS pathway and antioxidant requirement may be different.
5. As mentioned above, ROS-associated damage can be any of the major functional elements mentioned earlier, and the degree of damage is likely dependent on the site, dose of risk factors, and type of trigger that initiate an ROS-mediated response.

The above arguments highlight the need for more basic and functional studies to better understand the role of ROS in the initiation of the pathologic pathways ending in adverse birth outcomes and also the need to design more appropriate clinical trials based on the following: (1) risk exposure, (2) type of ROS response, and (3) type of antioxidant requirement.

OS in PTB and pPROM

There is a paucity of studies specifically addressing the mechanistic role of ROS in PTB and pPROM [29, 45, 46, 51–54, 56, 81, 124]. Numerous studies have examined either ROS markers or antioxidant gene expression; however, these studies have not sufficiently generated targets for intervention or projected biomarkers of these conditions. The Menon laboratory has recently been interested in understanding the factors that generate ROS, and we are trying to understand the mechanistic role of ROS in PTB and pPROM. The next section of this chapter introduces several new concepts based on our laboratory's recent published and unpublished data relevant to ongoing studies. Readers are warned that we are neither projecting PTB or pPROM biomarkers nor intervention targets. Our aim is to inspire readers to engage in future research leading to drawing pathways influenced by ROS along with inflammation.

Measurement of Lipid Peroxidation Marker F2-Isoprostanes in PTB and pPROM Amniotic Fluid Samples

Isoprostanes (IsoP) are eicosanoids formed by the oxidation of arachidonic acid, a ubiquitous polyunsaturated fatty acid [76, 125–128]. Unlike prostaglandins (PG), which are formed via the action of the cyclooxygenase enzymes, the most well-studied form, F2-IsoP, contains F-type prostane rings that are isomeric to PGF2α and are formed nonenzymatically as a result of the free-radical-mediated peroxidation of arachidonic acid. The biological functions of IsoP include vasoconstriction, induction of osteoclast and macrophage activation differentiation leading to bone resorption, inhibition of platelet aggregation, induction of hypertrophy in cardiac endothelial cells, induced endothelin release, and proliferation of vascular smooth muscle cells [76, 127]. In a case-control study, amniotic fluids were collected at the time of labor or during cesarean delivery [52]. Samples were subjected to gas chromatography, negative ion chemical ionization, and mass spectrometry for F2-IsoP. Primary analysis examined differences between prostanoid concentrations in PTB ($n=133$) compared with term births ($n=189$). Secondary stratified analyses compared eicosanoid concentrations in three epidemiological risk factors (race, cigarette smoking, and microbial invasion of amniotic cavity [MIAC]). Amniotic fluid F2-IsoP was significantly higher at term than

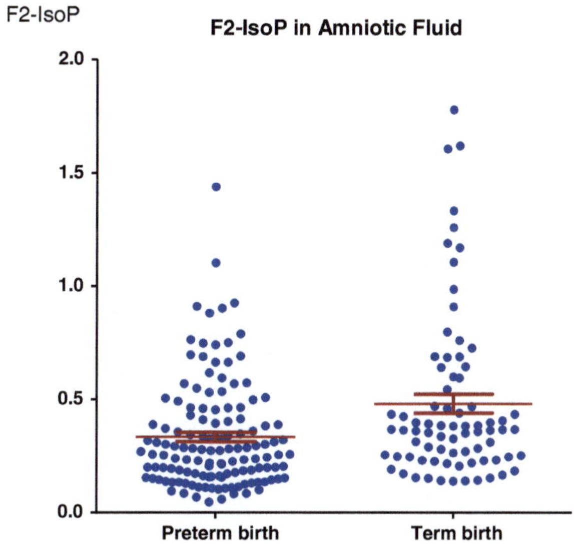

F2-IsoP concentrations were measured in amniotic fluid samples from spontaneous preterm birth (PTB) and normal term birth using GC/MS. F2-IsoP concentration was higher at term than PTB.

Fig. 5.1

in PTB [52] (Fig. 5.1). Regardless of gestational age status, F2-IsoP was threefold higher in smokers than nonsmokers. PTB with MIAC had significantly higher F2-isoprostane compared with PTB without MIAC. The concentrations of F2-IsoP in smokers from PTB were much higher than those found in cases with MIAC. These results prompt us to postulate that OS, as manifested by F2-IsoP, is a risk-specific response to smoking and infection but not necessarily a general effector of the labor. PGF2α, a proinflammatory uterotonic, was severalfold higher in PTB than normal term births suggesting that PG may play the role of the effector for PTB pathway, whereas underlying OS could be a major initiator of the labor process.

In the next phase of this study, we examined amniotic fluid samples from pPROM and compared these with data from PTB and term births. When F2-IsoP data were compared with PTB and term birth, pPROM had the highest concentration of F2-IsoP compared to gestational age-matched PTB or normal term birth. Similar to PTB and term births, smokers and MIAC also had the highest F2-IsoP concentrations. The concentrations were also higher in subjects who reported smoking during pregnancy than in cases with MIAC. These data are

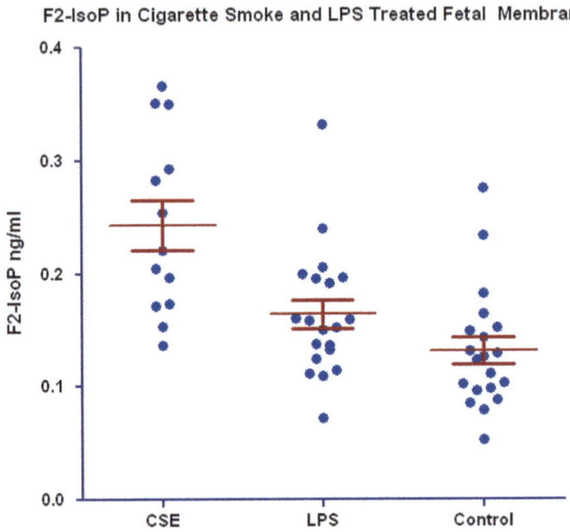

Fig. 5.2

indicative of measurable OS response due to lipid peroxidation in the intra-amniotic environment and OS's differential role in two pregnancy complications that otherwise have overlapping etiologies, have pathophysiologic pathways, and share several common inflammatory biomarkers. Although considerable overlap and interaction exist between the inflammatory and OS pathways, our results suggest that these pathways may uniquely be operative in different pregnancy complications and their activation is likely dependent on the type of risk exposure.

Cigarette Smoke Extract (CSE) Increases Lipid Peroxidation and F2-IsoP Production from Human Fetal Membranes In Vitro

The effect of cigarette smoke and infection, two risk factors for PTB and pPROM, on fetal membrane ROS was further verified using our fetal membrane organ explant model. CSE were prepared by bubbling smoke drawn from a single-lit commercial cigarette as explained in our prior reports. Each cigarette is reported to contain 26 mg of tar and 1.7 mg of nicotine. CSE was prepared fresh for each experiment. Normal term, not-in-labor, fetal membrane explants were exposed to culture media containing CSE and LPS (100 ng/ml) and stimulated for 24 h. CSE-exposed fetal membranes produced significantly higher F2-IsoP than unexposed controls [46] (Fig. 5.2). The concentration of F2-IsoP in LPS-treated membranes was considerably less than that in CSE-treated tissues. These data confirmed our in vivo analysis of amniotic fluid where we observed similar changes in response to different risk factors.

DNA Peroxidation and Telomere Attrition Associated with PTB and pPROM

Telomeres are DNA-protein complexes that cap the ends of chromosomes and preserve chromosomal stability throughout the cell cycle [14, 15]. When chromosomes undergo replication during cell division, the telomere is not fully replicated secondary to limitations of the DNA polymerase activity at the 59 end of the lagging strand, resulting in progressive telomere shortening with each cell cycle. Thus, telomere lengths serve as a valid marker of a cell's "biologic age" [129–132]. Emerging studies suggest that in adults, biochemical mediators of physiological and psychological stress result in OS, accelerate telomere shortening, and advance cellular aging [133–135]. Cellular senescence, as evidenced by shortened telomeres, is a predictor of mortality and a number of chronic diseases such as obesity, diabetes, and cardiovascular and inflammatory disorders [136–140]. Our study analyzed fetal leukocyte telomere length during gestation, comparing telomere length among products of pPROM, PTB, and normal term births to ascertain the risk of the three pregnancy outcomes in subjects with reduced telomere length. Telomere lengths were quantified in cord blood leukocytes ($n=133$) from three major groups using real-time quantitative PCR: (1) pPROM ($n=28$), (2) PTB ($n=69$), and (3) uncomplicated full-term births (controls, $n=35$). Telomere length was calculated using the empirically derived quadratic formula: telomere length (bp) = 37,631x22 85075x + 53,005. Reference standards of genomic DNA with known telomere length (from pooled "old," "middle-aged," and "young" subjects) were used to normalize the assay. Placental membrane specimens ($n=18$) were used to correlate fetal leukocyte and placental telomere lengths.

Initially, we calculated fetal leukocyte telomere length in pregnancies delivered with intact membranes. As anticipated, we observed a general trend that telomere length was inversely proportional to gestational age [45]. The longest telomeres (25,258 bp) were noted in a PTB sample at 174 days gestation and the shortest telomeres (5,022 bp) in a term pregnancy delivered at 282 days. These data should be interpreted with caution as fetal telomeres from PTB may be confounded by pathological complications leading to early delivery. Primary analyses compared telomere lengths among pPROM, PTB, and term birth. Overall, telomere length was not different between pPROM (996,263,124 bp) and term birth (901,162,497 bp; $p=0.31$). PTB had significantly longer telomeres (1,154,664,348 bp) than either pPROM ($p=0.05$) or term birth ($p<0.01$) groups. When the data were further stratified based on gestational age, early pPROM and PTB (<32 weeks) had similar evidence of infection, inflammation, and low birth weight. Fetal telomeres from PTB <32 weeks had longer telomeres (1,167,964,926 bp) than those of gestation age-matched pPROM (943,262,618 bp; $p=0.01$), whereas no difference was seen in telomere length between fetuses from pPROM (1,240,064,374 bp) and PTB <2 weeks (1,141,663,771 bp; $p=0.63$). Interestingly, pPROM <32 weeks had similar fetal telomere length (943,262,618 bp) to that detected in term births (901,162,497 bp; $p=0.67$). However, marginally longer telomeres were observed in pPROM >32 weeks (1,240,064,374 bp) compared with term births (901,162,497 bp; $p=0.07$). When data were compared based on infection status, we did not observe any statistically significant difference between the infection groups in pPROM and PTB, and this very well may be attributed to sample size. Regardless, pPROM had much lower telomere length than gestational-age-matched PTB which were similar to those found at term.

Analyses were performed to correlate matched fetal leukocyte and placental membrane telomere lengths. Telomere lengths in placental membranes were systematically longer than corresponding fetal leukocyte telomeres; however, the same trends in pPROM and PTB were maintained. Placental membrane telomeres were shorter in pPROM (922,763,389 bp) compared with PTB (1,444,463,320 bp; $p=0.05$), whereas no differences were seen between pPROM and term birth (1,184,264,456 bp; $p=0.32$). A strong Pearson's correlation was observed between placental membrane and fetal DNA telomere lengths ($r=0.77$; $p<0.01$) despite a small sample size.

The responses of telomeres to OS support our hypothesis and prior observations that specific subsets of PTB may be caused by OS [45, 46]. Due to their high content of guanines, which undergo oxidative modification to 8-oxo-dG bases, telomeres are highly sensitive to damage by OS. Senescent cells have four times the 8-oxo-dG content of healthy cells [141–143]. Furthermore, ROS, especially HO•, can introduce single-strand DNA breaks, which are less amenable to repair in telomeric than genomic DNA. In vitro studies also indicate that infection, inflammation, and inflammatory cytokines per se (e.g., TNF-α) downregulate hTERT enzyme activity (an enzyme that maintains telomere length) accelerating cell senescence [144–146]. We postulate that OS-mediated telomere shortening is a normal physiologic process leading to term delivery but is activated prematurely in pPROM prior to PTB.

Telomere Shortening in Amnion Cells Exposed to Cigarette Smoke

We recently performed a pilot study to test our system for telomere measurements using in vitro-cultured amnion cells from normal term cesarean sections at 38 weeks. In control cells, telomere length was 1.433 ± 1.369 (mean ± STD), and in CSE-treated cells, 0.6177 ± 0.2797 after 24 h. These lengths were further decreased over time as expected, and after 96 h in culture and simulation, telomere lengths were 1.092 ± 0.6564 in controls and 0.2273 ± 0.09817 in CSE-treated cells. Although we saw a severalfold decrease in telomere length, these data did not achieve statistical significance due to sample size. Additionally, we noticed DNA damage in these cells and generation of 8-oxoG lesions.

Lipid and DNA Peroxidation Indicate OS-Associated Pathology in Adverse Pregnancy Outcome

The above detailed data are strong indicators of OS response in PTB and pPROM and likely a driving force in normal term birth as well. OS response is much more prevalent in pPROM than PTB with intact membranes, especially in early pPROM cases (<34 weeks).

OS and Senescence

Emerging data from ongoing studies in our laboratory are indicative of OS-associated changes inducible in fetal tissues by PTB and pPROM risk factors, and we postulate that one of the changes that can result from OS damage is senescence of the fetal membranes or other intrauterine tissues. Senescence of the placenta, a condition associated with OS, is expected in normal term births, and this is likely a natural physiologic response that eventually leads to the birthing process. However, premature senescence at the maternal fetal interface due to ROS (indicated by lipid peroxidation, DNA damage, and telomere reduction) is a pathologic process that can cause fetal membrane extracellular matrix degradation, chorioamnionitis, and eventual rupture of the membranes. Apoptosis of the membranes has been well connected to PTB and pPROM, but senescence is not well documented. Senescence is an irreversible growth arrest of cells. Unlike apoptosis, these cells can persist, alter their function, and change the tissue environment. This change is often associated with senescence-associated secretory phenotype (SASP) characterized by inflammatory markers like cytokines and matrix metalloproteinases, an inflammatory feature underlying pPROM and a larger subset of PTB [147] that reduces fetal membrane tensile strength, thus weakening the membranes. Activation of this inflammatory process can also cause myometrial activation and cervical ripening. It is likely that both apoptosis and senescence are still factors that can associate with PTB and pPROM, and the delineation of pathways may very well depend on the dose of the stimulus and the extent of activation of downstream elements that can either cause apoptosis or senescence. When exposed to OS-inducer cigarette smoke, amnion cells showed an increased number of cells with senescence-associated β-gal staining (Fig. 5.3) compared with control, confirming risk-associated induction of senescence changes in fetal cells.

We therefore postulate that PTB and pPROM risk factors can cause OS damages at the maternal fetal interface, and this can result in pathologic changes as explained above. OS damages can be the initiator of preterm labor process or it can be secondary to inflammation. Depending on the type of risk factor, its dose, and the tissue type (location), either inflammation or OS can be the initiator of the phenotypic changes associated with PTB and pPROM. Further studies are

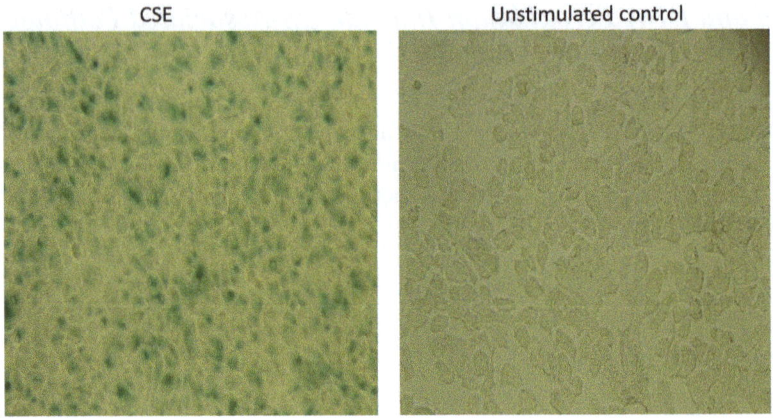

Fig. 5.3

required to understand the kinetics of OS and inflammation to identify the initiator to better understand the underlying pathology. Diagnosis of high-risk pregnancies cannot be generalized based on inflammation alone as most of current literature suggests. OS markers along with inflammatory indicators may provide much more valuable information for better diagnosis and identifying targets for intervention.

Summary

PTB is a clinically important and heterogeneous syndrome leading to neonatal compromise. The major categories of PTB are driven by many different clinical and biological causes. However, this chapter proposes that OS and the complex biology of ROS may specifically contribute significantly to the pathophysiology involved in each category. Herein we review how ROS interact with several major biological pathways including inflammation, apoptosis, autophagy, senescence, and collagen metabolism. Further, we propose that this interactive role underlies several of the processes involved in PTB by impacting the final common pathways involved in placental dysfunction, uterine contractions, and the premature rupture of membranes. This chapter further suggests that the apparent conflict in the association between exposure to OS and PTB seen in epidemiologic studies on the one hand and the apparent lack of effect in antioxidant therapy in clinical trials to improve clinical outcome on the other stems from the need for a more complete understanding of the underlying biology of ROS. Such an understanding may be derived from careful studies in animal models and humans.

References

1. Beck S, Wojdyla D, Say L, et al. The worldwide incidence of preterm birth: a systematic review of maternal mortality and morbidity. Bull World Health Organ. 2010;88:31–8.
2. Goldenberg RL, Culhane JF, Iams JD, Romero R. Epidemiology and causes of preterm birth. Lancet. 2008;371:75–84.
3. Menon R. Spontaneous preterm birth, a clinical dilemma: etiologic, pathophysiologic and genetic heterogeneities and racial disparity. Acta Obstet Gynecol Scand. 2008;87:590–600.
4. Romero R, Espinoza J, Kusanovic JP, et al. The preterm parturition syndrome. BJOG. 2006; 113 Suppl 3:17–42.
5. Menon R, Torloni MR, Voltolini C, et al. Biomarkers of spontaneous preterm birth: an overview of the literature in the last four decades. Reprod Sci. 2011;18:1046–70.
6. Savini I, Catani MV, Evangelista D, Gasperi V, Avigliano L. Obesity-associated oxidative stress: strategies finalized to improve redox state. Int J Mol Sci. 2013;14:10497–538.
7. Aseervatham GS, Sivasudha T, Jeyadevi R, Arul AD. Environmental factors and unhealthy lifestyle influence oxidative stress in humans-an overview. Environ Sci Pollut Res Int. 2013;20:4356–69.
8. Pohanka M. Role of oxidative stress in infectious diseases. A review. Folia Microbiol (Praha). 2013;58:503–13.
9. Zamora PL, Villamena FA. Pharmacological approaches to the treatment of oxidative stress-induced cardiovascular dysfunctions. Future Med Chem. 2013;5:465–78.
10. Vitale G, Salvioli S, Franceschi C. Oxidative stress and the ageing endocrine system. Nat Rev Endocrinol. 2013;9:228–40.
11. Popolo A, Autore G, Pinto A, Marzocco S. Oxidative stress in patients with cardiovascular disease and chronic renal failure. Free Radic Res. 2013;47:346–56.
12. Smith JA, Park S, Krause JS, Banik NL. Oxidative stress, DNA damage, and the telomeric complex as therapeutic targets in acute neurodegeneration. Neurochem Int. 2013;62:764–75.
13. Shah AA, Sinha AA. Oxidative stress and autoimmune skin disease. Eur J Dermatol. 2013;23:5–13.
14. Stadler K. Oxidative stress in diabetes. Adv Exp Med Biol. 2012;771:272–87.
15. Pimentel C, Batista-Nascimento L, Rodrigues-Pousada C, Menezes RA. Oxidative stress in Alzheimer's and Parkinson's diseases: insights from the yeast Saccharomyces cerevisiae. Oxid Med Cell Longev. 2012;2012:132146.
16. Tuuli MG, Longtine MS, Nelson DM. Review: Oxygen and trophoblast biology – a source of controversy. Placenta. 2011;32 Suppl 2:S109–18.
17. Moutquin JM. Classification and heterogeneity of preterm birth. BJOG. 2003;110 Suppl 20:30–3.
18. Agarwal A, Aponte-Mellado A, Premkumar BJ, Shaman A, Gupta S. The effects of oxidative stress on female reproduction: a review. Reprod Biol Endocrinol. 2012;10:49.
19. Al-Gubory KH, Fowler PA, Garrel C. The roles of cellular reactive oxygen species, oxidative stress and antioxidants in pregnancy outcomes. Int J Biochem Cell Biol. 2010;42:1634–50.
20. Dennery PA. Oxidative stress in development: nature or nurture? Free Radic Biol Med. 2010;49:1147–51.
21. Myatt L. Review: reactive oxygen and nitrogen species and functional adaptation of the placenta. Placenta. 2010;31 Suppl:S66–9.
22. Roberts JM, Myatt L, Spong CY, et al. Vitamins C and E to prevent complications of pregnancy-associated hypertension. N Engl J Med. 2010;362:1282–91.
23. Walsh SW. Maternal-placental interactions of oxidative stress and antioxidants in preeclampsia. Semin Reprod Endocrinol. 1998;16:93–104.
24. Clerici G, Slavescu C, Fiengo S, et al. Oxidative stress in pathological pregnancies. J Obstet Gynaecol. 2012;32:124–7.

25. Hsieh TT, Chen SF, Lo LM, Li MJ, Yeh YL, Hung TH. The association between maternal oxidative stress at mid-gestation and subsequent pregnancy complications. Reprod Sci. 2012;19:505–12.
26. Janssen BG, Munters E, Pieters N, et al. Placental mitochondrial DNA content and particulate air pollution during in utero life. Environ Health Perspect. 2012;120:1346–52.
27. Micle O, Muresan M, Antal L, Bodog F, Bodog A. The influence of homocysteine and oxidative stress on pregnancy outcome. J Med Life. 2012;5:68–73.
28. Orhan H, Onderoglu L, Yucel A, Sahin G. Circulating biomarkers of oxidative stress in complicated pregnancies. Arch Gynecol Obstet. 2003;267:189–95.
29. Perrone S, Tataranno ML, Negro S, et al. May oxidative stress biomarkers in cord blood predict the occurrence of necrotizing enterocolitis in preterm infants? J Matern Fetal Neonatal Med. 2012;25 Suppl 1:128–31.
30. Mert I, Oruc AS, Yuksel S, et al. Role of oxidative stress in preeclampsia and intrauterine growth restriction. J Obstet Gynaecol Res. 2012;38:658–64.
31. Gauster M, Desoye G, Totsch M, Hiden U. The placenta and gestational diabetes mellitus. Curr Diab Rep. 2012;12:16–23.
32. Fujimaki A, Watanabe K, Mori T, Kimura C, Shinohara K, Wakatsuki A. Placental oxidative DNA damage and its repair in preeclamptic women with fetal growth restriction. Placenta. 2011;32:367–72.
33. Sharp AN, Heazell AE, Crocker IP, Mor G. Placental apoptosis in health and disease. Am J Reprod Immunol. 2010;64:159–69.
34. Bridges JP, Gilbert JS, Colson D, et al. Oxidative stress contributes to soluble fms-like tyrosine kinase-1 induced vascular dysfunction in pregnant rats. Am J Hypertens. 2009; 22:564–8.
35. Anastasakis E, Papantoniou N, Daskalakis G, Mesogitis S, Antsaklis A. Screening for preeclampsia by oxidative stress markers and uteroplacental blood flow. J Obstet Gynaecol. 2008;28:285–9.
36. Costantine MM, Clark EA, Lai Y, et al. Association of polymorphisms in neuroprotection and oxidative stress genes and neurodevelopmental outcomes after preterm birth. Obstet Gynecol. 2012;120:542–50.
37. Banerjee S, Randeva H, Chambers AE. Mouse models for preeclampsia: disruption of redox-regulated signaling. Reprod Biol Endocrinol. 2009;7:4.
38. Rumbold A, Duley L, Crowther CA, Haslam RR. Antioxidants for preventing pre-eclampsia. Cochrane Database Syst Rev. 2008;CD004227.
39. Rumbold A, Crowther CA. Vitamin E supplementation in pregnancy. Cochrane Database Syst Rev. 2005;CD004069.
40. Mercer BM, Lewis R. Preterm labor and preterm premature rupture of the membranes. Diagnosis and management. Infect Dis Clin North Am. 1997;11:177–201.
41. Tambor V, Kacerovsky M, Lenco J, Bhat G, Menon R. Proteomics and bioinformatics analysis reveal underlying pathways of infection associated histologic chorioamnionitis in pPROM. Placenta. 2013;34:155–61.
42. Menon R, Fortunato SJ. Infection and the role of inflammation in preterm premature rupture of the membranes. Best Pract Res Clin Obstet Gynaecol. 2007;21:467–78.
43. Menon R, Fortunato SJ. The role of matrix degrading enzymes and apoptosis in rupture of membranes. J Soc Gynecol Investig. 2004;11:427–37.
44. Mercer BM, Goldenberg RL, Meis PJ, et al. The Preterm Prediction Study: prediction of preterm premature rupture of membranes through clinical findings and ancillary testing. The National Institute of Child Health and Human Development Maternal-Fetal Medicine Units Network. Am J Obstet Gynecol. 2000;183:738–45.
45. Menon R, Yu J, Basanta-Henry P, et al. Short fetal leukocyte telomere length and preterm prelabor rupture of the membranes. PLoS One. 2012;7:e31136.
46. Menon R, Fortunato SJ, Yu J, et al. Cigarette smoke induces oxidative stress and apoptosis in normal term fetal membranes. Placenta. 2011;32:317–22.

47. Kwiatkowski S, Torbe A, Dolegowska B, et al. Isoprostanes 8-iPF2alpha-III: risk markers of premature rupture of fetal membranes? Biomarkers. 2009;14:406–13.
48. Knudtson EJ, Smith K, Mercer BM, et al. Serum homocysteine levels after preterm premature rupture of the membranes. Am J Obstet Gynecol. 2004;191:537–41.
49. Woods Jr JR. Reactive oxygen species and preterm premature rupture of membranes-a review. Placenta. 2001;22(Suppl A):S38–44.
50. Ananth CV, Vintzileos AM. Epidemiology of preterm birth and its clinical subtypes. J Matern Fetal Neonatal Med. 2006;19:773–82.
51. Chai M, Barker G, Menon R, Lappas M. Increased oxidative stress in human fetal membranes overlying the cervix from term non-labouring and post labour deliveries. Placenta. 2012;33:604–10.
52. Menon R, Fortunato SJ, Milne GL, et al. Amniotic fluid eicosanoids in preterm and term births: effects of risk factors for spontaneous preterm labor. Obstet Gynecol. 2011;118: 121–34.
53. Lappas M, Permezel M. The anti-inflammatory and antioxidative effects of nicotinamide, a vitamin B(3) derivative, are elicited by FoxO3 in human gestational tissues: implications for preterm birth. J Nutr Biochem. 2011;22:1195–201.
54. Sakata M, Sado T, Kitanaka T, et al. Iron-dependent oxidative stress as a pathogenesis for preterm birth. Obstet Gynecol Surv. 2008;63:651–60.
55. Suh YJ, Kim YJ, Park H, Park EA, Ha EH. Oxidative stress-related gene interactions with preterm delivery in Korean women. Am J Obstet Gynecol. 2008;198:541–7.
56. Buhimschi IA, Buhimschi CS, Weiner CP. Protective effect of N-acetylcysteine against fetal death and preterm labor induced by maternal inflammation. Am J Obstet Gynecol. 2003; 188:203–8.
57. Gerschman R, Gilbert D, Nye SW, Dwyer P, Fenn WO. Oxygen poisoning and X-irradiation: a mechanism in common. 1954. Nutrition. 2001;17:162.
58. Gerschman R, Nye SW, Gilbert DL, Dwyer P, Fenn WO. Studies on oxygen poisoning: protective effect of beta-mercaptoethylamine. Proc Soc Exp Biol Med. 1954;85:75–7.
59. Finkel T. Signal transduction by reactive oxygen species. J Cell Biol. 2011;194:7–15.
60. Cadenas E, Sies H. Oxidative stress: excited oxygen species and enzyme activity. Adv Enzyme Regul. 1985;23:217–37.
61. Sies H. Oxidative stress: from basic research to clinical application. Am J Med. 1991; 91:31S–8.
62. Durackova Z. Some current insights into oxidative stress. Physiol Res. 2010;59:459–69.
63. Halliwell B, Whiteman M. Measuring reactive species and oxidative damage in vivo and in cell culture: how should you do it and what do the results mean? Br J Pharmacol. 2004; 142:231–55.
64. Vercesi AE, Kowaltowski AJ, Oliveira HC, Castilho RF. Mitochondrial Ca2+ transport, permeability transition and oxidative stress in cell death: implications in cardiotoxicity, neurodegeneration and dyslipidemias. Front Biosci. 2006;11:2554–64.
65. Bandy B, Davison AJ. Mitochondrial mutations may increase oxidative stress: implications for carcinogenesis and aging? Free Radic Biol Med. 1990;8:523–39.
66. Bedard K, Krause KH. The NOX family of ROS-generating NADPH oxidases: physiology and pathophysiology. Physiol Rev. 2007;87:245–313.
67. Poynter ME. Airway epithelial regulation of allergic sensitization in asthma. Pulm Pharmacol Ther. 2012;25:438–46.
68. Bevilacqua E, Gomes SZ, Lorenzon AR, Hoshida MS, Amarante-Paffaro AM. NADPH oxidase as an important source of reactive oxygen species at the mouse maternal-fetal interface: putative biological roles. Reprod Biomed Online. 2012;25:31–43.
69. Fujii S, Yoshimura T. Detection and imaging of endogenously produced nitric oxide with electron paramagnetic resonance spectroscopy. Antioxid Redox Signal. 2000;2:879–901.
70. Vande LK, Ciardelli R, Decordier I, Plas G, Haumont D, Kirsch-Volders M. Preterm newborns show slower repair of oxidative damage and paternal smoking associated DNA damage. Mutagenesis. 2012;27:573–80.

71. Giusti B, Vestrini A, Poggi C, et al. Genetic polymorphisms of antioxidant enzymes as risk factors for oxidative stress-associated complications in preterm infants. Free Radic Res. 2012;46:1130–9.
72. Sies H. Strategies of antioxidant defense. Eur J Biochem. 1993;215:213–9.
73. McKersie BD, Bowley SR, Harjanto E, Leprince O. Water-deficit tolerance and field performance of transgenic alfalfa overexpressing superoxide dismutase. Plant Physiol. 1996;111: 1177–81.
74. Sigler K, Chaloupka J, Brozmanova J, Stadler N, Hofer M. Oxidative stress in microorganisms – I. Microbial vs. higher cells – damage and defenses in relation to cell aging and death. Folia Microbiol (Praha). 1999;44:587–624.
75. Halliwell B, Gutteridge JM. Oxygen toxicity, oxygen radicals, transition metals and disease. Biochem J. 1984;219:1–14.
76. Morrow JD, Roberts LJ. The isoprostanes: unique bioactive products of lipid peroxidation. Prog Lipid Res. 1997;36:1–21.
77. Richter C, Park JW, Ames BN. Normal oxidative damage to mitochondrial and nuclear DNA is extensive. Proc Natl Acad Sci U S A. 1988;85:6465–7.
78. Adelman R, Saul RL, Ames BN. Oxidative damage to DNA: relation to species metabolic rate and life span. Proc Natl Acad Sci U S A. 1988;85:2706–8.
79. Fraga CG, Shigenaga MK, Park JW, Degan P, Ames BN. Oxidative damage to DNA during aging: 8-hydroxy-2′-deoxyguanosine in rat organ DNA and urine. Proc Natl Acad Sci U S A. 1990;87:4533–7.
80. Subash P, Gurumurthy P, Sarasabharathi A, Cherian KM. Urinary 8-OHdG: a marker of oxidative stress to DNA and total antioxidant status in essential hypertension with South Indian population. Indian J Clin Biochem. 2010;25:127–32.
81. Peter ST, Scholl TO, Schluter MD, et al. Oxidative stress early in pregnancy and pregnancy outcome. Free Radic Res. 2008;42:841–8.
82. Chen HI, Liou SH, Ho SF, et al. Oxidative DNA damage estimated by plasma 8-hydroxydeoxyguanosine (8-OHdG): influence of 4, 4′-methylenebis (2-chloroaniline) exposure and smoking. J Occup Health. 2007;49:389–98.
83. Wu LL, Chiou CC, Chang PY, Wu JT. Urinary 8-OHdG: a marker of oxidative stress to DNA and a risk factor for cancer, atherosclerosis and diabetics. Clin Chim Acta. 2004;339:1–9.
84. Matsubasa T, Uchino T, Karashima S, et al. Oxidative stress in very low birth weight infants as measured by urinary 8-OHdG. Free Radic Res. 2002;36:189–93.
85. Tsukahara H, Haruta T, Ono N, et al. Oxidative stress in childhood meningitis: measurement of 8-hydroxy-2′-deoxyguanosine concentration in cerebrospinal fluid. Redox Rep. 2000; 5:295–8.
86. Tarng DC, Huang TP, Wei YH, et al. 8-hydroxy-2′-deoxyguanosine of leukocyte DNA as a marker of oxidative stress in chronic hemodialysis patients. Am J Kidney Dis. 2000; 36:934–44.
87. Marczynski B, Kraus T, Rozynek P, Raithel HJ, Baur X. Association between 8-hydroxy-2′-deoxyguanosine levels in DNA of workers highly exposed to asbestos and their clinical data, occupational and non-occupational confounding factors, and cancer. Mutat Res. 2000; 468:203–12.
88. Chen L, Bowen PE, Berzy D, Aryee F, Stacewicz-Sapuntzakis M, Riley RE. Diet modification affects DNA oxidative damage in healthy humans. Free Radic Biol Med. 1999;26: 695–703.
89. Leinonen J, Lehtimaki T, Toyokuni S, et al. New biomarker evidence of oxidative DNA damage in patients with non-insulin-dependent diabetes mellitus. FEBS Lett. 1997;417:150–2.
90. Daube H, Scherer G, Riedel K, et al. DNA adducts in human placenta in relation to tobacco smoke exposure and plasma antioxidant status. J Cancer Res Clin Oncol. 1997;123:141–51.
91. Kaneko T, Tahara S, Matsuo M. Non-linear accumulation of 8-hydroxy-2′-deoxyguanosine, a marker of oxidized DNA damage, during aging. Mutat Res. 1996;316:277–85.

92. Martinon F. Signaling by ROS drives inflammasome activation. Eur J Immunol. 2010; 40:616–9.
93. Yanagisawa S, Koarai A, Sugiura H, et al. Oxidative stress augments toll-like receptor 8 mediated neutrophilic responses in healthy subjects. Respir Res. 2009;10:50.
94. Koarai A, Sugiura H, Yanagisawa S, et al. Oxidative stress enhances toll-like receptor 3 response to double-stranded RNA in airway epithelial cells. Am J Respir Cell Mol Biol. 2010;42:651–60.
95. Temma K, Shimoya K, Zhang Q, et al. Effects of 4-hydroxy-2-nonenal, a marker of oxidative stress, on the cyclooxygenase-2 of human placenta in chorioamnionitis. Mol Hum Reprod. 2004;10:167–71.
96. Chu WM. Tumor necrosis factor. Cancer Lett. 2013;328:222–5.
97. Bayir H, Kagan VE. Bench-to-bedside review: Mitochondrial injury, oxidative stress and apoptosis – there is nothing more practical than a good theory. Crit Care. 2008;12:206.
98. Tao GZ, Lehwald N, Jang KY, et al. Wnt/beta-catenin signaling protects mouse liver against oxidative stress-induced apoptosis through the inhibition of forkhead transcription factor FoxO3. J Biol Chem. 2013;288:17214–24.
99. Zhang Y, Li Q, Zhou D, Chen H. Genistein, a soya isoflavone, prevents azoxymethane-induced up-regulation of WNT/beta-catenin signalling and reduces colon pre-neoplasia in rats. Br J Nutr. 2013;109:33–42.
100. Verhasselt V, Goldman M, Willems F. Oxidative stress up-regulates IL-8 and TNF-alpha synthesis by human dendritic cells. Eur J Immunol. 1998;28:3886–90.
101. Levine B, Mizushima N, Virgin HW. Autophagy in immunity and inflammation. Nature. 2011;469:323–35.
102. Deng X, Zhang F, Rui W, et al. PM-induced oxidative stress triggers autophagy in human lung epithelial A549 cells. Toxicol In Vitro. 2013;27:1762–70.
103. Rodriguez-Vargas JM, Ruiz-Magana MJ, Ruiz-Ruiz C, et al. ROS-induced DNA damage and PARP-1 are required for optimal induction of starvation-induced autophagy. Cell Res. 2012;22:1181–98.
104. Munoz-Gamez JA, Rodriguez-Vargas JM, Quiles-Perez R, et al. PARP-1 is involved in autophagy induced by DNA damage. Autophagy. 2009;5:61–74.
105. Smith R, Maiti K, Aitken RJ. Unexplained antepartum stillbirth: a consequence of placental aging? Placenta. 2013;34:310–3.
106. Chen RJ, Chu CT, Huang SC, Chow SN, Hsieh CY. Telomerase activity in gestational trophoblastic disease and placental tissue from early and late human pregnancies. Hum Reprod. 2002;17:463–8.
107. Hirota Y, Cha J, Yoshie M, Daikoku T, Dey SK. Heightened uterine mammalian target of rapamycin complex 1 (mTORC1) signaling provokes preterm birth in mice. Proc Natl Acad Sci U S A. 2011;108:18073–8.
108. Radisavljevic ZM, Gonzalez-Flecha B. TOR kinase and Ran are downstream from PI3K/Akt in H2O2-induced mitosis. J Cell Biochem. 2004;91:1293–300.
109. Kim JH, Chu SC, Gramlich JL, et al. Activation of the PI3K/mTOR pathway by BCR-ABL contributes to increased production of reactive oxygen species. Blood. 2005;105:1717–23.
110. Dasgupta J, Kar S, Liu R, et al. Reactive oxygen species control senescence-associated matrix metalloproteinase-1 through c-Jun-N-terminal kinase. J Cell Physiol. 2010;225: 52–62.
111. Okamoto T, Akaike T, Sawa T, Miyamoto Y, van der Vliet A, Maeda H. Activation of matrix metalloproteinases by peroxynitrite-induced protein S-glutathiolation via disulfide S-oxide formation. J Biol Chem. 2001;276:29596–602.
112. Sedeek M, Gilbert JS, LaMarca BB, et al. Role of reactive oxygen species in hypertension produced by reduced uterine perfusion in pregnant rats. Am J Hypertens. 2008;21:1152–6.
113. Xu DX, Wang H, Zhao L, Ning H, Chen YH, Zhang C. Effects of low-dose lipopolysaccharide (LPS) pretreatment on LPS-induced intra-uterine fetal death and preterm labor. Toxicology. 2007;234:167–75.

114. Jiang XF, Zeng WY, Yang KX, Xu ZH, Xu AQ. Effect of N-acetylcysteine on lipopolysaccharide induced preterm labor in mice. Sichuan Da Xue Xue Bao Yi Xue Ban. 2007; 38:279–83.
115. Chang EY, Zhang J, Sullivan S, Newman R, Singh I. N-acetylcysteine attenuates the maternal and fetal proinflammatory response to intrauterine LPS injection in an animal model for preterm birth and brain injury. J Matern Fetal Neonatal Med. 2011;24:732–40.
116. Myatt L, Cui X. Oxidative stress in the placenta. Histochem Cell Biol. 2004;122:369–82.
117. Dennery PA. Effects of oxidative stress on embryonic development. Birth Defects Res C Embryo Today. 2007;81:155–62.
118. Thorne-Lyman AL, Fawzi WW. Vitamin A and carotenoids during pregnancy and maternal, neonatal and infant health outcomes: a systematic review and meta-analysis. Paediatr Perinat Epidemiol. 2012;26 Suppl 1:36–54.
119. Rossi AC, Mullin PM. Prevention of pre-eclampsia with low-dose aspirin or vitamins C and E in women at high or low risk: a systematic review with meta-analysis. Eur J Obstet Gynecol Reprod Biol. 2011;158:9–16.
120. Dror DK, Allen LH. Interventions with vitamins B6, B12 and C in pregnancy. Paediatr Perinat Epidemiol. 2012;26 Suppl 1:55–74.
121. Conde-Agudelo A, Romero R, Kusanovic JP, Hassan SS. Supplementation with vitamins C and E during pregnancy for the prevention of preeclampsia and other adverse maternal and perinatal outcomes: a systematic review and metaanalysis. Am J Obstet Gynecol. 2011; 204:503–12.
122. Rumbold A, Middleton P, Crowther CA. Vitamin supplementation for preventing miscarriage. Cochrane Database Syst Rev. 2005;CD004073.
123. Basaran A, Basaran M, Topatan B. Combined vitamin C and E supplementation for the prevention of preeclampsia: a systematic review and meta-analysis. Obstet Gynecol Surv. 2010;65:653–67.
124. Dhobale M, Joshi S. Altered maternal micronutrients (folic acid, vitamin B(12)) and omega 3 fatty acids through oxidative stress may reduce neurotrophic factors in preterm pregnancy. J Matern Fetal Neonatal Med. 2012;25:317–23.
125. Il'yasova D, Morrow JD, Ivanova A, Wagenknecht LE. Epidemiological marker for oxidant status: comparison of the ELISA and the gas chromatography/mass spectrometry assay for urine 2,3-dinor-5,6-dihydro-15-F2t-isoprostane. Ann Epidemiol. 2004;14:793–7.
126. Roberts LJ, Milne GL. Isoprostanes. J Lipid Res. 2009;50 Suppl:S219–23.
127. Milne GL, Yin H, Morrow JD. Human biochemistry of the isoprostane pathway. J Biol Chem. 2008;283:15533–7.
128. Morrow JD, Roberts LJ. Mass spectrometric quantification of F2-isoprostanes as indicators of oxidant stress. Methods Mol Biol. 2002;186:57–66.
129. Sanders JL, Newman AB. Telomere length in epidemiology: a biomarker of aging, age-related disease, both, or neither? Epidemiol Rev. 2013;35:112–31.
130. Horn T, Robertson BC, Gemmell NJ. The use of telomere length in ecology and evolutionary biology. Heredity (Edinb). 2010;105:497–506.
131. Bekaert S, De MT, Van OP. Telomere attrition as ageing biomarker. Anticancer Res. 2005;25:3011–21.
132. von Zglinicki T, Martin-Ruiz CM. Telomeres as biomarkers for ageing and age-related diseases. Curr Mol Med. 2005;5:197–203.
133. Weischer M, Bojesen SE, Cawthon RM, Freiberg JJ, Tybjaerg-Hansen A, Nordestgaard BG. Short telomere length, myocardial infarction, ischemic heart disease, and early death. Arterioscler Thromb Vasc Biol. 2012;32:822–9.
134. Hanna CW, Bretherick KL, Gair JL, Fluker MR, Stephenson MD, Robinson WP. Telomere length and reproductive aging. Hum Reprod. 2009;24:1206–11.
135. Vaziri H, West MD, Allsopp RC, et al. ATM-dependent telomere loss in aging human diploid fibroblasts and DNA damage lead to the post-translational activation of p53 protein involving poly(ADP-ribose) polymerase. EMBO J. 1997;16:6018–33.

136. Fyhrquist F, Saijonmaa O, Strandberg T. The roles of senescence and telomere shortening in cardiovascular disease. Nat Rev Cardiol. 2013;10:274–83.
137. Khan S, Chuturgoon AA, Naidoo DP. Telomeres and atherosclerosis. Cardiovasc J Afr. 2012;23:563–71.
138. Babizhayev MA, Yegorov YE. Smoking and health: association between telomere length and factors impacting on human disease, quality of life and life span in a large population-based cohort under the effect of smoking duration. Fundam Clin Pharmacol. 2011;25:425–42.
139. Demerath EW, Cameron N, Gillman MW, Towne B, Siervogel RM. Telomeres and telomerase in the fetal origins of cardiovascular disease: a review. Hum Biol. 2004;76:127–46.
140. Balasubramanyam M, Adaikalakoteswari A, Monickaraj SF, Mohan V. Telomere shortening & metabolic/vascular diseases. Indian J Med Res. 2007;125:441–50.
141. Haghdoost S, Sjolander L, Czene S, Harms-Ringdahl M. The nucleotide pool is a significant target for oxidative stress. Free Radic Biol Med. 2006;41:620–6.
142. Kawanishi S, Oikawa S. Mechanism of telomere shortening by oxidative stress. Ann N Y Acad Sci. 2004;1019:278–84.
143. Herrero A, Barja G. 8-oxo-deoxyguanosine levels in heart and brain mitochondrial and nuclear DNA of two mammals and three birds in relation to their different rates of aging. Aging (Milano). 1999;11:294–300.
144. Al-Zoubi M, Salem AF, Martinez-Outschoorn UE, et al. Creating a tumor-resistant microenvironment: cell-mediated delivery of TNF alpha completely prevents breast cancer tumor formation in vivo. Cell Cycle. 2013;12:480–90.
145. Shibata N, Watanabe T, Okitsu T, et al. Establishment of an immortalized human hepatic stellate cell line to develop antifibrotic therapies. Cell Transplant. 2003;12:499–507.
146. Matsushita H, Chang E, Glassford AJ, Cooke JP, Chiu CP, Tsao PS. eNOS activity is reduced in senescent human endothelial cells: preservation by hTERT immortalization. Circ Res. 2001;89:793–8.
147. Campisi J, di d'Adda FF. Cellular senescence: when bad things happen to good cells. Nat Rev Mol Cell Biol. 2007;8:729–40.

Chapter 6
Developmental Origins of Disease: The Role of Oxidative Stress

Rebecca A. Simmons

Introduction

It is becoming increasingly apparent that the in utero environment in which a fetus grows and develops may have long-term effects on subsequent health and survival [1, 2]. The landmark cohort study of 300,000 men by Ravelli and colleagues showed that exposure to the Dutch famine of 1944–1945 during the first half of pregnancy resulted in significantly higher obesity rates at age 19 [3]. Subsequent studies demonstrated a relationship between low birth weight and the later development of cardiovascular disease [4] and impaired glucose tolerance [5–7] in men in England. Those men who were smallest at birth (2.5 kg) were nearly seven times more likely to have impaired glucose tolerance or type 2 diabetes than were those who were heaviest at birth. In addition, the investigators found a similar relationship between lower birth weight and higher systolic blood pressure and triglyceride levels [8]. Subsequent studies in diverse populations through the world have demonstrated a significant correlation between low birth weight and the later development of type 2 diabetes [9–21].

Role of Oxidative Stress

Uteroplacental insufficiency, caused by such disorders as preeclampsia, maternal smoking, and abnormalities of uteroplacental development, is one of the most common causes of fetal growth retardation. In the face of uteroplacental insufficiency,

R.A. Simmons, M.D. (✉)
Department of Pediatrics Children's Hospital Philadelphia, University of Pennsylvania,
BRB II/III, Room 1308, 421 Curie Boulevard, Philadelphia, PA 19104, USA
e-mail: rsimmons@mail.med.upenn.edu

P.A. Dennery et al. (eds.), *Perinatal and Prenatal Disorders*,
Oxidative Stress in Applied Basic Research and Clinical Practice,
DOI 10.1007/978-1-4939-1405-0_6

the fetus adapts to an inadequate supply of substrates (such as glucose, amino acids, fatty acids, and oxygen) by metabolic changes, redistribution of blood flow, and changes in the production of fetal and placental hormones which control growth. The fetus' immediate metabolic response to placental insufficiency is catabolism: it consumes its own substrates to provide energy, and there are shifts from accretion to oxidative metabolism and breakdown of protein/glycogen for oxidative metabolism, or both. A more prolonged reduction in availability of substrates leads to a slowing in growth. This enhances the fetus' ability to survive by reducing the use of substrates and lowering the metabolic rate. Slowing of growth in late gestation leads to disproportion in organ size, since organs and tissues that are growing rapidly at the time are affected the most.

Multiple studies have now shown that intrauterine growth retardation is associated with increased oxidative stress in the human fetus [22–30]. A major consequence of limited nutrient availability is an alteration in the redox state in susceptible fetal tissues leading to oxidative stress. In particular, low levels of oxygen, evident in growth-retarded fetuses, will decrease the activity of complexes of the electron transport chain, which will generate increased levels of reactive oxygen species (ROS) [31–33]. Overproduction of ROS initiates many oxidative reactions that lead to oxidative damage not only in the mitochondria but also in cellular proteins, lipids, and nucleic acids. Increased ROS levels inactivate the iron-sulfur centers of the electron transport chain complexes, and tricarboxylic acid cycle aconitase, resulting in shutdown of mitochondrial energy production.

What Animal Models Can Tell Us

Animal models have a normal genetic background upon which environmental effects during gestation or early postnatal life can be tested for their role in inducing diabetes. For a comprehensive survey of the numerous animal models of fetal growth retardation, the reader is referred to two excellent reviews [34, 35]. The most commonly used animal models are caloric or protein restriction, glucocorticoid administration, or induction of uteroplacental insufficiency in the pregnant rodent. In the rat, maternal dietary protein restriction (approximately 40–50 % of normal intake) throughout gestation and lactation has been reported to impair glucose tolerance and induce hypertension in the adult offspring [36–41]. Offspring are significantly growth retarded, remain growth retarded throughout life, and in some cases develop mild β-cell secretory abnormalities [36–40] and in others insulin resistance [38, 41]. Aged rats develop hyperglycemia characterized by defects in insulin signaling in muscle, adipocytes, and liver [41–45].

Fetal overexposure to glucocorticoids either via maternal administration or by inhibition of placental 11beta-hydroxysteroid dehydrogenase type 2 (11βHSD2) in the rat results in fetal growth restriction and induces hypertension and glucose intolerance, and increased basal hypothalamic-pituitary-adrenal (HPA) activity after birth [46–49].

To extend these experimental studies of growth retardation, we developed a model of uteroplacental insufficiency (IUGR) induced by bilateral uterine artery ligation at day 18 of gestation (term is 22 days) in the rat that restricts fetal growth [50, 51]. Growth-retarded fetal rats have critical features of a metabolic profile characteristic of growth-retarded human fetuses: decreased levels of glucose, insulin, insulin-like growth factor 1 (IGF-I), amino acids, and oxygen [52–54]. By 6 months of age, IUGR rats develop diabetes with a phenotype remarkably similar to that observed in the humans with type 2 diabetes: progressive dysfunction in insulin secretion and insulin action. Thus, the studies in various animal models support the hypothesis that an abnormal intrauterine milieu can induce permanent changes in glucose homeostasis after birth and lead to type 2 diabetes in adulthood.

Cellular Mechanisms: Mitochondrial Dysfunction and Oxidative Stress

The intrauterine environment influences development of the fetus by modifying gene expression in both pluripotential cells and terminally differentiated cells. The long-range effects on the offspring (into adulthood) depend upon the cells undergoing differentiation, proliferation, and/or functional maturation at the time of the disturbance in maternal fuel economy.

A key adaptation enabling the fetus to survive in a limited energy environment may be the reprogramming of mitochondrial function [55–57]. However, these alterations in mitochondrial function can have deleterious effects, especially in cells that have a high energy requirement, such as the β-cell. The β-cell depends upon the normal production of ATP for nutrient-induced insulin secretion [58–65] and proliferation [66]. Thus, an interruption of mitochondrial function can have profound consequences for the β-cell.

Mitochondrial dysfunction can also lead to increased production of reactive oxygen species (ROS), which will lead to oxidative stress if the defense mechanisms of the cell are overwhelmed. β-cells are especially vulnerable to attacks by ROS because expression of antioxidant enzymes in pancreatic islets is very low [67, 68] and β-cells have a high oxidative energy requirement. Increased ROS impair glucose stimulated insulin secretion [66, 69, 70], decrease gene expression of key β-cell genes [71–77], and induce cell death [78–80].

We have found that uteroplacental insufficiency induces oxidative stress and marked mitochondrial dysfunction in the fetal β-cell [81]. ATP production is impaired and continues to deteriorate with age. The activities of complexes I and III of the electron transport chain progressively decline in IUGR islets. Mitochondrial DNA point mutations accumulate with age and are associated with decreased mtDNA content and reduced expression of mitochondrial-encoded genes in IUGR islets. Mitochondrial dysfunction results in impaired insulin secretion. These results demonstrate that IUGR induces mitochondrial dysfunction in the fetal β-cell, leading to increased production of ROS, which in turn damage mtDNA [81].

A self-reinforcing cycle of progressive deterioration in mitochondrial function leads to a corresponding decline in β-cell function. Finally, a threshold in mitochondrial dysfunction and ROS production is reached and diabetes ensues.

Mitochondrial dysfunction is not limited to the β-cell in the IUGR animal. IUGR animals exhibit marked insulin resistance early in life (prior to the onset of hyperglycemia), characterized by blunted whole-body glucose disposal in response to insulin and impaired insulin suppression of hepatic glucose output [82]. Basal hepatic glucose production is also increased [82]. Oxidation rates of pyruvate, glutamate, succinate, and α-ketoglutarate are significantly blunted in isolated hepatic mitochondria from IUGR pups (prior to the onset of diabetes) [82]. Rotenone-sensitive NADH-O_2 oxidoreductase activity is similar in control and IUGR mitochondria, showing that the defect responsible for decreased pyruvate, glutamate, and $\tilde{\alpha}$ ketoglutarate oxidation in the IUGR liver precedes the electron transport chain and involves pyruvate and α-ketoglutarate dehydrogenases. Increased levels of manganese superoxide dismutase (MnSOD) suggest that an antioxidant response has been mounted, and 4-hydroxynonenal (HNE) modification of pyruvate dehydrogenase E2 catalytic and E3-binding protein subunits suggests that HNE-induced inactivation of this key enzyme may play a role in the mechanism of injury. These results indicate that uteroplacental insufficiency impairs mitochondrial oxidative phosphorylation in the liver, and this derangement predisposes the IUGR rat to increased hepatic glucose production by suppressing pyruvate oxidation and increasing gluconeogenesis [82]. As increased hepatic glucose production and hepatic insulin resistance are associated with type 2 diabetes, the observed abnormalities in the IUGR liver likely play key roles in the pathophysiology of type 2 diabetes in this animal model.

Mitochondria in muscle of IUGR young adult rats, prior to the onset of hyperglycemia, exhibit significantly decreased rates of state 3 oxygen consumption with pyruvate, glutamate, α-ketoglutarate, and succinate [83]. Decreased pyruvate oxidation in IUGR mitochondria is associated with decreased ATP production, decreased pyruvate dehydrogenase activity, and increased expression of pyruvate dehydrogenase kinase 4 (PDK4). Such a defect in IUGR mitochondria leads to a chronic reduction in the supply of ATP available from oxidative phosphorylation. Impaired ATP synthesis in muscle compromises energy-dependent GLUT4 recruitment to the cell surface, glucose transport, and glycogen synthesis, which contributes to insulin resistance and hyperglycemia of type 2 diabetes [84].

Other animal models of fetal growth retardation also show mitochondrial abnormalities. Mitochondrial DNA content is reduced in the liver, pancreas, and skeletal muscle of male offspring of dams fed a low-protein diet during pregnancy and lactation [85, 86]. This was associated with reduced expression of mitochondrial DNA-encoded genes [86]. More recently, a targeted metabolomics study in a rabbit IUGR model (unilateral uterine artery ligation) revealed a significant increase in metabolites associated with oxidative stress in IUGR rabbit brain [87]. Thus, multiple models of fetal growth restriction in different species show oxidative stress in several tissues, suggesting that mitochondrial dysfunction is a central mechanism in the pathogenesis of fetal growth restriction and the later development of disease in adulthood.

A number of studies in humans further suggest that mitochondrial dysfunction may contribute to type 2 diabetes. Studies using ^{13}C and ^{31}P magnetic resonance spectroscopy (MRS) have shown decreases in mitochondrial activity and increases in intramyocellular fat content in young insulin-resistant offspring of parents with type 2 diabetes, a group that has a strong tendency to develop diabetes later in life [88]. Expression of genes involved in oxidative phosphorylation is reduced among patients with type 2 diabetes mellitus and insulin resistance [89], although this may be an effect rather than a cause of diabetes. However, it is not known whether mitochondrial defects are causal to the disease and what the contribution of mitochondrial dysfunction is to the development of diabetes in individuals who were born growth restricted. Thus, future studies need to be directed toward addressing this question.

Conclusions

The combined epidemiological, clinical, and animal studies clearly demonstrate that the intrauterine environment influences both growth and development of the fetus and the subsequent development of adult diseases. There are critical specific windows during development, often coincident with periods of rapid cell division, during which a stimulus or insult may have long-lasting consequences on tissue or organ function postnatally. Birth weight is only one marker of an adverse fetal environment, and confining studies to this population only may lead to erroneous conclusions regarding etiology. Studies using animal models of uteroplacental insufficiency suggest that mitochondrial dysfunction and oxidative stress play an important role in the pathogenesis of the fetal origins of adult disease.

References

1. Hales CN, Barker DJP. Type 2 diabetes mellitus: the thrifty phenotype hypothesis. Diabetologia. 1992;35:595–601.
2. Kermack WO. Death rates in Great Britain and Sweden. Lancet. 1934;1:698–703.
3. Ravelli GP, Stein ZA, Susser MW. Obesity in young men after famine exposure in utero and early infancy. N Engl J Med. 1976;295:349–53.
4. Barker DJP, Winter PD, Osmond C, Margetts B, Simmons SJ. Weight in infancy and death from ischaemic heart disease. Lancet. 1989;2:577–80.
5. Hales CN, Barker DJP, Clark PMS, Cox LJ, Fall C, Osmond C, Winter PD. Fetal and infant growth and impaired glucose tolerance at age 64. BMJ. 1991;303:1019–22.
6. Phipps K, Barker DJ, Hales CN, Fall CH, Osmond C, Clark PM. Fetal growth and impaired glucose tolerance in men and women. Diabetologia. 1993;36:225–8.
7. Fall CHD, Osmond C, Barker DJP, Clark PMS, Hales CN, Stirling Y, Meade TW. Fetal and infant growth and cardiovascular risk factors in women. BMJ. 1995;310:428–32.
8. Barker DJP, Hales CN, Fall CHD, Osmond C, Phipps K, Clark PMS. Type 2 diabetes mellitus, hypertension, and hyperlipidemia (syndrome X): relation to reduced fetal growth. Diabetologia. 1993;36:62–7.

9. Hales CN, Barker DJ. The Thrifty phenotype hypothesis. Br Med Bull. 2001;60:5–20.
10. Valdez R, Athens MA, Thompson GH, Bradshaw BS, Stern MP. Birthweight and adult health outcomes in a biethnic population in the USA. Diabetologia. 1994;37:624–31.
11. Curhan GC, Willett WC, Rimm EB, Spiegelman D, Ascherio AL, Stampfer MJ. Birthweight and adult hypertension, diabetes mellitus and obesity in US men. Circulation. 1996;94: 3246–50.
12. Lithell HO, McKeigue PM, Berglund L, Mohsen R, Lithell UBN, Leon DA. Relation of size at birth to non-insulin dependent diabetes and insulin concentrations in men aged 50–60 years. BMJ. 1996;312:406–10.
13. McKeigue PM, Lithell HO, Leon DA. Glucose tolerance and resistance to insulin-stimulated glucose uptake in men aged 70 years in relation to size at birth. Diabetologia. 1998;41: 1133–8.
14. Leger J, Levy-Marchal C, Bloch J, Pinet A, Chevenne D, Porquet D, Collin D, Czernichow P. Reduced final height and indications for insulin resistance in 20 year olds born small for gestational age: regional cohort study. BMJ. 1997;315:341–7.
15. Jaquet D, Gaboriau A, Czernichow P, Levy-Marchal C. Insulin resistance early in adulthood in subjects born with intrauterine growth retardation. J Clin Endocrinol Metab. 2000;85: 1401–6.
16. Egeland GM, Skjaerven R, Irgrens LM. Birth characteristics of women who develop gestational diabetes: population based study. BMJ. 2000;321:546–7.
17. Forsen T, Eriksson J, Tuomilehto J, Reunanen A, Osmond C, Barker D. The fetal and childhood growth of persons who develop type 2 diabetes. Ann Intern Med. 2000;133:176–82.
18. Rich-Edwards JW, Colditz GA, Stampfer MJ, Willett WC, Gillman MW, Hennekens CH, Speizer FE, Manson JE. Birthweight and the risk for type 2 diabetes mellitus in adult women. Ann Intern Med. 1999;130:278–84.
19. Eriksson J, Forsen T, Tuomilehto J, Osmond C, Barker D. Fetal and childhood growth and hypertension in adult life. Hypertension. 2000;36:790–4.
20. Bavdekar A, Sachdev HS, Fall CHD, Osmond C, Lakshmy R, Barker DJP, Biswas SKD, Ramji S, Prabhakaran D, Reddy KS. Relation of serial changes in childhood body-mass index to impaired glucose tolerance in young adulthood. N Engl J Med. 2004;350:865–75.
21. Bavdekar A, Yajnik CS, Fall CH, Bapat S, Pandit AN, Deshpande V, Bhave S, Kellingray SD, Joglekar C. Insulin resistance syndrome in 8-year-old Indian children:small at birth, big at 8 years, or both? Diabetes. 1999;48:2422–9.
22. Myatt L, Eis ALW, Brockman DE, Kossenjans W, Greer IA, Lyall F. Differential localization of superoxide dismutase isoforms in placental villous tissue of normotensive, pre-eclamptic, and intrauterine growth-restricted pregnancies. J Histochem Cytochem. 1997;45:1433–8.
23. Karowicz-Bilinska A, Suzin J, Sieroszewski P. Evaluation of oxidative stress indices during treatment in pregnant women with intrauterine growth retardation. Med Sci Monit. 2002; 8:CR211–6.
24. Ejima K, Nanri H, Toki N, Kashimura M, Ikeda M. Localization of thioredoxin reductase and thioredoxin in normal human placenta and their protective effect against oxidative stress. Placenta. 1999;20:95–101.
25. Kato H, Yoneyama Y, Araki T. Fetal plasma lipid peroxide levels in pregnancies complicated by preeclampsia. Gynecol Obstet Invest. 1997;43:158–61.
26. Bowen RS, Moodley J, Dutton MF, Theron AJ. Oxidative stress in pre-eclampsia. Acta Obstet Gynecol Scand. 2001;80:719–25.
27. Wang Y, Walsh SW. Increased superoxide generation is associated with decreased superoxide dismutase activity and mRNA expression in placental trophoblast cells in pre-eclampsia. Placenta. 2001;22:206–12.
28. Wang Y, Walsh SW. Placental mitochondria as a source of oxidative stress in pre-eclampsia. Placenta. 1998;19:581–6.
29. Guvendag Guven ES, Karcaaltincaba D, Kandemir O, Kiykac S, Mentese A. Cord blood oxidative stress markers correlate with umbilical artery pulsatility in fetal growth restriction. J Matern Fetal Neonatal Med. 2013;26(6):576–80.

30. Kimura C, Watanabe K, Iwasaki A, Mori T, Matsushita H, Shinohara K, Wakatsuki A. The severity of hypoxic changes and oxidative DNA damage in the placenta of early-onset preeclamptic women and fetal growth restriction. J Matern Fetal Neonatal Med. 2013;26(5): 491–6.
31. Esposti MD, McLennan H. Mitochondria and cells produce reactive oxygen species in virtual anaerobiosis: relevance to ceramide-induced apoptosis. FEBS Lett. 1998;430:338–42.
32. Chandel NS, Budinger GRS, Schumacker PT. Molecular oxygen modulates cytochrome c oxidase function. J Biol Chem. 1996;271:8672–18677.
33. Gorgias N, Maidatsi P, Tsolaki M, Alvanou A, Kiriazis G, Kaidoglou K, Giala M. Hypoxic pretreatment protects against neuronal damage of the rat hippocampus induced by severe hypoxia. Brain Res. 1996;714:215–25.
34. Fowden AL, Forhead AJ. Endocrine mechanisms of intrauterine programming. Reproduction. 2004;127:515–26.
35. McMillen C, Robinson JS. Developmental origins of the metabolic syndrome: prediction, plasticity, and programming. Physiol Rev. 2005;85:571–633.
36. Dahri S, Snoeck A, Reusens-Billen B, Remacle C, Hoet JJ. Islet function in off-spring of mothers on low-protein diet during gestation. Diabetes. 1991;40:115–20.
37. Snoeck A, Remacle C, Reusens B, Hoet JJ. Effect of a low protein diet during pregnancy on the fetal rat endocrine pancreas. Biol Neonate. 1990;57:107–18.
38. Ozanne SE, Wang CL, Coleman N, Smith GD. Altered muscle insulin sensitivity in the male offspring of protein-malnourished rats. Am J Physiol. 1996;271:E1128–34.
39. Berney DM, Desai M, Palmer DJ, Greenwald S, Brown A, Hales CN, Berry CL. The effects of maternal protein deprivation on the fetal rat pancreas: major structural changes and their recuperation. J Pathol. 1997;183:109–15.
40. Wilson MR, Hughes SJ. The effect of maternal protein deficiency during pregnancy and lactation on glucose tolerance and pancreatic islet function in adult rat offspring. J Endocrinol. 1997;154:177–85.
41. Burns SP, Desai M, Cohen RD, Hales CN, Iles RA, Germain JP, Going TCH, Bailey RA. Gluconeogenesis, glucose handling, and structural changes in livers of the adult offspring of rats partially deprived of protein during pregnancy and lactation. J Clin Invest. 1997; 100:1768–74.
42. Ozanne SE, Jensen CB, Tingey KJ, Storgaard H, Madsbad S, Vaag AA. Low birthweight is associated with specific changes in muscle insulin-signalling protein expression. Diabetologia. 2005;48:547–52.
43. Ozanne SE, Olsen GS, Hansen LL, Tingey KJ, Nave BT, Wang CL, Hartil K, Petry CJ, Buckley AJ, Mosthaf-Seedorf L. Early growth restriction leads to down regulation of protein kinase C zeta and insulin resistance in skeletal muscle. J Endocrinol. 2003;177:235–41.
44. Petry CJ, Dorling MW, Pawlak DB, Ozanne SE, Hales CN. Diabetes in old male offspring of rat dams fed a reduced protein diet. Int J Exp Diabetes Res. 2001;2:139–43.
45. Fernandez-Twinn DS, Wayman A, Ekizoglou S, Martin MS, Hales CN, Ozanne SE. Maternal protein restriction leads to hyperinsulinemia and reduced insulin-signaling protein expression in 21-mo-old female rat offspring. Am J Physiol Regul Integr Comp Physiol. 2005;288: R368–73.
46. Benediktsson R, Lindsay R, Noble J, Seckl JR, Edwards CRW. Glucocorticoid exposure *in utero*; a new model for adult hypertension. Lancet. 1993;341:339–41.
47. Lindsay RS, Lindsay RM, Edwards CRW, Seckl JR. Inhibition of 11β-hydroxysteroid dehydrogenase in pregnant rats and the programming of blood pressure in the offspring. Hypertension. 1996;27:1200–4.
48. Lindsay RS, Lindsay RM, Waddell B, Seckl JR. Programming of glucose tolerance in the rat: role of placental 11β-hydroxysteroid dehydrogenase. Diabetologia. 1996;39:1299–305.
49. Niyirenda MJ, Seckl JR. Intrauterine events and the programming of adulthood disease: the role of fetal glucocorticoid exposure. Int J Mol Med. 1998;2:607–14.
50. Simmons RA, Templeton L, Gertz S, Niu H. Intrauterine growth retardation leads to type II diabetes in adulthood in the rat. Diabetes. 2001;50:2279–86.

51. Boloker J, Gertz S, Simmons RA. Offspring of diabetic rats develop obesity and type II diabetes in adulthood. Diabetes. 2002;51:1499–506.
52. Ogata ES, Bussey M, Finley S. Altered gas exchange, limited glucose, branched chain amino acids, and hypoinsulinism retard fetal growth in the rat. Metabolism. 1986;35:950–77.
53. Simmons RA, Gounis AS, Bangalore SA, Ogata ES. Intrauterine growth retardation: fetal glucose transport is diminished in lung but spared in brain. Pediatr Res. 1991;31:59–63.
54. Unterman T, Lascon R, Gotway M, Oehler D, Gounis A, Simmons RA, Ogata ES. Circulating levels of insulin-like growth factor binding protein-1 (IGFBP-1) and hepatic mRNA are increased in the small for gestational age fetal rat. Endocrinology. 1990;127:2035–7.
55. Sakai C, Tomitsuka E, Miyagishi M, Harada S, Kita K. Type II Fp of human mitochondrial respiratory complex II and its role in adaptation to hypoxia and nutrition-deprived conditions. Mitochondrion. 2013;13:602–9.
56. Pejznochova M, Tesarova M, Hansikova H, Magner M, Honzik T, Vinsova K, Hajkova Z, Havlickova V, Zeman J. Mitochondrial DNA content and expression of genes involved in mtDNA transcription, regulation and maintenance during human fetal development. Mitochondrion. 2010;10:321–9.
57. Lattuada D, Colleoni F, Martinelli A, Garretto A, Magni R, Radaelli T, Cetin I. Higher mitochondrial DNA content in human IUGR placenta. Placenta. 2008;29:1029–33.
58. Panten U, Zielman S, Langer J, Zunkler BJ, Lenzen S. Regulation of insulin secretion by energy metabolism in pancreatic β-cell mitochondria. Biochem J. 1984;219:189–96.
59. Newgard CB, McGarry JD. Metabolic coupling factors in pancreatic β-cell signal transduction. Annu Rev Biochem. 1995;64:689–719.
60. Schuit F. Metabolic fate of glucose in purified islet cells. Glucose regulated anaplerosis in β-cells. J Biol Chem. 1997;272:18572–9.
61. Mertz RJ, Worley III JF, Spencer B, Johnson JH, Dukes ID. Activation of stimulus-secretion coupling in pancreatic β-cells by specific products of glucose metabolism. J Biol Chem. 1996;271:4838–45.
62. Ortsater H, Liss P, Akerman KEO. Contribution of glycolytic and mitochondrial pathways in glucose-induced changes in islet respiration and insulin secretion. Pflugers Arch Eur J Physiol. 2002;444:506–12.
63. Antinozzi PA, Ishihara H, Newgard CB, Wollheim CB. Mitochondrial metabolism sets the maximal limit of fuel-stimulated insulin secretion in a model pancreatic beta cell. A survey of four fuel secretagogues. J Biol Chem. 2002;277:11746–55.
64. Malaisse WJ, Hutton JC, Carpinelli AR, Herchuelz A, Senner A. The stimulus-secretion coupling of amino acid-induced insulin release. Metabolism and cationic effects of leucine. Diabetes. 1980;29:431–7.
65. Lenzen S, Schmidt W, Rustenbeck I, Panten U. 2-Ketoglutarate generation in pancreatic β-cell mitochondria regulates insulin secretory action of amino acids and 2-keto acids. Biosci Rep. 1986;6:163–9.
66. Noda M, Yamashita S, Takahashi N, Eto K, Shen LM, Izumi K, Daniel S, Tsubamoto Y, Nemoto T, Lino M, Kasai H, Sharp GW, Kadowaki T. Switch to anaerobic glucose metabolism with NADH accumulation in the beta-cell model of mitochondrial diabetes. Characteristics of betaHC9 cells deficient in mitochondrial DNA transcription. J Biol Chem. 2002;277:41817–26.
67. Lenzen S, Drinkgern J, Tiedge M. Low antioxidant enzyme gene expression in pancreatic islets compared with various other mouse tissues. Free Radic Biol Med. 1996;20:463–6.
68. Tiedge M, Lortz S, Drinkgern J, Lenzen S. Relationship between antioxidant enzyme gene expression and antioxidant defense status of insulin-producing cells. Diabetes. 1997;46:1733–42.
69. Maechler P, Jornot L, Wollheim CB. Hydrogen peroxide alters mitochondrial activation and insulin secretion in pancreatic beta cells. J Biol Chem. 1999;274:27905–13.
70. Sakai K, Matsumoto K, Nishikawa T, Suefuji M, Nakamaru K, Hirashima Y, Kawashima J, Shirotani T, Ichinose I, Brownlee M, Araki E. Mitochondrial reactive oxygen species reduce insulin secretion by pancreatic β-cells. Biochem Biophys Res Commun. 2003;300:216–22.
71. Kaneto H, Xu G, Fujii N, Kim S, Bonner-Weir S, Weir GC. Involvement of c-Jun N-terminal kinase in oxidative stress-mediated suppression of insulin gene expression. J Biol Chem. 2002; 277:30010–8.

72. Kaneto HH, Xu G, Fujii N, Kim S, Bonner-Weir S, Weir GC. Involvement of protein kinase C beta 2 in c-myc induction by high glucose in pancreatic beta-cells. J Biol Chem. 2002; 277:3680–5.
73. Kaneto H, Xu G, Song KH, Suzuma K, Bonner-Weir S, Sharma A, Weir GC. Activation of the hexosamine pathway leads to deterioration of pancreatic beta-cell function through the induction of oxidative stress. J Biol Chem. 2001;276:31099–104.
74. Kaneto H, Kajimoto Y, Fujitani Y, Matsuoka T, Sakamoto K, Matsuhisa M, Yamasaki Y, Hori M. Oxidative stress induces p21 expression in pancreatic islet cells: possible implication in beta-cell dysfunction. Diabetologia. 1999;42:1093–7.
75. Jonas JC, Laybutt DR, Steil GM, Trivedi N, Pertusa JG, Van de Casteele M, Weir GC, Henquin JC. High glucose stimulates early response gene c-Myc expression in rat pancreatic beta cells. J Biol Chem. 2001;276:35375–81.
76. Jonas JC, Sharma A, Hasenkamp W, Ilkova H, Patane G, Laybutt R, Bonner-Weir S, Weir GC. Chronic hyperglycemia triggers loss of pancreatic beta cell differentiation in an animal model of diabetes. J Biol Chem. 1999;274:14112–21.
77. Efanova IB, Zaitsev SV, Zhivotovsky B, Kohler M, Efendic S, Orrenius S, Berggren PO. Glucose and tolbutamide induce apoptosis in pancreatic β-cells. J Biol Chem. 1998;273: 22501–7.
78. Moran A, Zhang HJ, Olsonm LK, Harmon JS, Poitoust V, Robertson RP. Differentiation of glucose toxicity from β-cell exhaustion during the evolution of defective insulin gene expression in the pancreatic islet cell line, HIT-T15. J Clin Invest. 2000;99:534–9.
79. Donath MY, Gross DJ, Cerasi E, Kaiser N. Hyperglycemia-induced β-cell apoptosis in pancreatic islets of Psammomys obesus during development of diabetes. Diabetes. 1999;48:738–44.
80. Silva JP, Kohler M, Graff C, Oldfors A, Magnuson MA, Berggren PO, Larsson NG. Impaired insulin secretion and β-cell loss in tissue specific knockout mice with mitochondrial diabetes. Nat Genet. 2000;26:336–40.
81. Simmons RA, Suponitsky-Kroyter I, Selak MA. Progressive accumulation of mitochondrial DNA mutations and decline in mitochondrial function lead to beta-cell failure. J Biol Chem. 2005;280:28785–91.
82. Peterside IE, Selak MA, Simmons RA. Impaired oxidative phosphorylation in hepatic mitochondria of growth retarded rats alters glucose metabolism. Am J Physiol. 2003;285: E1258–64.
83. Selak MA, Storey BT, Peterside IE, Simmons RA. Impaired oxidative phosphorylation in skeletal muscle contributes to insulin resistance and hyperglycemia. Am J Physiol. 2003; 285:E130–7.
84. Boyle KE, Newsom SA, Janssen RC, Lappas M, Friedman JE. Skeletal muscle MnSOD, mitochondrial complex II, and SIRT3 enzyme activities are decreased in maternal obesity during human pregnancy and gestational diabetes mellitus. J Clin Endocrinol Metab. 2013;98: E1601–9.
85. Park HK, Jin CJ, Cho YM, do Park J, Shin CS, Park KS, Kim SY, Cho BY, Lee HK. Changes of mitochondrial DNA content in the male offspring of protein-malnourished rats. Ann N Y Acad Sci. 2004;1011:205–16.
86. Park KS, Kim SK, Kim MS, Cho EY, Lee JH, Lee KU, Pak YK, Lee HK. Fetal and early postnatal protein malnutrition cause long-term changes in rat liver and muscle mitochondria. J Nutr. 2003;133:3085–90.
87. van Vliet E, Eixarch E, Illa M, Arbat-Plana A, González-Tendero A, Hogberg HT, Zhao L, Hartung T, Gratacos E. Metabolomics reveals metabolic alterations by intrauterine growth restriction in the fetal rabbit brain. PLoS One. 2013;8:e64545.
88. Petersen KF, Dufour S, Befroy D, Garcia R, Shulman GI. Impaired mitochondrial activity in the insulin-resistant offspring of patients with type 2 diabetes. N Engl J Med. 2004;350:664–7.
89. Mootha VK, Lindgren CM, Eriksson KF, Subramanian A, Sihag S, Lehar J, Puigserver P, Carlsson E, Ridderstrale M, Laurila E, Houstis N, Daly MJ, Patterson N, Mesirov JP, Golub TR, Tamayo P, Spiegelman B, Lander ES, Hirschhorn JN, Altshuler D, Groop LC. PGC-1alpha-responsive genes involved in oxidative phosphorylation are coordinately downregulated in human diabetes. Nat Genet. 2003;34:267–73.

Chapter 7
Down Syndrome as a Special Case of Oxidatively Induced Developmental Dysregulation

Marzia Perluigi and D. Allan Butterfield

Down Syndrome

Down syndrome (DS) is the most common genetic cause of intellectual disability that arises from the triplication of the entire portion or part of chromosome 21 (trisomy 21). DS occurs in 1 every 700–800 live births, and the only known risk factor is the increased meiotic maternal nondisjunction errors occurring with age [1]. Although genetic defects are the crucial determinants of the majority of clinical presentations of the disease such as craniofacial abnormalities, small brain size, accelerated aging, and cognitive decline, other factors play a crucial role in determining the severity of pleiotropic phenotypes. Trisomy 21 is associated with a genetic instability that affects not only chromosome 21 (Chr21) but also other chromosomes, and it is essentially responsible of two types of phenotypes: (1) those present in every DS individual and (2) those that occur only in a subset of DS individuals. Further, for any given phenotype, there is wide variability in expression that in turn results in a complex set of clinical features [2]. One of the hypotheses to explain such variability is "gene dosage hypothesis," which states that the increased dosage of Chr21 genes

M. Perluigi
Department of Biochemical Sciences,
Sapienza University of Rome, 00185 Rome, Italy

D.A. Butterfield (✉)
Department of Chemistry, Sanders-Brown Center on Aging,
and Center of Membrane Sciences, University of Kentucky,
Lexington, KY 40506-0055, USA
e-mail: dabcns@email.uky.edu

P.A. Dennery et al. (eds.), *Perinatal and Prenatal Disorders*,
Oxidative Stress in Applied Basic Research and Clinical Practice,
DOI 10.1007/978-1-4939-1405-0_7

is the direct cause of the phenotypical alterations of DS [3–5]. This scenario becomes much more complex when considering the effects of overexpression of trisomic genes on disomic genes, which, in turn, may gain aberrant expression and contribute to some clinical manifestations [5]. In this view, "the amplified developmental instability hypothesis" suggests that the variability of phenotypic features may actually involve a relatively small number of genes which have the ability to impact global gene expression thus leading to deleterious effects on development [6].

By combining these two different theories, it is likely that, in this deregulated scenario, the effects caused by some dosage-sensitive genes are extremely amplified and result in a plethora of different phenotypic traits according to the "number and dose" of genes involved.

Life expectancy in the DS population is shorter compared with normal individuals; however, improvements in medical care and identification and treatment of psychiatric disorders have significantly improved the quality of life of individuals with DS [7, 8]. Numerous developmental defects are associated with DS [9, 10], including skeletal defects, reduced brain size, craniofacial dysmorphic features, defects of atrioventricular septum, and abnormalities of the gastrointestinal system [10]. Brain development and intellectual disabilities are the most striking features of trisomy [11, 12], and language, learning, and memory appear to be severely affected [13]. The comprehension of degenerative phenomena related to accelerated aging and neurodegeneration is one of the major challenges of research in this field, both at molecular and clinical levels.

Growing data support the link between the DS phenotype and an increased risk of development of AD [14]. The incidence of dementia among DS patients is 8 % in the age range 35–49, 55 % in the age range 50–59, and 75 % above the age of 60 years, but AD neuropathology is present in all DS individuals by the age of 40 [15]. Neuropathological features of AD-like dementia are the presence of senile plaques (SPs) and neurofibrillary tangles (NFTs) and also cholinergic and serotonergic reduction [16, 17]. However, although Aβ plaques are present in DS patients early in life, even in the fetus, it is only very later on that DS individuals may develop AD.

It is reasonable to speculate that neurodegenerative processes may in part overlap between these two disorders, and identification of common, as well as divergent pathways, may help to elucidate the role of specific molecular events within a complex puzzle. Among the proposed "unifying molecular mechanisms" for both DS and AD, OS is a major candidate and the genetics of DS are a rich environment for the amplification of the cascade of reactions initiated by free radicals.

Oxidative Stress in DS: Implication for Development of Alzheimer Disease

A growing number of studies have demonstrated that OS is involved in the biology of DS (reviewed in [18]). OS represents an imbalance between the production of ROS and the ability of the cell to detoxify the reactive intermediates or to repair the

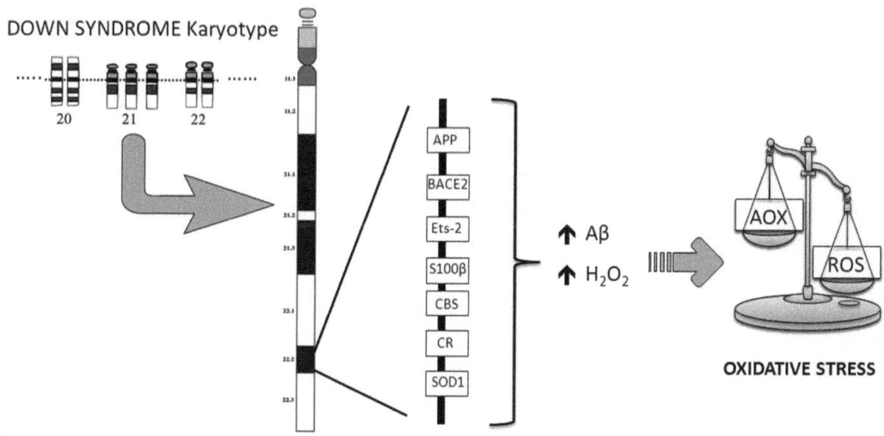

Fig. 7.1 Oxidative stress hypothesis in Down syndrome. Increased oxidative stress likely results by the overexpression of some of the genes encoded by Chr21. Among these, amyloid precursor protein (APP), copper-zinc superoxide dismutase (SOD1), beta secretase (BACE2), cystathionine beta synthase (CBS), carbonyl reductase (CR), and transcription factor Ets-2 can directly or indirectly lead to OS

resulting damage [19, 20]. Increasing levels of OS-induced by-products have been measured in brain tissue from DS [20, 21]. The most striking link between OS and the pathology of DS is the fact that increased conditions of OS are exacerbated by the overexpression of some of the genes encoded by Chr21 including SOD1, APP, BACE2, Ets-2, S100b, carbonyl reductase, amyloid precursor protein, among others (see Fig. 7.1).

SOD1 is one of the major enzymes involved in the antioxidant defense and catalyzes the dismutation of $O_2^{\bullet-}$ to molecular oxygen (O_2) and H_2O_2, which in turn is metabolized by catalase (CAT) and by (selenium-containing) glutathione peroxidase (GPX) to water [22]. However, brain levels of both CAT and GPX are generally lower compared with other tissues, and this may in part explain why neuronal tissues are particularly susceptible to oxidative damage. Further, the triplication of Chr21 leads to an imbalance in the ratio of SOD-1 to CAT and GPX, resulting in the accumulation of H_2O_2 [15]. Interestingly, all DS tissues, in addition to the brain, display an altered SOD-1/GPX activity ratio [23]. SOD-1 was found at levels approximately 50 % higher than normal in a variety of DS cells and tissues, including erythrocytes, B and T lymphocytes, and fibroblasts. Indeed, the erythrocytes of DS children, adolescents, and adults exhibited systemic increases in SOD-1, SOD-1/GPX, or the SOD-1/(GPX+CAT) activity ratio. A key role of SOD-1 was also demonstrated by Shin et al. which found that transgenic mice overexpressing wild-type human SOD1 (Tg-SOD1) showed mitochondrial swelling, vacuolization, learning, and memory deficits [24].

In line with these studies, Busciglio and Yankner [25] reported that neurons of DS patients have higher intracellular ROS levels also associated with elevated levels of lipid peroxidation. However, a proteomics study from Gulesserian et al. [26] suggested that OS in fetal DS is not merely a consequence of SOD-1 overexpression but appeared to be caused principally by low levels of antioxidant enzymes, such as glutathione transferases and thioredoxin peroxidases.

It is worth mentioning that elevated levels of OS could be also caused by increased production of amyloid beta-peptide (Aβ) [27–29] due to the overexpression of the amyloid precursor protein (*APP*) gene. Indeed, accumulation of Aβ(1–42) peptide, a characteristic hallmark of AD pathology which correlates with age, was observed in postmortem brain from DS persons [30]. Mehta et al. [31] measured the levels of Aβ(1–42) and Aβ(1–40) and found that both were higher in DS plasma than controls. The ratio of Aβ42/Aβ40 was lower in DS than in controls, and a significant negative correlation between age and Aβ40 in DS and controls was observed, and between age and Aβ42 levels in DS but not in controls. Recently, the same group demonstrated that among adults with DS, decreasing levels of plasma Aβ42, a decline in the Aβ42/Aβ40 ratio, or increasing levels of Aβ40 may be putative markers of conversion to AD, possibly reflecting compartmentalization of Aβ peptides in the brain [32]. Intriguingly, recent studies from Anandatheerthavarada et al. [33] demonstrated also that full-length APP may be neurotoxic, mostly at the mitochondrial level. The authors hypothesized that increased APP expression caused a progressive accumulation of transmembrane-arrested APP which in turn affected mitochondria ultimately resulting in impairment of energy metabolism. Additional studies supported this novel hypothesis by showing that mice overexpressing wild-type human APP develop neuronal pathology similar to AD, but without robust Aβ deposition in the hippocampus [34]. These findings support the notion that overexpression of APP may promote mitochondrial dysfunction in DS independent of aberrant Aβ deposition.

Related to Aβ metabolism, another gene encoded by Chr21 is the *β*-site APP-cleaving 2 enzyme (BACE2). BACE2, homologous to BACE1, is a *β*-secretase involved in the amyloidogenic pathway of APP proteolysis. Though initially it was hypothesized that the co-overexpression of APP and BACE2 could contribute to Alzheimer-like neuropathology in DS, recent evidence showed that co-overexpression of BACE2 and APP did not increase amyloid-*β* peptide concentration in brain of Tg mice. It is likely that in vivo effects of APP are not exacerbated by BACE2 co-overexpression but, on the contrary, may confer protective effects in specific behavioral and cognitive domains in transgenic mice [35].

Chr21 also maps the gene for S100β, an astroglial-derived Ca^{2+}-binding protein acting as a neurotrophic factor on neurons and glial cells. S100β is a neurite growth-promoting factor that is important for brain development, which promotes neuronal growth and differentiation, intracellular calcium signal transduction, and proliferation and morphogenesis in astrocytes [36]. Both DS and AD are characterized by an elevated astrocytic S100β expression, and the pattern of elevated tissue S100β levels across brain regions in AD reflects that of neuritic plaques [37, 38]. This strong correlation found between numbers of activated S100β positive astrocytes and the numerical density of amyloid plaques, together with the finding that the numbers of activated S100β-positive astrocytes (but not nonactivated $S100\beta^{+}$ cells) increases in parallel with the density of classic plaques, supports the idea that S100β is an important element in the accumulation of plaques in DS and AD. It is likely that chronic overexpression of S100β promotes increased neuronal and neuritic βAPP expression with consequent increased amyloid deposition, as well as abnormal growth of neurites in β-amyloid plaques, as seen in middle-aged DS patients [37]. However, this proinflammatory role of S100β is not restricted to the brain. Indeed, S100β stimulates both NO production and iNOS protein transcription and expression in rat peritoneal macrophages [39].

Ets-2 is a member of the Ets family of transcription factors that has been proposed to have important functions in cancer, bone development, and immune responses. Ets-2 was shown to be upregulated by OS, and it is involved in differentiation, maturation, and signaling cascade [40]. Previous studies have demonstrated that increased expression of Ets-2 is associated with pathologic markers. Interestingly, though Ets-2 overexpression is involved in neuronal apoptotic cell death, it is conceivable that this protein plays a role in the reduced incidence of solid tumors occurring in DS individuals [41].

The overexpression of Ets-2 has been hypothesized to be an important contributor to the increased susceptibility of DS cells to apoptotic stimuli that might, at least in part, be responsible for the thymic and splenic hypoplasia and conceivably other pathophysiological features shared between Ets-2 transgenic mice and individuals with DS [42].

By mapping Chr21, another candidate gene that may be involved in OS is the enzyme carbonyl reductase (CBR). Carbonyls are cytotoxic metabolic intermediates, which are detoxified by oxidation catalyzed by aldehyde dehydrogenase (ALDH) or by reduction to their corresponding alcohols by CBR. Protein levels of these enzymes were found to be increased in different brain regions of both DS and AD patients [43]. Further, CBR is an oxidatively modified protein in the brain of subjects with mild cognitive impairment (MCI), arguably a prodromal phase of AD [44].

There is evidence of a link between 1-carbon/transsulfuration (1C-TS) metabolism and DS. There are at least six genes encoding enzymes important for 1C-TS metabolism located on human Chr21, including the gene for cystathionine beta synthase (CBS) [45]. CBS catalyzes the condensation of serine and homocysteine to form cystathionine. CBS plays a critical role in linking the folate cycle and the methionine cycle and in regulating homocysteine levels [46]. In addition, CBS can convert cysteine to hydrogen sulfide, which researchers are beginning to recognize as an important neuromodulator in the brain [47]. There is evidence that CBS protein levels and enzyme activity are increased in persons with DS [48]. Elevated CBS activity can lower homocysteine levels, which in turn perturb the balance of 1C-TS metabolism and lead to elevated—perhaps toxic—levels of hydrogen sulfide. These metabolic alterations might play a role in the cognitive disability seen in DS [49]. Accordingly, CBS is reportedly a risk factor for AD [50].

Elevated OS markers have been measured in peripheral and CNS specimens of DS patients and animal models of the disease [25] (summarized in Table 7.1). Levels of TBARS, total protein carbonyls, and advanced glycation end products

Table 7.1 Markers of oxidative stress in brain and body fluids from DS individuals

Markers of oxidative damage in DS	Brain	Amniotic fluid	Urine
Allantoin and 2,3-dinor-iPF2α-III			No change Tolun [54]
Isoprostane 8,12-*iso*-iPF2α			Pratico [27]
HNE-bound proteins and protein carbonyls		Perluigi [60]	
TBARS, AGEs, carbonyls	Odetti [51]		
8OHdG, 3NT	Nunomura [52]		
AGEs, dityrosine, H_2O_2			Campo [53]
HNE-bound proteins	Cenini et al. [71]		

These markers have been found increased compared with their age-matched controls except for allantoin and 2,3-dinor-iPF2α-III

(AGEs) were increased in the cortex from DS fetal brain compared with controls [51]. Accumulation of 8-hydroxy-2-deoxyguanosine (8OHdG), oxidized proteins, and nitrotyrosine was also observed in the cytoplasm of cerebral neurons in DS [52]. At the systemic level, the amount of isoprostane 8,12-*iso*-iPF2α (iPF2α), a specific marker of lipid peroxidation, has been found to be elevated in urine samples from adults with DS [27]. In addition, levels of AGEs, dityrosine, H_2O_2, and nitrite/nitrate were significantly higher in urine samples of DS compared with age-matched controls [53, 54].

Recently, an epidemiologic study showed that the levels of allantoin and 2,3-dinor-iPF2α-III are not increased in DS individuals compared with controls [54]. These results emphasize a controversy within the field of DS. Although the prevailing concept considers oxidative stress a major contributor of several morbidities associated with DS, results from human studies are conflicting as to whether or not such oxidative stress exists and can be reliably measured at the systemic level has yet to be clarified.

Most of the results reported have been collected on animal models of the disease, including Ts65Dn mice and Ts1Cje mice. The Ts65Dn mouse carries a small chromosome derived primarily from mouse chromosome 16, causing dosage imbalance for approximately half of the human chromosome 21 orthologs; these mice have cerebellar pathology with direct parallels to DS [55]. The Ts1Cje mouse, containing a translocated chromosome 16, is dosage imbalanced for 67 % of the genes triplicated in Ts65Dn [56]. Ts1Cje mice do not express the SOD1 gene and show some DS-related abnormalities such as craniofacial alterations and spatial learning deficits [57], but different from Ts65Dn mice.

Ishihara et al. [58] showed increased level of ROS and mitochondrial dysfunction in primary cultured astrocytes and neurons from Ts1Cje transgenic mice. These results support the "gene-dosage" hypothesis, which in this case is suitable to explain the major part of OS-induced intracellular damage observed in this animal model of DS. The authors applied a redox proteomics approach, and they were able to identify those proteins that were modified by lipid peroxidation-derived products [58]. ATP synthase mitochondrial F1 complex b subunit, α-enolase, and triose phosphate isomerase 1 were identified as proteins modified by 3-hydroperoxy-9Z,11E-octadecadienoic acid (13-HPODE). Neurofilament light polypeptide, internexin neuronal intermediate filament α, neuron-specific enolase, peroxiredoxin 6, phosphoglycerate kinase 1, and triose phosphate isomerase were shown to be HNE-modified proteins. Based on these data, the dysfunction of these proteins as a consequence of oxidative damage may disturb ATP production and the neuronal cytoskeleton system and antioxidant system function. Interestingly, previous studies from the Butterfield laboratory already identified by redox proteomics some of these proteins modified by hydroxynonenal, a reactive product of lipid peroxidation, in AD and MCI brain [44, 59–62]. These data, together with the findings on Tg mice, suggest that these brain proteins might contribute to cognitive dysfunction and neurodegenerative processes occurring in both DS and AD. As discussed above, DS and AD neuropathology share common features, and redox proteomics studies contributed to illuminate putative pathways altered in both pathologies.

In order to better understand the role of OS, we have analyzed the amniotic fluid (AF) from women carrying DS pregnancy compared with that from women carrying healthy fetuses. Most prior studies were performed on DS fetal brains or DS mouse models, but few data were available on AF, which is a more reliable index of the physiological condition of the fetus. Indeed, in analogy with CSF, AF could be a powerful source for the identification of disease biomarkers to be coupled with current genetic screening. The composition of AF is complex and includes fetal and maternal proteins, amino acids, carbohydrates, hormones, lipids, and electrolytes. AF is in direct contact with multiple organs of the fetus and provides both physical and biochemical support for the developing fetus. AF contains high concentrations of proteins that are directly secreted from the fetus [63]. Recent development of advanced proteomics platforms allowed analysis of AF, in order to better understand its complex biological function and to discover disease-specific biomarkers for fetal aneuploidies and pregnancy-related complications [64].

In detail, we evaluated a set of oxidative stress biomarkers in AF from women carrying DS fetuses, and we found that the levels of protein carbonyl and lipid peroxidation were increased, coupled with reduction of GSH and Trx levels, and induction of the heat shock protein (HSP) response. By a redox proteomics approach, we identified selective proteins that showed increased oxidation in DS AF compared with that from mothers carrying healthy fetuses. The identified proteins are involved in iron homeostasis (ceruloplasmin and transferin), lipid metabolism (Zinc-alpha2-glycoprotein, retinol-binding protein 4, and Apolipoprotein A1), and inflammation (Complement C9, Alpha-1B-glycoprotein, Collagen alpha-1 V chain) with critical relevance in the clinical outcome of DS [60].

As already discussed, another key player in the oxidative stress hypothesis of neurodegeneration is Aβ peptide. A consistent Aβ deposition and neurofibrillary tangle formation was evidenced in DS brain [40] that correlates with age. Although plaque deposition occurs very early in DS individuals, even during fetal development, it is only very later on that DS individuals may develop AD [65]. Clinical features of dementia in DS population, over 50 years of age, appeared many years after the first significant insoluble Aβ accumulation or plaque deposition and also after the first signs of neurofibrillary tangle pathology [66]. Thus, other molecular pathways, in addition to Aβ metabolism and tangles formation, likely are involved to cause consistent neuronal dysfunction and cognitive decline. Synaptic dysfunction may be a consequence of APP overexpression or increased Aβ [67]. Neuroinflammation [68], endosomal dysfunction [69], and oxidative damage [70] may play a crucial role in DS as well as in AD pathology [66].

Recently the Butterfield group [71] showed that DS brains with neuropathological hallmarks of AD have more oxidative, but not nitrosative, stress than those with DS but without significant AD pathology, as compared with similarly aged-matched non-DS controls. They also found that soluble and insoluble Aβ and oligomers increase as a function of age in DS frontal cortex. Of the oxidative stress markers, HNE-bound proteins were increased overall in DS. Protein carbonyls were correlated with Aβ40 levels. These results suggest that oxidative damage, but not nitrosative stress, may contribute to the onset and progression of AD pathogenesis in DS. Further studies are needed to better understand aging-related neurodegenerative phenomena in DS, which from one side contribute to development of AD but also paradoxically result in AD neuropathology but without dementia.

Recent studies reported a quite surprising trend of oxidative stress damage in DS. While increased OS conditions are detectable as early as during pregnancy [72] and increase with age in young DS, adults with DS do not show a significant increased oxidative damage to DNA from peripheral lymphocytes [73]. These data are in contrast with other findings supporting the correlation of increased oxidative damage with increasing age. This discrepancy may be due to the type of samples analyzed, that is, peripheral lymphocytes which display a specific functionality and which could be able to activate compensatory mechanisms that the brain does not. Further, results obtained by measuring different markers of oxidative stress do not always correlate with each other, because the susceptibility of lipids, proteins, and nucleic acid to accumulate oxidative damage is different together with the specific susceptibility of different cells to repair the damage.

Young DS individual are exposed to chronic oxidative stress and the "surviving" cells are those able to activate defense mechanisms that counteract increasing oxidative stress conditions over the lifespan [73]. It is likely that newborn DS have to challenge high levels of ROS that are responsible of the pathogenesis of many of the pathological manifestations. Living in a more oxidant environment leads to the selection of more resistant cellular phenotypes that otherwise show multiple dysfunctions. Studies performed by Head et al. [65] demonstrated by PET that compensatory increases in metabolic rate and activation of plasticity mechanisms in vulnerable brain regions in DS occurred prior to the development of dementia.

It seems likely that some trisomic genes may interact with each other and are responsible for learning and memory deficits during development, but with increasing age, their interaction may become beneficial and possibly protective [63, 65]. The molecular mechanisms that drive dysfunction versus protection need to be further elucidated.

Summary Comments

The cause of increased OS conditions in DS has to be searched by mapping Chr21, where a number of genes, directly or indirectly, lead to overproduction of ROS, enhanced oxidative damage, and possibly to ROS-induced cell death. Among the most relevant contributory factors to OS, SOD1, APP, BACE2, Ets-2, S100b, carbonyl reductase are encoded on Chr21. Thus, it is likely that deregulation of several intracellular pathways occurs early in DS as demonstrated by studies performed on fetal brain and AF from DS pregnancy showing increased markers of OS, including TBARS, protein oxidation, AGEs, 8OHdG, isoprostanes, among others.

In particular, the development of advanced proteomics platforms contributed to better understand the role of OS in DS neuropathology allowing identification of intracellular targets that have selective oxidative damage. These studies demonstrated that dysfunction of energy metabolism, neuronal cytoskeleton system, and antioxidant system function contributes to neuronal loss observed in DS individuals. We suggest that OS is an early event that contributes to exacerbation of neurodegenerative phenomena in DS brain from early in life.

Posttranslational modifications of proteins—caused by OS—are critical to their functions, as they can affect activity, stability, aggregation, and turnover. Thus, future studies should employ selective proteomics approaches that aim to characterize specific posttranslational modifications more than simply looking at difference in protein expression levels. By following this approach, novel putative markers may emerge and may also contribute to better understand the mechanism of disease pathogenesis and progression. Similar approaches can also be employed to test the efficacy of novel therapeutic approaches that may slow or decrease OS-induced neuronal damage.

Role of Mitochondria in DS Neuropathology

Several studies highlighted the central role of mitochondrial impairment in neurodegenerative processes [74]. Mitochondria are subcellular organelles essential for ATP production and maintenance of calcium homoeostasis and also crucially involved in apoptotic signaling. As an essential link to oxidative stress, mitochondrial dysfunctions are observed whenever redox imbalances occur, due to the main role of mitochondria in oxygen metabolism, and this is the case of DS. Mitochondrial abnormalities, such as abnormal shape, reduced levels of microtubules, among others, in DS were reported in cultured cerebellar neurons from trisomy 16 (Ts16) mice [75]. Previous findings demonstrated a decreased functionality of mitochondrial enzymes, including monoamine oxidase, cytochrome oxidase, and isocitrate dehydrogenase [76]. A number of reports demonstrated that the accumulation of mitochondrial DNA (mtDNA) mutations may contribute to degenerative phenomena associated with both human aging and cognitive decline [77]. Mutations in mtDNA may be responsible of increased production of free radicals and also cause decreased ATP levels. Druzhyna et al. [78] demonstrated increased mtDNA oxidative damage associated with reduced activity of some repair systems in fibroblasts from DS patients. Reasonably, increased oxidative stress levels were caused by increased superoxide formation, as demonstrated in Ts16 neurons compared to control neurons. This stress condition was also evident in Ts16 neurons treated with rotenone (complex I inhibitor) that blocked $O_2^{\bullet-}$ production only in diploid neurons, but not in Ts16 neurons. Similar results were also obtained by treating Ts16 neurons and diploid neurons with carbonyl cyanide p-trifluoromethoxyphenylhydrazone, used to uncouple mitochondrial oxidative phosphorylation, which caused irreversible deficiency in the energy metabolism in Ts16 neurons, but not in diploid control neurons. These findings support the hypothesis that the Ts16 neurons display increased $O_2^{\bullet-}$ basal generation due to deficient complex I and impaired mitochondrial energy metabolism that severely contribute to neuronal cell death [79]. Accordingly, data published by Valenti et al. [80] showed a selective deficit in the catalytic efficiency of Complex I in DS-HSFs (DS human fetal skin fibroblasts). Impairment of Complex I was associated with a decrease in cAMP-dependent phosphorylation of the 18 kDa subunit of the complex, due to a decrease in PKA activity. Furthermore, the authors found a threefold increase in cellular levels of ROS, in particular $O_2^{\bullet-}$, mainly produced by DS-HSF mitochondria. The levels of H_2O_2 and calcium uptake were not significantly different in the Ts16 mitochondria, while decreased pyruvate dehydrogenase levels were detected, similar to the pattern found in Parkinson disease [81].

Mitochondrial functionality was also evaluated by the study of Conti et al. [82], who analyzed the expression profile of several genes on Chr21 using oligonucleotide microarrays in hearts of human DS fetuses compared with normal fetuses. Results from this work demonstrated that a dosage-dependent upregulation of Chr21 genes also affected those genes responsible for mitochondrial function and for the extracellular matrix organization in the fetal heart of trisomic subjects [82]. Further, Roat et al. [83] reported the in vivo alteration of mitochondrial function in blood cells from DS patients indexed by increased loss of mitochondrial membrane potential. Mitochondrial function is also regulated by the methyl status, and we discussed above the genes encoded by Chr21 that participate in recycling of methionine/homocysteine in the methyl cycle sequence of reactions. Methylation is a necessary event in mitochondria and relies on the availability and uptake of the methyl donor S-adenosylmethionine. Indeed, mitochondrial dysfunctions have been described in DS, but they have never been correlated to a possible mitochondrial methyl unbalance. Infantino et al. [84] recently showed that the mitochondrial levels of S-adenosylmethionine were reduced in DS compared to control cells consistent with a methyl imbalance in mitochondria functionality.

Summary Comments

It is well recognized that mitochondrial dysfunction is a crucial event for neurodegeneration. Mitochondria are known to play a central role in many cell functions including ATP generation, intracellular Ca^{2+} homeostasis, ROS formation as by-products of oxidative phosphorylation, and apoptosis. Neurons are particularly dependent on mitochondria because of their high-energy demands, and it is likely that neurons are intolerant of mitochondrial dysfunction.

Altered mitochondrial activity and oxidative stress have long been associated with DS. DS cortical neurons exhibit higher production of ROS and lipid peroxidation; both DS neurons and astrocytes display an abnormal pattern of protein processing consistent with chronic energy deficits; altered mitochondrial activity has been reported in DS fibroblasts and mitochondrial DNA mutations were found in DS brain tissue.

This scenario is further complicated by the fact that a number of proteins that cause neurodegeneration, including Aβ, interact with mitochondria or affect mitochondrial function. Taken together, DS neurons are threatened by increased ROS levels, decreased mitochondrial functionality, decreased ATP production, and increased Aβ load. All these events may participate in a self-sustaining vicious cycle thus ultimately making neurons highly susceptible to death.

Conclusions and Future Perspectives

The findings highlighted in this review that trisomy affects gene/protein expression of trisomic, as well as some disomic genes, ultimately leading to increased OS and impaired mitochondrial function. These alterations are early events in DS pathophysiological development as demonstrated by studies performed on fetal brain and amniotic fluid. OS conditions are caused, and further amplified, not only from overexpression of some of the genes encoded by Chr21 but also in the presence of reduced antioxidant system efficiency. Redox homeostasis also is disturbed by overproduction of Aβ, which accumulates into plaques as a function of age in DS as well as in AD.

Indeed, Aβ neurotoxicity contributes to neuronal loss and cognitive dysfunctions observed both in DS and in AD and amplifies oxidative damage to macromolecules. OS is detrimental because it modifies several intracellular regulatory pathways such as cell growth/death, gene expression, protein function, energy production, among many others.

That OS contributes to neurodegeneration is well established, but in the case of DS and AD, genetic similarities, due to the fact that some of the genes responsible for familial form of AD are encoded by Chr21, provide the basis to gain insights into specific dysregulated pathways. A better understanding of such molecular events would allow testing the possibility of using antioxidant nutrients to prevent/slow free radical-mediated damage and possibly modulate some of the complications of DS. Unfortunately, a very recent paper by Lott et al. [85] showed that a 2-year randomized, double-blind, placebo-controlled trial with daily oral antioxidant supplementation (900 IU of alpha-tocopherol, 200 mg of ascorbic acid, and 600 mg of alpha-lipoic acid), though safe and tolerable for individuals with DS and dementia, did not result in improvement in cognitive functioning nor a stabilization of cognitive decline compared with control group.

These results are not in agreement with those obtained by Lockrow et al. [86], who observed that a long-term supplementation with vitamin E effectively reduced the levels of ROS in the adult Ts65Dn brain. Treated mice also exhibited improved performance on a spatial working memory task and showed an attenuation of cholinergic neuron pathology in the basal forebrain.

These contrasting results likely reflect the "biological discrepancy" between human and animal studies. Transgenic mice, though useful models to study the molecular basis of a disease and test the efficacy of drug treatment, do not show all the features of human disease. More importantly, when testing the protective effects of antioxidants, antioxidants supplementation should be administered at the very early stage of the disease, before persistent oxidative damage occurs. Indeed, many individuals with AD have significant AD pathology about 20 years prior to presentation of prominent symptoms and subsequent disease diagnosis.

An intriguing field or research for the development of novel therapeutic strategies in DS is the putative compensatory mechanisms that may be activated in the presence of genetic instability in the DS population. In fact, overexpression of several genes on Chr21, including APP, DYRK1A, SOD1, and RCAN1, in addition to impairing neuronal growth and synapse maintenance, may participate to the activation of compensatory pathways during aging. DS may be considered a kind of prodromal AD, and better understanding of common as well as different pathways of neurodegeneration occurring in both these neuropathologies is urgently needed. Studies on human and animal models of both DS and AD are currently ongoing in our laboratory.

References

1. Hassold T, Hunt P. To err (meiotically) is human: the genesis of human aneuploidy. Nat Rev Genet. 2001;2:280–91.
2. Dierssen M, Herault Y, Estivill X. Aneuploidy: from a physiological mechanism of variance to Down syndrome. Physiol Rev. 2009;89:887–920.
3. Antonarakis SE, Lyle R, Chrast R, Scott HS. Differential gene expression studies to explore the molecular pathophysiology of Down syndrome. Brain Res Brain Res Rev. 2001;36:265–74.
4. Antonarakis SE, Lyle R, Dermitzakis ET, Reymond A, Deutsch S. Chromosome 21 and Down syndrome: from genomics to pathophysiology. Nat Rev Genet. 2004;5:725–38.
5. Iannello RC, Crack PJ, de Haan JB, Kola I. Oxidative stress and neural dysfunction in Down syndrome. J Neural Transm Suppl. 1999;57:257–67.
6. Shapiro BL. Down syndrome – a disruption of homeostasis. Am J Med Genet. 1983;14:241–69.
7. Roizen NJ, Patterson D. Down's syndrome. Lancet. 2003;361:1281–9.
8. Bittles AH, Glasson EJ. Clinical, social, and ethical implications of changing life expectancy in Down syndrome. Dev Med Child Neurol. 2004;46:282–6.
9. Antonarakis SE, Epstein CJ. The challenge of Down syndrome. Trends Mol Med. 2006;12:473–9.
10. Delabar JM, Aflalo-Rattenbac R, Creau N. Developmental defects in trisomy 21 and mouse models. ScientificWorldJournal. 2006;6:1945–64.

11. Devenny DA, Krinsky-McHale SJ, Sersen G, Silverman WP. Sequence of cognitive decline in dementia in adults with Down's syndrome. J Intellect Disabil Res. 2000;44(Pt 6):654–65.
12. Devenny DA, Wegiel J, Schupf N, Jenkins E, Zigman W, Krinsky-McHale SJ, Silverman WP. Dementia of the Alzheimer's type and accelerated aging in Down syndrome. Sci Aging Knowledge Environ. 2005;2005:dn1.
13. Carlesimo GA, Marotta L, Vicari S. Long-term memory in mental retardation: evidence for a specific impairment in subjects with Down's syndrome. Neuropsychologia. 1997;35:71–9.
14. Bush A, Beail N. Risk factors for dementia in people with Down syndrome: issues in assessment and diagnosis. Am J Ment Retard. 2004;109:83–97.
15. Zana M, Janka Z, Kalman J. Oxidative stress: a bridge between Down's syndrome and Alzheimer's disease. Neurobiol Aging. 2007;28:648–76.
16. Capone GT. Down syndrome: advances in molecular biology and the neurosciences. J Dev Behav Pediatr. 2001;22:40–59.
17. Krasuski JS, Alexander GE, Horwitz B, Rapoport SI, Schapiro MB. Relation of medial temporal lobe volumes to age and memory function in nondemented adults with Down's syndrome: implications for the prodromal phase of Alzheimer's disease. Am J Psychiatry. 2002;159:74–81.
18. Perluigi M, Butterfield DA. Oxidative stress and Down syndrome: a route toward Alzheimer-like dementia. Curr Gerontol Geriatr Res. 2012;2012:724904.
19. Lott IT. Antioxidants in Down syndrome. Biochim Biophys Acta. 2012;1822:657–63.
20. Reynolds GP, Cutts AJ. Free radical damage in Down's syndrome brain. Biochem Soc Trans. 1993;21:221S.
21. Brooksbank BW, Martinez M, Balazs R. Altered composition of polyunsaturated fatty acyl groups in phosphoglycerides of Down's syndrome fetal brain. J Neurochem. 1985;44:869–74.
22. Benzi G, Moretti A. Are reactive oxygen species involved in Alzheimer's disease? Neurobiol Aging. 1995;16:661–74.
23. de Haan JB, Cristiano F, Iannello RC, Kola I. Cu/Zn-superoxide dismutase and glutathione peroxidase during aging. Biochem Mol Biol Int. 1995;35:1281–97.
24. Shin JH, London J, Le Pecheur M, Hoger H, Pollak D, Lubec G. Aberrant neuronal and mitochondrial proteins in hippocampus of transgenic mice overexpressing human Cu/Zn superoxide dismutase 1. Free Radic Biol Med. 2004;37:643–53.
25. Busciglio J, Yankner BA. Apoptosis and increased generation of reactive oxygen species in Down's syndrome neurons in vitro. Nature. 1995;378:776–9.
26. Gulesserian T, Engidawork E, Fountoulakis M, Lubec G. Antioxidant proteins in fetal brain: superoxide dismutase-1 (SOD-1) protein is not overexpressed in fetal Down syndrome. J Neural Transm Suppl. 2001;61:71–84.
27. Pratico D, Iuliano L, Amerio G, Tang LX, Rokach J, Sabatino G, Violi F. Down's syndrome is associated with increased 8,12-iso-iPF2alpha-VI levels: evidence for enhanced lipid peroxidation in vivo. Ann Neurol. 2000;48:795–8.
28. Butterfield DA, Drake J, Pocernich C, Castegna A. Evidence of oxidative damage in Alzheimer's disease brain: central role for amyloid beta-peptide. Trends Mol Med. 2001;7: 548–54.
29. Butterfield DA, Galvan V, Lange MB, Tang H, Sowell RA, Spilman P, Fombonne J, Gorostiza O, Zhang J, Sultana R, Bredesen DE. In vivo oxidative stress in brain of Alzheimer disease transgenic mice: requirement for methionine 35 in amyloid beta-peptide of APP. Free Radic Biol Med. 2010;48:136–44.
30. Head E, Lott IT. Down syndrome and beta-amyloid deposition. Curr Opin Neurol. 2004;17: 95–100.
31. Mehta PD, Capone G, Jewell A, Freedland RL. Increased amyloid beta protein levels in children and adolescents with Down syndrome. J Neurol Sci. 2007;254:22–7.
32. Schupf N, Zigman WB, Tang MX, Pang D, Mayeux R, Mehta P, Silverman W. Change in plasma Ass peptides and onset of dementia in adults with Down syndrome. Neurology. 2010; 75:1639–44.

33. Anandatheerthavarada HK, Biswas G, Robin MA, Avadhani NG. Mitochondrial targeting and a novel transmembrane arrest of Alzheimer's amyloid precursor protein impairs mitochondrial function in neuronal cells. J Cell Biol. 2003;161:41–54.
34. Simon AM, Schiapparelli L, Salazar-Colocho P, Cuadrado-Tejedor M, Escribano L, Lopez de Maturana R, Del Rio J, Perez-Mediavilla A, Frechilla D. Overexpression of wild-type human APP in mice causes cognitive deficits and pathological features unrelated to Abeta levels. Neurobiol Dis. 2009;33:369–78.
35. Azkona G, Levannon D, Groner Y, Dierssen M. In vivo effects of APP are not exacerbated by BACE2 co-overexpression: behavioural characterization of a double transgenic mouse model. Amino Acids. 2010;39:1571–80.
36. Griffin WS, Sheng JG, McKenzie JE, Royston MC, Gentleman SM, Brumback RA, Cork LC, Del Bigio MR, Roberts GW, Mrak RE. Life-long overexpression of S100beta in Down's syndrome: implications for Alzheimer pathogenesis. Neurobiol Aging. 1998;19:401–5.
37. Royston MC, McKenzie JE, Gentleman SM, Sheng JG, Mann DM, Griffin WS, Mrak RE. Overexpression of s100beta in Down's syndrome: correlation with patient age and with beta-amyloid deposition. Neuropathol Appl Neurobiol. 1999;25:387–93.
38. Van Eldik LJ, Griffin WS. S100 beta expression in Alzheimer's disease: relation to neuropathology in brain regions. Biochim Biophys Acta. 1994;1223:398–403.
39. Esposito G, De Filippis D, Cirillo C, Sarnelli G, Cuomo R, Iuvone T. The astroglial-derived S100beta protein stimulates the expression of nitric oxide synthase in rodent macrophages through p38 MAP kinase activation. Life Sci. 2006;78:2707–15.
40. Lott IT, Head E, Doran E, Busciglio J. Beta-amyloid, oxidative stress and Down syndrome. Curr Alzheimer Res. 2006;3:521–8.
41. Sanij E, Hatzistavrou T, Hertzog P, Kola I, Wolvetang EJ. Ets-2 is induced by oxidative stress and sensitizes cells to H(2)O(2)-induced apoptosis: implications for Down's syndrome. Biochem Biophys Res Commun. 2001;287:1003–8.
42. Wolvetang EJ, Wilson TJ, Sanij E, Busciglio J, Hatzistavrou T, Seth A, Hertzog PJ, Kola I. ETS2 overexpression in transgenic models and in Down syndrome predisposes to apoptosis via the p53 pathway. Hum Mol Genet. 2003;12:247–55.
43. Balcz B, Kirchner L, Cairns N, Fountoulakis M, Lubec G. Increased brain protein levels of carbonyl reductase and alcohol dehydrogenase in Down syndrome and Alzheimer's disease. J Neural Transm Suppl. 2001;61:193–201.
44. Reed T, Perluigi M, Sultana R, Pierce WM, Klein JB, Turner DM, Coccia R, Markesbery WR, Butterfield DA. Redox proteomic identification of 4-hydroxy-2-nonenal-modified brain proteins in amnestic mild cognitive impairment: insight into the role of lipid peroxidation in the progression and pathogenesis of Alzheimer's disease. Neurobiol Dis. 2008;30:107–20.
45. Munke M, Kraus JP, Ohura T, Francke U. The gene for cystathionine beta-synthase (CBS) maps to the subtelomeric region on human chromosome 21q and to proximal mouse chromosome 17. Am J Hum Genet. 1988;42:550–9.
46. Hensley K, Venkova K, Christov A. Emerging biological importance of central nervous system lanthionines. Molecules. 2010;15:5581–94.
47. Whiteman M, Armstrong JS, Chu SH, Jia-Ling S, Wong BS, Cheung NS, Halliwell B, Moore PK. The novel neuromodulator hydrogen sulfide: an endogenous peroxynitrite 'scavenger'? J Neurochem. 2004;90:765–8.
48. Ichinohe A, Kanaumi T, Takashima S, Enokido Y, Nagai Y, Kimura H. Cystathionine beta-synthase is enriched in the brains of Down's patients. Biochem Biophys Res Commun. 2005;338:1547–50.
49. Kamoun P, Belardinelli MC, Chabli A, Lallouchi K, Chadefaux-Vekemans B. Endogenous hydrogen sulfide overproduction in Down syndrome. Am J Med Genet A. 2003;116A:310–1.
50. Beyer K, Lao JI, Carrato C, Rodriguez-Vila A, Latorre P, Mataro M, Llopis MA, Mate JL, Ariza A. Cystathionine beta synthase as a risk factor for Alzheimer disease. Curr Alzheimer Res. 2004;1:127–33.
51. Odetti P, Angelini G, Dapino D, Zaccheo D, Garibaldi S, Dagna-Bricarelli F, Piombo G, Perry G, Smith M, Traverso N, Tabaton M. Early glycoxidation damage in brains from Down's syndrome. Biochem Biophys Res Commun. 1998;243:849–51.

52. Nunomura A, Perry G, Pappolla MA, Friedland RP, Hirai K, Chiba S, Smith MA. Neuronal oxidative stress precedes amyloid-beta deposition in Down syndrome. J Neuropathol Exp Neurol. 2000;59:1011–7.
53. Campos C, Guzman R, Lopez-Fernandez E, Casado A. Evaluation of urinary biomarkers of oxidative/nitrosative stress in adolescents and adults with Down syndrome. Biochim Biophys Acta. 2011;1812:760–8.
54. Tolun AA, Scarbrough PM, Zhang H, McKillop JA, Wang F, Kishnani PS, Millington DS, Young SP, Il'yasova D. Systemic oxidative stress, as measured by urinary allantoin and F(2)-isoprostanes, is not increased in Down syndrome. Ann Epidemiol. 2012;22:892–4.
55. Liu C, Belichenko PV, Zhang L, Fu D, Kleschevnikov AM, Baldini A, Antonarakis SE, Mobley WC, Yu YE. Mouse models for Down syndrome-associated developmental cognitive disabilities. Dev Neurosci. 2011;33:404–13.
56. Olson LE, Roper RJ, Baxter LL, Carlson EJ, Epstein CJ, Reeves RH. Down syndrome mouse models Ts65Dn, Ts1Cje, and Ms1Cje/Ts65Dn exhibit variable severity of cerebellar phenotypes. Dev Dyn. 2004;230:581–9.
57. Sago H, Carlson EJ, Smith DJ, Kilbridge J, Rubin EM, Mobley WC, Epstein CJ, Huang TT. Ts1Cje, a partial trisomy 16 mouse model for Down syndrome, exhibits learning and behavioral abnormalities. Proc Natl Acad Sci U S A. 1998;95:6256–61.
58. Ishihara K, Amano K, Takaki E, Ebrahim AS, Shimohata A, Shibazaki N, Inoue I, Takaki M, Ueda Y, Sago H, Epstein CJ, Yamakawa K. Increased lipid peroxidation in Down's syndrome mouse models. J Neurochem. 2009;110:1965–76.
59. Butterfield DA, Bader Lange ML, Sultana R. Involvements of the lipid peroxidation product, HNE, in the pathogenesis and progression of Alzheimer's disease. Biochim Biophys Acta. 2010;1801:924–9.
60. Perluigi M, di Domenico F, Fiorini A, Cocciolo A, Giorgi A, Foppoli C, Butterfield DA, Giorlandino M, Giorlandino C, Schinina ME, Coccia R. Oxidative stress occurs early in Down syndrome pregnancy: a redox proteomics analysis of amniotic fluid. Proteomics Clin Appl. 2011;5:167–78.
61. Perluigi M, Sultana R, Cenini G, Di Domenico F, Memo M, Pierce WM, Coccia R, Butterfield DA. Redox proteomics identification of 4-hydroxynonenal-modified brain proteins in Alzheimer's disease: role of lipid peroxidation in Alzheimer's disease pathogenesis. Proteomics Clin Appl. 2009;3:682–93.
62. Sultana R, Perluigi M, Allan Butterfield D. Lipid peroxidation triggers neurodegeneration: a redox proteomics view into the Alzheimer disease brain. Free Radic Biol Med. 2013;62: 157–69.
63. Underwood MA, Gilbert WM, Sherman MP. Amniotic fluid: not just fetal urine anymore. J Perinatol. 2005;25:341–8.
64. Kolialexi A, Tounta G, Mavrou A, Tsangaris GT. Proteomic analysis of amniotic fluid for the diagnosis of fetal aneuploidies. Expert Rev Proteomics. 2011;8:175–85.
65. Head E, Lott IT, Patterson D, Doran E, Haier RJ. Possible compensatory events in adult Down syndrome brain prior to the development of Alzheimer disease neuropathology: targets for nonpharmacological intervention. J Alzheimers Dis. 2007;11:61–76.
66. Head E, Doran E, Nistor M, Hill M, Schmitt FA, Haier RJ, Lott IT. Plasma amyloid-beta as a function of age, level of intellectual disability, and presence of dementia in Down syndrome. J Alzheimers Dis. 2011;23:399–409.
67. Nistor M, Don M, Parekh M, Sarsoza F, Goodus M, Lopez GE, Kawas C, Leverenz J, Doran E, Lott IT, Hill M, Head E. Alpha- and beta-secretase activity as a function of age and beta-amyloid in Down syndrome and normal brain. Neurobiol Aging. 2007;28:1493–506.
68. Head E, Azizeh BY, Lott IT, Tenner AJ, Cotman CW, Cribbs DH. Complement association with neurons and beta-amyloid deposition in the brains of aged individuals with Down syndrome. Neurobiol Dis. 2001;8:252–65.
69. Cataldo AM, Petanceska S, Peterhoff CM, Terio NB, Epstein CJ, Villar A, Carlson EJ, Staufenbiel M, Nixon RA. App gene dosage modulates endosomal abnormalities of Alzheimer's disease in a segmental trisomy 16 mouse model of Down syndrome. J Neurosci. 2003;23:6788–92.

70. Nunomura A, Perry G, Aliev G, Hirai K, Takeda A, Balraj EK, Jones PK, Ghanbari H, Wataya T, Shimohama S, Chiba S, Atwood CS, Petersen RB, Smith MA. Oxidative damage is the earliest event in Alzheimer disease. J Neuropathol Exp Neurol. 2001;60:759–67.
71. Cenini G, Dowling ALS, Beckett TL, Barone E, Mancuso, C, Murphy MP, LeVine H II, Lott IT, Schmitt FA, Butterfield DA, Head E. Association between frontal cortex oxidative damage and beta-amyloid as a function of age in Down Syndrome. Biochim Biophys Acta. 2012; 1822;130–138.
72. Perrone S, Longini M, Bellieni CV, Centini G, Kenanidis A, De Marco L, Petraglia F, Buonocore G. Early oxidative stress in amniotic fluid of pregnancies with Down syndrome. Clin Biochem. 2007;40:177–80.
73. Zana M, Szecsenyi A, Czibula A, Bjelik A, Juhasz A, Rimanoczy A, Szabo K, Vetro A, Szucs P, Varkonyi A, Pakaski M, Boda K, Rasko I, Janka Z, Kalman J. Age-dependent oxidative stress-induced DNA damage in Down's lymphocytes. Biochem Biophys Res Commun. 2006;345:726–33.
74. Pallardo FV, Degan P, d'Ischia M, Kelly FJ, Zatterale A, Calzone R, Castello G, Fernandez-Delgado R, Dunster C, Lloret A, Manini P, Pisanti MA, Vuttariello E, Pagano G. Multiple evidence for an early age pro-oxidant state in Down Syndrome patients. Biogerontology. 2006;7:211–20.
75. Bersu ET, Ahmad FJ, Schwei MJ, Baas PW. Cytoplasmic abnormalities in cultured cerebellar neurons from the trisomy 16 mouse. Brain Res Dev Brain Res. 1998;109:115–20.
76. Prince J, Jia S, Bave U, Anneren G, Oreland L. Mitochondrial enzyme deficiencies in Down's syndrome. J Neural Transm Park Dis Dement Sect. 1994;8:171–81.
77. Howell N, Elson JL, Chinnery PF, Turnbull DM. mtDNA mutations and common neurodegenerative disorders. Trends Genet. 2005;21:583–6.
78. Druzhyna N, Nair RG, LeDoux SP, Wilson GL. Defective repair of oxidative damage in mitochondrial DNA in Down's syndrome. Mutat Res. 1998;409:81–9.
79. Schuchmann S, Heinemann U. Increased mitochondrial superoxide generation in neurons from trisomy 16 mice: a model of Down's syndrome. Free Radic Biol Med. 2000;28:235–50.
80. Valenti D, Manente GA, Moro L, Marra E, Vacca RA. Deficit of complex I activity in human skin fibroblasts with chromosome 21 trisomy and overproduction of reactive oxygen species by mitochondria: involvement of the cAMP/PKA signalling pathway. Biochem J. 2011; 435:679–88.
81. Li H, Dryhurst G. Oxidative metabolites of 5-S-cysteinyldopamine inhibit the pyruvate dehydrogenase complex. J Neural Transm. 2001;108:1363–74.
82. Conti A, Fabbrini F, D'Agostino P, Negri R, Greco D, Genesio R, D'Armiento M, Olla C, Paladini D, Zannini M, Nitsch L. Altered expression of mitochondrial and extracellular matrix genes in the heart of human fetuses with chromosome 21 trisomy. BMC Genomics. 2007;8:268.
83. Roat E, Prada N, Ferraresi R, Giovenzana C, Nasi M, Troiano L, Pinti M, Nemes E, Lugli E, Biagioni O, Mariotti M, Ciacci L, Consolo U, Balli F, Cossarizza A. Mitochondrial alterations and tendency to apoptosis in peripheral blood cells from children with Down syndrome. FEBS Lett. 2007;581:521–5.
84. Infantino V, Castegna A, Iacobazzi F, Spera I, Scala I, Andria G, Iacobazzi V. Impairment of methyl cycle affects mitochondrial methyl availability and glutathione level in Down's syndrome. Mol Genet Metab. 2011;102:378–82.
85. Lott IT, Doran E, Nguyen VQ, Tournay A, Head E, Gillen DL. Down syndrome and dementia: a randomized, controlled trial of antioxidant supplementation. Am J Med Genet A. 2011; 155A:1939–48.
86. Lockrow J, Prakasam A, Huang P, Bimonte-Nelson H, Sambamurti K, Granholm AC. Cholinergic degeneration and memory loss delayed by vitamin E in a Down syndrome mouse model. Exp Neurol. 2009;216:278–89.

Chapter 8
Retinopathy of Prematurity and Oxygen

Anna-Lena Hård, Ann Hellström and Lois Smith

Introduction

Retinopathy of prematurity (ROP) is a potentially blinding disease, the end stage of which was first described by Terry in 1942 [1]. During the 1940s, a dramatic increase in ROP-induced blindness occurred in highly developed countries [2], but the incidence decreased after the identification of oxygen supplementation as a risk factor. In recent years, improved neonatal care and increased survival of very immature babies has led to a new epidemic of severe ROP. This epidemic differs between countries and even between institutions in the same country [3]. With highly developed care, the most immature babies, i.e., those born before 28 gestational weeks, risk sight-threatening ROP, while with less advanced care and unmonitored oxygen use, more mature infants develop severe ROP. With poor or no care, immature infants die and ROP is not seen [4]. ROP is the result of early neonatal impairment of retinal vessel growth, which renders the peripheral retina avascular and hypoxic. Many of the risk factors associated with ROP are associated with excessive levels of reactive oxygen species (ROS), including hyperoxia, hypoxia, hyperglycemia, and disturbed metabolism. Infants who develop ROP have higher levels of markers of oxidative damage [5].

A.-L. Hård • A. Hellström
Section of Pediatric Ophthalmology, The Queen Silvia Children's Hospital,
The Sahlgrenska Academy at University of Gothenburg, S-416 85 Göteborg, Sweden

L. Smith, M.D., Ph.D. (✉)
Department of Ophthalmology, Children's Hospital Boston, Harvard Medical School,
300 Longwood Avenue, Boston, MA 02115, USA
e-mail: lois.smith@childrens.harvard.edu

P.A. Dennery et al. (eds.), *Perinatal and Prenatal Disorders*,
Oxidative Stress in Applied Basic Research and Clinical Practice,
DOI 10.1007/978-1-4939-1405-0_8

The Neonatal Period of the Very Preterm Infant

At birth, the very preterm infant leaves the maternal environment where oxygen, nutrients, hormones, and other factors are provided through the placenta where a close crosstalk regulates this transfer. After birth, the metabolic rate of the preterm infant becomes higher than that of a fetus, which is surrounded by its mother and has a metabolic rate similar to hers. In mammals, small size is associated with increased metabolism presumably to compensate for heat loss through a relatively larger surface area. Preterm infants undergo a prolonged adaptation to an extrauterine metabolism appropriate for size resulting in increased calorie demands compared to those of a fetus of the same age [6]. Oxygen is needed for metabolism and oxygen demands also increase after birth. Despite much research, there is still uncertainty about the requirements regarding both oxygen and nutrition of the very preterm neonate. The immaturity of vital organ systems makes the results of interventions hard to predict.

The first weeks of extrauterine life of very immature babies is characterized by varying oxygenation with frequent desaturation episodes [7], varying glycemia with hyperglycemic episodes [8], undernutrition, and growth retardation [9].

Normal Fetal Retinal Vascularization

Retinal vascularization starts around 12 weeks of gestation (WG) at the optic nerve [10]. The area immediately anterior to the vasculature is densely populated with astrocyte precursor cells, which first appear at the optic nerve in the inner retina at 8 weeks and extend to the retinal periphery at 28 weeks with reduced density by 32 weeks [11]. Astrocytes are intimately associated with the blood vessels during retinal vascularization which initially occurs by vasculogenesis, i.e., blood vessel formation from vascular precursor cells which are restricted to the central two thirds of the retina [10]. The expansion of the vascular network, to meet the needs for oxygen and nutrients of the developing retina, then proceeds through angiogenesis, i.e., budding from already formed vessels. Increased metabolic demands of the retina as it matures in a radial fashion result in a "physiological hypoxia" which induces expression of vascular endothelial growth factor (VEGF), which stimulates vascular growth [12]. In the retina, minimum levels of nutrient-dependent insulin-like growth factor-I (IGF-I) are required for VEGF-mediated blood vessel growth [13].

Retinopathy of Prematurity

ROP is considered a two-phase disease with an initial phase of vasoobliteration and impaired vessel growth followed by a second phase of pathologic neovascularization. This order of events is found in animal models of the disease where mice [14] and rats [15] are most commonly used. Human ROP has been classified into five stages: demarcation line (stage 1), ridge (stage 2), ridge with extraretinal neovascularization (stage 3), partial retinal detachment (stage 4), and total retinal detachment (stage 5). The latter proceeds to the end stage of ROP, retrolental fibroplasia (RLF) described by Terry [1], where the retina is dragged anteriorly and is visible behind the crystalline lens. In extremely preterm infants with ROP, the demarcation line and ridge usually appear around 2 months after birth, and in those with more severe disease, vasoproliferative stage 3 becomes visible with ophthalmoscopy some weeks later [16]. The two phases of the disease are not as evident in humans as in rodents, and it has not been defined whether ROP stage 1 and 2 belongs to phase I or II. Vascular leakage has been found in stage 2 ROP indicating neovascularization [17]. Perhaps, all ROP visible with ophthalmoscopy should be regarded as phase II ROP. Plus disease implies increased central vessel dilation and tortuosity and is an ominous sign of disease progression. Pre-plus also denotes vessel dilation and tortuosity but is less severe than plus disease. The extent of vascularization from the optic nerve head to the periphery is expressed with regard to zone where zone I is the most central and zone III most peripheral. More posterior (closer to the optic nerve) ROP implies a larger peripheral avascular area and greater risk of sight-threatening ROP. Aggressive posterior (AP) ROP is a form of the disease with posterior location, prominence of plus disease, ill-defined nature with regard to stage, and a tendency for rapid progression [18].

Ophthalmoscopically detectable ROP is first seen at around 30 weeks PMA [16]. At this age, the initial period of growth retardation, unstable oxygenation, and hyperglycemia is commonly followed by a period of more or less intense general catch-up growth.

Oxygen as Risk Factor for ROP

Oxygen supplementation is the earliest identified and best known risk factor for ROP; controlling oxygen reduces the incidence of ROP. Animal models of ROP are based on oxygen-induced retinopathy (OIR). However, there are other important risk factors, especially the degree of immaturity and factors related to metabolism and growth.

Optimum oxygen saturation for each postmenstrual age is unknown. Oxygen as a risk factor for ROP has more recently been studied in terms of higher or lower oxygen saturation targets in peripheral blood without measuring actual saturation or retinal oxygenation [19]. Much of our knowledge of the pathophysiology of the disease comes from cell cultures and animal studies with exposure of normal newborn animals to hyperoxia for a period when retinal vessel growth would normally occur but is impeded by hyperoxia exposure, followed by a return to room air of the now incompletely vascularized retina. Manipulations of genes, nutrition, and oxygen protocols in these animal models have increased our knowledge of ROP. According to its most simple description, ROP is a disease caused by initial retinal vessel growth retardation and loss, later hypoxia and pathologic neovascularization.

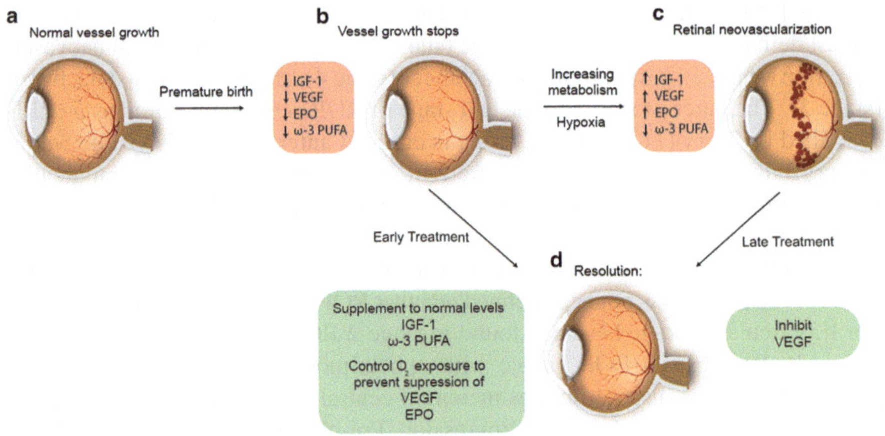

Fig. 8.1 Progression of retinopathy of prematurity. *A* In utero, in the infant, the oxygen tension is low. *B* After birth until ~30 weeks postmenstrual age, retinal vascularization is inhibited due to hyperoxia which suppresses production of VEGF and Epo. In addition after preterm birth, serum IGF-I levels fall due to loss of in utero sources and inability to produce IGF-I due to liver immaturity. Blood vessel growth stops, and as the retina matures and metabolic demand increases, there is resulting hypoxia (Phase I). *C* The hypoxic retina stimulates expression of oxygen-regulated factors VEGF and Epo, which stimulate retinal neovascularization. IGF-I levels rise slowly from low levels after preterm birth to level that VEGF activates MAPK and Akt (Phase II). *D* Resolution of retinopathy can be achieved by preventing phase I by limiting oxygen to prevent loss of VEGF and Epo (early treatment), or VEGF can be suppressed in phase II after neovascularization by suppressing VEGF with an antibody or by careful oxygen supplementation and/or by increasing IGF-I to in utero levels (Adapted from Smith [96], by permission of the Association for Research in Vision and Ophthalmology)

However, factors other than oxygen are involved because high oxygen levels are neither necessary nor sufficient to induce ROP since some infants who have never received oxygen treatment developed ROP and not all infants with long periods of oxygen supplementation develop the disease [20].

Attempts to prevent ROP have mainly focused on avoidance of hyperoxia, which is thought to stop vascular growth and destroy retinal vessels in the first phase. Avascular retina then leads to hypoxia with upregulation of vascular endothelial growth factor (VEGF) and uncontrolled vessel growth in the second phase (Fig. 8.1). During the first epidemic of ROP, uncontrolled use of pure oxygen made even rather mature babies blind. The ensuing oxygen restriction to fraction of inspired oxygen (FIO_2) of no more than 50 % was estimated to result in an excess of 16 deaths per case of blindness prevented [21]. More than half a century later, the question of the best oxygen supplementation practice is still not solved and we know very little about actual retinal oxygenation.

Measuring Oxygenation

Although sensitive retinal oximetry methods have been developed, their use in clinical studies of preterm infants has not been reported. In studies aiming to evaluate oxygen manipulation as prevention for ROP, different blood oxygen saturation SpO_2 target ranges have usually been compared. Blood oxygen saturation is measured using pulse oximeter, a technology from the 1990s, which provides continuous monitoring and targeting within a specified range. However, the accuracy of pulse oximetry technology varies.

Signal extraction technology (SET) as used in the Masimo SET pulse oximeter has been shown to reduce the number of false alarms and missed events during motion and low perfusion [22]. In one study, changes in clinical practice in combination with use of Masimo SET but not without it led to significant reduction in severe ROP, highlighting the importance of using the best technology [23]. Also, software characteristics can influence results [24]. Recent evidence suggests that oxygen delivery to preterm infants during ventilation may be systematically underestimated [25]. In the brain, increase in FiO_2 to assist recovery from desaturation may cause hyper-oxygenation for 5–15 min [26], which could also affect the retina.

Retinal Vessel Alternations in the First Phase of OIR and ROP

At very preterm birth, both vasculogenesis and angiogenesis are ongoing processes in the retina. Immature retinal vessels have increased susceptibility to hyperoxia during a sensitive period, which in mice is confined to the first few weeks of life.

Retinal endothelial cell death by apoptosis has been regarded as the major cause of vessel loss in hyperoxia [27, 28]. In a canine cell culture study, hyperoxia inhibited differentiation and proliferation of canine retinal angioblasts and proliferation and migration of endothelial cells thus probably affecting both angiogenesis and vasculogenesis. Apoptosis of endothelial cells was seen immediately after hyperoxia exposure but contributed less (18 %) to the decrease in endothelial cell number than inhibition of proliferation (75 %) [29] Fig. 8.1.

Animal Models

Rodents have commonly been used for studies of oxygen-induced retinopathy (OIR) in the neonatal period. In mice and rats, oxygen exposure inhibits retinal vascularization which normally takes place during the first weeks of life resulting in cessation of radial development of vessels and loss of central vascular beds in rats [30], mice [14] and dogs [31]. In the beagle dog study by McLeod, vasoconstriction after exposure to 100 % oxygen from 4 to 8 days of age resulted in arterial constriction from a diameter of 35.6 to 14.4 μm, vein constriction from 97.9 to 27.9 μm, and capillary constriction from 15.6 to 4.7 μm [31]. In addition, the area of vascularized retina after 4 days of continuous hyperoxia was reduced to 51 mm^2, 60 % less than that of control animals breathing air (127 mm^2). Retinal thickness was also reduced by 11 % in hyperoxia exposed puppies, and this reduction was mainly confined to the inner retinal layers. In the mouse model by Smith et al. [14], reduced perfusion of the fine central branching capillaries was found probably due to vasoconstriction. This perfusion defect was reversible initially but became persistent with time. However, in the mouse model, retinal vascularization normalizes after a period of pathologic vascularization and no retinal detachments occur.

Humans

In preterm infants with ROP, delayed retinal vessel growth is evident at ophthalmoscopy. With well-regulated oxygen supplementation, it has not been clear if vasoobliteration, as found in animal OIR, also occurs in humans.

Screening for ROP starts at the earliest at about 4 weeks of age or at a postmenstrual age (PMA) of 31 weeks [32] since ROP is seldom detected earlier. Thus, the first phase of ROP is rarely observed. Zepeda-Romero et al. [33] recently reported the results of fluorescein angiograms (FAs) taken at 2 weeks of age in ten infants with GA 26–32 weeks. Capillary loss and non-perfused areas were found in the two infants who later developed Type I ROP. Other early findings were arteriovenous shunts, which were surrounded by capillary non-perfusion, rosary-bead-like hyperfluorescence and tortuosity, and leakage from distal arterioles. In a study by Velia et al. [17] with serial FAs every other week from a PMA of 31 weeks, capillary bed loss in the junction between vascularized and avascular retina was common, and shunts and circumferential vessel formation was always close to the junction or to non-perfused areas. In another study, FA performed at the time of treatment for aggressive posterior ROP in infants with GA 24–25 weeks demonstrated areas of non-perfusion posterior to the demarcation line. Lack of capillary development in the vascularized retina was one explanation suggested by the authors [34]. Thus, areas of non-perfusion in the vascularized retina are a feature also of human ROP. Whether it is caused by vasoobliteration or lack of capillary development remains unknown. Interestingly, in a recent study from India, rather mature infants (GA 28–35 weeks) subjected to very high uncontrolled oxygen concentrations developed AP-ROP. In babies with immature vascularization in zone II or III at first examinations 2 days to 3.5 weeks after birth, loss of vascularized retina from zone II and even zone III to zone I was demonstrated. FA revealed large geographic areas of vasoobliteration posterior to the shunt vessels within otherwise vascularized retina. Thus, loss of already formed vessels occurs after excessive oxygen exposure also in human preterm infants [35].

Oxidative Stress in the First Phase of ROP

Retinal vessel loss in the first phase of ROP is generally attributed to hyperoxia-induced oxidative stress with formation of reactive oxygen species (ROS) and vasoobliteration. Reactive oxygen species are produced during normal cellular metabolism, and in low to moderate concentrations, they have beneficial effects in the defense against infectious agents and in the function of a number of cellular signaling systems [36]. In physiologic conditions, ROS are scavenged by intrinsic antioxidant defense mechanisms.

Excessive levels of ROS are produced in mitochondria in response to both hyperoxia [37] and hypoxia [38]. In high concentrations, ROS oxidize lipids, proteins, and polysaccharides and can damage DNA and RNA [39, 40]. Preterm infants have poor antioxidant defense systems and are especially vulnerable to oxidative stress [41]. At a postnatal age (PNA) of 4–6 weeks, higher levels of markers of oxidative damage were found in infants who developed ROP than in those who did not [5].

Retinal oxidative stress with production of ROS also occurs in diabetic retinopathy [42] which in many ways is similar to ROP, pathologic neovascularization subsequent to retinal hypoxia. Hyperglycemic episodes are common in preterm infants especially in those with BW < 1,000 g and hyperglycemia has been recognized as a risk factor for ROP [43, 44].

Oxidative Stress in the Second Phase of ROP

Mitochondrial reactive oxygen species are generated in hypoxia but less than in hyperoxia. However, the relationship between hypoxia and ROS production is controversial [45].

Oxygen Treatment in the First Phase of ROP

In the delivery room, resuscitation with 100 % oxygen has been the rule until randomized trials showed that for full-term infants, room air was more effective than pure oxygen [46–48]. In a study of preterm infants with GA ≤ 28 weeks, Escrig et al. compared the use of 30 and 90 % FiO_2 during delivery room resuscitation. Infants in the high oxygen group had a tendency to an increased incidence of ROP although not statistically significant ($p < 0.069$) [49]. In another randomized study comparing room air and 100 % oxygen for resuscitation of infants with GA <32 weeks, no difference in ROP incidence was found [50]. Thus, there are no conclusive results regarding the effect on ROP incidence of various fractions of inspired oxygen at resuscitation. Based on other variables, treatment recommendations for resuscitation of infants with GA <32 weeks were released in 2010 [51]: Neither room air nor pure oxygen was recommended but a mixture that resulted in a "normal" oxygen saturation as defined by Dawson [52] and measured using pulse oximetry.

A few observational studies of higher and lower SpO_2 targets and ROP have been published. Tin et al. found that babies with GA <28 weeks in neonatal intensive care units (NICU) targeting an oxygen saturation of 88–98 % for at least the first 8 weeks of life needed treatment for ROP four times as often as babies in NICUs targeting 70–90 % saturation [53]. Interestingly, babies in the high saturation target group were more likely to have a weight below the third centile at discharge. No difference in survival or cerebral palsy was found.

In a US national survey of infants with BW <1,500 g, NICUs with a SpO_2 max >98 % in the two first postnatal weeks reported ROP treatment in 5.5 % of infants vs. 3 % in units with saturation targets ≤98 % ($p<0.05$). With SpO_2targets after 2 weeks of age >92 %, 3.3 % of infants needed ROP treatment compared to 1.3 % with a SpO_2 target ≤92 % ($p<0.00001$). ROP ≥ stage 3 was found in 5.5 % when max SpO_2 was >92 % vs. 2.4 % when it was ≤92 % ($p<0.0005$) after 2 weeks of age [54].

Recently, SpO_2 target ranges of 85–89 % and 91–95 % have been compared in two large multicentre double-blind, randomized controlled studies: the Surfactant, Positive Pressure, and Pulse Oximetry Randomized Trial (SUPPORT) performed by the SUPPORT Study Group of Eunice Kennedy Shriver NICHD Neonatal Research Network and the multicenter Benefits of Oxygen saturation Targeting (BOOST)-II study. The latter was planned to include 1,200 infants in each of Australia and the UK and 320 in New Zealand [55]. In addition, a prospective meta-analysis including the SUPPORT, BOOST-II, and the Canadian Oxygen Trial (COT) (NCT00637169) was organized to get a large sample size ($n=5,230$) to study outcome of oxygen saturation targets in infants with GA <28 weeks, primarily in terms of mortality and major disability by 2 years of corrected age [55].

In the SUPPORT study, 1,316 infants born with GA between 24 and 27 + 6/7 weeks were included and the Masimo Radical Pulse Oximeter was used [56]. Infants in the low oxygen target group had increased mortality before discharge (19.9 % of infants vs. 16.2 %), but survivors had reduced incidence (8.6 % vs. 17.9 %) of severe ROP, i.e., type 1 ROP as defined in the ETROP study [57].

In the UK and Australian BOOST-II studies, infants were initially managed with the use of Masimo oximeters similar to those used in the SUPPORT trial, but in early 2009, all oximeters were fitted with a revised calibration algorithm with improved SpO_2 targeting. In December 2010, a joint safety analysis of survival at 36 weeks PMA was undertaken including 2,315 infants in the UK, Australian, and New Zealand BOOST-II trials and the 1,316 infants of the SUPPORT trial. Reduced survival was found among the infants assigned to the lower SpO_2 target (85–89 %) compared to those assigned to the higher target (91–95 %) with mortalities of 17.3 % vs. 14.4 % ($p=0.015$). For the 1,055 infants in the UK and Australian trials managed after the change of calibration algorithm, mortality differences were greater (21.8 % vs. 13.3 %; $p<0.001$) and the recruitment of infants in these trials was closed [19].

Thus, the optimal target range of oxygen saturation is still unknown. In addition, it might be that the need for oxygen is different at different developmental stages since retinal hyperoxia is thought to be a problem in the first phase of ROP and hypoxia in the second. Accordingly, Chen et al. performed a systematic review and meta-analysis of ten publications to report the association between severe ROP and high or low target oxygen saturation as measured by pulse oximetry. They found that both low oxygen saturation (70–96 %) in the first several postnatal weeks and high oxygen saturation (94–99 %) at ≥32 weeks PMA were associated with decreased risk for progression to severe ROP [58]. In addition to the degree of oxygenation, long duration of mechanical ventilation has been identified as a ROP risk factor in several studies [59].

Fluctuating Oxygenation

Other oxygen-related variables such as fluctuations in oxygen levels during the first few weeks of life have also been found to be associated with increased ROP risk [60–62] which is supported by the results of experimental studies [15, 63, 64]. In rat pups, clustered brief episodes of hypoxia produced more severe retinopathy than dispersed episodes [65]. Frequent intermittent episodes of hypoxia during the first 8 weeks of life have been associated with later ROP needing treatment. Using pulse oximetry with more frequent sampling than usually applied, Di Fiore et al. showed that for all infants in their study ($n=79$, GA 24 to 27+6/7 weeks), few hypoxic episodes occurred during the first week of life with a progressive increase in weeks 2–4 and a decrease in weeks 6–8. Interestingly, a higher incidence of hypoxemic events was associated with increased risk of severe ROP, while a higher incidence of hyperoxemic events was associated with reduced risk [66].

It might be that variations in oxygenation increase ROP risk more than hyperoxia per se and one might speculate that oxygenation variability is increased with higher rather than lower oxygen saturation targets contributing to higher ROP incidence.

Other Risk Factors in Relation to Oxygen

Even in the early 1950s, when oxygen treatment had been identified as an important risk factor for ROP, it was clear that it was not the only cause of the disease. The degree of immaturity is by far the most important risk factor and others such as poor weight gain, low serum IGF-I levels, lack of antioxidant defense, and low levels of omega 3 long chain polyunsaturated fatty acids are complexly interrelated with each other and with oxygen. Retinal vessel development is driven by "physiological hypoxia" as it progresses from central to peripheral retina [67]. Since the vasculature provides not only oxygen but also nutrients, hormones, and many other factors necessary for metabolism and growth, it seems likely that factors related to nutrition and metabolism also influence retinal vessel growth.

More than 50 years ago, the Swedish scientist Bo Hellström reported that undernutrition in mouse pups resulted not only in poor weight gain but also in a slightly reduced growth of retinal vessels. Exposure to 98–100 % oxygen almost completely stopped vessel outgrowth irrespective of the state of nutrition.

In newborn mice that were starving during 10 days of oxygen exposure and after return to room air for 10 days, 2 of 14 developed pathologic neovascularization, while all 12 pups that were undernourished during oxygen exposure but received a normal diet thereafter had extraretinal neovascularization [68]. Thus, poor neonatal growth followed by a later catch up, which is common in preterm infants, appeared to be associated with an increased risk of retinopathy. This phenomenon was also noted by Lubchenco, also more than 50 years ago [69]. Interestingly, in a rat OIR model, starvation affected the outgrowth of the vasculature more than the outgrowth of the neural retina, resulting in a smaller ratio of vascularized to total retina. In addition, rat pups in larger litters receiving less milk per pup developed more neovascularization than better nourished pups in a small litter. A smaller area of normal vascularization was associated with higher incidence and severity of abnormal neovascularization. This agrees with other reports suggesting that the size of the peripheral avascular zone determines the incidence of neovascularization in rats exposed to variable oxygen [15, 70]. Acidosis in the neonatal period results in acidosis-induced retinopathy (AIR) which is similar to OIR and likewise results in more neovascularization in undernourished than normally fed rat pups [71]. Recently, OIR in mice with poor weight gain showed a delayed onset and prolonged course of neovascularization and VEGF-expression [72].

Insulin-like growth factor-1 (IGF-1) is a nutrient-dependent growth factor, which is essential for growth and development during the third trimester when fetal serum levels normally increase two- to threefold [73]. After very preterm birth, with loss of the maternal to fetal supply of nutrients and the crosstalk between mother and fetus in the placenta, serum IGF-1 levels drop dramatically [74]. Liver IGF-1 synthesis is mainly regulated by energy and amino acid availability and low serum levels are a natural consequence of the insufficient nutrient intake of most very preterm babies. Low neonatal serum IGF-1 levels have been identified as an important risk factor for ROP and other morbidities in preterm infants [75]. In the retina, low serum IGF-1 suppresses VEGF-mediated vessel growth [13] reflecting the need for sufficient nutrients and IGF-1 for vascular growth.

Interestingly, hyperoxia causes a decrease in metabolism and oxygen consumption in cells and tissues [76, 77]. Few studies have examined an association between oxygenation and weight gain in preterm infants, but Tin et al. reported that weight centiles fell more between birth and discharge with higher (88–98 %) than lower (70–90 %) oxygen saturation targets [53]. Thus, both hyperoxia and poor nutrition may contribute to poor weight gain which is a strong risk factor for ROP [78, 79].

Possible Preventive Strategies to Reduce Oxidative Stress and ROP

In addition to optimizing nutrition and other factors important for proper growth, minimizing oxidative stress either by preventing the production of free radicals or improving antioxidant capacity is likely to reduce ROP.

Prevention of ROS

Oxygen

Attempts to prevent ROP through decreased oxygen saturation targets have been mentioned above. Clinical efforts to control oxygen fluctuations with rigorous oxygen control, and in one study, avoidance of hyperoxia and episodes of hypoxia-hyperoxia resulted in reduced incidence of proliferative ROP from 12.5 % in 1997 to 2.5 % in 2001 and reduction in the need for laser treatment from 4.5 to 0 % [80]. Thus, ROP prevention by better oxygen control appears possible even in settings with advanced care. In parts of the world where oxygen supplementation is unmonitored and more mature infants develop ROP, better oxygen control would certainly prevent some ROP [35].

Several reports indicate that alleviating the retinal hypoxia of phase II ROP by supplemental oxygen may reduce pathologic neovascularization [81–83]. The Supplemental Therapeutic Oxygen for Prethreshold Retinopathy of prematurity (STOP-ROP) [84] investigated whether severe ROP could be prevented by higher oxygen saturation levels to prevent the hypoxic stimulus to the avascular retina. Infants with oxygen saturation >94 % in room air were excluded and included in the High-Oxygen-Percentage Retinopathy of prematurity (HOPE-ROP) [85] study. In the STOP-ROP study, infants with prethreshold disease without plus disease progressed less often (32 % vs. 46 %) to threshold disease if receiving supplemental oxygen. Infants with >94 % oxygen saturation (HOPE-ROP) at prethreshold disease progressed less often to threshold than those in the STOP-ROP cohort for unknown reasons.

Metabolism

Avoiding hyperglycemia may reduce oxidative stress and ROP. Increasing early protein intake may be one intervention leading to less hyperglycemia [86].

Increasing Antioxidant Capacity

Omega 3 Polyunsaturated Fatty Acids (PUFAs)

In animal OIR, increased intake of fish oil rich in omega-3 PUFAs increases vessel regrowth after hyperoxic injury as well as directly inhibits pathological vessels [87].

Preterm infants with parenteral nutrition for at least 2 weeks after birth, who were administered SMOFlipid (Fresenius Kabi) containing medium-chain triglycerides (30 %), lipids from soy bean oil (30 %), olive oil (25 %), and fish oil (15 %) containing omega 3 polyunsaturated fatty acids DHA and EPA, had increased total antioxidant potential compared to infants receiving standard Intralipid 20 %

(Fresenius Kabi) based on soy bean oil. In a recent Polish study, infants with BW <1,250 g born before the 32nd week of gestation, needing total parenteral nutrition from day one, received a lipid emulsion consisting of equal parts of 20 % Clinoleic (soy bean and olive oil) (Baxter SA, Norfolk UK) and 10 % Omegaven (fish oil) (Fresenius Kabi AG) (Omegaven). A historic control group received Clinoleic only. The frequency of retinopathy in the two groups was similar, 32.5 vs. 36.3 %. However, regression of disease was more common in the Omegaven-treated group and fewer needed laser treatment than in the historical control cohort [88].

SOD

In a rat OIR model, retinal activities of superoxide dismutase (SOD), an important antioxidant, were reduced in pups raised in hyperoxia, compared to room air-raised controls, and supplementation with SOD resulted in increased SOD activity, reduced vaso-attenuation, and decreased avascular area [89].

Vitamins

Vitamin E, a free radical scavenger, was the first intervention studied for ROP prevention [90]. High doses of vitamin E were later found to prevent ROP and blindness but increased the risk of sepsis [91]. However, current enteral and parenteral nutrition options result in less vitamin E provided than recommended for very low birth weight infants [92].

Low vitamin A levels have been found in preterm infants, but no protective effect of vitamin A supplementation against ROP has been proven [93].

IGF-1

In cell culture, IGF-1 has been found to prevent oxidative stress-induced apoptosis [94] indicating that this might be one important mechanism for IGF-1 in preventing OIR [95] and ROP.

Conclusions

Oxygen is one of many risk factors for ROP and proper oxygenation levels are still not known. However, if current knowledge is applied, and oxygen closely monitored, and high oxygenation levels and episodes of hypoxia-hyperoxia avoided, the burden of ROP-induced blindness can be substantially reduced worldwide. To further reduce ROP incidence, other risk factors also have to be addressed such as lack of nutrition and growth factors. Research must also continue to identify and explore other unknown factors.

References

1. Terry TL. Fibroblastic overgrowth of persistent tunica vasculosa lentis in infants born prematurely: II. Report of cases-clinical aspects. Trans Am Ophthalmol Soc. 1942;40:262–84.
2. Campbell K. Intensive oxygen therapy as a possible cause of retrolental fibroplasia; a clinical approach. Med J Aust. 1951;2:48–50.
3. Zin AA, Moreira ME, Bunce C, Darlow BA, Gilbert CE. Retinopathy of prematurity in 7 neonatal units in Rio de Janeiro: screening criteria and workload implications. Pediatrics. 2010; 126:e410–7.
4. Gilbert C. Retinopathy of prematurity: a global perspective of the epidemics, population of babies at risk and implications for control. Early Hum Dev. 2008;84:77–82.
5. Ates O, Alp HH, Caner I, Yildirim A, Tastekin A, et al. Oxidative DNA damage in retinopathy of prematurity. Eur J Ophthalmol. 2009;19:80–5.
6. Singer D, Muhlfeld C. Perinatal adaptation in mammals: the impact of metabolic rate. Comp Biochem Physiol A Mol Integr Physiol. 2007;148:780–4.
7. Miller MJ, Martin RJ. Apnea of prematurity. Clin Perinatol. 1992;19:789–808.
8. Beardsall K, Vanhaesebrouck S, Ogilvy-Stuart AL, Vanhole C, Palmer CR, et al. Prevalence and determinants of hyperglycemia in very low birth weight infants: cohort analyses of the NIRTURE study. J Pediatr. 2010;157:715–19.e711–13.
9. Martin CR, Brown YF, Ehrenkranz RA, O'Shea TM, Allred EN, et al. Nutritional practices and growth velocity in the first month of life in extremely premature infants. Pediatrics. 2009; 124:649–57.
10. Chan-Ling T, McLeod DS, Hughes S, Baxter L, Chu Y, et al. Astrocyte-endothelial cell relationships during human retinal vascular development. Invest Ophthalmol Vis Sci. 2004;45: 2020–32.
11. Chu Y, Hughes S, Chan-Ling T. Differentiation and migration of astrocyte precursor cells and astrocytes in human fetal retina: relevance to optic nerve coloboma. FASEB J. 2001;15: 2013–5.
12. Stone J, Itin A, Alon T, Pe'er J, Gnessin H, et al. Development of retinal vasculature is mediated by hypoxia-induced vascular endothelial growth factor (VEGF) expression by neuroglia. J Neurosci. 1995;15:4738–47.
13. Hellstrom A, Perruzzi C, Ju M, Engstrom E, Hard AL, et al. Low IGF-I suppresses VEGF-survival signaling in retinal endothelial cells: direct correlation with clinical retinopathy of prematurity. Proc Natl Acad Sci U S A. 2001;98:5804–8.
14. Smith LE, Wesolowski E, McLellan A, Kostyk SK, D'Amato R, et al. Oxygen-induced retinopathy in the mouse. Invest Ophthalmol Vis Sci. 1994;35:101–11.
15. Penn JS, Tolman BL, Lowery LA. Variable oxygen exposure causes preretinal neovascularization in the newborn rat. Invest Ophthalmol Vis Sci. 1993;34:576–85.
16. Austeng D, Kallen KB, Hellstrom A, Tornqvist K, Holmstrom GE. Natural history of retinopathy of prematurity in infants born before 27 weeks' gestation in Sweden. Arch Ophthalmol. 2010;128:1289–94.
17. Velia P, Antonio B, Patrizia P, Carmen G, Anna M, et al. Fluorescein angiography and retinal vascular development in premature infants. J Matern Fetal Neonatal Med. 2012;25 Suppl 3:53–6.
18. International Committee for the Classification of Retinopathy of P. The International Classification of Retinopathy of Prematurity revisited. Arch Ophthalmol. 2005;123:991–9.
19. Stenson B, Brocklehurst P, Tarnow-Mordi W. Increased 36-week survival with high oxygen saturation target in extremely preterm infants. N Engl J Med. 2011;364:1680–2.
20. Lucey JF, Dangman B. A reexamination of the role of oxygen in retrolental fibroplasia. Pediatrics. 1984;73:82–96.
21. Bolton DP, Cross KW. Further observations on cost of preventing retrolental fibroplasia. Lancet. 1974;1:445–8.
22. Sahni R, Gupta A, Ohira-Kist K, Rosen TS. Motion resistant pulse oximetry in neonates. Arch Dis Child Fetal Neonatal Ed. 2003;88:F505–8.

23. Castillo A, Deulofeut R, Critz A, Sola A. Prevention of retinopathy of prematurity in preterm infants through changes in clinical practice and SpO technology. Acta Paediatr. 2011;100: 188–92.
24. Johnston ED, Boyle B, Juszczak E, King A, Brocklehurst P, et al. Oxygen targeting in preterm infants using the Masimo SET Radical pulse oximeter. Arch Dis Child Fetal Neonatal Ed. 2011;96:F429–33.
25. De Felice C, Bechelli S, Tonni G, Latini G, Hansmann G. Systematic underestimation of oxygen delivery in ventilated preterm infants. Neonatology. 2010;98:18–22.
26. Baerts W, Lemmers PM, van Bel F. Cerebral oxygenation and oxygen extraction in the preterm infant during desaturation: effects of increasing FiO(2) to assist recovery. Neonatology. 2011; 99:65–72.
27. Alon T, Hemo I, Itin A, Pe'er J, Stone J, et al. Vascular endothelial growth factor acts as a survival factor for newly formed retinal vessels and has implications for retinopathy of prematurity. Nat Med. 1995;1:1024–8.
28. Gu X, El-Remessy AB, Brooks SE, Al-Shabrawey M, Tsai NT, et al. Hyperoxia induces retinal vascular endothelial cell apoptosis through formation of peroxynitrite. Am J Physiol Cell Physiol. 2003;285:C546–54.
29. Uno K, Merges CA, Grebe R, Lutty GA, Prow TW. Hyperoxia inhibits several critical aspects of vascular development. Dev Dyn. 2007;236:981–90.
30. Penn JS, Thum LA, Naash MI. Oxygen-induced retinopathy in the rat. Vitamins C and E as potential therapies. Invest Ophthalmol Vis Sci. 1992;33:1836–45.
31. McLeod DS, Brownstein R, Lutty GA. Vaso-obliteration in the canine model of oxygen-induced retinopathy. Invest Ophthalmol Vis Sci. 1996;37:300–11.
32. Ophthalmology AAOPSo, American Academy Of O, American Association For Pediatric O, Strabismus, American Association Of Certified O. Screening examination of premature infants for retinopathy of prematurity. Pediatrics. 2013;131:189–95.
33. Zepeda-Romero LC, Oregon-Miranda AA, Lizarraga-Barron DS, Gutierrez-Camarena O, Meza-Anguiano A, et al. Early retinopathy of prematurity findings identified with fluorescein angiography. Graefes Arch Clin Exp Ophthalmol. 2013;251:2093–7.
34. Schulenburg WE, Tsanaktsidis G. Variations in the morphology of retinopathy of prematurity in extremely low birthweight infants. Br J Ophthalmol. 2004;88:1500–3.
35. Shah PK, Narendran V, Kalpana N. Aggressive posterior retinopathy of prematurity in large preterm babies in South India. Arch Dis Child Fetal Neonatal Ed. 2012;97:F371–5.
36. Valko M, Leibfritz D, Moncol J, Cronin MT, Mazur M, et al. Free radicals and antioxidants in normal physiological functions and human disease. Int J Biochem Cell Biol. 2007;39:44–84.
37. Li J, Gao X, Qian M, Eaton JW. Mitochondrial metabolism underlies hyperoxic cell damage. Free Radic Biol Med. 2004;36:1460–70.
38. Pearlstein DP, Ali MH, Mungai PT, Hynes KL, Gewertz BL, et al. Role of mitochondrial oxidant generation in endothelial cell responses to hypoxia. Arterioscler Thromb Vasc Biol. 2002;22:566–73.
39. Saugstad OD. Mechanisms of tissue injury by oxygen radicals: implications for neonatal disease. Acta Paediatr. 1996;85:1–4.
40. Sarker AH, Watanabe S, Seki S, Akiyama T, Okada S. Oxygen radical-induced single-strand DNA breaks and repair of the damage in a cell-free system. Mutat Res. 1995;337:85–95.
41. Sullivan JL, Newton RB. Serum antioxidant activity in neonates. Arch Dis Child. 1988; 63:748–50.
42. Yang Y, Hayden MR, Sowers S, Bagree SV, Sowers JR. Retinal redox stress and remodeling in cardiometabolic syndrome and diabetes. Oxid Med Cell Longev. 2010;3:392–403.
43. Garg R, Agthe AG, Donohue PK, Lehmann CU. Hyperglycemia and retinopathy of prematurity in very low birth weight infants. J Perinatol. 2003;23:186–94.
44. Ertl T, Gyarmati J, Gaal V, Szabo I. Relationship between hyperglycemia and retinopathy of prematurity in very low birth weight infants. Biol Neonate. 2006;89:56–9.
45. Chandel NS, Budinger GR. The cellular basis for diverse responses to oxygen. Free Radic Biol Med. 2007;42:165–74.

46. Davis PG, Tan A, O'Donnell CP, Schulze A. Resuscitation of newborn infants with 100 % oxygen or air: a systematic review and meta-analysis. Lancet. 2004;364:1329–33.
47. Rabi Y, Rabi D, Yee W. Room air resuscitation of the depressed newborn: a systematic review and meta-analysis. Resuscitation. 2007;72:353–63.
48. Saugstad OD, Ramji S, Soll RF, Vento M. Resuscitation of newborn infants with 21 % or 100 % oxygen: an updated systematic review and meta-analysis. Neonatology. 2008;94: 176–82.
49. Escrig R, Arruza L, Izquierdo I, Villar G, Saenz P, et al. Achievement of targeted saturation values in extremely low gestational age neonates resuscitated with low or high oxygen concentrations: a prospective, randomized trial. Pediatrics. 2008;121:875–81.
50. Wang CL, Anderson C, Leone TA, Rich W, Govindaswami B, et al. Resuscitation of preterm neonates by using room air or 100 % oxygen. Pediatrics. 2008;121:1083–9.
51. Kattwinkel J, Perlman JM, Aziz K, Colby C, Fairchild K, et al. Part 15: neonatal resuscitation: 2010 American Heart Association Guidelines for Cardiopulmonary Resuscitation and Emergency Cardiovascular Care. Circulation. 2010;122:S909–19.
52. Dawson JA, Kamlin CO, Vento M, Wong C, Cole TJ, et al. Defining the reference range for oxygen saturation for infants after birth. Pediatrics. 2010;125:e1340–7.
53. Tin W, Milligan DW, Pennefather P, Hey E. Pulse oximetry, severe retinopathy, and outcome at one year in babies of less than 28 weeks gestation. Arch Dis Child Fetal Neonatal Ed. 2001;84:F106–10.
54. Anderson CG, Benitz WE, Madan A. Retinopathy of prematurity and pulse oximetry: a national survey of recent practices. J Perinatol. 2004;24:164–8.
55. Askie LM, Brocklehurst P, Darlow BA, Finer N, Schmidt B, et al. NeOProM: Neonatal Oxygenation Prospective Meta-analysis Collaboration study protocol. BMC Pediatr. 2011; 11:6.
56. Carlo WA, Finer NN, Walsh MC, Rich W, Gantz MG, et al. Target ranges of oxygen saturation in extremely preterm infants. N Engl J Med. 2010;362:1959–69.
57. group Etfropc. Revised indications for the treatment of retinopathy of prematurity. Arch Ophthalmol. 2003;121:1684.
58. Chen ML, Guo L, Smith LE, Dammann CE, Dammann O. High or low oxygen saturation and severe retinopathy of prematurity: a meta-analysis. Pediatrics. 2010;125:e1483–92.
59. Incidence of and risk factors for neonatal morbidity after active perinatal care: extremely preterm infants study in Sweden (EXPRESS). Acta Paediatr. 2010;99: 978–92
60. Saito Y, Omoto T, Cho Y, Hatsukawa Y, Fujimura M, et al. The progression of retinopathy of prematurity and fluctuation in blood gas tension. Graefes Arch Clin Exp Ophthalmol. 1993;231:151–6.
61. Cunningham S, Fleck BW, Elton RA, McIntosh N. Transcutaneous oxygen levels in retinopathy of prematurity. Lancet. 1995;346:1464–5.
62. York JR, Landers S, Kirby RS, Arbogast PG, Penn JS. Arterial oxygen fluctuation and retinopathy of prematurity in very-low-birth-weight infants. J Perinatol. 2004;24:82–7.
63. Penn JS, Henry MM, Tolman BL. Exposure to alternating hypoxia and hyperoxia causes severe proliferative retinopathy in the newborn rat. Pediatr Res. 1994;36:724–31.
64. Penn JS, Henry MM, Wall PT, Tolman BL. The range of PaO2 variation determines the severity of oxygen-induced retinopathy in newborn rats. Invest Ophthalmol Vis Sci. 1995;36: 2063–70.
65. Coleman RJ, Beharry KD, Brock RS, Abad-Santos P, Abad-Santos M, et al. Effects of brief, clustered versus dispersed hypoxic episodes on systemic and ocular growth factors in a rat model of oxygen-induced retinopathy. Pediatr Res. 2008;64:50–5.
66. Di Fiore JM, Bloom JN, Orge F, Schutt A, Schluchter M, et al. A higher incidence of intermittent hypoxemic episodes is associated with severe retinopathy of prematurity. J Pediatr. 2010;157:69–73.
67. Chan-Ling T, Gock B, Stone J. The effect of oxygen on vasoformative cell division. Evidence that 'physiological hypoxia' is the stimulus for normal retinal vasculogenesis. Invest Ophthalmol Vis Sci. 1995;36:1201–14.

68. Hellstrom BE. Experimental approach to the pathogenesis of retrolental fibroplasia. V. The influence of the state of nutrition on oxygen-induced changes in the mouse eye. Acta Paediatr. 1956;45:43–57.
69. Lubchenco LO, Boyd E, Dressler MS. Growth of prematurely born infants with and without retrolental fibroplasia. AMA Am J Dis Child. 1953;86:466–8.
70. Reynaud X, Dorey CK. Extraretinal neovascularization induced by hypoxic episodes in the neonatal rat. Invest Ophthalmol Vis Sci. 1994;35:3169–77.
71. Zhang S, Leske DA, Lanier WL, Holmes JM. Postnatal growth retardation exacerbates acidosis-induced retinopathy in the neonatal rat. Curr Eye Res. 2001;22:133–9.
72. Stahl A, Chen J, Sapieha P, Seaward MR, Krah NM, et al. Postnatal weight gain modifies severity and functional outcome of oxygen-induced proliferative retinopathy. Am J Pathol. 2010;177:2715–23.
73. Lassarre C, Hardouin S, Daffos F, Forestier F, Frankenne F, et al. Serum insulin-like growth factors and insulin-like growth factor binding proteins in the human fetus. Relationships with growth in normal subjects and in subjects with intrauterine growth retardation. Pediatr Res. 1991;29:219–25.
74. Lineham JD, Smith RM, Dahlenburg GW, King RA, Haslam RR, et al. Circulating insulin-like growth factor I levels in newborn premature and full-term infants followed longitudinally. Early Hum Dev. 1986;13:37–46.
75. Hellstrom A, Engstrom E, Hard AL, Albertsson-Wikland K, Carlsson B, et al. Postnatal serum insulin-like growth factor I deficiency is associated with retinopathy of prematurity and other complications of premature birth. Pediatrics. 2003;112:1016–20.
76. Kimura RE, Thulin GE, Wender D, Warshaw JB. Decreased oxidative metabolism in neonatal rat lung exposed to hyperoxia. J Appl Physiol. 1983;55:1501–5.
77. Schoonen WG, Wanamarta AH, van der Klei-van Moorsel JM, Jakobs C, Joenje H. Respiratory failure and stimulation of glycolysis in Chinese hamster ovary cells exposed to normobaric hyperoxia. J Biol Chem. 1990;265:1118–24.
78. Hellstrom A, Ley D, Hansen-Pupp I, Niklasson A, Smith L, et al. New insights into the development of retinopathy of prematurity – importance of early weight gain. Acta Paediatr. 2010;99:502–8.
79. Hellstrom A, Hard AL, Engstrom E, Niklasson A, Andersson E, et al. Early weight gain predicts retinopathy in preterm infants: new, simple, efficient approach to screening. Pediatrics. 2009;123:e638–45.
80. Chow LC, Wright KW, Sola A. Can changes in clinical practice decrease the incidence of severe retinopathy of prematurity in very low birth weight infants? Pediatrics. 2003;111: 339–45.
81. Chan-Ling T, Gock B, Stone J. Supplemental oxygen therapy. Basis for noninvasive treatment of retinopathy of prematurity. Invest Ophthalmol Vis Sci. 1995;36:1215–30.
82. Zhang W, Yokota H, Xu Z, Narayanan SP, Yancey L, et al. Hyperoxia therapy of pre-proliferative ischemic retinopathy in a mouse model. Invest Ophthalmol Vis Sci. 2011;52:6384–95.
83. Gu X, Samuel S, El-Shabrawey M, Caldwell RB, Bartoli M, et al. Effects of sustained hyperoxia on revascularization in experimental retinopathy of prematurity. Invest Ophthalmol Vis Sci. 2002;43:496–502.
84. Supplemental Therapeutic Oxygen for Prethreshold Retinopathy Of Prematurity (STOP-ROP), a randomized, controlled trial. I: primary outcomes. Pediatrics. 2000;105:295–310.
85. McGregor ML, Bremer DL, Cole C, McClead RE, Phelps DL, et al. Retinopathy of prematurity outcome in infants with prethreshold retinopathy of prematurity and oxygen saturation >94 % in room air: the high oxygen percentage in retinopathy of prematurity study. Pediatrics. 2002;110:540–4.
86. Mahaveer A, Grime C, Morgan C. Increasing early protein intake is associated with a reduction in insulin-treated hyperglycemia in very preterm infants. Nutr Clin Pract. 2012; 27:399–405.

87. Connor KM, SanGiovanni JP, Lofqvist C, Aderman CM, Chen J, et al. Increased dietary intake of omega-3-polyunsaturated fatty acids reduces pathological retinal angiogenesis. Nat Med. 2007;13:868–73.
88. Pawlik D, Lauterbach R, Turyk E. Fish-oil fat emulsion supplementation may reduce the risk of severe retinopathy in VLBW infants. Pediatrics. 2011;127:223–8.
89. Niesman MR, Johnson KA, Penn JS. Therapeutic effect of liposomal superoxide dismutase in an animal model of retinopathy of prematurity. Neurochem Res. 1997;22:597–605.
90. Owens WC, Owens EU. Retrolental fibroplasia in premature infants; studies on the prophylaxis of the disease; the use of alpha tocopheryl acetate. Am J Ophthalmol. 1949;32:1631–7.
91. Brion LP, Bell EF, Raghuveer TS (2003) Vitamin E supplementation for prevention of morbidity and mortality in preterm infants. Cochrane Database Syst Rev. CD003665
92. Porcelli PJ, Weaver Jr RG. The influence of early postnatal nutrition on retinopathy of prematurity in extremely low birth weight infants. Early Hum Dev. 2010;86:391–6.
93. Tyson JE, Wright LL, Oh W, Kennedy KA, Mele L, et al. Vitamin A supplementation for extremely-low-birth-weight infants. National Institute of Child Health and Human Development Neonatal Research Network. N Engl J Med. 1999;340:1962–8.
94. Li Y, Shelat H, Geng YJ. IGF-1 prevents oxidative stress induced-apoptosis in induced pluripotent stem cells which is mediated by microRNA-1. Biochem Biophys Res Commun. 2012;426: 615–9.
95. Vanhaesebrouck S, Daniels H, Moons L, Vanhole C, Carmeliet P, et al. Oxygen-induced retinopathy in mice: amplification by neonatal IGF-I deficit and attenuation by IGF-I administration. Pediatr Res. 2009;65:307–10.
96. Smith LE. Invest Ophthalmol Vis Sci. 2008;49(12):5177–82.

Chapter 9
Necrotizing Enterocolitis and Oxidative Stress

Peter J. Giannone, Craig A. Nankervis, Michael R. Stenger, Brandon Schanbacher, and John Anthony Bauer

Introduction

Necrotizing enterocolitis (NEC) is one of the most common gastrointestinal conditions affecting newborn infants. Despite vigorous research the pathogenesis of NEC remains unclear. Oxidative stress has been shown to contribute to the development of other diseases involving premature infants such as bronchopulmonary dysplasia (BPD) and retinopathy of prematurity (ROP). More recently, oxidative stress has implicated in the pathogenesis of NEC. This chapter will briefly review the clinical presentation and management of NEC, discuss the potential role of oxidative stress in the pathogenesis of NEC, and finally review possible therapeutic targets in the hope of preventing NEC in high-risk infants.

Necrotizing Enterocolitis (NEC) and Prematurity

NEC is diagnosed in approximately 1–5 % of all neonatal intensive care (NICU) admissions with approximately 5,000–7,000 new cases of NEC occurring in the USA each year [1–3]. The risk of this condition is highest for the most prematurely born infants as approximately 5–10 % of all infants <1,500 g at birth and born prior

P.J. Giannone, M.D. (✉) • B. Schanbacher, MS • J.A. Bauer, Ph.D.
Division of Neonatology, Department of Pediatrics, Kentucky Children's Hospital, University of Kentucky College of Medicine, 800 Rose Street, Lexington, KY 40536, USA
e-mail: peter.giannone@uky.edu

C.A. Nankervis, M.D. • M.R. Stenger, M.D.
Department of Pediatrics, The Ohio State University College of Medicine and Nationwide Children's Hospital & Research Institute, Columbus, OH 43205, USA

P.A. Dennery et al. (eds.), *Perinatal and Prenatal Disorders*,
Oxidative Stress in Applied Basic Research and Clinical Practice,
DOI 10.1007/978-1-4939-1405-0_9

to 30 weeks gestational age develop NEC [4]. Although NEC can occur in near term and late preterm infants, low birth weight premature infants are the most commonly affected and represent 90–95 % of the cases. As medical care, in particular surfactant replacement therapy, has led to the increased survival of extremely premature infants, the incidence of NEC has been increasing [5].

The medical and economic impact of NEC remains substantial [6, 7]. The risk of death or lifelong disability and the long-term social and financial impact on the family and other healthcare resources make NEC a significant morbidity associated with preterm birth [6]. As the preterm birth rate increases worldwide and as the survivability of extremely premature infants is enhanced, it is likely that the medical significance of NEC will also rise over the next decades [7, 8]. Consequently, there is a clear need to further define this condition and to develop rational strategies to affect its incidence and/or progression.

Several preventative strategies have been developed over the years based on clinical and experimental data [1]. Cautious feeding regimens, the use of maternal breast milk, passive immunization, and the use of probiotics have each been shown to reduce the incidence of NEC in small studies [9]. As there is no medical treatment for NEC, finding novel preventative strategies is therefore important. Also, since NEC is primarily a disease of the premature newborn, until we are able to markedly decrease the incidence of preterm birth, NEC will likely remain a significant problem within this population.

Clinical Course, Medical Diagnosis, and Consequences of NEC

Key aspects of this medical condition in this high-risk patient population (e.g., the preterm infant) are the initially subtle symptoms, a likely chronic/progressive phase of intestinal pathology that may last days or weeks, and an often "suddenly apparent" intestinal crisis requiring emergent resection of the severely diseased portion of bowel. The bedside identification, diagnosis, and clinical care for preterm and other infants with NEC are therefore challenging, especially since the early signs and symptoms of NEC are often nonspecific in nature and are similar to the presenting symptoms of more common entities encountered in the neonatal intensive care setting [12]. The timing of presentation, typically in the first few weeks of life, is related to gestational age with less mature infants presenting later compared to more mature infants. Indeed, one of the most devastating and demoralizing aspects of NEC for families and clinicians alike is that often these small fragile infants develop NEC and become critically ill when they are over their acute cardiorespiratory problems associated with prematurity, are taking enteral feedings, and are starting to progress toward discharge [10].

The early signs and symptoms of NEC are often nonspecific and include apnea, bradycardia, lethargy, irritability, and temperature instability. Unfortunately, due to the nonspecific nature of the early signs of NEC, delays in the diagnosis of NEC are

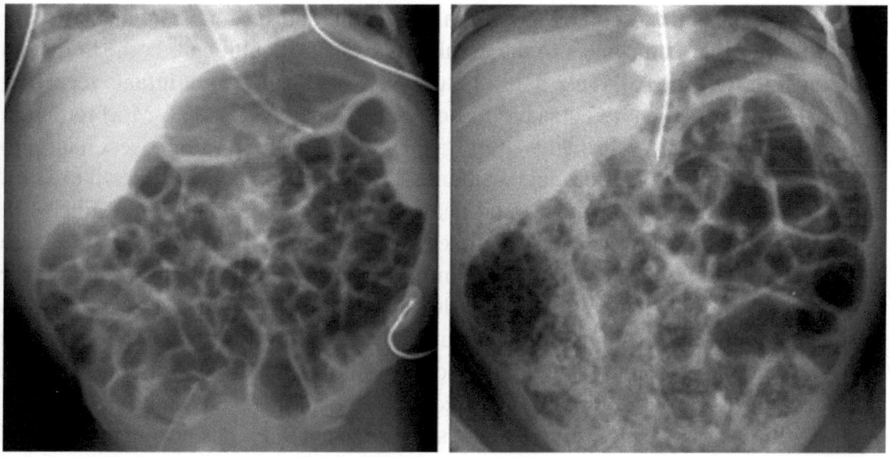

Fig. 9.1 Radiologic images of preterm infant abdomen showing a normal bowel (*left*) and a bowel showing large dilated loops (*right*)

not uncommon [3]. Abdominal findings include abdominal distension, abdominal tenderness, decreased bowel sounds, feeding intolerance, bile- or blood-stained emesis or gastric aspirates, bloody stools, and abdominal wall erythema. Laboratory findings include metabolic acidosis, thrombocytopenia, leukopenia, or leukocytosis with a relative increased number of immature cells, glucose instability, hyponatremia, coagulopathy, and elevated markers of inflammation (particularly C – reactive protein). Radiographic findings include abdominal distension, ileus, and pneumatosis intestinalis and, in more advanced stages, evidence of intestinal perforation(s).

In 1978, Bell and colleagues proposed a combination of clinical and radiographic criteria which could be used to describe the severity of NEC which are still useful to this day [11]. Stage I NEC designates infants with clinical symptoms of NEC, including bile stained gastric aspirates, bloody stools, and bowel stasis; this is considered "suspected NEC" [11]. Approximately one half of infants with stage I NEC advance to stage II NEC, also known as "medical NEC" which is confirmed by radiographic findings including ileus, bowel dilation, and intramural gas or pneumatosis intestinalis [11, 12]. Approximately 70 % of these infants advance to stage III NEC, or "surgical NEC" requiring surgical intervention. Shown in Fig. 9.1 are representative abdominal radiological images of preterm infants with relatively normal findings (left panel), versus evidence of intestinal air trapping and dilation (e.g., pneumatosis) (right panel); see figure legend for details.

The medical management of NEC is supportive and includes bowel rest, broad spectrum antibiotics, parenteral nutrition, and cardiorespiratory support as clinically indicated [13]. Surgical intervention is performed when there is evidence of bowel perforation and/or bowel necrosis [14]. Surgical treatment involves either resection of diseased bowel or peritoneal drainage. Unfortunately, both therapies are associated with very high morbidity and mortality [11, 12].

The clinical outcome of NEC is highly variable and has substantial morbidity and mortality that contributes to the escalating costs of health care both short term and longer term [15, 16]. Only half of extremely low birth weight infants receiving surgical resection for NEC survive to hospital discharge [12, 17]. Morbidities in survivors include poor growth, short gut syndrome, liver failure secondary to prolonged hyperalimentation, and strictures [18]. Development of intestinal strictures may be due to abnormal wound healing within the premature intestine [19]. Furthermore, neurodevelopmental outcomes in infants who develop NEC are concerning and may be related to the timing of surgical interventions required for bowel resection [6]. Among extremely low birth weight infants, surgical NEC is associated with significant growth delay and adverse neurodevelopmental outcomes at 18–22 months corrected age compared with age-matched controls [20]. Infants who develop NEC but do not go on to require surgical intervention do not seem to have additional risks, suggesting that the surgical event itself has lasting detrimental consequences in these vulnerable patients [6].

Mechanisms of NEC: Disease Initiation and/or Progression

Despite over 30 years of investigation, the pathogenesis of NEC remains enigmatic but is likely multifactorial in nature. Retrospective surveys have identified numerous risk factors for NEC, including catheterization of umbilical vessels [21], hypoxemia and hypotension [22], intrauterine growth retardation [23, 24], alterations in wound healing capabilities of the immature intestine [25], and prematurity [26]. Preterm birth remains the single most identifiable risk factor, especially for extremely low birth weight (ELBW) infants (born weighing < 1,000 g) or gestational age <30 weeks [16, 27]. Formula feeding has been implicated in the pathogenesis of NEC. Indeed, the exclusive use of breast milk for feeding premature infants is associated with a twofold reduction in NEC when compared to exclusive formula use. These two observations are consistent with the basic concept that the premature newborn has an immature and underdeveloped intestinal tract that is unprepared for the postnatal challenges of a mature and fully developed intestine.

Shown in Fig. 9.2 is a pathological illustration of the progressive nature of NEC with respect to intestinal tissue damage that occurs. Representative images of surgically resected small intestine from NEC cases (formalin fixed, paraffin embedded, and H/E stained sections of intestinal tissue) are presented, showing full-thickness intestinal cross sections. The panels shown include visibly normal intestinal morphology tissue (which was harvested from the boundary of the resected segment found to be uninvolved and visibly healthy), as well as sequential "grades" of intestinal injury, illustrating the pathological progression from modest injury to the mucosal villous layer to the severe disruption of the mucosa and complete loss of luminal integrity. A key aspect of the pathophysiological process in this complex tissue is the likely involvement of several cell types (epithelial, vascular, immune cells), the ultimate event of mucosal damage and destruction, and potentially

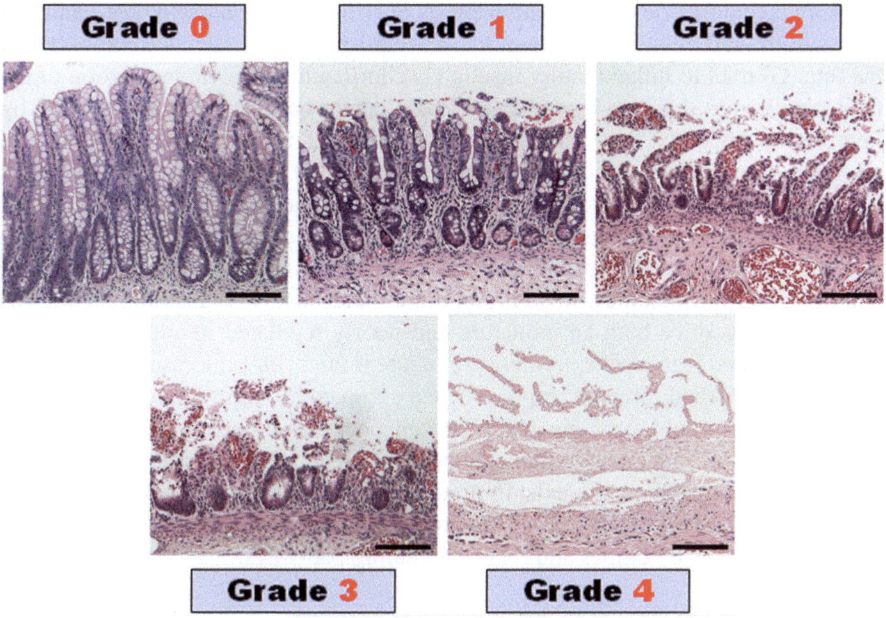

Fig. 9.2 Illustration of grading system for necrotizing enterocolitis. Grade 0 is comprised of normal healthy villi. Grade 1 shows evidence of cell sloughing from the villus tips. Grade 2 shows cell sloughing progressing midway down the villus. Grade 3 is characterized by loss of villi and some swelling within the submucosa. Grade 4 is transmural necrosis with total villi loss and separation of the mucosa from the submucosa

intestinal perforation due to severe and full-thickness necrosis. The histopathology of NEC is characterized by destruction of the mucosal layer in initial stages and by transmural necrosis of the intestinal wall in advanced stages of the disease [2, 28]. Therefore, it appears that the epithelial cells in the mucosa may be the first cells to undergo cell death, either by apoptosis or necrosis, prior to transmural necrosis [29]. Determining the mechanisms involved in early cell death of the mucosal layer may lead to possible therapeutic targets that prevent large sections of intestinal necrosis from requiring surgical intervention.

Early theories on the pathogenesis of NEC implicated blood flow limitations and ischemic damage to the premature intestinal tract as the primary insult. Data supporting this concept include animal models of vascular occlusion which create lesions similar to human NEC and other evidence that ischemia and hypoxic conditions are features of involved intestinal segments. Whether blood flow limitation leads to intestinal stasis and injury or whether intestinal distension can lead to blood flow limitation is uncertain. Others have suggested that the increase in metabolic demand that occurs following feeding, coupled with immature vascular reactivity and control of blood flow distribution, may lead to regions of hypoperfused intestine and thus promote ischemic injury.

More recent theories on the pathogenesis of NEC have implicated host-pathogen interactions and related inflammatory mechanisms. These include exposure of the fetal GI tract to inflammatory insults via chorioamnionitis (inflammation of the fetal membranes due to a bacterial infection), abnormal intestinal colonization by microbes postnatally that lead to mucosal damage, inflammatory responses of an innate immune system locally within the immature GI tract to these microbes particularly by Toll-like receptors, and eventually a large systemic inflammatory responses likely due to translocation of bacteria or signaling from the GI tract leading to cardiovascular collapse. These possible mechanisms are supported by both animal and clinical studies. In addition, these mechanisms are also rational in this clinical setting since both an immature and poorly regulated immune system is a well-established feature of extremely premature infants. In some cases, studies have shown a diminished immune response when compared to term infants or older age groups (which could promote inappropriate bacterial colonization), whereas in other studies an exaggerated inflammatory response to LPS and other bacterial constituents have been shown to exist in preterm infants, as well as a reduced capacity to limit an inflammatory response by appropriate braking mechanisms via release of anti-inflammatory cytokines and related signaling pathways.

A Role for Oxidative Stress

Recently, oxidative stress has been implicated in the pathogenesis of NEC and could represent a final common pathway leading to tissue due to a variety of inciting mechanisms. Tissue ischemia and related oxygen limitation is a well-recognized cause of redox imbalance and reactive oxygen intermediates. Likewise, abnormal or uncontrolled bacterial colonization leading to immune cell infiltration and activation or the exposure of intestinal epithelium to various cytokines and other factors which stimulate inflammatory gene expression are key processes known to cause reactive intermediates and oxidative stress in tissues. Factors in the newborn intestine modulating endothelial function [30, 31], cytokine and oxidant production [32, 33], infections [34], protective and toxic effects of nitric oxide (NO) [31, 35, 36], and apoptosis [29, 37] have all been postulated as contributing to the initiation and/or progression of NEC. Furthermore, interactions of epithelial cell injury and vascular endothelial alterations have been suggested as important factors involved in the pathogenesis of NEC, and epithelial barrier function may be related to coincident local changes in vascular integrity [38, 39]. Regardless of the upstream mechanisms, enterocyte death leading to an attendant breach of intestinal barrier function and integrity are the hallmark pathological outcomes. Whether the limitations are related to an underdeveloped epithelium with immature barrier function and poor restitution capabilities, an immature vasculature, and/or an immature enteric nervous system remains unclear.

NOS 2, Oxidants and Intestinal Cell Death

Endogenous NO production occurs via the metabolism of L-arginine catalyzed by a family of enzymes known as nitric oxide synthase (NOS). There are three isoforms of NOS: inducible NOS (iNOS or NOS2), endothelial NOS (eNOS), and neuronal NOS (nNOS) with iNOS being the isoform most commonly implicated in NO dysregulation [40–42]. Under physiologic conditions, relatively low concentrations of NO are produced in the intestine by eNOS and nNOS, where NO serves as a mediator of intestinal tone and contractility [43]. Additionally, when produced by iNOS at low levels, NO acts as an intercellular transduction mediator, [44] an intracellular effector of enzymes, and as a toxicant for immune defense. These divergent roles suggest that strict control of intra- and intercellular levels of NO are critical and NO dysregulation may have drastic consequences.

NO produced by eNOS is quickly dispersed into red blood cells and then converted into nitrate. NO produced by iNOS within the intestinal tissue cells cannot be converted readily into nitrate. Thus, during times of inflammation increased NO production can react with other molecules, such as superoxide, and produce various reactive nitrogen species (peroxynitrite and others). For example, during inflammation, superoxide anion destroys NO, reduces its efficacy as a signal transduction agent, and promotes the formation of peroxynitrite, a highly reactive nitrogen species known to nitrate protein tyrosine residues and cause cellular oxidative damage, including DNA strand breaks [45]. This nitration reaction produces a chemically stable oxidative modification of tyrosine residues in proteins (3-nitrotyrosine) and is somewhat selective. Thus, accumulation of 3-NT serves as a biomarker for reactive nitrogen species [46]. The reaction of NO with superoxide anion is extremely rapid and occurs at a diffusion-limited rate [47]. Excess or uncontrolled production of reactive nitrogen species can shift the actions of available NO from a useful cellular signaling molecule to a toxic free radical [48]. Giannone et al. and others have demonstrated protein nitration in both experimental animal models and in tissue resected from infants with NEC [36, 49]. Therefore, the consequences of inflammation and the formation of reactive nitrogen species may be the putative causes of enterocyte death in NEC, especially in the early progression of the disease (i.e., transition from stage 1 to stage 2).

Increased levels of NOS2 and 3-NT have been found in intestinal tissue from both experimental animal models of NEC as well as in tissue resected from infants with NEC. In newborn mice, lipopolysaccharide (LPS) has been shown to markedly enhance intestinal iNOS production. This implies that abnormal bacterial colonization within the newborn intestine may increase baseline iNOS production. These cytopathic effects of NO dysregulation associated with LPS in the newborn intestine have been implicated in barrier dysfunction. It is postulated that bacterial translocation of abnormal bacteria further activates the newborn intestine's innate system (i.e., Toll-like receptors) further propagating the inflammatory cascade leading to intestinal injury. When these mice were administered aminoguandine, an iNOS inhibitor, the bacterial translocation was inhibited when exposed to LPS. In addition, when iNOS knockout mice were exposed to LPS, they had significantly less bacterial translocation than their wild-type littermates exposed to the same dose of LPS.

Increased levels of tissue 3-NT have been linked to cell death both in vitro and in vivo. In tissue resected from infants with NEC, high levels of iNOS and 3-NT were found in areas with significant apoptosis and necrosis along the crypt/villus area of the intestinal mucosa. Maintenance of the intestinal barrier involves a balance between apoptosis at the villus tip, cell proliferation at the crypt, and migration of proliferating cells up to the tip. Increased cell death due to nitrative or other forms of cell death may overwhelm the ability of the intestinal mucosa to replenish the cells and thus lead to barrier dysfunction and further propagation of intestinal injury.

Aside from nonspecific oxidation of proteins caused by reactive nitrogen species, some enzymes can be nitrated at key tyrosine residues in a quasi-specific manner leading to enzymatic dysfunction. For example, several proliferative protein kinase pathways involve tyrosine protein phosphorylation, including both EGFR-meditated cell proliferation and SRC kinase-mediated proliferation. Disruption of these pathways due to nitration can lead to decrease crypt cell proliferation and an inability of the mucosa to replenish cells entering apoptosis, thus leading to barrier dysfunction. Finally, decreased immature intestinal cell migration has been seen in the presence of increased NO production in animal models of NEC as well as in vitro models of cell migration. Further studies of reactive nitrogen species in the pathogenesis of NEC, as well as short-term and long-term consequences to gut development, are needed.

Poly(ADP-Ribose) Polymerase – 1 (PARP-1) as a Respondent to Reactive Nitrogen Species

Nitration has been found in human NEC specimens [36], and it is known that reactive nitrogen species also cause DNA damage [50]. Following DNA damage due to oxidation, PARP-1 is a critical enzyme activated to facilitate DNA repair [51]. This enzyme uses NAD+ (nicotinamide adenine dinucleotide) as a substrate and attaches poly ADP-ribose (PAR) units to itself and other acceptor proteins [52]. This poly(ADP-ribosyl)ation allows the acceptor proteins to selectively influence important cellular responses that enhance DNA repair and transcription of inflammatory mediators such as NF-*k*B [53]. When the PAR polymer becomes too large, it will result in PARP-1 deactivation. In this case, poly(ADP-ribose) glycohydrolase (PARG) may reactivate these proteins by removing the PAR that is being continually added to PARP-1 or other acceptor proteins, allowing PARP-1 to remain activated and continue to increase PAR production [54]. However, in the presence of severe cellular oxidative stress and DNA damage, over-activation of PARP-1 may ensue. This may lead to cell death by two possible mechanisms. First, high PAR turnover via PARP-1 activation may deplete the cells of NAD+/ATP, killing the cells by metabolic catastrophe [55]. Second, in the presence of adequate cell energy stores, increased PARP-1 activation can lead to apoptosis via apoptotic inducing factor (AIF) from the mitochondria or by caspase-dependent mechanisms [56]. If the cell already has low NAD+ stores, as in highly proliferative cells like the enterocyte, the cell may be more likely to enter a necrotic cell death pathway when faced with marked DNA damage. Necrotic cell death leads to the release of proinflammatory mediators, such as high-mobility group protein B1 (HMGB1), into the local environment, and further propagation of tissue injury may ensue. Thus, the activity of PARP-1 in an injured or activated cell serves as a checkpoint or governor between the fate of cellular repair and the fate of cell death (via either apoptosis or necrosis), and cellular energetics play an important role in this fate [57]. PARP-1 is increased in the rat NEC model, and it is associated with tissue regions with high abundance of protein nitration [58]. These data suggest that PARP-1 activation may indeed be a downstream consequence of reactive nitrogen species formation in this setting. Enhancing cellular NAD+ stores and attenuating PARP-1 activation with nicotinamide decreased intestinal necrosis seen in NEC [58].

Red Blood Cell Transfusion and NEC

While a convincing link has not been established, there are multiple retrospective studies suggesting an association between recent exposure to transfusion and the development of NEC. In 1987 McGrady et al. reported a significant increase in NEC cases from a single center during a 3-month period involving 31 % of very low birth weight (VLBW) infants [59]. After investigation, RBC transfusion was found to be strongly associated with NEC with an odds ratio (OR) for NEC after transfusion of 15.1. In 1998, Bednarek et al. reported a similar association in 6 NICUs in the Boston area with different transfusion practices [60]. VLBW infants in the highest transfusing NICUs had a higher prevalence of NEC (7 %) compared with a 2 % prevalence in the low-transfusing centers ($p<0.05$). A recent meta-analysis identified 11 retrospective case-control studies and 1 cohort study evaluating the temporal association of blood transfusion in the previous 48 h with the development of transfusion-associated NEC (TANEC) [61]. The authors concluded that recent exposure to transfusion was associated with NEC in both adjusted and unadjusted analyses. Infants who developed TANEC had lower birth weights, were of lower gestational age, had higher odds of patent ductus arteriosus (PDA), and were more likely to be ventilated [61]. While the true mechanism of transfusion associated NEC remains unclear, several potential mechanisms have been proposed, including once by the authors demonstrated in Fig. 9.3.

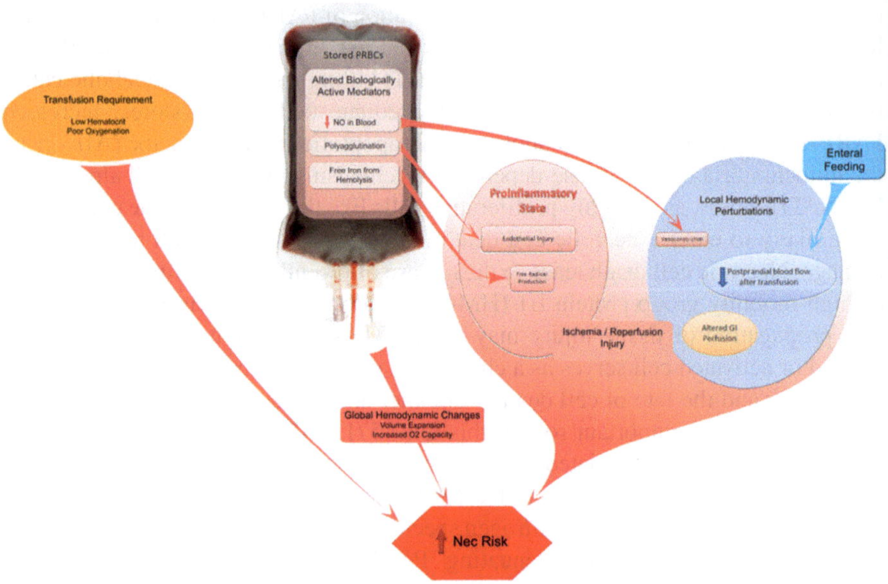

Fig. 9.3 Illustration of a potential mechanism for transfusion-associated NEC

The Effects of Storage on Packed Red Blood Cells

The standard transfusion practice in many NICUs is to limit donor exposure by assigning a specific unit of donated red blood cells (RBCs) for each transfused patient that is used exclusively over the infant's course up to the expiration date of the unit. While this decreases the risk of viral transmission through transfusions of blood from multiple donors, this approach increases the rates of transfusion of older RBCs. Recent observational studies have shown that prolonged RBC storage is associated with increased mortality and organ dysfunction in critically ill adults and children [62, 63]. The storage of red blood cells in preservative medium is associated with harmful metabolic, biochemical, and molecular changes to the erythrocytes including oxidative injury to cytoskeleton proteins and membrane phospholipids [64]. During the preparation of packed red blood cell units, most of the plasma proteins capable of binding and sequestering iron, such as transferrin and albumin, are removed. The extracellular medium surrounding red blood cells in just-expired pediatric packed cell preparations was found to be rich in iron, likely due to oxidative damage to the erythrocyte membrane [65]. The iron released under these conditions is associated with methemoglobin formation and appears to be redox active and therefore potentially harmful [66]. Josephson et al. found that the median age of blood storage of RBCs was longer for infants that developed NEC compared to transfused patients who did not develop NEC [67]. However, a recent randomized trial found no improvement in survival or rates of complications including NEC among critically ill premature infants receiving fresh RBC transfusions (stored ≤ 7 days) compared with standard RBC transfusion practices [68].

Blood Transfusions and Oxidative Stress

Previous studies have demonstrated an association between the number of transfusions a premature infant receives and their risk for developing bronchopulmonary dysplasia (BPD) and retinopathy of prematurity (ROP) [69, 70]. One explanation for these links and for the relationship between blood transfusions and NEC may be the release of oxygen-derived free radicals. Blood transfusions may deliver excess ferrous iron, which through the Haber-Weiss reaction may enable formation of the highly reactive hydroxyl radical (·OH) from superoxide [71]:

$$O_2^{-} + Fe^{3+} \rightarrow O_2 + Fe^{2+}$$

$$2O_2^{-} + 2H^{+} \rightarrow O_2 + H_2O_2$$

$$Fe^{2+} + H_2O_2 \rightarrow Fe^{3+} + OH^{-} + {}^{\cdot}OH$$

The reactive hydroxy radical causes lipid peroxidation which may result in cell injury in a premature infant with limited antioxidant defenses [72]. Blood transfusions were found to significantly increase plasma non-transferrin bound iron ("free" iron with the potential to promote the generation of reactive oxygen species) in premature infants but not in full term infants [73]. Under normal conditions, protection against free iron is provided by ceruloplasmin (which converts the pro-oxidant ferrous iron into the ferric state) and transferrin (which binds the ferric iron), but in premature infants the concentrations of these binding proteins are low [73, 74]. Consequently, blood transfusions were associated with lipid peroxidation in premature infants as measured by urinary malondialdehyde (MDA) and levels of MDA in bronchoalveolar lavage fluid [71, 75]. Biomarkers of oxidative stress have also been linked to NEC, as levels of non-protein-bound iron and total hydroperoxides in cord blood were found to be significantly higher in infants who developed NEC [76].

Potential Interventions for Transfusion Associated NEC

Ideally, one approach to prevent transfusion-associated NEC would be to limit late transfusions. While no standard national guidelines for transfusions in the NICU exist, protocols may exist in individual units. A transfusion-reduction program instituted by the NICUs of Intermountain Healthcare in 2009 was accompanied by a trend toward a lower rate of NEC ($P<0.09$) [77]. In addition, erythropoietin (Epo) has been shown to be protective against NEC [78]. Finally, Singh et al. found that iron supplementation, another therapeutic intervention for treatment of anemia, was also associated with a lower risk for NEC [79]. Previous concerns about excess iron and oxidative injury in premature infants with reduced antioxidant defenses have been limited after two recent studies showed that iron supplementation to stable growing premature infants does not induce oxidative stress [80, 81].

The issue of enteral feeding practices in relationship with blood transfusions was first addressed by Perciaccante and Young in an abstract which showed that making infants *Nil per os* (NPO) 4 h before transfusion was associated with a reduction in the occurrence of TANEC to zero compared to 39 % in the previous time period [82]. More recently, El-Dib et al. completed a case-control study involving 50 infants born < 32 weeks gestation and found a reduction in TANEC after making patients NPO immediately before and during the transfusion [83]. Clearly, well-designed prospective studies are needed to convincingly establish the association between NEC and blood transfusion and to evaluate the impact of withholding feeds prior and during a blood transfusion.

Challenges and New Prospects for NEC Therapy

Progress in treating NEC after initial diagnosis remains extremely limited. In fact, at this time there are no therapies for NEC treatment, aside from surgical resection of damaged intestinal tissue. Progress in this setting has therefore been mainly in the area prevention of NEC. Several emerging possibilities for NEC therapy have been investigated in preclinical and clinical settings.

Enteral Antibiotic Administration

LPS appears to exacerbate nitrative injury in the immature epithelium of the neonate. Oral administration of antibiotics has been used in an attempt to decrease abnormal colonization [84]. Meta-analysis has shown a decrease in the incidence of NEC but no decrease in mortality in NEC cases [85]. Therefore, with the concerns of resistance development by pathogenic bacteria and altering the normal flora, further studies are needed to evaluate the long-term safety of this strategy.

Probiotics

Probiotics are live microbial food supplements (bacteria or yeast) that beneficially affect the host animal by improving its intestinal microbial balance [86]. These beneficial, commensal organisms play a central role in the maintenance of intestinal barrier function and intestinal homeostasis. As abnormal colonization of the premature gut with pathogenic bacteria has been implicated in the pathogenesis of NEC, the ability of probiotics to alter the intestinal flora of infants by favoring the growth of beneficial commensal bacteria makes their use an attractive potential strategy for the prevention of NEC.

Role of Intestinal Flora

Commensal bacteria regulate appropriate bacterial colonization by increasing the mucosal barrier to bacteria and bacterial products [87, 88]. These bacteria play a role in the development of intestinal tolerance to luminal antigens, regulate the expression of certain protective barrier genes as well as genes involved in digestion and angiogenesis, and can induce specific mucus genes. Other functions of commensal bacteria include regulation of intestinal permeability, modulation of intestinal inflammation, degradation protein and carbohydrates, and increased production of anti-inflammatory cytokines [89, 90]. These varied functions may be particularly important in the immature intestine with inadequate defense mechanisms against pathogenic bacteria and the potential for an exaggerated inflammatory response [84]. Therefore, the establishment of a stable and diverse intestinal bacterial community plays a vital role in maintenance of intestinal barrier function and immune regulation.

Probiotics for the Prevention of NEC

The potential benefits of probiotics include more than simply modifying the gut flora. Indeed, probiotics may limit some of the pathologic features thought to predispose premature infants to NEC including an exaggerated innate immune response, local perturbations in intestinal perfusion leading to ischemia, and an immature intestinal barrier. Interaction between commensal bacteria and the intestinal epithelium allow for appropriate regulation of the intestinal immune and inflammatory response. When the immature intestinal barrier is breeched by pathogenic bacteria, activation and upregulation of innate immune receptors in the epithelium, such as Toll-like receptor 4, can occur. The innate immune receptors recognize these bacteria as harmful and initiate the inflammatory cascade via NF-κB. Commensal bacteria prevent the nuclear ubiquination or degradation of the NF-κB/IκB complex, thus preventing the nuclear translocation and activation of NF-κB [91]. IκB increases with maturation; thus, immature enterocytes have less IκB to inhibit NF-κB and thus are particularly susceptible to uncontrolled inflammatory response when presented with a pathogen [92]. Therefore, commensal bacteria may play an important role in the premature enterocyte to prevent activation of NF-κB, thus modulating the inflammatory response and potentially protect the newborn intestine from injury.

Treatment of human intestinal epithelial cells primed with TNF-α, IL-1β, and IFNγ with a probiotic led to increased nitric oxide (NO) production through the induction of inducible nitric oxide synthase (iNOS). This controlled induction of NOS and subsequent low level production of NO may be a potential mechanism whereby probiotics serve to protect the intestinal tract [93]. In immature mice, Lin et al. demonstrated that a probiotic blocked inflammatory signaling via generation of reactive oxygen species, regulated apoptosis, and promoted cytoprotective responses [94, 95]. D'Souza et al. noted downregulation of caveolin-1, eNOS, and nNOS and upregulation of iNOS in the terminal ileum of formula fed neonatal rats. Following probiotic supplementation, superoxide dismutase and glutathione peroxidase were upregulated [96]. Later this group demonstrated that probiotic supplementation decreased proinflammatory cytokine production and downregulated genes involved in oxidative stress and toll-like receptor pathways in the terminal ileum of formula-fed neonatal rats, although the effect was attenuated in hypoxia/hyperoxia [97]. In formula-fed neonatal rat pups, Fordjour et al. demonstrated preservation of bowel integrity and induction of IGF-1 and EGF by formula supplemented with probiotics [98]. Utilizing immature and mature human intestinal xenografts and primary enterocytes isolated from resected NEC tissue, Ganguli et al. noted an attenuated response to inflammatory stimuli following exposure to probiotic conditioned media due to the secretion of an anti-inflammatory probiotic factor, possibly a glycan or glycolipid [99].

The possibility of decreasing the incidence of NEC by the administration of probiotics has been investigated in human infants. Several groups have noted a decrease in the incidence of NEC following probiotic supplementation [100–104]. In contrast, other groups have failed to demonstrate a reduction in the incidence of NEC following probiotic administration [105, 106]. Despite these negative findings, several recent meta-analyses have concluded that probiotic supplementation reduces the incidence of NEC in preterm infants [107, 108]. While the available data are insufficient to support the routine use of probiotics to prevent NEC, these studies do support the need for a large randomized, controlled trial to further investigate the role of probiotics in the prevention of NEC in premature infants.

Growth Factors

Growing clinical evidence supports the use of EGF as a predictive marker of NEC and its use for prevention and treatment of NEC [109]. In addition, experimental data indicate potential mechanisms of EGF prevention against NEC. These consist of reduction of inflammation, improvement of barrier function, and regulation of epithelial apoptosis, all of which appear to be directly affected by nitrative injury. HB-EGF is one such growth factor that appears to be a promising therapeutic agent in regard to preventing and possibly treating NEC. Further studies are necessary to identify the optimal dose, timing, and route of administration of members of the EGF family, including HB-EGF, to NEC patients.

Glucocorticoids

Glucocorticoids have been found to accelerate maturation of the immature intestinal epithelium as well as decrease inflammation and abnormal bacterial colonization [84, 110–112]. However, due to concerns regarding the incidence of spontaneous intestinal perforations as well as poor neurodevelopmental outcomes noted following postnatal steroid exposure in premature infants, further studies are needed to further evaluate the safety and efficacy of this strategy.

NOS2 Inhibitors

NOS2 inhibitors (L-NIL, aminoguanidine) have been found to decrease markers of nitrative stress and as well as tissue injury in experimental animal models of NEC [36, 113]. NOS inhibition in adults with acute states of disease, such as sepsis, unfortunately has not been as promising. Current iNOS inhibitors may not be as selective as hoped and thus also inhibit the beneficial effects of NO. Perhaps in the future there may be more specific iNOS inhibitors able to attenuate the detrimental effects of overproduction of NO vs. its required effects for homeostasis at lower levels. Other options would include therapeutics targeting more focal downstream effects of NO dysregulation.

PARP Antagonism

There are three potential mechanisms by which PARP-1 may contribute to tissue injury and cell death. First, NAD+ consumption by PARP-1 may deplete the cells of energy metabolites leading to cellular necrosis [55, 114]. Second, in the presence of severe DNA damage but adequate NAD+ stores, PARP-1 activation may lead to apoptosis by either caspase-dependent mechanisms or apoptosis-inducing factor (AIF) release from mitochondria that causes caspase-independent apoptotic cell death [115]. Lastly, PARP-1 appears to be part of the NF-*k*B transcriptosome and thus contributes to the synthesis of inflammatory mediators [116]. These potential pathways by which PARP-1 may contribute to the intestinal injury and intestinal cell death seen in NEC need to be tested mechanistically. A substantial number of pharmacologic studies have shown the benefit of various classes of PARP inhibitors in different disease models of inflammation and oxidation, neurodegeneration, and vascular disease, including myocardial infarction, septic shock, and colitis [117, 118]. Some of these inhibitors have entered human trials [118, 119]. PARP inhibition appears to be a viable candidate for medical therapy in NEC and warrants further investigation.

Summary

NEC is a devastating intestinal condition that has severe consequences in preterm infants. While there are several clinical strategies that have been shown to reduce the incidence of NEC, no medical approach has been established for NEC therapy.

The identification of various risk factors for the development of NEC has shed light on important contributors and mechanisms of the disease process. Studies have suggested alterations of intestinal blood flow, abnormal bacterial colonization, and/or limited growth factor presence in the immature neonatal intestine as contributors to NEC risk and its progression. These mechanisms may act alone or exist in concert, and this may explain the variable timing and presentation of NEC symptoms and outcomes. Each of these mechanisms is a known contributor to activation of inflammatory processes and related innate immunity mechanisms. In turn, intestinal inflammation can promote local oxidative stress and may be a common pathway for NEC-related tissue injury. As the mechanisms of NEC are better defined, newer opportunities for rational therapeutic strategies will continue to emerge. Further studies to elucidate the unifying and common processes in this heterogeneous and often unpredictable disease state are clearly needed.

References

1. Pietz J, et al. Prevention of necrotizing enterocolitis in preterm infants: a 20-year experience. Pediatrics. 2007;119(1):e164–70.
2. Gibbs K, Lin J, Holzman IR. Necrotising enterocolitis: the state of the science. Indian J Pediatr. 2007;74(1):67–72.
3. Lin PW, Stoll BJ. Necrotising enterocolitis. Lancet. 2006;368(9543):1271–83.
4. Luig M, Lui K. Epidemiology of necrotizing enterocolitis – Part I: Changing regional trends in extremely preterm infants over 14 years. J Paediatr Child Health. 2005;41(4):169–73.
5. Luig M, Lui K. Epidemiology of necrotizing enterocolitis – Part II: Risks and susceptibility of premature infants during the surfactant era: a regional study. J Paediatr Child Health. 2005;41(4):174–9.
6. Rees CM, Pierro A, Eaton S. Neurodevelopmental outcomes of neonates with medically and surgically treated necrotizing enterocolitis. Arch Dis Child Fetal Neonatal Ed. 2007; 92(3): F193–8.
7. Holzman IR, Brown DR. Necrotizing enterocolitis: a complication of prematurity. Semin Perinatol. 1986;10(3):208–16.
8. Yost CC. Neonatal necrotizing enterocolitis: diagnosis, management, and pathogenesis. J Infus Nurs. 2005;28(2):130–4.
9. Henry MC, Moss RL. Current issues in the management of necrotizing enterocolitis. Semin Perinatol. 2004;28(3):221–33.
10. Cikrit D, et al. Necrotizing enterocolitis: factors affecting mortality in 101 surgical cases. Surgery. 1984;96(4):648–55.
11. Bell MJ, et al. Neonatal necrotizing enterocolitis. Therapeutic decisions based upon clinical staging. Ann Surg. 1978;187(1):1–7.
12. Bell MJ, et al. Surgical mortality and morbidity in extremely low-birth-weight infants. Am J Dis Child. 1983;137(7):682–4.
13. Bell EF. Preventing necrotizing enterocolitis: what works and how safe? Pediatrics. 2005;115(1):173–4.
14. Henry MC, Lawrence Moss R. Surgical therapy for necrotizing enterocolitis: bringing evidence to the bedside. Semin Pediatr Surg. 2005;14(3):181–90.
15. Arnold LD. The cost-effectiveness of using banked donor milk in the neonatal intensive care unit: prevention of necrotizing enterocolitis. J Hum Lact. 2002;18(2):172–7.
16. Stoll BJ, et al. Epidemiology of necrotizing enterocolitis: a case control study. J Pediatr. 1980;96(3 Pt 1):447–51.

17. Stevenson DK, et al. Late morbidity among survivors of necrotizing enterocolitis. Pediatrics. 1980;66(6):925–7.
18. Horwitz JR, et al. Complications after surgical intervention for necrotizing enterocolitis: a multicenter review. J Pediatr Surg. 1995;30(7):994–8; discussion 998–9.
19. Vieten D, et al. Proliferative response in necrotising enterocolitis is insufficient to prevent disease progression. Pediatr Surg Int. 2006;22(1):50–6.
20. Hintz SR, et al. Neurodevelopmental and growth outcomes of extremely low birth weight infants after necrotizing enterocolitis. Pediatrics. 2005;115(3):696–703.
21. McCormack CJ, Emmens RW, Putnam TC. Evaluation of factors in high risk neonatal necrotizing enterocolitis. J Pediatr Surg. 1987;22(6):488–91.
22. Bunton GL, et al. Necrotizing enterocolitis. Controlled study of 3 years' experience in a neonatal intensive care unit. Arch Dis Child. 1977;52(10):772–7.
23. Baschat AA, et al. Predictors of neonatal outcome in early-onset placental dysfunction. Obstet Gynecol. 2007;109(2 Pt 1):253–61.
24. Yu VY, et al. Perinatal risk factors for necrotizing enterocolitis. Arch Dis Child. 1984;59(5): 430–4.
25. Vieten D, et al. Impaired mucosal regeneration in neonatal necrotising enterocolitis. Pediatr Surg Int. 2005;21(3):153–60.
26. Hsueh W, et al. Neonatal necrotizing enterocolitis: clinical considerations and pathogenetic concepts. Pediatr Dev Pathol. 2003;6(1):6–23.
27. De Curtis M, et al. A case control study of necrotizing enterocolitis occurring over 8 years in a neonatal intensive care unit. Eur J Pediatr. 1987;146(4):398–400.
28. Anand RJ, et al. The role of the intestinal barrier in the pathogenesis of necrotizing enterocolitis. Shock. 2007;27(2):124–33.
29. Jilling T, et al. Intestinal epithelial apoptosis initiates gross bowel necrosis in an experimental rat model of neonatal necrotizing enterocolitis. Pediatr Res. 2004;55(4):622–9.
30. Nowicki PT. Ischemia and necrotizing enterocolitis: where, when, and how. Semin Pediatr Surg. 2005;14(3):152–8.
31. Nowicki PT, et al. Endothelial nitric oxide synthase in human intestine resected for necrotizing enterocolitis. J Pediatr. 2007;150(1):40–5.
32. Halpern MD, et al. Up-regulation of IL-18 and IL-12 in the ileum of neonatal rats with necrotizing enterocolitis. Pediatr Res. 2002;51(6):733–9.
33. Kelly N, et al. The role of the glutathione antioxidant system in gut barrier failure in a rodent model of experimental necrotizing enterocolitis. Surgery. 2004;136(3):557–66.
34. Hammerman C, Kaplan M. Probiotics and neonatal intestinal infection. Curr Opin Infect Dis. 2006;19(3):277–82.
35. Potoka DA, et al. Role of nitric oxide and peroxynitrite in gut barrier failure. World J Surg. 2002;26(7):806–11.
36. Zamora R, et al. Nitrosative stress in an animal model of necrotizing enterocolitis. Free Radic Biol Med. 2005;39(11):1428–37.
37. Hackam DJ, et al. Utility of radiographic hepatic injury grade in predicting outcome for children after blunt abdominal trauma. J Pediatr Surg. 2002;37(3):386–9.
38. Feng J, El-Assal ON, Besner GE. Heparin-binding epidermal growth factor-like growth factor decreases the incidence of necrotizing enterocolitis in neonatal rats. J Pediatr Surg. 2006; 41(1):144–9; discussion 144–9.
39. Upperman JS, et al. Mechanisms of nitric oxide-mediated intestinal barrier failure in necrotizing enterocolitis. Semin Pediatr Surg. 2005;14(3):159–66.
40. Moncada S, Higgs A. The L-arginine-nitric oxide pathway. N Engl J Med. 1993;329(27): 2002–12.
41. Sessa WC. The nitric oxide synthase family of proteins. J Vasc Res. 1994;31(3):131–43.
42. Stuehr DJ, Marletta MA. Mammalian nitrate biosynthesis: mouse macrophages produce nitrite and nitrate in response to Escherichia coli lipopolysaccharide. Proc Natl Acad Sci U S A. 1985; 82(22):7738–42.
43. Takahashi T. Pathophysiological significance of neuronal nitric oxide synthase in the gastrointestinal tract. J Gastroenterol. 2003;38(5):421–30.

44. Stamler JS, Singel DJ, Loscalzo J. Biochemistry of nitric oxide and its redox-activated forms. Science. 1992;258(5090):1898–902.
45. Beckman JS, Koppenol WH. Nitric oxide, superoxide, and peroxynitrite: the good, the bad, and ugly. Am J Physiol. 1996;271(5 Pt 1):C1424–37.
46. Pacher P, et al. Role of poly(ADP-ribose) polymerase activation in endotoxin-induced cardiac collapse in rodents. Biochem Pharmacol. 2002;64(12):1785–91.
47. Huie RE, Padmaja S. The reaction of no with superoxide. Free Radic Res Commun. 1993;18(4):195–9.
48. Pryor WA, Squadrito GL. The chemistry of peroxynitrite: a product from the reaction of nitric oxide with superoxide. Am J Physiol. 1995;268(5 Pt 1):L699–722.
49. Giannone PJ, et al. Effects of prenatal lipopolysaccharide exposure on epithelial development and function in newborn rat intestine. J Pediatr Gastroenterol Nutr. 2006;43(3):284–90.
50. Ohshima H, Sawa T, Akaike T. 8-nitroguanine, a product of nitrative DNA damage caused by reactive nitrogen species: formation, occurrence, and implications in inflammation and carcinogenesis. Antioxid Redox Signal. 2006;8(5–6):1033–45.
51. Bonicalzi ME, et al. Regulation of poly(ADP-ribose) metabolism by poly(ADP-ribose) glycohydrolase: where and when? Cell Mol Life Sci. 2005;62(7–8):739–50.
52. Diefenbach J, Burkle A. Introduction to poly(ADP-ribose) metabolism. Cell Mol Life Sci. 2005;62(7–8):721–30.
53. Meyer-Ficca ML, et al. Poly(ADP-ribose) polymerases: managing genome stability. Int J Biochem Cell Biol. 2005;37(5):920–6.
54. Cuzzocrea S, et al. Role of poly(ADP-ribose) glycohydrolase in the development of inflammatory bowel disease in mice. Free Radic Biol Med. 2007;42(1):90–105.
55. Szabo C. Poly(ADP-ribose) polymerase activation by reactive nitrogen species – relevance for the pathogenesis of inflammation. Nitric Oxide. 2006;14(2):169–79.
56. Kolthur-Seetharam U, et al. Control of AIF-mediated cell death by the functional interplay of SIRT1 and PARP-1 in response to DNA damage. Cell Cycle. 2006;5(8):873–7.
57. Dantzer F, et al. Poly(ADP-ribose) polymerase-1 activation during DNA damage and repair. Methods Enzymol. 2006;409:493–510.
58. Giannone PJ, et al. Poly(ADP-ribose) polymerase-1: a novel therapeutic target in necrotizing enterocolitis. Pediatr Res. 2011;70(1):67–71.
59. McGrady GA, et al. An outbreak of necrotizing enterocolitis. Association with transfusions of packed red blood cells. Am J Epidemiol. 1987;126(6):1165–72.
60. Bednarek FJ, et al. Variations in blood transfusions among newborn intensive care units. SNAP II Study Group. J Pediatr. 1998;133(5):601–7.
61. Mohamed A, Shah PS. Transfusion associated necrotizing enterocolitis: a meta-analysis of observational data. Pediatrics. 2012;129(3):529–40.
62. Gauvin F, et al. Association between length of storage of transfused red blood cells and multiple organ dysfunction syndrome in pediatric intensive care patients. Transfusion. 2010;50(9):1902–13.
63. Koch CG, et al. Duration of red-cell storage and complications after cardiac surgery. N Engl J Med. 2008;358(12):1229–39.
64. Chaudhary R, Katharia R. Oxidative injury as contributory factor for red cells storage lesion during twenty eight days of storage. Blood Transfus. 2012;10(1):59–62.
65. Collard K, White D, Copplestone A. The effect of maximum storage on iron status, oxidative stress and antioxidant protection in paediatric packed cell units. Blood Transfus. 2012; 12:1–9.
66. Ferrali M, et al. Iron release and membrane damage in erythrocytes exposed to oxidizing agents, phenylhydrazine, divicine and isouramil. Biochem J. 1992;285(Pt 1):295–301.
67. Josephson CD, et al. Do red cell transfusions increase the risk of necrotizing enterocolitis in premature infants? J Pediatr. 2010;157(6):972–978 e1-3.
68. Fergusson DA, et al. Effect of fresh red blood cell transfusions on clinical outcomes in premature, very low-birth-weight infants: the ARIPI randomized trial. JAMA. 2012;308(14): 1443–51.

69. Cooke RW, et al. The apparent role of blood transfusions in the development of retinopathy of prematurity. Eur J Pediatr. 1993;152(10):833–6.
70. Korhonen P, et al. Frequency and risk factors in bronchopulmonary dysplasia in a cohort of very low birth weight infants. Early Hum Dev. 1999;54(3):245–58.
71. Wardle SP, et al. Effect of blood transfusion on lipid peroxidation in preterm infants. Arch Dis Child Fetal Neonatal Ed. 2002;86(1):F46–8.
72. Davis JM, Auten RL. Maturation of the antioxidant system and the effects on preterm birth. Semin Fetal Neonatal Med. 2010;15(4):191–5.
73. Hirano K, et al. Blood transfusion increases radical promoting non-transferrin bound iron in preterm infants. Arch Dis Child Fetal Neonatal Ed. 2001;84(3):F188–93.
74. Lackmann GM, Schnieder C, Bohner J. Gestational age-dependent reference values for iron and selected proteins of iron metabolism in serum of premature human neonates. Biol Neonate. 1998;74(3):208–13.
75. Collard KJ, Godeck S, Holley JE. Blood transfusion and pulmonary lipid peroxidation in ventilated premature babies. Pediatr Pulmonol. 2005;39(3):257–61.
76. Perrone S, et al. May oxidative stress biomarkers in cord blood predict the occurrence of necrotizing enterocolitis in preterm infants? J Matern Fetal Neonatal Med. 2012;25 Suppl 1:128–31.
77. Baer VL, et al. Implementing a program to improve compliance with neonatal intensive care unit transfusion guidelines was accompanied by a reduction in transfusion rate: a pre-post analysis within a multihospital health care system. Transfusion. 2011;51(2):264–9.
78. Ledbetter DJ, Juul SE. Erythropoietin and the incidence of necrotizing enterocolitis in infants with very low birth weight. J Pediatr Surg. 2000;35(2):178–81; discussion 182.
79. Singh R, et al. Association of necrotizing enterocolitis with anemia and packed red blood cell transfusions in preterm infants. J Perinatol. 2011;31(3):176–82.
80. Braekke K, et al. Oxidative stress markers and antioxidant status after oral iron supplementation to very low birth weight infants. J Pediatr. 2007;151(1):23–8.
81. Miller SM, McPherson RJ, Juul SE. Iron sulfate supplementation decreases zinc protoporphyrin to heme ratio in premature infants. J Pediatr. 2006;148(1):44–8.
82. Perciaccante J, Young T, Necrotizing enterocolitis associated with packed red blood cell transfusions in premature neonates. Honolulu: Pediatric Academic Society; 2008. p. 5839.8.
83. El-Dib M, et al. Red blood cell transfusion, feeding and necrotizing enterocolitis in preterm infants. J Perinatol. 2011;31(3):183–7.
84. Petrosyan M, et al. Current concepts regarding the pathogenesis of necrotizing enterocolitis. Pediatr Surg Int. 2009;25(4):309–18.
85. Bury RG, Tudehope D. Enteral antibiotics for preventing necrotising enterocolitis in low birthweight or preterm infants. Cochrane Database Syst Rev. 2000;(2):CD000405.
86. Stenger MR, et al. Probiotics and prebiotics for the prevention of necrotizing enterocolitis. Curr Infect Dis Rep. 2011;13(1):13–20.
87. Martin CR, Walker WA. Probiotics: role in pathophysiology and prevention in necrotizing enterocolitis. Semin Perinatol. 2008;32(2):127–37.
88. Salminen S, Isolauri E. Opportunities for improving the health and nutrition of the human infant by probiotics. Nestle Nutr Workshop Ser Pediatr Program. 2008;62:223–33; discussion 233–7.
89. Emami CN, et al. Role of the host defense system and intestinal microbial flora in the pathogenesis of necrotizing enterocolitis. Surg Infect (Larchmt). 2009;10(5):407–17.
90. Sharma R, Tepas 3rd JJ. Microecology, intestinal epithelial barrier and necrotizing enterocolitis. Pediatr Surg Int. 2010;26(1):11–21.
91. Petrof EO, et al. Probiotics inhibit nuclear factor-kappaB and induce heat shock proteins in colonic epithelial cells through proteasome inhibition. Gastroenterology. 2004;127(5): 1474–87.
92. Claud EC, et al. Developmentally regulated IkappaB expression in intestinal epithelium and susceptibility to flagellin-induced inflammation. Proc Natl Acad Sci U S A. 2004;101(19): 7404–8.

93. Korhonen R, et al. Induction of nitric oxide synthesis by probiotic Lactobacillus rhamnosus GG in J774 macrophages and human T84 intestinal epithelial cells. Inflammation. 2001; 25(4):223–32.
94. Lin PW, et al. Lactobacillus rhamnosus blocks inflammatory signaling in vivo via reactive oxygen species generation. Free Radic Biol Med. 2009;47(8):1205–11.
95. Lin PW, et al. The probiotic Lactobacillus GG may augment intestinal host defense by regulating apoptosis and promoting cytoprotective responses in the developing murine gut. Pediatr Res. 2008;64(5):511–6.
96. D'Souza A, et al. Effects of probiotics, prebiotics, and synbiotics on messenger RNA expression of caveolin-1, NOS, and genes regulating oxidative stress in the terminal ileum of formula-fed neonatal rats. Pediatr Res. 2010;67(5):526–31.
97. D'Souza A, et al. Cytokines and Toll-like receptor signaling pathways in the terminal ileum of hypoxic/hyperoxic neonatal rats: benefits of probiotics supplementation. Am J Transl Res. 2012;4(2):187–97.
98. Fordjour L, et al. Comparative effects of probiotics, prebiotics, and synbiotics on growth factors in the large bowel in a rat model of formula-induced bowel inflammation. J Pediatr Gastroenterol Nutr. 2010;51(4):507–13.
99. Ganguli K, et al. Probiotics prevent necrotizing enterocolitis by modulating enterocyte genes that regulate innate immune-mediated inflammation. Am J Physiol Gastrointest Liver Physiol. 2013;304(2):G132–41.
100. Awad H, et al. Comparison between killed and living probiotic usage versus placebo for the prevention of necrotizing enterocolitis and sepsis in neonates. Pak J Biol Sci. 2010;13(6): 253–62.
101. Bin-Nun A, et al. Oral probiotics prevent necrotizing enterocolitis in very low birth weight neonates. J Pediatr. 2005;147(2):192–6.
102. Hoyos AB. Reduced incidence of necrotizing enterocolitis associated with enteral administration of Lactobacillus acidophilus and Bifidobacterium infantis to neonates in an intensive care unit. Int J Infect Dis. 1999;3(4):197–202.
103. Li LQ, et al. Role of probiotics in the prevention of neonatal necrotizing enterocolitis: a case-control study. Zhongguo Dang Dai Er Ke Za Zh. 2006;8(6):464–6.
104. Lin HC, et al. Oral probiotics reduce the incidence and severity of necrotizing enterocolitis in very low birth weight infants. Pediatrics. 2005;115(1):1–4.
105. Dani C, et al. Probiotics feeding in prevention of urinary tract infection, bacterial sepsis and necrotizing enterocolitis in preterm infants. A prospective double-blind study. Biol Neonate. 2002;82(2):103–8.
106. Luoto R, et al. Incidence of necrotizing enterocolitis in very-low-birth-weight infants related to the use of Lactobacillus GG. Acta Paediatr. 2010;99(8):1135–8.
107. Deshpande G, et al. Updated meta-analysis of probiotics for preventing necrotizing enterocolitis in preterm neonates. Pediatrics. 2010;125(5):921–30.
108. Wang Q, Dong J, Zhu Y. Probiotic supplement reduces risk of necrotizing enterocolitis and mortality in preterm very low-birth-weight infants: an updated meta-analysis of 20 randomized, controlled trials. J Pediatr Surg. 2012;47(1):241–8.
109. Coursodon CF, Dvorak B. Epidermal growth factor and necrotizing enterocolitis. Curr Opin Pediatr. 2012;24(2):160–4.
110. Buchmiller TL, et al. Effect of prenatal dexamethasone administration: fetal rabbit intestinal nutrient uptake and disaccharidase development. J Surg Res. 1994;57(2):274–9.
111. Israel EJ, et al. Prevention of necrotizing enterocolitis in the rat with prenatal cortisone. Gastroenterology. 1990;99(5):1333–8.
112. Israel EJ, et al. Cortisone strengthens the intestinal mucosal barrier in a rodent necrotizing enterocolitis model. Adv Exp Med Biol. 1991;310:375–80.
113. Giannone PJ, et al. Prenatal lipopolysaccharide increases postnatal intestinal injury in a rat model of necrotizing enterocolitis. J Pediatr Gastroenterol Nutr. 2009;48(3):276–82.

114. Zhou HZ, et al. Poly(ADP-ribose) polymerase-1 hyperactivation and impairment of mitochondrial respiratory chain complex I function in reperfused mouse hearts. Am J Physiol Heart Circ Physiol. 2006;291(2):H714–23.
115. van Wijk SJ, Hageman GJ. Poly(ADP-ribose) polymerase-1 mediated caspase-independent cell death after ischemia/reperfusion. Free Radic Biol Med. 2005;39(1):81–90.
116. Hassa PO, et al. Acetylation of poly(ADP-ribose) polymerase-1 by p300/CREB-binding protein regulates coactivation of NF-kappaB-dependent transcription. J Biol Chem. 2005; 280(49):40450–64.
117. Booz GW. PARP inhibitors and heart failure – translational medicine caught in the act. Congest Heart Fail. 2007;13(2):105–12.
118. Genovese T, et al. Treatment with a novel poly(ADP-ribose) glycohydrolase inhibitor reduces development of septic shock-like syndrome induced by zymosan in mice. Crit Care Med. 2004;32(6):1365–74.
119. Jagtap P, Szabo C. Poly(ADP-ribose) polymerase and the therapeutic effects of its inhibitors. Nat Rev Drug Discov. 2005;4(5):421–40.

Chapter 10
Oxidative Stress and the Perinatal Circulation

Robin H. Steinhorn and Stephen Wedgwood

Introduction

The lung is the human organ that is exposed to the highest concentrations of atmospheric oxygen, and as a result, it has developed an intricate array of responses to changes in oxygen concentrations. The perinatal lung is particularly interesting in this regard, because the fetus must survive and grow in a hypoxic environment, and then rapidly adapt to an oxygen-rich atmosphere. Respiratory failure affects 2 % of live births and contributes significantly to neonatal morbidity and mortality [1], and therapeutic oxygen is commonly required. This review will focus on the role of reactive oxygen species as mediators of responses to neonatal respiratory failure, hypoxia, and hyperoxia.

Overview of Reactive Oxygen Species

The vascular effects of both hypoxia and hyperoxia are largely mediated through the formation of reactive oxygen species (ROS), small molecules derived from molecular oxygen. ROS are important vascular signaling molecules, regulators of vascular tone and function, and a potential source of vascular injury through

R.H. Steinhorn, M.D. (✉)
Department of Pediatrics, Children's Hospital, University of California Davis Medical Center, Sacramento, CA, USA
e-mail: robin.steinhorn@ucdmc.ucdavis.edu

S. Wedgwood
Department of Pediatrics, University of California Davis Medical Center, Sacramento, CA, USA

P.A. Dennery et al. (eds.), *Perinatal and Prenatal Disorders*, Oxidative Stress in Applied Basic Research and Clinical Practice, DOI 10.1007/978-1-4939-1405-0_10

their interaction with proteins, DNA, RNA, and lipids [2, 3]. Multiple enzymatic oxidase systems can contribute to generation of ROS in the vessel wall, and each system has specific roles in vascular physiology and pathophysiology. Antioxidant systems also regulate vascular signaling pathways by scavenging ROS. In the endothelium, mitochondria, xanthine oxidase, cytochrome P450, and cyclooxygenase can generate ROS that influence vascular function, while under stress conditions, an uncoupling of endothelial nitric oxide synthase (eNOS) can also contribute [4, 5]. In vascular smooth muscle, NADPH oxidases (Nox) and mitochondria can function as significant sources of ROS generation [6–10].

ROS-generating systems donate an electron to molecular oxygen to generate superoxide anion (O_2^-), the precursor for hydrogen peroxide and other reactive species. Superoxide dismutases (SODs) generate hydrogen peroxide (H_2O_2) through dismutation of superoxide, although spontaneous dismutation can occur at low pH even in the absence of SOD. H_2O_2 induces vasoconstriction, which is blocked by scavengers such as catalase and glutathione peroxidase. In the presence of SOD, superoxide is relatively short lived, and its negative charge renders it relatively impermeable to biological membranes. By contrast, H_2O_2 has a longer half life and is uncharged and freely diffusible through aquaporins in membranes. As such, H_2O_2 is an important intracellular and intercellular signaling molecule. Superoxide can also react with other radicals such as nitric oxide (NO^-) to form peroxynitrite ($ONOO^-$), an exceptionally reactive oxidant that can promote vascular dysfunction by nonspecific nitration of proteins resulting in significant consequences for their functions. In the presence of iron, H_2O_2 can generate hydroxyl radicals through the Fenton reaction. However, these highly reactive radicals are unlikely to participate in signaling pathways and are therefore exclusively involved in cell damage responses.

Perinatal Lung Development

As the human fetal lung develops, lung vascular growth and branching is closely coupled with the growth and branching of the airway epithelium [11]. Formation of the pulmonary vasculature is dependent on vasculogenesis, or de novo formation of blood vessels, and angiogenesis, the formation of new vessels from preexisting ones. During the final stages of vascular development, the pulmonary capillaries surround the thinning alveolar walls, providing the increased alveolar and capillary surface areas necessary for efficient gas exchange at birth. This choreographed structural organization requires the tight regulation of vascularization and alveolarization. After birth, vascular development continues with expansion of capillary volume and surface area, driven by angiogenesis from preexisting vessels and intussusceptive growth [12]. Antenatal or postnatal events that affect the developmental program of the fetal or newborn lung may contribute to defective pulmonary vascular development.

The fetal lung is programmed to develop in a low-oxygen intrauterine environment that favors multiple growth factor signaling pathways. The VEGF and NO signaling pathways are among those critical for antenatal and postnatal lung vascular growth. Vascular endothelial growth factor (VEGF) is expressed in vascular endothelial and smooth muscle cells and in airway epithelium in the fetal lung and is central to vascular development of the perinatal lung. Two distinct transmembrane tyrosine kinase receptors, VEGF receptor 1 (VEGFR-1) and VEGF receptor 2 (VEGFR-2), are expressed on vascular endothelium. Experimental inactivation of VEGF or its receptor genes results in embryonic lethality characterized by deficient organization of endothelial cells and vascularization, and targeted inactivation in the early postnatal period increases mortality and impairs lung vascular development [13]. Inhibition of VEGF receptors in adult rodents also produces abnormalities of lung architecture, suggesting that VEGF signaling remains important after infancy for maintenance of normal pulmonary vasculature and alveolar structure.

Lung VEGF expression is regulated by members of the hypoxia-inducible factor (HIF) family of transcription factors [14, 15]. HIFs are heterodimers consisting of oxygen-sensitive α subunits (HIF-1α, HIF-2α) and constitutively expressed β subunits. Hypoxia stabilizes the α subunit leading to nuclear accumulation and activation of multiple target genes [16]. Conversely, elevated levels of oxygen target the protein for proteasomal degradation thereby decreasing target gene expression. Deletion of HIF-1 or HIF-2 results in embryonic lethality, and an important recent study demonstrated that SMC-specific knockout of HIF-1α in mice results in pulmonary hypertension and increased phosphorylation of myosin light chain under normoxia and following hypoxic exposure [17].

Nitric oxide (NO) is a well-known pulmonary vasodilator but also plays a key role in lung vascular growth during fetal and neonatal life. For example, the lungs of late fetal and neonatal eNOS-deficient mice have striking abnormalities of vascularization [18] and are more susceptible to failed vascular growth following exposure to mild hypoxia [19]. Recent studies also suggest that VEGF-induced lung angiogenesis is in part mediated by NO. For instance, inhibition of VEGF receptors decreased lung eNOS protein expression and NO production, and lung vascular growth could be restored by treatment with inhaled NO [20, 21].

The perinatal period is also distinguished by specific circulatory patterns required for placental respiration. Fetal blood flow bypasses the lungs via the foramen ovale and the ductus arteriosus, thereby directing oxygenated blood to the systemic circulation. Multiple mechanisms maintain high pulmonary vascular resistance and low pulmonary blood flow in the fetus, including low-oxygen tension, low basal production of vasodilator products (such as PgI_2 and NO), and increased production of vasoconstrictors (such as endothelin-1). Maternal hyperoxygenation has no effect on human fetal pulmonary blood flow prior to 26 weeks gestation but produces significant increases by mid-third trimester, suggesting a developmentally regulated capacity of the pulmonary circulation to sense and respond to changes in oxygen tension [22]. The low-oxygen environment of the fetus maintains low production of vasoactive mediators such as nitric oxide and prostacyclin. For example, maternal hyperoxygenation activates endothelial nitric oxide synthase and increases pulmonary blood flow to postnatal levels in fetal lambs [23].

At birth, the lung adapts to replace the placenta as the organ of gas exchange. This is facilitated by a dramatic decrease in pulmonary vascular resistance, regulated by complex physiological and biochemical processes, and resulting in an eight- to ten-fold increase in pulmonary blood flow [24]. Hypoxic pulmonary vasoconstriction is reversed at birth by the sudden increase in lung oxygenation when the newborn takes its first breath. If this process is altered by abnormal lung vascular development and/or vascular dysfunction, pulmonary hypertension and its attendant complications will result. In addition, the responsiveness to hypoxia is retained into adulthood, and pulmonary hypertension can be triggered by respiratory failure, other hypoxic lung disease, travel to high altitude, or intermittent obstructive sleep apnea.

Neonatal Pulmonary Vascular Disease

Persistent pulmonary hypertension of the newborn (PPHN) describes the failure of normal pulmonary vascular adaptation at birth and is characterized by elevated pulmonary vascular resistance and right-to-left extrapulmonary shunting of deoxygenated blood that produces severe hypoxemia [25]. The PPHN syndrome complicates the course of approximately 10 % of term and preterm infants with respiratory failure and carries a significant risk of death, pulmonary morbidity, and neurodevelopmental impairment. Early, severe PPHN is associated with pathologic remodeling in utero, with findings of medial thickening of small pulmonary arteries and abnormal extension of vascular smooth muscle to nonmuscular arteries (Fig. 10.1) [26]. The extent

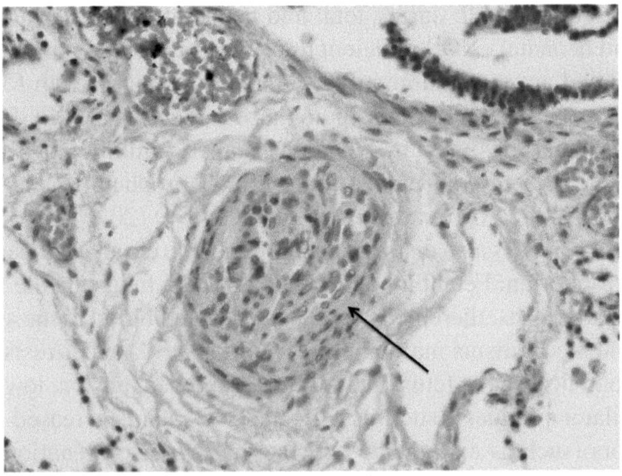

Fig. 10.1 Representative histology of a pulmonary vessel from an infant that died with asphyxia and severe PPHN, demonstrating dramatic smooth muscle cell thickening around pulmonary arteries (*arrow*) and mild adventitial proliferation. (**b**) Lung photomicrograph with elastin staining from an infant with BPD-associated pulmonary hypertension and organizing pneumonia. A thickened medial layer, double elastic lamina, and modest proliferation of the adventitia are noted (*arrow*). Both examples indicate a lack of the intimal proliferation that characterizes adult pulmonary hypertension

of pulmonary vascular remodeling correlates with the severity of the disease, although the in utero abnormalities that alter the development and adaptation of the pulmonary circulation remain poorly understood. Levels of eNOS expression are decreased in umbilical venous endothelial cell cultures from human infants with meconium staining who develop PPHN [27]. Circulating levels of cGMP are decreased [28], and endothelin (ET-1) levels are elevated in PPHN infants [29, 30].

Chronic intrauterine hypoxia can occur due to high altitude, cigarette smoking, anemia, placental insufficiency, and numerous other causes. Chronic fetal hypoxia is commonly associated with fetal intrauterine growth restriction and increases the risk of pulmonary hypertension [31]. During chronic hypoxia, sustained constriction of pulmonary arteries causes pulmonary arterial hypertension leading to right ventricular hypertrophy. In addition, acute perinatal asphyxia or alveolar hypoxia arising from lung parenchymal disorders such as meconium aspiration syndrome, respiratory distress syndrome, and pneumonia can cause structurally normal pulmonary vessels to constrict.

Bronchopulmonary dysplasia (BPD) occurs most frequently in extremely preterm infants born before 28 weeks gestation and is characterized by alveolar simplification, loss of small pulmonary arteries, and decreased capillary density [16]. Pulmonary hypertension (PH) complicates the course of up to a third of infants with BPD. Over time, BPD-associated PH contributes to ongoing hypoxemia, which induces further vascular remodeling and right ventricular hypertrophy.

Preterm birth exposes the lung to ambient oxygen concentrations that are severalfold higher than fetal levels, and supplemental oxygen is frequently required to treat respiratory failure. The preterm lung is ill equipped to cope with increased ambient oxygen concentrations, and exposure to hyperoxia during this developmentally sensitive period increases lipid and protein oxidation products [32] and disrupts normal parenchymal and vascular lung development [16]. While multiple studies support an association between oxidative stress and BPD, a causal relationship has yet to be definitively established. However, these conditions could rapidly degrade HIF and promote abnormal lung development after preterm birth through disrupting expression of downstream targets such as VEGF and mitochondrial SOD (Fig. 10.2). Mitochondrial SOD (MnSOD) is essential for survival in an aerobic environment, and its deletion leads to early neonatal death and increased sensitivity to oxidant stress [33, 34]. In support of this theory, short-term hyperoxic ventilation of preterm lambs increased PHD2 expression and disrupted pulmonary expression of HIF and its downstream target VEGF [35]. Furthermore, decreased VEGF is evident in the lungs of premature neonates who died with BPD [36], and VEGF expression is decreased in tracheal aspirates from premature infants who later develop BPD [37]. These data suggest that hyperoxia may interrupt the expression of critical factors that promote normal lung vascular and parenchymal development.

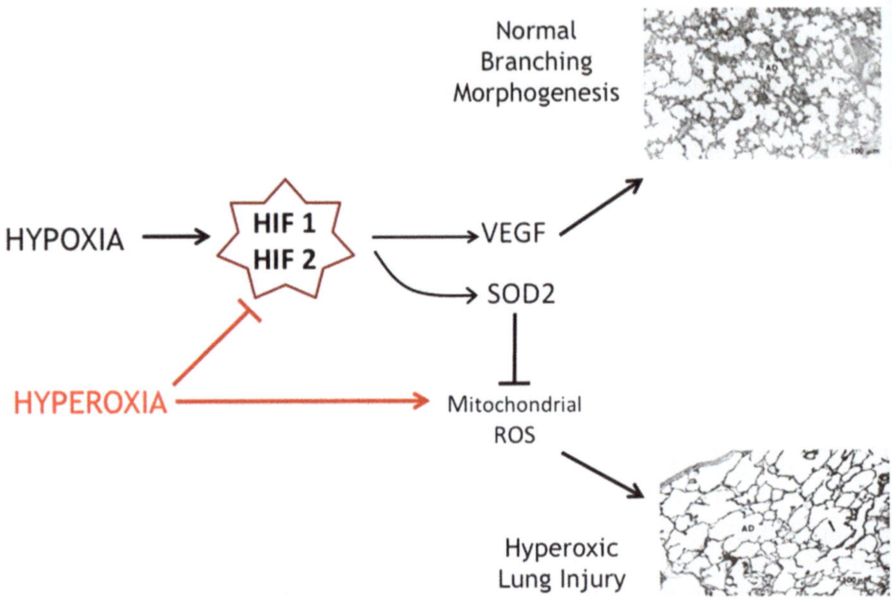

Fig. 10.2 Hyperoxic conditions activate prolyl hydroxylases, which degrade hypoxia-inducible factors (HIF1 and HIF2), key regulators of vascular angiogenesis in fetal life. Reduced HIF activity likely promotes abnormal lung development after preterm birth by disrupting expression of downstream targets such as VEGF and mitochondrial SOD

Experimental Models of Neonatal Pulmonary Vascular Disease

Understanding the mechanisms that produce abnormal lung vascular and parenchymal development and function is important in improving early detection and treatment strategies for infants with pulmonary hypertension. As it is not feasible to study the processes in the human infant, much of our current knowledge is derived from animal models.

Chronic Intrauterine Pulmonary Hypertension

In fetal lambs, ligation or compression of the ductus arteriosus rapidly induces fetal and neonatal pulmonary hypertension. Similar to newborns that die of PPHN, these lambs have medial hypertrophy within the small pulmonary arteries, complete muscularization of normally partially muscularized pulmonary arteries, and extension of muscle to non-muscularized arteries. PPHN lambs exhibit impaired NOS expression and activity [38], decreased responsiveness to NO, decreased expression of soluble guanylate cyclase [39], increased PDE5 expression and activity [40], and increased

ET-1 levels [41]. Recent studies found elevated production of 5-hydroxytryptamine in PPHN lambs [42], which causes fetal pulmonary vasoconstriction via mechanisms involving Rho kinase [43]. Further, Rho kinase activity is elevated in pulmonary artery endothelial cells (PAEC) isolated from PPHN lambs [44] in an ET-1-dependent fashion [45]. Together these data suggest that dysregulation of multiple redox-sensitive pathways contributes to impaired pulmonary vasodilation in PPHN lambs.

Hypoxia-Induced Pulmonary Hypertension

Hypoxia can occur antenatally due to maternal hypoxia or placental dysfunction or postnatally due to acute or chronic cardiopulmonary disease. Pulmonary arterial smooth muscle cells undergo calcium-dependent contraction in response to acute hypoxia, indicating that the O_2 responsiveness is intrinsic to vascular cells. While controversial, evidence suggests that hypoxia activates oxidant signaling in pulmonary vascular cells and that the acute vasoconstriction response is redox-sensitive. During chronic hypoxia, sustained constriction of pulmonary arteries causes pulmonary arterial hypertension, leading to right ventricular hypertrophy. If this condition persists for more than a few weeks, the walls of pulmonary arteries remodel through a process that involves hypertrophy and hyperplasia of smooth muscle cells and in some cases the endothelium [46]. After the vessels have remodeled, the pulmonary hypertension can become refractory to normoxia and vasodilating drugs, most likely because reorganization of the extracellular matrix restricts the ability of the vessel to dilate.

Pregnancy and delivery of sheep at high altitude yields offspring with pulmonary hypertension, coupled with increased constrictor reactivity of isolated pulmonary vessels. This effect persists even if the lamb is returned to sea level and is associated with higher sensitivity to endothelin-1, elevated expression of eNOS and PDE5, and decreased heme oxygenase-1 expression and carbon monoxide production [47]. Exposure of neonatal mice to 2 weeks of chronic hypoxia (12 % O_2) in an airflow chamber causes pulmonary hypertension and vascular remodeling that is mediated in part by ET-1 [48]. Pulmonary hypertension develops in newborn piglets exposed to hypoxia for 3 days and worsens after 10 days of hypoxia [49]. These findings are associated with uncoupled NOS activity and reduced NO bioavailability, associated with excessive oxidant stress [50]. These data suggest that chronic hypoxia may induce pulmonary hypertension via mechanisms similar to those associated with antenatal ductal ligation in lambs.

Hyperoxic Lung Injury

The increase in oxygen tension at birth is one of the most important stimuli to facilitate the fetal to newborn transition, hence high oxygen concentrations and commonly used to treat hypoxemia and reverse pulmonary vasoconstriction in neonates with pulmonary hypertension. Until very recently, hyperoxic gas mixtures were also recommended whenever an infant required resuscitative measures. While the use of hyperoxic gas mixtures has short-term benefits, there is an emerging understanding that hyperoxia may also greatly exaggerate oxidative stress in multiple cellular compartments of the normal and diseased pulmonary vasculature.

Healthy newborn lambs ventilated with 100 % oxygen for the first 30 minutes of life show a more rapid decrease in pulmonary vascular resistance than those ventilated with 21 % oxygen [51]. However, when weaned to 21 % oxygen and studied 4 h later, the high oxygen group display impaired pulmonary vasodilator responses to endogenous and exogenous NO [51]. Further studies in a newborn lamb model of acute perinatal asphyxia confirm that vasodilator responses are decreased following resuscitation with 100 % O_2 and suggest that elevated vascular oxidant stress is involved (Fig. 10.3) [52].

In PPHN lambs born after antenatal ductal ligation, the use of 100 % oxygen does not enhance the decrease in pulmonary vascular resistance relative to ventilation with 21 % or 50 % oxygen, confirming the reduced responsiveness to oxygen in remodeled pulmonary vessels. However, hyperoxia impairs the vasodilator response to inhaled NO and produces striking increases in the activity of cGMP-specific phosphodiesterase [53]. These data suggest that short-term pulmonary vascular benefits of ventilation with high levels of oxygen need to be weighed against longer-lasting adverse effects on vascular reactivity.

Given the additional potential for harm due to oxidant-induced injury, the routine use of high oxygen concentrations for initial delivery room resuscitation is no longer recommended [54]. However, the optimal concentration of oxygen to treat hypoxemic respiratory failure and/or pulmonary hypertension in the NICU setting remains controversial and poorly studied. In clinical practice, only limited information (i.e., calculated alveolar PO_2 and measured arterial PO_2) is typically available to evaluate the relative effect of higher oxygen mixtures on pulmonary vasodilation. The relationship between PaO_2 and PVR has been studied in animal models. In young healthy calves, PVR rises steeply when the PaO_2 falls below 50 mmHg but decreases minimally when the PaO_2 is >50 mmHg [55]. Equivalent studies in neonatal lambs ventilated with 10, 21, 50, or 100 % oxygen show a strikingly similar relationship between PaO_2 and PVR in both healthy lambs and lambs with PPHN. While both the normal and remodeled vasculature constrict vigorously to hypoxia, increasing the $FiO_2 > 50$ % or $PaO_2 > 50–60$ mmHg produces little additional reduction of PVR in both groups [51, 56]. These findings suggest that the vasodilatory effects of supplemental oxygen reach a plateau at about 50 % oxygen or a PaO_2 of 50–60 mmHg and are similar to previous clinical observations in children with bronchopulmonary dysplasia and pulmonary hypertension [57].

Chronic exposure to hyperoxia induces additional factors that may impair normal lung development. Mice and rats are born in the saccular stage of lung development and do not begin alveolarization until postnatal day 5 (P5) and have been used to model the effects of hyperoxia in the immature lung. Chronic exposure to hyperoxia (60 % O_2 and greater) induces pulmonary hypertension and vascular remodeling in newborn rats [58] and mice [59]. Similar to premature infants who die with BPD

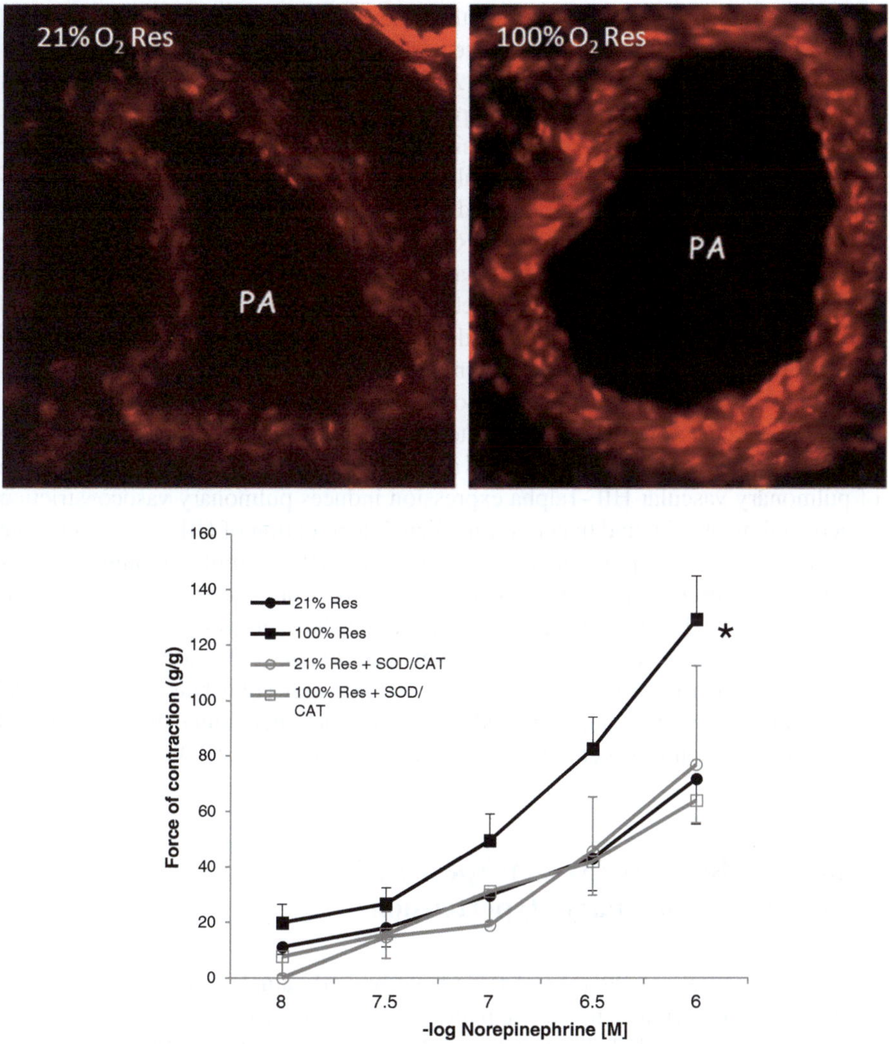

Fig. 10.3 Term lambs were subjected to acute asphyxia and then ventilated with 100 % oxygen or 21 % oxygen for 30 min. Representative sections of the lung were imaged with the superoxide probe dihydroethidium (DHE). The DHE signal was more than twofold higher after 100 % oxygen resuscitation than with 21 % oxygen. The bottom panel shows that the force of contraction of small pulmonary arterial rings to norepinephrine is greater after resuscitation with 100 % oxygen. Contractility is restored to control levels after treatment with superoxide dismutase and catalase. *PA* pulmonary artery

[36, 60], lung VEGF expression is decreased in various rodent models of neonatal hyperoxic lung injury [61–63]. Disrupted pulmonary VEGF leads to abnormal vascular and alveolar development, lung hypoplasia, and pulmonary hypertension [15, 64, 65], and pharmacologic inhibition of VEGF receptors in neonatal rats produces similar impairments in lung alveolarization and vascular growth [20, 66, 67].

Recent research has investigated the role of microRNAs (miRNAs), small conserved noncoding RNAs that regulate posttranscriptional gene expression, in normal and abnormal lung development. Abnormal regulation of miRNAs has been implicated in the development of multiple pathological conditions, and their importance in pulmonary arterial hypertension is becoming more apparent [68]. For instance, microRNA-21 (miR-21) is linked to hypoxia, BMP signaling, and inflammatory signaling and has been proposed as a central regulator in adult-onset pulmonary hypertension [69]. Inhibitors of miR-21 reduced pulmonary vascular remodeling in hypoxia-exposed mice, suggesting that miRNAs could become therapeutic targets [70]. Several studies have monitored changes in lung expression of miRNAs and their target genes in neonatal rodent models of BPD [71, 72]. Continuing studies may provide further insight into the mechanisms of dysregulated gene expression that contribute to the pathophysiology of neonatal pulmonary hypertension and BPD.

The importance of a hypoxic intrauterine environment suggests a potential role for HIFs in the regulation of normal fetal and neonatal lung development. Knockdown of pulmonary vascular HIF-1alpha expression induces pulmonary vasoconstriction in neonatal mice [17], and hyperoxia-mediated degradation of HIFs may contribute to decreased VEGF expression in rodent models of BPD. Similar to human infants that develop BPD [32], markers of oxidative stress are increased in hyperoxia-exposed mice [73] and rats [74]. Additional studies have also implicated alterations in the thioredoxin system by hyperoxia, which could alter HIF and VEGF signaling [75, 76]. Together these data suggest that hyperoxia may contribute to BPD and PH by several distinct mechanisms including oxygen-mediated attenuation of HIF and VEGF signaling and oxygen-derived reactive oxygen species (ROS).

Sources of Reactive Oxygen Species (ROS) in Neonatal Pulmonary Hypertension

The animal models of PPHN and BPD have proved valuable in determining how oxidant stress contributes to disease pathogenesis. Increased superoxide levels have been found in the endothelium and vascular smooth muscle of PPHN pulmonary arteries [77, 78] (Fig. 10.4). Vascular superoxide anions react with NO in a diffusion-limited reaction to form the highly reactive intermediate peroxynitrite (ONOO), which reduces NO bioavailability, constricts the neonatal pulmonary vasculature [79], and inactivates other enzymatic pathways via the formation of nitrotyrosine (Fig. 10.4). In addition, increased production of H_2O_2 in PPHN pulmonary arteries [80, 81] may contribute to decreased eNOS expression [82], impaired cGMP production [83], and elevated PDE5 activity [83]. Figure 10.5 shows impaired NO signaling due to elevated ROS in PPHN. In addition to inhibiting vasodilation, ROS can stimulate vascular SMC growth and contribute to pulmonary vascular remodeling [84, 85]. The following section will discuss potential sources of ROS generation in animal models of pulmonary hypertension.

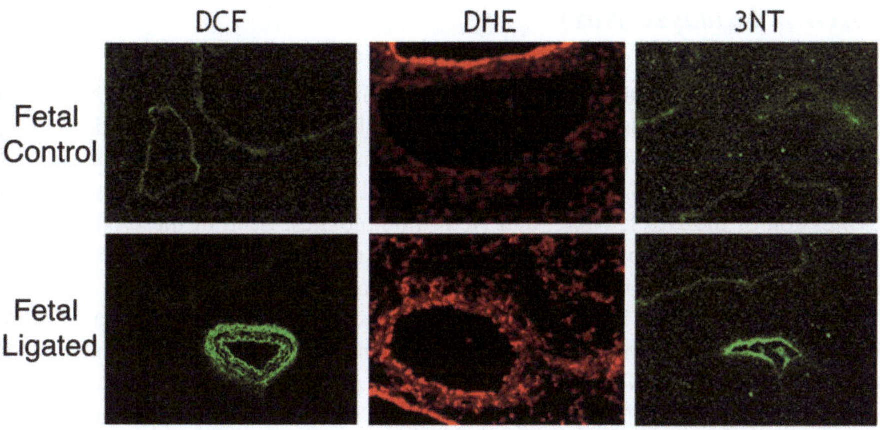

Fig. 10.4 Vascular remodeling is associated with increased ROS in PPHN pulmonary arteries. Frozen sections from control and PPHN lamb lungs were incubated with the H_2O_2-sensitive dye 2″, 7″-dichlorodihydrofluorescein diacetate (Adapted from [81]), with the superoxide-sensitive dye dihydroethidium (Adapted from [77]), or fixed and incubated with an antibody to 3-nitrotyrosine residues (3-NT) (Adapted from [134]), and visualized by fluorescence microscopy

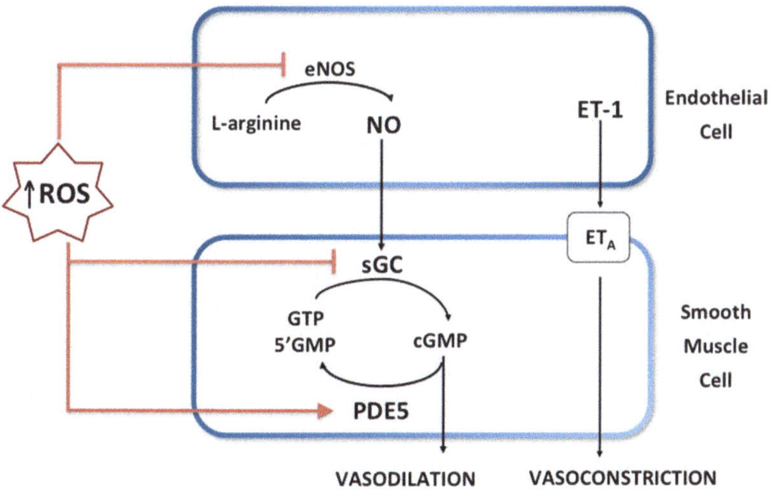

Fig. 10.5 Elevated levels of reactive oxygen species (*ROS*) induces vasoconstriction and pulmonary vascular remodeling in PPHN via multiple mechanisms. In endothelial cells, ROS inhibit eNOS activity resulting in decreased levels of bioavailable NO. In smooth muscle cells, ROS inactivate sGC and activate PDE5 resulting in decreased levels of cGMP. Combined with increased endothelin, these abnormalities promote vascular constriction

NADPH Oxidases (Nox)

Nox enzymes are membrane proteins that transfer electrons from NADPH to molecular oxygen, producing superoxide intracellularly or extracellularly depending on the isoform and subcellular location of the enzyme. Seven Nox enzymatic subtypes have been identified in a wide range of cell types, which include vascular cell expression of the Nox1, Nox2, and Nox4 homologues. Each of these Nox family members is regulated by specific physiological mechanisms, and dysregulation of expression or activity may contribute to pulmonary vascular dysfunction.

Nox1 is expressed in vascular smooth muscle, endothelium, and adventitia [86, 87] and has been localized to membranes, including plasma membranes, caveolae, and endosomes. Nox1 activity requires the association with other protein subunits including $p22^{phox}$, Noxo1, Noxa1, and Rac [87, 88]. Overexpression of Nox1 in vascular smooth muscle increases production of reactive oxygen species, causes eNOS uncoupling, and decreases nitric oxide bioavailability [89]. Nox1 expression is increased in mouse lung cell lines exposed to 72 h hyperoxia, while hyperoxia-induced ROS generation and lung injury is attenuated in Nox1-deficient mice [90]. Increased pulmonary Nox1 expression was reported in neonatal mice with PH after hyperoxia [91] and in neonatal piglets exposed to hypoxia [92], and Nox1 may be involved in pulmonary vascular remodeling in monocrotaline-treated rats [93]. Nox1 levels are unchanged in the lungs of fetal lambs with chronic intrauterine pulmonary hypertension [81], although expression and activity of Nox1 following delivery and hyperoxic ventilation remains under investigation.

The Nox2 isoform is expressed in phagocytic cells as well as in cells comprising the vascular wall and is activated by pathways very similar to Nox1. It requires assembly of protein subunits including $p22^{phox}$, $p47^{phox}$, $p67^{phox}$, and Rac for activation. When assembled in the plasma membrane, Nox2 secretes superoxide into the extracellular space and potentially inhibits NO signaling through formation of peroxynitrite (Fig. 10.6), but when endocytotic vesicles arising from the plasma membrane are formed, the superoxide is secreted into the lysosome [94, 95]. Similar to Nox1, Nox2 produces superoxide that is associated with vasoconstriction, and increased Nox2 subunit expression correlates with increased superoxide levels and impaired pulmonary vasorelaxation in both lamb and piglet models of neonatal pulmonary hypertension [77, 96]. Nox2-deficient mice display attenuated hypoxia-induced ROS, vascular remodeling, and pulmonary hypertension [46] but are not different from controls when exposed to hyperoxia [90]. The Nox inhibitor apocynin improves oxygenation and decreases ROS in ventilated PPHN lambs [97]. In addition, serum-induced PASMC proliferation requires the Nox subunit Rac1, indicating that Nox2 may be important in vascular remodeling [98].

Nox4 is more abundantly expressed in vascular cells relative to Nox1 and Nox2, and has been shown to localize to the mitochondria, endoplasmic reticulum, and nucleus [99–101]. Nox4 was initially thought to require only $p22^{phox}$ for activity and to be constitutively active [102], although recent data indicate that polymerase delta interacting protein 2 (Poldip2) interacts with Nox4 to enhance its activity [103]. Nox4 activity appears to be regulated primarily by expression, and Nox4 has been

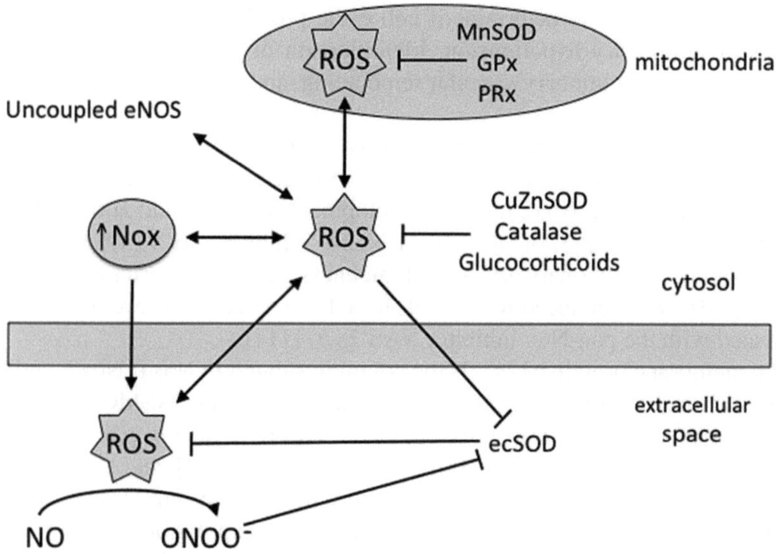

Fig. 10.6 Diagram depicting the interactions between cellular ROS generators and scavengers. In mitochondria, ROS levels are regulated by enzymes including manganese superoxide dismutase (*MnSOD*), glutathione peroxidase (*GPx*), and peroxiredoxin (*PRx*). In the cytosol, NADPH oxidases (*Nox*), xanthine oxidase, and uncoupled endothelial nitric oxide synthase (*eNOS*) generate ROS, while copper/zinc superoxide dismutase (*CuZnSOD*) and catalase scavenge ROS. Nox also contribute to ROS in the extracellular space, while extracellular superoxide dismutase (*ecSOD*) scavenges extracellular superoxide. Increased extracellular superoxide decreases bioavailable NO in the formation of peroxynitrite (*ONOO*), and this vasoconstrictor is removed in the presence of a decomposition catalyst

shown to generate both superoxide and H_2O_2 depending upon the stimulus and cell type [104]. Nox4 mediates PASMC proliferation in response to transforming growth factor (TGF)-β1 [105], urotensin II [106] and hypoxia [107]. Increased Nox4 expression correlates with increased H_2O_2 in PPHN pulmonary arteries, while Nox4 knockdown attenuates ROS levels in SMC isolated from PPHN pulmonary arteries relative to controls [81]. Nox4-derived H_2O_2 may contribute to impaired vasodilation in PPHN lambs via multiple mechanisms including decreased eNOS and sGC expression and increased PDE5 activity [81] (Fig. 10.5). Nox4 has also been implicated in the regulation of cellular migration, growth, and differentiation [108], and increased Nox4 activity may also contribute to pulmonary vascular remodeling in PPHN lambs. An understanding of the downstream targets of ROS-induced PASMC proliferation is just emerging. Cyclin D1 regulates the transition from G_0/G_1 to S phase in the cell cycle, resulting in activation of genes necessary for cell cycle progression. Cyclin D1 expression is increased in PPHN lungs and PASMC [81]. Nox4 siRNA decreases cyclin D1 expression in PPHN PASMC, and intratracheal catalase decreases cyclin D1 expression in the lungs of ventilated PPHN lambs [81]. Together these data indicate a link between increased ROS generation, activation of cell cycle promoters, and pulmonary vascular remodeling in pulmonary hypertension. Further studies are warranted to determine if cell cycle proteins are dysregulated in other models of pulmonary hypertension. Identification of such proteins may improve early detection of pulmonary vascular remodeling, and their targeting may attenuate or reverse this process.

Nox4 may also contribute to hypoxia-induced pulmonary hypertension [109], although data from Nox4 knockout models in mice are just emerging. Novel Nox pharmacologic inhibitors are in early development, which would allow for targeted therapy. Administration of the Nox1/Nox4 inhibitor GKT137831 attenuates hypoxia-induced pulmonary artery wall thickness and right ventricular hypertrophy in mice [110], while reduced ROS levels have been detected in several animal models treated with the pan-Nox inhibitor VAS 2870 [111].

The membrane protein $p22^{phox}$ is the common subunit to Nox1, Nox2, and Nox4, and $p22^{phox}$ expression is elevated in the lungs, PA, and PASMC isolated from PPHN lambs [81]. The transcription factor NFκB regulates $p22^{phox}$ and Nox4 transcription in human aortic smooth muscle cells [112, 113], and NFκB activity is elevated in PA and PASMC of fetal lambs with chronic intrauterine pulmonary hypertension [81]. NFκB inhibition decreases Nox4 expression in PPHN PASMC [81] and attenuates monocrotaline-induced pulmonary arterial hypertension in rats [114, 115], suggesting a potential role for NFκB in triggering pulmonary vascular ROS generation. Hyperoxia selectively activates NFκB in fetal, but not adult, lung fibroblasts [116], suggesting that NFκB could also play a role in the development of neonatal pulmonary hypertension due to hyperoxic lung injury. NFκB activity is regulated through its interaction with the regulatory protein IκB. NFκB is activated by ROS via the phosphorylation of IκB, which targets IκB for protein degradation [117] and allows NFκB to translocate into the nucleus and regulate the transcription of target genes. Together these data suggest the possibility of a feed-forward mechanism in PPHN whereby ROS generated by Nox isoforms increases the activity of key transcription factors such as NFκB, resulting in sustained Nox subunit expression. NFκB could prove to be an attractive target to reduce the amplification of ROS in hypertensive diseases.

Mitochondrial Electron Transport Chain

During normal oxidative phosphorylation, electrons are transferred to molecular oxygen at the terminal cytochrome oxidase in the mitochondrial electron transport chain, generating H_2O. However some electrons are captured by O_2 at more proximal sites, resulting in the formation of superoxide radical. Superoxide generated at Complex I, II, or III can result in oxidant stress in the mitochondrial matrix or intermembrane space [118]. In some conditions such as atherosclerosis, the mitochondria become dysfunctional and the leak of electrons is enhanced [119]. Mitochondrial ROS appear to trigger increased Nox activity in hypoxic pulmonary arteries [120], and increased oxidative stress in the mitochondrial matrix is associated with increased Nox4 expression in pulmonary artery smooth muscle cells isolated from

PPHN lambs [81, 121]. Emerging evidence indicates cross talk between the mitochondria and Nox isoforms [122], and further investigation of the underlying mechanisms may produce novel therapies for pulmonary hypertensive diseases. In addition, selective inhibition of mitochondrial oxidant stress reduces vascular oxidant stress and hypertension in systemic vessels [123], suggesting that effective antioxidant therapy will require targeting of specific subcellular compartments. Indeed, hypoxia decreases ROS levels in the mitochondrial matrix while increasing ROS derived from mitochondrial complex III in the mitochondrial intermembrane space and cytosol [124, 125]. Conversely PPHN and hyperoxia both increase oxidant stress in the mitochondrial matrix and cytosol in fetal sheep PASMC [121, 126], suggesting that precise antioxidant targeting within cellular subcompartments may be required to treat hypertensive diseases with different etiologies.

Endothelial Nitric Oxide Synthase (eNOS)

Endothelial NOS converts L-arginine and molecular oxygen to L-citrulline and the vasodilator NO using O_2 as well as electrons from NADPH. The activity of eNOS is normally regulated by availability of substrate as well as several cofactors including calcium calmodulin, HSP90, and tetrahydrobiopterin (BH_4), and mechanisms that inhibit eNOS activity or attenuate downstream NO signaling can induce vasoconstriction. Elevated ROS decrease eNOS expression and reduce available tetrahydrobiopterin in PPHN lambs [80, 127] and reduce downstream responses to NO by decreasing sGC expression [80] and increasing cGMP-specific phosphodiesterase activity [83].

Impaired eNOS expression and activity may also contribute to abnormal lung and vascular development that produce BPD. Substantial reductions in total NOS activity and expression of all three NOS isoforms have been observed in baboon and rodent models of hyperoxic lung injury [128, 129]. NO not only mediates the downstream effects of VEGF during lung development but may also upregulate VEGF expression [130]. Intrauterine growth restriction (IUGR) increases the risk for BPD [131], although a mechanistic role for oxidant stress has not yet been established. Still, PAECs isolated from a lamb model of IUGR exhibit decreased eNOS expression and phosphorylation, decreased NO production, and attenuated tube formation and migration [132]. eNOS-deficient mice display pulmonary hypertension and vascular remodeling when exposed to mild hypoxia [133], further highlighting the central role of eNOS in maintaining normal vascular tone and development.

Peroxynitrite formation is elevated in PPHN lambs [134] and inhibits NOS activity via mechanisms that include decreased association with HSP90 [135, 136]. eNOS becomes a source of ROS when the enzyme becomes "uncoupled," resulting in incomplete reduction of molecular oxygen with the formation of superoxide. Endothelial NOS uncoupling is evident in hypoxic piglets [137] and can occur via several mechanisms, including degradation or oxidation of cofactors such as tetrahydrobiopterin and HSP90 or by inactivation of the enzyme through increased peroxynitrite levels [134, 138]. Increased Nox activity may be an important trigger for eNOS uncoupling [89, 139], while eNOS uncoupling can promote mitochondrial dysfunction and ROS generation via increased peroxynitrite [140]. Together these data suggest that abnormal regulation of ROS can promote oxidant production from additional sources, thus amplifying and sustaining a pathological state.

Antioxidants

Cells also regulate ROS levels through a wide variety of enzymatic scavengers that are developmentally regulated [141, 142]. The expression of these ROS scavengers is specific to certain subcellular compartments and most antioxidant enzymes are selective for a single type of ROS molecule. Superoxide dismutase (SOD) degrades superoxide to H_2O_2, and H_2O_2 produced by SOD is regulated by its rate of degradation by enzymes such as catalase, glutathione peroxidase, and peroxiredoxins [143].

There are three known forms of superoxide dismutase: Cu/ZnSOD (SOD1), MnSOD (SOD2), and extracellular superoxide dismutase (ecSOD or SOD3). Cu/ZnSOD is expressed in the cytosol and intermembrane space of the mitochondria, and ecSOD is secreted to the extracellular space where it binds to the extracellular matrix. Lung expression of all three SOD isoforms is greater in the lungs of 8-week-old mice relative to 1-week-old mice [91], suggesting that the newborn lung is more susceptible to increased oxidant stress due to limited antioxidant capacity. MnSOD is localized to the mitochondria and is responsible for protecting against excessive mitochondrial superoxide generation. Mice with homozygous deletion of the MnSOD gene die from oxidative stress shortly after birth [144], and mice lacking one allele of MnSOD develop hypertension with aging and in response to a high-salt diet [145]. MnSOD levels are reduced in pulmonary artery endothelial cells of PPHN lambs [146] but were unexpectedly higher in PPHN PASMC [81, 121], highlighting the complex cell-specific regulatory mechanisms.

ecSOD is predominantly synthesized by the vascular smooth muscle cells and comprises a significant component of the total SOD activity in the blood vessel wall [147]. It is most highly expressed in the lung [148, 149] and is present in high concentrations between the endothelium and smooth muscle surrounding blood vessels, the same domain that NO must pass through to stimulate smooth muscle relaxation. This suggests that high concentrations of ecSOD in this region are especially important in maintaining low superoxide concentrations and preserving NO function [150]. ecSOD activity is decreased in PPHN lungs and PASMC [126], potentially via a mechanism involving Nox4-derived H_2O_2 that oxidizes copper at the enzyme active site [81]. Decreased ecSOD activity is predicted to decrease bioavailable NO via the formation of peroxynitrite, while protein nitration inhibits ecSOD activity [151] indicating a potential feed-forward mechanism of enzyme inhibition. Therapeutic intervention to maintain ecSOD activity is therefore predicted to be beneficial in the treatment of cardiovascular disease. Overexpression of ecSOD ameliorates pulmonary hypertension in rats [152], protects lung development [153], and attenuates pulmonary vascular remodeling in hypoxic mice [154]. The H_2O_2 scavenger catalase also enhances ecSOD activity and decreases PA superoxide in PPHN lambs [126]. The development of novel proteins such as a chimeric SOD2/3 [155] may allow for more sustained pulmonary vascular antioxidant activity. Treatment with a peroxynitrite decomposition catalyst attenuates hyperoxia-induced decreases in VEGF expression and enhances alveolar formation in neonatal rats [156] suggesting that this approach may also increase ecSOD activity and improve NO signaling in PPHN.

Catalase functions to catalyze the decomposition of hydrogen peroxide to water and oxygen. However, mice deficient in catalase develop normally, indicating that other complementary antioxidant systems must be present [157]. The glutathione peroxidase (GPx) and peroxiredoxin (PRx) systems utilize reduced glutathione to scavenge H_2O_2 and are critical for minimizing oxidant stress and for regulating redox signaling pathways. GPx levels are decreased in the lungs of patients with IPAH [158], although genetic deletion of GPx-1 does not affect the increase in aortic pressure or vascular hypertrophy induced by angiotensin II, and GPx-1 levels are unchanged in the lungs of PPHN lambs [80, 159]. By contrast, deletion of peroxiredoxin 1 induces hemolytic anemia and a significant decrease in lifespan [160], while deletion of peroxiredoxin 3 leads to oxidant-mediated lung inflammation and an enhanced susceptibility to LPS challenge [161].

Therapeutic Considerations

While the data presented above strongly suggest a potential role for antioxidant therapy in the treatment of neonatal pulmonary hypertension, clinical trials of antioxidant therapy for a wide range of diseases including BPD have had only limited success [162]. Antioxidant therapy may be ineffective once the disease has progressed beyond a critical stage, suggesting that early detection strategies may improve treatment outcomes. ROS scavenging may also interfere with normal signaling pathways in the developing lung. Furthermore, oxidant stress may be localized to specific subcellular compartments in different diseases, which may limit the efficacy of nonspecific antioxidants. Thus, highly targeted therapies may be required to maximize the potential of antioxidants in the treatment of hypertensive diseases.

Inhaled NO therapy improves pulmonary hypertension resulting from impaired eNOS signaling, but could potentially increase peroxynitrite formation, resulting in nitration and inhibition of endogenous eNOS activity [136]. Intratracheal antioxidants decrease ROS, increase eNOS expression, and normalize tetrahydrobiopterin levels in PPHN lambs [97, 127], suggesting that antioxidants could enhance the vascular effects of inhaled NO. Intratracheal recombinant human SOD (rhSOD) also reduces ONOO-mediated protein nitration [134], decreases PDE5 activity, and increases cGMP in the pulmonary arteries of ventilated PPHN lambs [163], suggesting that antioxidant therapy may improve NO signaling at multiple points in the pathway. The SOD mimetic MnTE-2-PyP attenuates hypoxia-induced pulmonary vascular remodeling and pulmonary hypertension in mice [164], and the SOD mimetic M40403 improves NO-mediated relaxation in PAs isolated from hypoxic piglets [137].

Administration of intratracheal catalase to ventilated PPHN lambs improves oxygenation, increases ecSOD activity, and decreases PA superoxide levels [126]. Further, intratracheal catalase decreases PDE5 activity and increases cGMP in the PAs of ventilated PPHN lambs [121]. These data suggest that H_2O_2 scavenging improves NO signaling in PPHN, possibly by increasing ecSOD activity (Fig. 10.7). Accordingly, treatment with catalase improves NO-mediated vasodilation in PAs

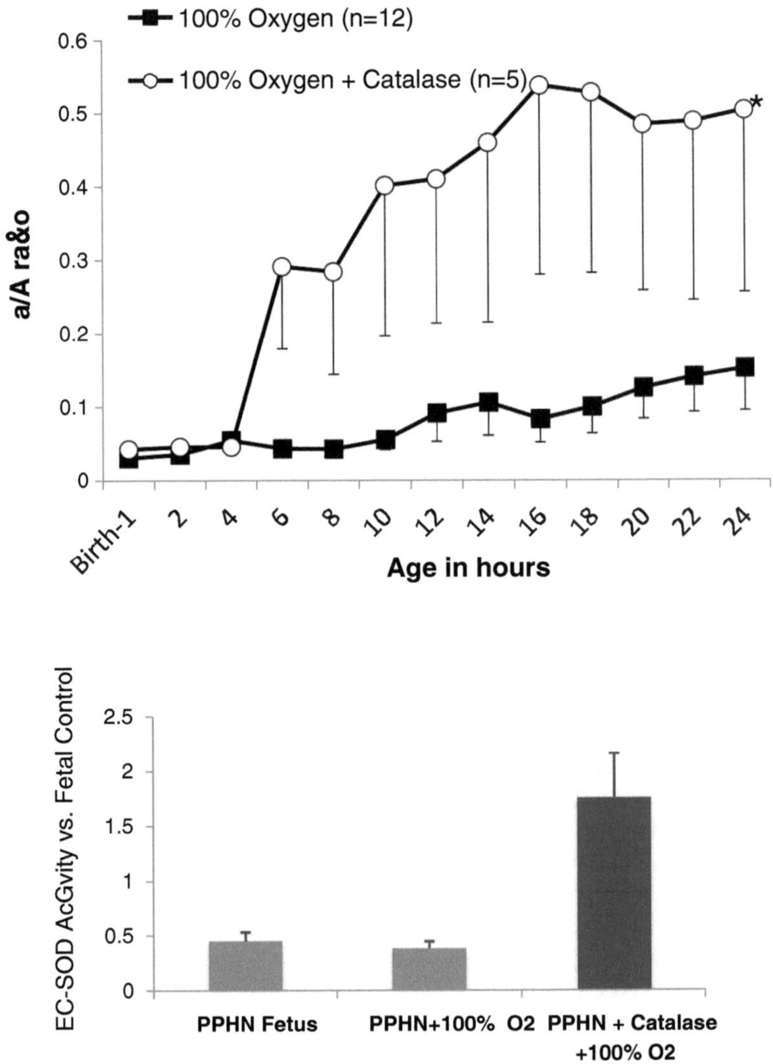

Fig. 10.7 PPHN lambs were ventilated with 100 % O_2, and five were treated with 15,000 U/kg of PEG-catalase (open symbols). Catalase-treated lambs showed sustained improvement in oxygenation at the end of the 24 h protocol vs. O_2-treated controls. The bottom panel shows that catalase increased activity of EC-SOD, which reduced superoxide levels as measured by DHE

isolated from PPHN lambs [80] and from hypoxic piglets [137]. Anti-inflammatory glucocorticoids are used to treat neonates with MAS [165], and hydrocortisone improves oxygenation, decreases PA ROS, decreases PDE5 activity, and increases cGMP levels in PPHN lambs [50]. Figure 10.6 illustrates the localized interaction between antioxidants and ROS.

Overexpression of GTP cyclohydrolase, the enzyme catalyzing the rate-limiting step in BH_4 synthesis, attenuates hypoxic pulmonary hypertension [166], which could be due to improved NOS function. GTP-cyclohydrolase expression is diminished in PPHN lambs and restored by treatment with antioxidants [127]. Clinical trials with oral BH_4 have been conducted in adults with pulmonary hypertension [167], and future studies may determine the efficacy of this approach in the treatment of children. Other approaches to increase NOS activity include supplementation with L-citrulline. eNOS generates NO from the oxidation of L-arginine, and L-citrulline is formed as a by-product. L-citrulline is converted back to L-arginine by enzymes that colocalize with eNOS in the endothelium, and L-citrulline may be a more potent activator of eNOS by providing a supply of L-arginine in close proximity to eNOS. Oral supplementation with L-citrulline attenuates pulmonary hypertension and increases NO production in newborn piglets exposed to hypoxia [168]. Moreover, hyperoxia decreases plasma L-arginine and L-citrulline levels in a rat model of BPD, while supplementation with L-citrulline preserves lung alveolar and vascular development and attenuates pulmonary hypertension and vascular remodeling [169]. L-arginine can also be converted to urea and L-ornithine by arginase enzymes expressed in the lung, and increased arginase activity decreases NO production from eNOS by competing for the same substrate. Hypoxia induces human PASMC proliferation via arginase [170], while hypoxic mice deficient in MAP kinase signaling display elevated arginase expression, exaggerated pulmonary hypertension and vascular remodeling, and decreased eNOS expression relative to wild type mice [171]. Decreased L-arginine promotes eNOS uncoupling [172], suggesting that therapies including L-citrulline supplementation and arginase inhibition may attenuate eNOS uncoupling and stimulate NO production in hypertensive diseases.

Augmenting cGMP concentrations through other routes may also prevent or reverse pulmonary vascular remodeling due to oxidant stress. Novel activators of sGC such as cinaciguat (BAY 58–2667) are functional in the enzyme's oxidized, NO-resistant state [173, 174], and phosphodiesterase inhibitors such as sildenafil are a logical approach to overcome increased PDE5 activity. Sildenafil treatment of neonatal rats exposed to hyperoxia normalizes lung alveolar and vascular development and attenuates pulmonary hypertension and vascular remodeling [175]. Furthermore, sildenafil treatment after initiation of hyperoxia restores HIF-1 expression [176] and significantly reduces medial wall thickness and right ventricular hypertrophy, suggesting hyperoxia-induced vascular remodeling is reversible [177]. Iron is an important regulator of ROS levels, and an iron chelator prevents hypoxia-induced pulmonary hypertension and pulmonary vascular remodeling in rats [68]. Future studies are needed to determine if similar approaches are effective in the neonatal pulmonary vasculature.

Table 10.1 Gaps in knowledge, research strategies, and their therapeutic implications in improving treatment outcomes in infants with pulmonary hypertension

Gaps in knowledge	Research strategies	Therapeutic implications
Few biomarkers and early predictors of disease	Biomarker identification in animal models; development of maternal and neonatal genetic and metabolite screening	Earlier identification and clinical intervention for patients at highest risk
Limited understanding of oxygen toxicity	Animal studies to monitor changes in ROS levels, gene expression, and metabolite levels in response to hypoxia and hyperoxia	Clinical approaches that improve oxygen delivery while minimizing ROS toxicity
Disease-specific identity and subcellular location of ROS molecules	Redox sensitive probes targeted to specific ROS and subcellular locations	Development of successful therapies that target specific antioxidants to specific locations/cell types
Precise roles of pro-oxidants and antioxidants in disease	Genetically modified animal models and knock-out and knock-in mice	Development of targeted molecular inhibitors and activators of specific genes

Summary and Conclusions

Oxidant stress plays multiple roles in the susceptibility and pathogenesis of pulmonary hypertension in preterm infants and newborns. Increased oxidant stress can arise due to exposure to a variety of injurious stimuli, including hyperoxia and hypoxia, which activate ROS generators, and/or inactivate endogenous antioxidants. The mechanisms differ based on the cell type and underlying disease process, which will have implications for the most effective therapeutic approach. As noted in Table 10.1, further investigation into the specific mechanisms involved, along with the development of novel antioxidant and non-antioxidant therapies, may improve the outcomes for infants with pulmonary hypertension.

References

1. Angus DC, Linde-Swirble WT, Clermont G, Griffin MF, Clark RH. Epidemiology of neonatal respiratory failure in the United States. Am J Respir Crit Care Med. 2001;164:1154–60.
2. Rhoades R, Packer C, Roepke D, Jin N, Meiss R. Reactive oxygen species alter contractile properties of pulmonary arterial smooth muscle. Can J Physiol Pharmacol. 1990;68:1581–9.
3. Demiryürek A, Wadsworth R. Superoxide in the pulmonary circulation. Pharmacol Ther. 1999;84:355–65.
4. Kukreja RC, Kontos HA, Hess ML, Ellis EF. PGH synthase and lipoxygenase generate -superoxide in the presence of NADH or NADPH. Circ Res. 1986;59:612–9.
5. Mueller CF, Laude K, McNally JS, Harrison DG. ATVB in focus: redox mechanisms in blood vessels. Arterioscler Thromb Vasc Biol. 2005;25:274–8.
6. Archer SL, Gomberg-Maitland M, Maitland ML, Rich S, Garcia JG, Weir EK. Mitochondrial metabolism, redox signaling, and fusion: a mitochondria-ROS-HIF-1alpha-Kv1.5 O2-sensing

pathway at the intersection of pulmonary hypertension and cancer. Am J Physiol Heart Circ Physiol. 2008;294:H570–8.
7. Hilenski LL, Clempus RE, Quinn MT, Lambeth JD, Griendling KK. Distinct subcellular localizations of Nox1 and Nox4 in vascular smooth muscle cells. Arterioscler Thromb Vasc Biol. 2004;24:677–83.
8. Lyle AN, Griendling KK. Modulation of vascular smooth muscle signaling by reactive oxygen species. Physiology (Bethesda). 2006;21:269–80.
9. Waypa GB, Schumacker PT. Oxygen sensing in hypoxic pulmonary vasoconstriction: using new tools to answer an age-old question. Exp Physiol. 2008;93:133–8.
10. Wolin MS, Ahmad M, Gupte SA. Oxidant and redox signaling in vascular oxygen sensing mechanisms: basic concepts, current controversies, and potential importance of cytosolic NADPH. Am J Physiol Lung Cell Mol Physiol. 2005;289:L159–73.
11. Hall SM, Hislop AA, Pierce CM, Haworth SG. Prenatal origins of human intrapulmonary arteries: formation and smooth muscle maturation. Am J Respir Cell Mol Biol. 2000;23: 194–203.
12. Djonov V, Schmid M, Tschanz SA, Burri PH. Intussusceptive angiogenesis: its role in embryonic vascular network formation. Circ Res. 2000;86:286–92.
13. Gerber HP, Hillan KJ, Ryan AM, Kowalski J, Keller GA, Rangell L, Wright BD, Radtke F, Aguet M, Ferrara N. VEGF is required for growth and survival in neonatal mice. Development. 1999;126:1149–59.
14. Rajatapiti P, van der Horst IW, de Rooij JD, Tran MG, Maxwell PH, Tibboel D, Rottier R, de Krijger RR. Expression of hypoxia-inducible factors in normal human lung development. Pediatr Dev Pathol. 2008;11:193–9.
15. Compernolle V, Brusselmans K, Acker T, Hoet P, Tjwa M, Beck H, Plaisance S, Dor Y, Keshet E, Lupu F, Nemery B, Dewerchin M, Van Veldhoven P, Plate K, Moons L, Collen D, Carmeliet P. Loss of HIF-2alpha and inhibition of VEGF impair fetal lung maturation, whereas treatment with VEGF prevents fatal respiratory distress in premature mice. Nat Med. 2002;8:702–10. Epub 2002 Jun 10.
16. Thebaud B, Abman SH. Bronchopulmonary dysplasia: where have all the vessels gone? Roles of angiogenic growth factors in chronic lung disease. Am J Respir Crit Care Med. 2007;175:978–85.
17. Kim YM, Barnes EA, Alvira CM, Ying L, Reddy S, Cornfield DN. Hypoxia-inducible factor-1alpha in pulmonary artery smooth muscle cells lowers vascular tone by decreasing myosin light chain phosphorylation. Circ Res. 2013;112:1230–3.
18. Han RN, Babaei S, Robb M, Lee T, Ridsdale R, Ackerley C, Post M, Stewart DJ. Defective lung vascular development and fatal respiratory distress in endothelial NO synthase- deficient mice: a model of alveolar capillary dysplasia? Circ Res. 2004;94:1115–23. Epub 2004 Mar 11.
19. Balasubramaniam V, Tang JR, Maxey A, Plopper CG, Abman SH. Mild hypoxia impairs alveolarization in the endothelial nitric oxide synthase-deficient mouse. Am J Physiol Lung Cell Mol Physiol. 2003;284:L964–71.
20. Tang JR, Markham NE, Lin YJ, McMurtry IF, Maxey A, Kinsella JP, Abman SH. Inhaled nitric oxide attenuates pulmonary hypertension and improves lung growth in infant rats after neonatal treatment with a VEGF receptor inhibitor. Am J Physiol Lung Cell Mol Physiol. 2004;287:L344–51.
21. Balasubramaniam V, Maxey AM, Morgan DB, Markham NE, Abman SH. Inhaled NO restores lung structure in eNOS-deficient mice recovering from neonatal hypoxia. Am J Physiol Lung Cell Mol Physiol. 2006;291:L119–27.
22. Rasanen J, Wood DC, Debbs RH, Cohen J, Weiner S, Huhta JC. Reactivity of the human fetal pulmonary circulation to maternal hyperoxygenation increases during the second half of pregnancy: a randomized study. Circulation. 1998;97:257–62.
23. Tiktinsky MH, Morin III FC. Increasing oxygen tension dilates fetal pulmonary circulation via endothelium-derived relaxing factor. Am J Physiol Heart Circ Physiol. 1993;265:H376–80.
24. Dawes GS, Mott JC, Widdicombe JG, Wyatt DG. Changes in the lungs of the newborn lamb. J Physiol. 1953;121:141–62.

25. Steinhorn RH. Neonatal pulmonary hypertension. Pediatr Crit Care Med. 2010;11:S79–84.
26. Haworth S, Reid L. Persistent fetal circulation: newly recognized structural features. J Pediatr. 1976;88:614–20.
27. Esterlita M, Villanueva T, Zaher FM, Svinarick DM, Konduri GG. Decreased gene expression of endothelial nitric oxide synthase in newborns with persistent pulmonary hypertension. Pediatr Res. 1998;44:338–43.
28. Christou H, Adatia I, Van Marter LJ, Kane JW, Thompson JE, Stark AR, Wessel DL, Kourembanas S. Effect of inhaled nitric oxide on endothelin-1 and cyclic guanosine 5′-monophosphate plasma concentrations in newborn infants with persistent pulmonary hypertension. J Pediatr. 1997;130:603–11.
29. Kumar P, Kazzi NJ, Shankaran S. Plasma immunoreactive endothelin-1 concentrations in infants with persistent pulmonary hypertension of the newborn. Am J Perinatol. 1996;13: 335–41.
30. Rosenberg AA, Kennaugh J, Koppenhafer SL, Loomis M, Chatfield BA, Abman SH. Elevated immunoreactive endothelin-1 levels in newborn infants with persistent pulmonary hypertension. J Pediatr. 1993;123:109–14.
31. Niermeyer S. Cardiopulmonary transition in the high altitude infant. High Alt Med Biol. 2003;4:225–39.
32. Saugstad OD. Oxygen and oxidative stress in bronchopulmonary dysplasia. J Perinat Med. 2010;38:571–7.
33. Scortegagna M, Ding K, Oktay Y, Gaur A, Thurmond F, Yan LJ, Marck BT, Matsumoto AM, Shelton JM, Richardson JA, Bennett MJ, Garcia JA. Multiple organ pathology, metabolic abnormalities and impaired homeostasis of reactive oxygen species in Epas1–/– mice. Nat Genet. 2003;35:331–40.
34. Asikainen TM, Huang TT, Taskinen E, Levonen AL, Carlson E, Lapatto R, Epstein CJ, Raivio KO. Increased sensitivity of homozygous Sod2 mutant mice to oxygen toxicity. Free Radic Biol Med. 2002;32:175–86.
35. Grover TR, Asikainen TM, Kinsella JP, Abman SH, White CW. Hypoxia-inducible factors HIF-1alpha and HIF-2alpha are decreased in an experimental model of severe respiratory distress syndrome in preterm lambs. Am J Physiol Lung Cell Mol Physiol. 2007;292: L1345–51.
36. Bhatt AJ, Pryhuber GS, Huyck H, Watkins RH, Metlay LA, Maniscalco WM. Disrupted pulmonary vasculature and decreased vascular endothelial growth factor, Flt-1, and TIE-2 in human infants dying with bronchopulmonary dysplasia. Am J Respir Crit Care Med. 2001;164:1971–80.
37. Lassus P, Turanlahti M, Heikkila P, Andersson LC, Nupponen I, Sarnesto A, Andersson S. Pulmonary vascular endothelial growth factor and Flt-1 in fetuses, in acute and chronic lung disease, and in persistent pulmonary hypertension of the newborn. Am J Respir Crit Care Med. 2001;164:1981–7.
38. Shaul PW, Yuhanna IS, German Z, Chen Z, Steinhorn RH, Morin 3rd FC. Pulmonary endothelial NO synthase gene expression is decreased in fetal lambs with pulmonary hypertension. Am J Physiol. 1997;272:L1005–12.
39. Steinhorn RH, Russell JA, Morin 3rd FC. Disruption of cGMP production in pulmonary arteries isolated from fetal lambs with pulmonary hypertension. Am J Physiol. 1995;268: H1483–9.
40. Black SM, Johengen MJ, Soifer SJ. Coordinated regulation of genes of the nitric oxide and endothelin pathways during the development of pulmonary hypertension in fetal lambs. Pediatr Res. 1998;44:821–30.
41. Ivy DD, Ziegler JW, Dubus MF, Fox JJ, Kinsella JP, Abman SH. Chronic intrauterine pulmonary hypertension alters endothelin receptor activity in the ovine fetal lung. Pediatr Res. 1996;39:435–42.
42. Delaney C, Gien J, Roe G, Isenberg N, Kailey J, Abman SH. Serotonin contributes to high pulmonary vascular tone in a sheep model of persistent pulmonary hypertension of the newborn. Am J Physiol Lung Cell Mol Physiol. 2013;304:L894–901.

43. Delaney C, Gien J, Grover TR, Roe G, Abman SH. Pulmonary vascular effects of serotonin and selective serotonin reuptake inhibitors in the late-gestation ovine fetus. Am J Physiol Lung Cell Mol Physiol. 2011;301:L937–44.
44. Gien J, Seedorf GJ, Balasubramaniam V, Tseng N, Markham N, Abman SH. Chronic intrauterine pulmonary hypertension increases endothelial cell Rho kinase activity and impairs angiogenesis in vitro. Am J Physiol Lung Cell Mol Physiol. 2008;295:L680–7.
45. Gien J, Tseng N, Seedorf G, Roe G, Abman SH. Endothelin-1 impairs angiogenesis in vitro through Rho-kinase activation after chronic intrauterine pulmonary hypertension in fetal sheep. Pediatr Res. 2013;73:252–62.
46. Liu J, Zelko I, Erbynn E, Sham J, Folz R. Hypoxic pulmonary hypertension: role of superoxide and NADPH oxidase (gp91phox). Am J Physiol Lung Cell Mol Physiol. 2006;290:L2–10.
47. Herrera EA, Pulgar VM, Riquelme RA, Sanhueza EM, Reyes RV, Ebensperger G, Parer JT, Valdez EA, Giussani DA, Blanco CE, Hanson MA, Llanos AJ. High-altitude chronic hypoxia during gestation and after birth modifies cardiovascular responses in newborn sheep. Am J Physiol Regul Integr Comp Physiol. 2007;292:R2234–40.
48. Ambalavanan N, Bulger A, Murphy-Ullrich J, Oparil S, Chen Y. Endothelin-A receptor blockade prevents and partially reverses neonatal hypoxic pulmonary vascular remodeling. Pediatr Res. 2005;57:631–6.
49. Fike CD, Kaplowitz MR, Rehorst-Paea LA, Nelin LD. L-Arginine increases nitric oxide production in isolated lungs of chronically hypoxic newborn pigs. J Appl Physiol. 2000; 88:1797–803.
50. Perez M, Lakshminrusimha S, Wedgwood S, Czech L, Gugino SF, Russell JA, Farrow KN, Steinhorn RH. Hydrocortisone normalizes oxygenation and cGMP regulation in lambs with persistent pulmonary hypertension of the newborn. Am J Physiol Lung Cell Mol Physiol. 2012;302:L595–603.
51. Lakshminrusimha S, Russell JA, Steinhorn RH, Swartz DD, Ryan RM, Gugino SF, Wynn KA, Kumar VH, Mathew B, Kirmani K, Morin 3rd FC. Pulmonary hemodynamics in neonatal lambs resuscitated with 21 %, 50 %, and 100 % oxygen. Pediatr Res. 2007;62:313–8.
52. Lakshminrusimha S, Steinhorn RH, Wedgwood S, Savorgnan F, Nair J, Mathew B, Gugino SF, Russell JA, Swartz DD. Pulmonary hemodynamics and vascular reactivity in asphyxiated term lambs resuscitated with 21 and 100 % oxygen. J Appl Physiol. 2011;111:1441–7.
53. Lakshminrusimha S, Swartz DD, Gugino SF, Ma CX, Wynn KA, Ryan RM, Russell JA, Steinhorn RH. Oxygen concentration and pulmonary hemodynamics in newborn lambs with pulmonary hypertension. Pediatr Res. 2009;66:539–44.
54. Kattwinkel J, Perlman JM, Aziz K, Colby C, Fairchild K, Gallagher J, Hazinski MF, Halamek LP, Kumar P, Little G, McGowan JE, Nightengale B, Ramirez MM, Ringer S, Simon WM, Weiner GM, Wyckoff M, Zaichkin J. Part 15: neonatal resuscitation: 2010 American Heart Association Guidelines for Cardiopulmonary Resuscitation and Emergency Cardiovascular Care. Circulation. 2010;122:S909–19.
55. Rudolph AM, Yuan S. Response of the pulmonary vasculature to hypoxia and H+ ion concentration changes. J Clin Invest. 1966;45:399–411.
56. Lakshminrusimha S, Porta NF, Farrow KN, Chen B, Gugino SF, Kumar VH, Russell JA, Steinhorn RH. Milrinone enhances relaxation to prostacyclin and iloprost in pulmonary arteries isolated from lambs with persistent pulmonary hypertension of the newborn. Pediatr Crit Care Med. 2009;10:106–12.
57. Mourani PM, Ivy DD, Gao D, Abman SH. Pulmonary vascular effects of inhaled nitric oxide and oxygen tension in bronchopulmonary dysplasia. Am J Respir Crit Care Med. 2004;170:1006–13.
58. Koppel R, Han RN, Cox D, Tanswell AK, Rabinovitch M. Alpha 1-antitrypsin protects neonatal rats from pulmonary vascular and parenchymal effects of oxygen toxicity. Pediatr Res. 1994;36:763–70.
59. Warner BB, Stuart LA, Papes RA, Wispe JR. Functional and pathological effects of prolonged hyperoxia in neonatal mice. Am J Physiol. 1998;275:L110–7.

60. Abman SH. Impaired vascular endothelial growth factor signaling in the pathogenesis of neonatal pulmonary vascular disease. Adv Exp Med Biol. 2010;661:323–35.
61. Maniscalco WM, Watkins RH, D'Angio CT, Ryan RM. Hyperoxic injury decreases alveolar epithelial cell expression of vascular endothelial growth factor (VEGF) in neonatal rabbit lung. Am J Respir Cell Mol Biol. 1997;16:557–67.
62. Hosford GE, Olson DM. Effects of hyperoxia on VEGF, its receptors, and HIF-2alpha in the newborn rat lung. Am J Physiol Lung Cell Mol Physiol. 2003;285:L161–8.
63. Thebaud B, Ladha F, Michelakis ED, Sawicka M, Thurston G, Eaton F, Hashimoto K, Harry G, Haromy A, Korbutt G, Archer SL. Vascular endothelial growth factor gene therapy increases survival, promotes lung angiogenesis, and prevents alveolar damage in hyperoxia-induced lung injury: evidence that angiogenesis participates in alveolarization. Circulation. 2005;112:2477–86.
64. Gien J, Seedorf GJ, Balasubramaniam V, Markham N, Abman SH. Intrauterine pulmonary hypertension impairs angiogenesis in vitro: role of vascular endothelial growth factor nitric oxide signaling. Am J Respir Crit Care Med. 2007;176:1146–53.
65. Grover TR, Parker TA, Zenge JP, Markham NE, Kinsella JP, Abman SH. Intrauterine hypertension decreases lung VEGF expression and VEGF inhibition causes pulmonary hypertension in the ovine fetus. Am J Physiol Lung Cell Mol Physiol. 2003;284:L508–17.
66. Jakkula M, Le Cras TD, Gebb S, Hirth KP, Tuder RM, Voelkel NF, Abman SH. Inhibition of angiogenesis decreases alveolarization in the developing rat lung. Am J Physiol Lung Cell Mol Physiol. 2000;279:L600–7.
67. Le Cras TD, Markham NE, Tuder RM, Voelkel NF, Abman SH. Treatment of newborn rats with a VEGF receptor inhibitor causes pulmonary hypertension and abnormal lung structure. Am J Physiol Lung Cell Mol Physiol. 2002;283:L555–62.
68. Sessa R, Hata A. Role of microRNAs in lung development and pulmonary diseases. Pulm Circ. 2013;3:315–28.
69. Parikh VN, Jin RC, Rabello S, Gulbahce N, White K, Hale A, Cottrill KA, Shaik RS, Waxman AB, Zhang YY, Maron BA, Hartner JC, Fujiwara Y, Orkin SH, Haley KJ, Barabasi AL, Loscalzo J, Chan SY. MicroRNA-21 integrates pathogenic signaling to control pulmonary hypertension: results of a network bioinformatics approach. Circulation. 2012;125:1520–32.
70. Yang S, Banerjee S, Freitas A, Cui H, Xie N, Abraham E, Liu G. miR-21 regulates chronic hypoxia-induced pulmonary vascular remodeling. Am J Physiol Lung Cell Mol Physiol. 2012;302:L521–9.
71. Bhaskaran M, Xi D, Wang Y, Huang C, Narasaraju T, Shu W, Zhao C, Xiao X, More S, Breshears M, Liu L. Identification of microRNAs changed in the neonatal lungs in response to hyperoxia exposure. Physiol Genomics. 2012;44:970–80.
72. Dong J, Carey WA, Abel S, Collura C, Jiang G, Tomaszek S, Sutor S, Roden AC, Asmann YW, Prakash YS, Wigle DA. MicroRNA-mRNA interactions in a murine model of hyperoxia-induced bronchopulmonary dysplasia. BMC Genomics. 2012;13:204.
73. Perveen S, Patel H, Arif A, Younis S, Codipilly CN, Ahmed M. Role of EC-SOD overexpression in preserving pulmonary angiogenesis inhibited by oxidative stress. PLoS One. 2012;7:e51945.
74. Pan L, Fu JH, Xue XD, Xu W, Zhou P, Wei B. Melatonin protects against oxidative damage in a neonatal rat model of bronchopulmonary dysplasia. World J Pediatr. 2009;5:216–21.
75. Tipple TE, Welty SE, Nelin LD, Hansen JM, Rogers LK. Alterations of the thioredoxin system by hyperoxia: implications for alveolar development. Am J Respir Cell Mol Biol. 2009;41:612–9.
76. Farrell MR, Rogers LK, Liu Y, Welty SE, Tipple TE. Thioredoxin-interacting protein inhibits hypoxia-inducible factor transcriptional activity. Free Radic Biol Med. 2010;49:1361–7.
77. Brennan LA, Steinhorn RH, Wedgwood S, Mata-Greenwood E, Roark EA, Russell JA, Black SM. Increased superoxide generation is associated with pulmonary hypertension in fetal lambs: a role for NADPH oxidase. Circ Res. 2003;92:683–91.
78. Chandrasekar I, Eis A, Konduri GG. Betamethasone attenuates oxidant stress in endothelial cells from fetal lambs with persistent pulmonary hypertension. Pediatr Res. 2008;63:67–72.

79. Belik J, Pan J, Jankov R, Tanswell A. Peroxynitrite inhibits relaxation and induces pulmonary artery muscle contraction in the newborn rat. Free Radic Biol Med. 2004;37:1384–92.
80. Wedgwood S, Steinhorn RH, Bunderson M, Wilham J, Lakshminrusimha S, Brennan LA, Black SM. Increased hydrogen peroxide downregulates soluble guanylate cyclase in the lungs of lambs with persistent pulmonary hypertension of the newborn. Am J Physiol Lung Cell Mol Physiol. 2005;289:L660–6.
81. Wedgwood S, Lakshminrusimha S, Czech L, Schumacker PT, Steinhorn RH. Increased p22phox/Nox4 expression triggers remodeling through hydrogen peroxide signaling in persistent pulmonary hypertension of the newborn. Antioxid Redox Signal. 2013;18:1765–76.
82. Wedgwood S, Black SM. Endothelin-1 decreases endothelial NOS expression and activity through ETA receptor-mediated generation of hydrogen peroxide. Am J Physiol Lung Cell Mol Physiol. 2005;288:L480–7.
83. Farrow KN, Groh BS, Schumacker PT, Lakshminrusimha S, Czech L, Gugino SF, Russell JA, Steinhorn RH. Hyperoxia increases phosphodiesterase 5 expression and activity in ovine fetal pulmonary artery smooth muscle cells. Circ Res. 2008;102:226–33.
84. Rao G, Berk B. Active oxygen species stimulate vascular smooth muscle cell growth and proto-oncogene expression. Circ Res. 1992;18:775–94.
85. Sundaresan M, Yu ZX, Ferrans VJ, Irani K, Finkel T. Requirement for generation of H2O2 for platelet-derived growth factor signal transduction. Science. 1995;270:296–9.
86. Lassegue B, Clempus RE. Vascular NAD(P)H oxidases: specific features, expression, and regulation. Am J Physiol Regul Integr Comp Physiol. 2003;285:R277–97.
87. Csanyi G, Taylor WR, Pagano PJ. NOX and inflammation in the vascular adventitia. Free Radic Biol Med. 2009;47:1254–66.
88. Brown DI, Griendling KK. Nox proteins in signal transduction. Free Radic Biol Med. 2009;47:1239–53.
89. Dikalova AE, Gongora MC, Harrison DG, Lambeth JD, Dikalov S, Griendling KK. Upregulation of Nox1 in vascular smooth muscle leads to impaired endothelium-dependent relaxation via eNOS uncoupling. Am J Physiol Heart Circ Physiol. 2010;299: H673–9.
90. Carnesecchi S, Deffert C, Pagano A, Garrido-Urbani S, Metrailler-Ruchonnet I, Schappi M, Donati Y, Matthay MA, Krause KH, Barazzone Argiroffo C. NADPH oxidase-1 plays a crucial role in hyperoxia-induced acute lung injury in mice. Am J Respir Crit Care Med. 2009;180:972–81.
91. Berkelhamer SK, Kim GA, Radder JE, Wedgwood S, Czech L, Steinhorn RH, Schumacker PT. Developmental differences in hyperoxia-induced oxidative stress and cellular responses in the murine lung. Free Radic Biol Med. 2013;61C:51–60.
92. Dennis KE, Aschner JL, Milatovic D, Schmidt JW, Aschner M, Kaplowitz MR, Zhang Y, Fike CD. NADPH oxidases and reactive oxygen species at different stages of chronic hypoxia- induced pulmonary hypertension in newborn piglets. Am J Physiol Lung Cell Mol Physiol. 2009;297:L596–607.
93. Veit F, Pak O, Egemnazarov B, Roth M, Kosanovic D, Sommer N, Seimetz M, Ghofrani H, Seeger W, Grimminger F, Brandes RP, Schermuly R, Weissmann N. Function of NADPH oxidase 1 in pulmonary arterial smooth muscle cells after monocrotaline-induced pulmonary vascular remodeling. Antioxid Redox: Signal; 2013.
94. Chamseddine AH, Miller Jr FJ. Gp91phox contributes to NADPH oxidase activity in aortic fibroblasts but not smooth muscle cells. Am J Physiol Heart Circ Physiol. 2003;285: H2284–9.
95. Petry A, Djordjevic T, Weitnauer M, Kietzmann T, Hess J, Gorlach A. NOX2 and NOX4 mediate proliferative response in endothelial cells. Antioxid Redox Signal. 2006;8:1473–84.
96. Fike CD, Slaughter JC, Kaplowitz MR, Zhang Y, Aschner JL. Reactive oxygen species from NADPH oxidase contribute to altered pulmonary vascular responses in piglets with chronic hypoxia-induced pulmonary hypertension. Am J Physiol Lung Cell Mol Physiol. 2008; 295:L881–8.

97. Wedgwood S, Lakshminrusimha S, Farrow KN, Czech L, Gugino SF, Soares F, Russell JA, Steinhorn RH. Apocynin improves oxygenation and increases eNOS in persistent pulmonary hypertension of the newborn. Am J Physiol Lung Cell Mol Physiol. 2012;302:L616–26.
98. Patil S, Bunderson M, Wilham J, Black SM. Important role for Rac1 in regulating reactive oxygen species generation and pulmonary arterial smooth muscle cell growth. Am J Physiol Lung Cell Mol Physiol. 2004;287:L1314–22.
99. Chen K, Kirber MT, Xiao H, Yang Y, Keaney Jr JF. Regulation of ROS signal transduction by NADPH oxidase 4 localization. J Cell Biol. 2008;181:1129–39.
100. Van Buul JD, Fernandez-Borja M, Anthony EC, Hordijk PL. Expression and localization of NOX2 and NOX4 in primary human endothelial cells. Antioxid Redox Signal. 2005; 7:308–17.
101. Kuroda J, Nakagawa K, Yamasaki T, Nakamura K, Takeya R, Kuribayashi F, Imajoh-Ohmi S, Igarashi K, Shibata Y, Sueishi K, Sumimoto H. The superoxide-producing NAD(P)H oxidase Nox4 in the nucleus of human vascular endothelial cells. Genes Cells. 2005;10: 1139–51.
102. Serrander L, Cartier L, Bedard K, Banfi B, Lardy B, Plastre O, Sienkiewicz A, Forro L, Schlegel W, Krause K. NOX4 activity is determined by mRNA levels and reveals a unique pattern of ROS generation. Biochem J. 2007;406:105–14.
103. Lyle AN, Deshpande NN, Taniyama Y, Seidel-Rogol B, Pounkova L, Du P, Papaharalambus C, Lassegue B, Griendling KK. Poldip2, a novel regulator of Nox4 and cytoskeletal integrity in vascular smooth muscle cells. Circ Res. 2009;105:249–59.
104. Dikalov S, Dikalova A, Bikineyeva A, Schmidt H, Harrison D, Griendling K. Distinct roles of Nox1 and Nox4 in basal and angiotensin II-stimulated superoxide and hydrogen peroxide production. Free Radic Biol Med. 2008;45:1340–51.
105. Sturrock A, Cahill B, Norman K, Huecksteadt TP, Hill K, Sanders K, Karwande SV, Stringham JC, Bull DA, Gleich M, Kennedy TP, Hoidal JR. Transforming growth factor-beta1 induces Nox4 NAD(P)H oxidase and reactive oxygen species-dependent proliferation in human pulmonary artery smooth muscle cells. Am J Physiol Lung Cell Mol Physiol. 2006; 290:L661–73.
106. Djordjevic T, BelAiba RS, Bonello S, Pfeilschifter J, Hess J, Gorlach A. Human urotensin II is a novel activator of NADPH oxidase in human pulmonary artery smooth muscle cells. Arterioscler Thromb Vasc Biol. 2005;25:519–25. Epub 2004 Dec 23.
107. Diebold I, Petry A, Hess J, Görlach A. The NADPH oxidase subunit NOX4 is a new target gene of the hypoxia-inducible factor-1. Mol Biol Cell. 2010;21:2087–96.
108. Lassegue B, Griendling KK. NADPH oxidases: functions and pathologies in the vasculature. Arterioscler Thromb Vasc Biol. 2009;30:653–61.
109. Mittal M, Roth M, Konig P, Hofmann S, Dony E, Goyal P, Selbitz AC, Schermuly RT, Ghofrani HA, Kwapiszewska G, Kummer W, Klepetko W, Hoda MA, Fink L, Hanze J, Seeger W, Grimminger F, Schmidt HH, Weissmann N. Hypoxia-dependent regulation of non-phagocytic NADPH oxidase subunit NOX4 in the pulmonary vasculature. Circ Res. 2007;101:258–67.
110. Green DE, Murphy TC, Kang BY, Kleinhenz JM, Szyndralewiez C, Page P, Sutliff RL, Hart CM. The Nox4 inhibitor GKT137831 attenuates hypoxia-induced pulmonary vascular cell proliferation. Am J Respir Cell Mol Biol. 2012;47:718–26.
111. Altenhofer S, Kleikers PW, Radermacher KA, Scheurer P, Rob Hermans JJ, Schiffers P, Ho H, Wingler K, Schmidt HH. The NOX toolbox: validating the role of NADPH oxidases in physiology and disease. Cell Mol Life Sci. 2012;69:2327–43.
112. Manea A, Tanase L, Raicu M, Simionescu M. Transcriptional regulation of NADPH oxidase isoforms, Nox1 and Nox4, by nuclear factor-kappaB in human aortic smooth muscle cells. Biochem Biophy Res Commun. 2010;396:901–7.
113. Manea A, Manea S, Gafencu A, Raicu M. Regulation of NADPH oxidase subunit p22(phox) by NF-kB in human aortic smooth muscle cells. Arch Physiol Biochem. 2007;113.
114. Kimura S, Egashira K, Chen L, Nakano K, Iwata E, Miyagawa M, Tsujimoto H, Hara K, Morishita R, Sueishi K, Tominaga R, Sunagawa K. Nanoparticle-mediated delivery of

nuclear factor kappaB decoy into lungs ameliorates monocrotaline-induced pulmonary arterial hypertension. Hypertension. 2009;53:877–83.
115. Sawada H, Mitani Y, Maruyama J, Jiang B, Ikeyama Y, Dida F, Yamamoto H, Imanka-Yoshida K, Shimpo H, Mizoguchi A, Maruyama K, Komada Y. A nuclear factor-kappaB inhibitor pyrrolidine dithiocarbamate ameliorates pulmonary hypertension in rats. Chest. 2007;132:1265–74.
116. Wright CJ, Zhuang T, La P, Yang G, Dennery PA. Hyperoxia-induced NF-kappaB activation occurs via a maturationally sensitive atypical pathway. Am J Physiol Lung Cell Mol Physiol. 2009;296:L296–306.
117. Traenckner E, Wilk S, Baeuerle P. A proteasome inhibitor prevents activation of NF-{kappa}B and stabilizes a newly phosphorylated form of I{kappa}B-{alpha} that is still bound to NF-{kappa}B. EMBO J. 1994;15:5433–41.
118. Guzy RD, Schumacker PT. Oxygen sensing by mitochondria at complex III: the paradox of increased reactive oxygen species during hypoxia. Exp Physiol. 2006;91:807–19.
119. Madamanchi NR, Runge MS. Mitochondrial dysfunction in atherosclerosis. Circ Res. 2007;100:460–73.
120. Rathore R, Zheng YM, Niu CF, Liu QH, Korde A, Ho YS, Wang YX. Hypoxia activates NADPH oxidase to increase [ROS]i and [Ca2+]i through the mitochondrial ROS-PKC epsilon signaling axis in pulmonary artery smooth muscle cells. Free Radic Biol Med. 2008;45:1223–31.
121. Farrow K, Wedgwood S, Lee K, Czech L, Gugino S, Lakshminrusimha S, Schumacker P, Steinhorn R. Mitochondrial oxidant stress increases PDE5 activity in persistent pulmonary hypertension of the newborn. Respir Physiol Neurobiol. 2010;174:272–81.
122. Dikalov S. Cross talk between mitochondria and NADPH oxidases. Free Radic Biol Med. 2011;51:1289–301.
123. Dikalova AE, Bikineyeva AT, Budzyn K, Nazarewicz RR, McCann L, Lewis W, Harrison DG, Dikalov SI. Therapeutic targeting of mitochondrial superoxide in hypertension. Circ Res. 2010;107:106–16.
124. Waypa G, Marks J, Guzy R, Mungai P, Schriewer J, Dokic D, Schumacker P. Hypoxia triggers subcellular compartmental redox signaling in vascular smooth muscle cells. Circ Res. 2010;106:526–35.
125. Waypa GB, Marks JD, Guzy RD, Mungai PT, Schriewer JM, Dokic D, Ball MK, Schumacker PT. Superoxide generated at mitochondrial complex III triggers acute responses to hypoxia in the pulmonary circulation. Am J Respir Crit Care Med. 2013;187:424–32.
126. Wedgwood S, Lakshminrusimha S, Fukai T, Russell J, Schumacker P, Steinhorn R. Hydrogen peroxide regulates extracellular superoxide dismutase activity and expression in neonatal pulmonary hypertension. Antioxid Redox Signal. 2011;15:1497–506.
127. Farrow KN, Lakshminrusimha S, Reda WJ, Wedgwood S, Czech L, Gugino SF, Davis JM, Russell JA, Steinhorn RH. Superoxide dismutase restores eNOS expression and function in resistance pulmonary arteries from neonatal lambs with persistent pulmonary hypertension. Am J Physiol Lung Cell Mol Physiol. 2008;295:L979–87.
128. Afshar S, Gibson LL, Yuhanna IS, Sherman TS, Kerecman JD, Grubb PH, Yoder BA, McCurnin DC, Shaul PW. Pulmonary NO synthase expression is attenuated in a fetal baboon model of chronic lung disease. Am J Physiol Lung Cell Mol Physiol. 2003;284:L749–58.
129. Lin YJ, Markham NE, Balasubramaniam V, Tang JR, Maxey A, Kinsella JP, Abman SH. Inhaled nitric oxide enhances distal lung growth after exposure to hyperoxia in neonatal rats. Pediatr Res. 2005;58:22–9.
130. Lopez E, Boucherat O, Franco-Montoya ML, Bourbon JR, Delacourt C, Jarreau PH. Nitric oxide donor restores lung growth factor and receptor expression in hyperoxia-exposed rat pups. Am J Respir Cell Mol Biol. 2006;34:738–45.
131. Check J, Gotteiner N, Liu X, Su E, Porta N, Steinhorn R, Mestan KK. Fetal growth restriction and pulmonary hypertension in premature infants with bronchopulmonary dysplasia. J Perinatol. 2013;33(7):553–7.

132. Rozance PJ, Seedorf GJ, Brown A, Roe G, O'Meara MC, Gien J, Tang JR, Abman SH. Intrauterine growth restriction decreases pulmonary alveolar and vessel growth and causes pulmonary artery endothelial cell dysfunction in vitro in fetal sheep. Am J Physiol Lung Cell Mol Physiol. 2011;301:L860–71.
133. Fagan KA, Fouty BW, Tyler RC, Morris Jr KG, Hepler LK, Sato K, LeCras TD, Abman SH, Weinberger HD, Huang PL, McMurtry IF, Rodman DM. The pulmonary circulation of homozygous or heterozygous eNOS-null mice is hyperresponsive to mild hypoxia. J Clin Invest. 1999;103:291–9.
134. Lakshminrusimha S, Russell J, Wedgwood S, Gugino S, Kazzaz J, Davis J, Steinhorn RH. Superoxide dismutase improves oxygenation and reduces oxidation in neonatal pulmonary hypertension. Am J Respir Crit Care Med. 2006;174:1370–7.
135. Teng RJ, Wu TJ, Bisig CG, Eis A, Pritchard KA, Konduri GG. Nitrotyrosine impairs angiogenesis and uncouples eNOS activity of pulmonary artery endothelial cells isolated from developing sheep lungs. Pediatr Res. 2011;69:112–7.
136. Wedgwood S, McMullan DM, Bekker JM, Fineman JR, Black SM. Role for endothelin-1-induced superoxide and peroxynitrite production in rebound pulmonary hypertension associated with inhaled nitric oxide therapy. Circ Res. 2001;89:357–64.
137. Fike CD, Dikalova A, Slaughter JC, Kaplowitz MR, Zhang Y, Aschner JL. Reactive oxygen species-reducing strategies improve pulmonary arterial responses to nitric oxide in piglets with chronic hypoxia-induced pulmonary hypertension. Antioxid Redox Signal. 2013;18:1727–38.
138. Lassegue B, Griendling KK. Reactive oxygen species in hypertension; an update. Am J Hypertens. 2004;17:852–60.
139. Landmesser U, Dikalov S, Price SR, McCann L, Fukai T, Holland SM, Mitch WE, Harrison DG. Oxidation of tetrahydrobiopterin leads to uncoupling of endothelial cell nitric oxide synthase in hypertension. J Clin Invest. 2003;111:1201–9.
140. Sud N, Wells SM, Sharma S, Wiseman DA, Wilham J, Black SM. Asymmetric dimethylarginine inhibits HSP90 activity in pulmonary arterial endothelial cells: role of mitochondrial dysfunction. Am J Physiol Cell Physiol. 2008;294:C1407–18.
141. Asikainen TM, White CW. Pulmonary antioxidant defenses in the preterm newborn with respiratory distress and bronchopulmonary dysplasia in evolution: implications for antioxidant therapy. Antioxid Redox Signal. 2004;6:155–67.
142. Clerch LB, Massaro D. Rat lung antioxidant enzymes: differences in perinatal gene expression and regulation. Am J Physiol. 1992;263:L466–70.
143. Rhee SG, Chae HZ, Kim K. Peroxiredoxins: a historical overview and speculative preview of novel mechanisms and emerging concepts in cell signaling. Free Radic Biol Med. 2005;38:1543–52.
144. Fukui T, Lassegue B, Kai H, Alexander RW, Griendling KK. Cytochrome b-558 alpha-subunit cloning and expression in rat aortic smooth muscle cells. Biochim Biophys Acta. 1995;1231:215–9.
145. Rodriguez-Iturbe B, Sepassi L, Quiroz Y, Ni Z, Wallace DC, Vaziri ND. Association of mitochondrial SOD deficiency with salt-sensitive hypertension and accelerated renal senescence. J Appl Physiol. 2007;102:255–60.
146. Afolayan AJ, Eis A, Teng RJ, Bakhutashvili I, Kaul S, Davis JM, Konduri GG. Decreases in manganese superoxide dismutase expression and activity contribute to oxidative stress in persistent pulmonary hypertension of the newborn. Am J Physiol Lung Cell Mol Physiol. 2012;303:L870–9.
147. Fukai T, Folz RJ, Landmesser U, Harrison DG. Extracellular superoxide dismutase and cardiovascular disease. Cardiovasc Res. 2002;55:239–49.
148. Oury TD, Chang L, Marklund S, Day BJ, Crapo JD. Immunocytochemical localization of extracellular superoxide dismutase in human lung. Lab Invest. 1994;70:889–98.
149. Oury TD, Day BJ, Crapo JD. Extracellular superoxide dismutase in vessels and airways of humans and baboons. Free Radic Biol Med. 1996;20:957–65.

150. Qin Z, Reszka KJ, Fukai T, Weintraub NL. Extracellular superoxide dismutase (ecSOD) in vascular biology: an update on exogenous gene transfer and endogenous regulators of ecSOD. Transl Res. 2008;151:68–78.
151. Mamo LB, Suliman HB, Giles BL, Auten RL, Piantadosi CA, Nozik-Grayck E. Discordant extracellular superoxide dismutase expression and activity in neonatal hyperoxic lung. Am J Respir Crit Care Med. 2004;170:313–8. Epub 2004 Apr 29.
152. Kamezaki F, Tasaki H, Yamashita K, Tsutsui M, Koide S, Nakata S, Tanimoto A, Okazaki M, Sasaguri Y, Adachi T, Otsuji Y. Gene transfer of extracellular superoxide dismutase ameliorates pulmonary hypertension in rats. Am J Respir Crit Care Med. 2008;177:219–26.
153. Ahmed MN, Suliman HB, Folz RJ, Nozik-Grayck E, Golson ML, Mason SN, Auten RL. Extracellular superoxide dismutase protects lung development in hyperoxia-exposed newborn mice. Am J Respir Crit Care Med. 2003;167:400–5. Epub 2002 Oct 3.
154. Nozik-Grayck E, Suliman H, Majka S, Albietz J, Van Rheen Z, Roush K, Stenmark K. Lung EC-SOD overexpression attenuates hypoxic induction of Egr-1 and chronic hypoxic pulmonary vascular remodeling. Am J Physiol Lung Cell Mol Physiol. 2008;295:L422–30.
155. Clarke MB, Wright R, Irwin D, Bose S, Van Rheen Z, Birari R, Stenmark KR, McCord JM, Nozik-Grayck E. Sustained lung activity of a novel chimeric protein, SOD2/3, after intratracheal administration. Free Radic Biol Med. 2010;49:2032–9.
156. Masood A, Belcastro R, Li J, Kantores C, Jankov RP, Tanswell AK. A peroxynitrite decomposition catalyst prevents 60 % O2-mediated rat chronic neonatal lung injury. Free Radic Biol Med. 2010;49:1182–91.
157. Ho YS, Xiong Y, Ma W, Spector A, Ho DS. Mice lacking catalase develop normally but show differential sensitivity to oxidant tissue injury. J Biol Chem. 2004;279:32804–12. Epub 2004 Jun 3.
158. Masri F, Comhair S, Dostanic-Larson I, Kaneko F, Dweik R, Arroliga A, Erzurum S. Deficiency of lung antioxidants in idiopathic pulmonary arterial hypertension. Clin Transl Sci. 2008;1:99–106.
159. Ardanaz N, Yang XP, Cifuentes ME, Haurani MJ, Jackson KW, Liao TD, Carretero OA, Pagano PJ. Lack of glutathione peroxidase 1 accelerates cardiac-specific hypertrophy and dysfunction in angiotensin II hypertension. Hypertension. 2010;55:116–23.
160. Neumann CA, Krause DS, Carman CV, Das S, Dubey DP, Abraham JL, Bronson RT, Fujiwara Y, Orkin SH, Van Etten RA. Essential role for the peroxiredoxin Prdx1 in erythrocyte antioxidant defence and tumour suppression. Nature. 2003;424:561–5.
161. Li L, Shoji W, Takano H, Nishimura N, Aoki Y, Takahashi R, Goto S, Kaifu T, Takai T, Obinata M. Increased susceptibility of MER5 (peroxiredoxin III) knockout mice to LPS-induced oxidative stress. Biochem Biophys Res Commun. 2007;355:715–21.
162. Davis JM, Parad RB, Michele T, Allred E, Price A, Rosenfeld W, North American Recombinant Human CuZnSOD Study Group. Pulmonary outcome at 1 year corrected age in premature infants treated at birth with recombinant human CuZn superoxide dismutase. Pediatrics. 2003;111:469–76.
163. Farrow KN, Lakshminrusimha S, Czech L, Groh BS, Gugino S, Davis JM, Russell JA, Steinhorn RH. SOD and inhaled nitric oxide normalize phosphodiesterase 5 expression and activity in neonatal lambs with persistent pulmonary hypertension. Am J Physiol Lung Cell Mol Physiol. 2010;299:L109–16.
164. Villegas LR, Kluck D, Field C, Oberley-Deegan RE, Woods C, Yeager ME, El Kasmi KC, Savani RC, Bowler RP, Nozik-Grayck E. Superoxide dismutase mimetic, MnTE-2-PyP, attenuates chronic hypoxia-induced pulmonary hypertension, pulmonary vascular remodeling, and activation of the NALP3 inflammasome. Antioxid Redox Signal. 2013;18:1753–64.
165. Tripathi S, Saili A. The effect of steroids on the clinical course and outcome of neonates with meconium aspiration syndrome. J Trop Pediatr. 2007;53:8–12.
166. Khoo JP, Zhao L, Alp NJ, Bendall JK, Nicoli T, Rockett K, Wilkins MR, Channon KM. Pivotal role for endothelial tetrahydrobiopterin in pulmonary hypertension. Circulation. 2005;111:2126–33.

167. Robbins IM, Hemnes AR, Gibbs JS, Christman BW, Howard L, Meehan S, Cabrita I, Gonzalez R, Oyler T, Zhao L, Du RH, Mendes LA, Wilkins MR. Safety of sapropterin dihydrochloride (6r-bh4) in patients with pulmonary hypertension. Exp Lung Res. 2011;37:26–34.
168. Ananthakrishnan M, Barr FE, Summar ML, Smith HA, Kaplowitz M, Cunningham G, Magarik J, Zhang Y, Fike CD. L-Citrulline ameliorates chronic hypoxia-induced pulmonary hypertension in newborn piglets. Am J Physiol Lung Cell Mol Physiol. 2009;297:L506–11.
169. Vadivel A, Aschner JL, Rey-Parra GJ, Magarik J, Zeng H, Summar M, Eaton F, Thebaud B. L-citrulline attenuates arrested alveolar growth and pulmonary hypertension in oxygen-induced lung injury in newborn rats. Pediatr Res. 2010;68:519–25.
170. Chen B, Calvert AE, Cui H, Nelin LD. Hypoxia promotes human pulmonary artery smooth muscle cell proliferation through induction of arginase. Am J Physiol Lung Cell Mol Physiol. 2009;297:L1151–9.
171. Jin Y, Calvert TJ, Chen B, Chicoine LG, Joshi M, Bauer JA, Liu Y, Nelin LD. Mice deficient in Mkp-1 develop more severe pulmonary hypertension and greater lung protein levels of arginase in response to chronic hypoxia. Am J Physiol Heart Circ Physiol. 2010;298: H1518–28.
172. Schneider R, Raff U, Vornberger N, Schmidt M, Freund R, Reber M, Schramm L, Gambaryan S, Wanner C, Schmidt HH, Galle J. L-Arginine counteracts nitric oxide deficiency and improves the recovery phase of ischemic acute renal failure in rats. Kidney Int. 2003;64: 216–25.
173. Coggins MP, Bloch KD. Nitric oxide in the pulmonary vasculature. Arterioscler Thromb Vasc Biol. 2007;27:1877–85.
174. Chester M, Tourneux P, Seedorf G, Grover TR, Gien J, Abman SH. Cinaciguat, a soluble guanylate cyclase activator, causes potent and sustained pulmonary vasodilation in the ovine fetus. Am J Physiol Lung Cell Mol Physiol. 2009;297:L318–25.
175. Ladha F, Bonnet S, Eaton F, Hashimoto K, Korbutt G, Thebaud B. Sildenafil improves alveolar growth and pulmonary hypertension in hyperoxia-induced lung injury. Am J Respir Crit Care Med. 2005;172:750–6.
176. Park HS, Park JW, Kim HJ, Choi CW, Lee HJ, Kim BI, Chun YS. Sildenafil alleviates bronchopulmonary dysplasia in neonatal rats by activating the hypoxia-inducible factor signaling pathway. Am J Respir Cell Mol Biol. 2013;48:105–13.
177. de Visser YP, Walther FJ, el Laghmani H, Boersma H, van der Laarse A, Wagenaar GT. Sildenafil attenuates pulmonary inflammation and fibrin deposition, mortality and right ventricular hypertrophy in neonatal hyperoxic lung injury. Respir Res. 2009;10:30.

Chapter 11
Use of Oxygen in the Resuscitation of Neonates

Máximo Vento, Helmut Hummler, Jennifer Dawson, Javier Escobar, and Julia Kuligowski

Introduction

Aerobic Metabolism (Summarized in Fig. 11.1)

Oxygen (O_2) is one of the most abundant elements in nature and probably the most widely used drug in neonatology [130]. Evolution from unicellular to multicellular organisms required a substantial increase in the amount of energy needed, and this could only be provided by mitochondrial combustion of glucose, amino acids, and free fatty acids in the presence of O_2 [104]. Basic nutritional elements broken down into acetyl coenzyme A are metabolized in the tricarboxylic acid cycle (Kreb's cycle). During this process, highly energized electrons are liberated and transported by specific dinucleotides (NADH, NADPH, FADH) to the electron transport chain (ETC) located in the inner membrane of the mitochondria. These reducing equivalents are used to maintain the electrochemical gradient that drives adenosine triphosphate (ATP) synthesis. Oxygen is the final acceptor of electrons at the complex IV of the respiratory chain. The process involving complete reduction of oxygen to water comprises a four-electron process without production of reactive oxygen

M. Vento, M.D., Ph.D. (✉)
Division of Neonatology and Health Research Institute La Fe, University and Polytechnic Hospital La Fe, Valencia, Spain
e-mail: maximo.vento@uv.es

H. Hummler, M.D.
Division of Neonatology, Universitätsklinikum Ulm, Ulm, Germany

J. Dawson, RN, Ph.D.
Division of Neonatology, The Royal Women's Hospital, Melbourne, Australia

J. Escobar, MSci, Ph.D. • J. Kuligowski, MSc, Ph.D.
Neonatal Research Group Laboratory, Health Research Institute La Fe, Valencia, Spain

P.A. Dennery et al. (eds.), *Perinatal and Prenatal Disorders*,
Oxidative Stress in Applied Basic Research and Clinical Practice,
DOI 10.1007/978-1-4939-1405-0_11

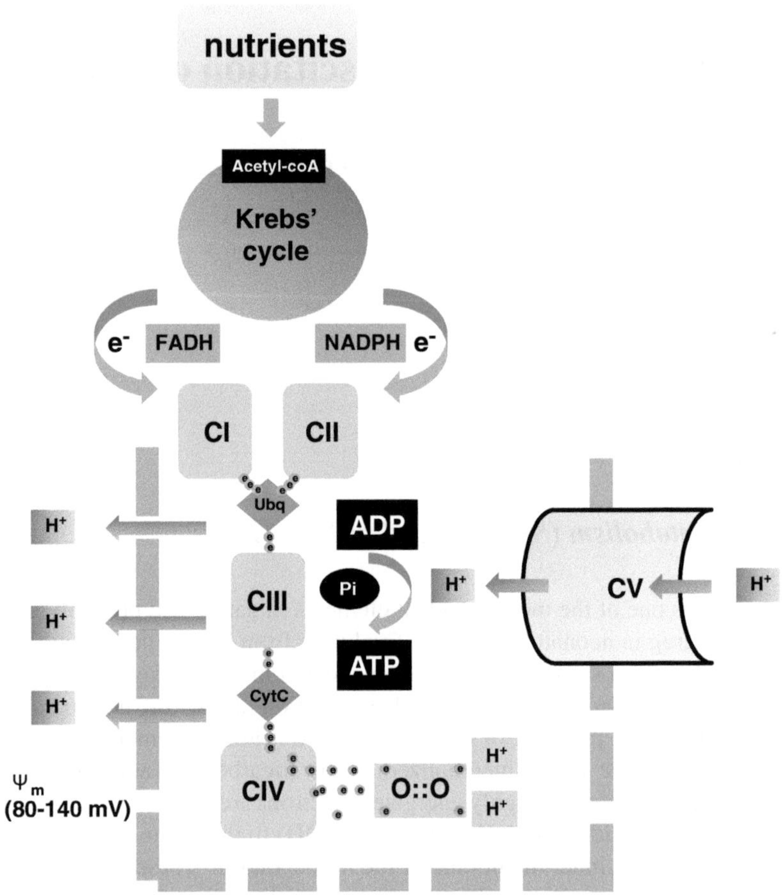

Fig. 11.1 In the presence of oxygen, acetyl-coA (the common merging metabolite derived from ingested nutrients) entering into the inner mitochondrial space will liberate highly energized electrons in the tricarboxylic cycle (Kreb's cycle) which will be thereafter transported to the electron transport chain (ETC). Energy is used to extrude protons and establish a transmembrane potential (Ψ_m). In the final step, ATP synthase will intrude protons again in the inner mitochondrial space, and the energy provided by Ψ_m will be employed to synthesize adenosine triphosphate (*ATP*) from adenosine diphosphate (*ADP*). Oxygen will be reduced with four electrons. This process is known as oxidative phosphorylation. For further explanation, see text

species (ROS) [130]. Because O_2 has a high standard oxidation-reduction (redox) potential, it is an ideal electron acceptor – and is therefore a sink for the capture of energy for intracellular use. The efficiency of aerobic metabolism is approximately 18–20-fold greater than anaerobic. Hence, while anaerobic metabolization of one molecule of glucose will produce 2–4 mol of ATP, aerobic combustion will produce 36 mol of ATP [121]. The proportion of carbohydrates, lipids, and proteins utilized to build up ATP is cell specific and organ specific. Hence, brain and erythrocytes use almost exclusively glucose (although the brain may use ketone bodies early in fetal to neonatal transition), whereas the energy for cardiac contraction derives fundamentally from free fatty acid oxidation [113]. The mitochondrial electrochemical gradient or proton motive force (Δp_m) is generated and utilized by the oxidative phosphorylation (OxPhos) system, composed by the electron transport chain (ETC) and ATP synthase. ETC transfers highly energized electrons to the final acceptor oxygen. For this purpose, the different proteins in the ETC complex intervene in a sequential manner. Briefly, the nonprotein carrier ubiquinone transfers electrons from complex I (NADH dehydrogenase) and II (succinate dehydrogenase) to complex III (bc_1 complex) followed by electron transfer to complex IV (cytochrome *c* oxidase, CcO) via the electron carrier cytochrome *c* (Cytc). CcO catalyzes the final and proposed rate-limiting step in electron transfer: the donation of electrons to O_2, allowing the conversion of $O_2 + H^+$ to H_2O. Along this process, energy extracted at the different complexes (I, III, and IV) is used to pumping of H^+ across the inner mitochondrial membrane. This process generates the mitochondrial membrane potential ($\Delta\Psi_m$), which represents the charge difference across the inner mitochondrial membrane. Extruded protons are captured by ATP synthase (complex V) which drives the conversion of adenosine diphosphate (ADP) + inorganic phosphate (Pi) to ATP. Remarkably, OxPhos provides the majority of ATP synthesized in our economy and is especially relevant in aerobic-dependent tissues such as brain [96]. Regulation of OxPhos depends on both the availability of substrate (NADH, O_2, ADP, Pi) and its end product ATP. ATP and ADP are allosteric inhibitors of cytochrome *c* oxidase (CcO), respectively, and this control mechanism adjusts ETC activity to energy demand [51]. Another relevant mechanism for respiratory control is Δp_m. Hence, in resting mitochondria when most of the ADP has already been converted into ATP, synthesis of ATP slows down, and Δp_m increases and inhibits ETC proton pumps preventing further proton pumping at higher Δp_m levels. This feedback mechanism helps maintaining $\Delta\Psi_m$ at physiological levels (80–140 mV) in which ATP production is efficient, and generation of reactive oxygen species (ROS) is minimal [96].

Reactive Oxygen Species, Redox Regulation, and Antioxidant Enzymes

Reactive Oxygen Species (ROS) Formation (Summarized in Fig. 11.2)

Reactive oxygen species (ROS) is a phrase used to describe a variety of molecules derived from incomplete reduction of molecular dyoxygen [9]. Under physiological conditions, more than 90 % of O_2 undergoes a complete reduction with four electrons by CcO to form H_2O. However, a small percentage of electrons will "leak" causing a "partial" reduction of oxygen. The most common incomplete reduction consistent in a monovalent reduction of oxygen elicits the production of anion superoxide [$\cdot O_2^-$]. This ROS production takes place proximal to CcO by inhibition of electron transfer through the Q cycle for complex III. Superoxide is released in the matrix

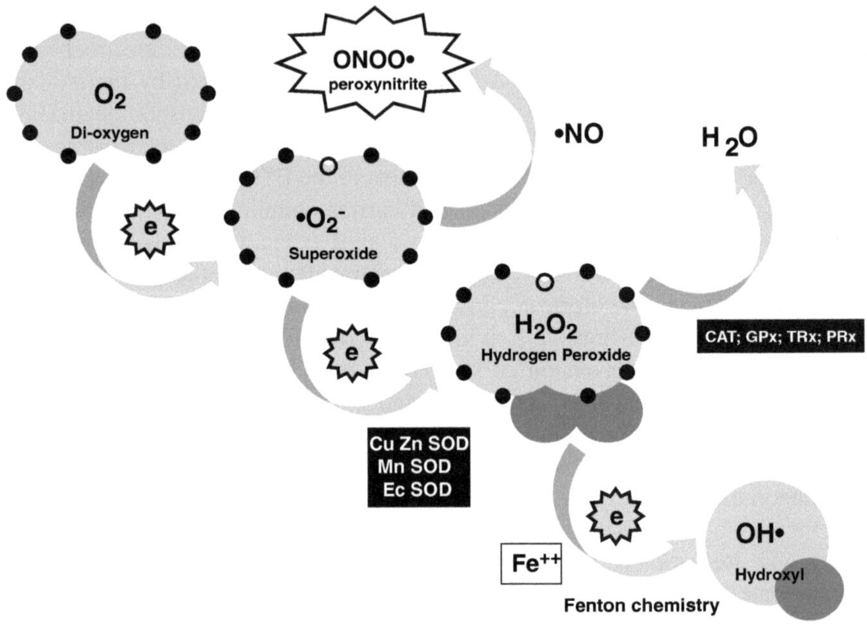

Fig. 11.2 Stepwise reduction of oxygen leads to the formation of reactive oxygen species such as superoxide anion (monovalent reduction), hydrogen peroxide (divalent reduction) and hydroxyl radical (trivalent reduction). Hydroxyl radical formation is enhanced in the presence of transition metals such as iron, copper, and manganese in the so-called Fenton's chemistry. The antioxidant system neutralizes chemical reactivity of free radicals

(complex I) and intermembrane space sides (complex III). Additional monovalent reduction of anion superoxide will lead to the generation of hydrogen peroxide (H_2O_2), while a third monovalent reduction will produce the highly reactive hydroxyl radical (·OH). Other ROS such as peroxyl, hydroperoxyl, and alkoxyl radicals are also produced however in minor quantities [66]. ROS are also relevant in the regulation of nitric oxide bioavailability and, therefore, indirectly influence airway and vascular reactivity. Hence, anion superoxide can sequester great amounts of ·NO producing the very reactive peroxynitrite ($\cdot ONOO^-$) and altering pulmonary vascular reactivity [60]. ROS generation is highly dependent on keeping $\Delta\Psi_m$ within a physiological range (80–140 mV). However, hyperpolarization ($\Delta\Psi_m$ >140 mV) causes an exponential increase in ROS generation at both complexes I and III [37].

The production of ROS under physiological and also pathologic conditions is very highly dependent on the concentration of oxygen in the tissue. ROS can be extremely aggressive when acting as free radicals causing direct structural and/or functional damage and/or interfere with essential redox regulatory elements. Contrarily, ROS can also be relatively stable molecules and act as signaling molecules in physiological processes. A free radical can be defined as any molecule capable of independent existence with one or more unpaired electrons in the outer shell (e.g., anion superoxide, hydroxyl radical). When reacting with other free radicals, they form covalent bonds to share one electron; however, the resulting molecule easily decomposes leading to the formation of toxic products. Free radicals may also react with non-radical molecules in typical chain reactions causing damage to DNA, proteins, and lipids or by promoting the formation of adducts with DNA. The range of molecular damage produced by free radicals is rather remarkable and encompasses, for instance, lipid peroxidation and nitration, protein oxidation and nitration, protein-thiol depletion, nucleic acid hydroxylation and nitration, DNA strand breakage, and DNA adduct formation. Multiple pathways have been involved in ROS-induced apoptosis and/or necrosis. Remarkably, under very stressful conditions (ischemia-reperfusion, inflammation, or hyperoxia) damage to proteins, lipids, and nucleic acids can inevitably lead to cell death. Nitration or carbonylation of proteins can alter their structure and function causing marked cellular dysfunction. Oxidation of cell membrane phospholipids can activate sphingomyelinase and release of ceramide, which activates apoptosis. In addition, intense damage to DNA can also lead to necrosis or apoptosis. ROS can also activate multiple signal transduction pathways relevant to cell function but also to triggering apoptosis. Hence, ROS can activate mitogen-activated protein kinases (MAP-kinases) such as c-Jun N-terminal kinase to release Bcl-2-related proteins. These proteins will activate Bax which in turn will translocate to the mitochondria and initiate release of cytochrome c and other pro-death mediators into the cytosol [4, 5, 37].

Enzyme complexes linked to other structures and functions different from mitochondrial respiration can also generate ROS. Hence, cytochrome P_{450} monooxygenase system, xanthine oxidoreductase, nitric oxide synthases, heme oxygenases, or several others involved in the inflammatory process are also capable of producing significant amounts of ROS. Moreover, in the presence of "free" metals such as iron, copper, and manganese, generation of ROS can be exponentially increased [96].

Interestingly, ROS have a relevant role in cell physiology. Beneficial effects of ROS occur at low/moderate concentrations and imply an ample array of cellular responses. Thus, various ROS-mediated actions can protect cells against ROS-induced oxidative stress and reestablish or maintain redox homeostasis (see section "Redox regulation"). Among the physiological functions controlled by redox-responsive signaling pathways, the most relevant involve the redox-regulated production of NO, the oxidative burst produced as response to infectious agents by phagocytic NAD(P)H oxidase, or the ROS production as sensing elements for regulating tissue oxygen needs, cell adhesion, immune responses, or regulation of induced apoptosis [118].

Redox Regulation

Recently, a seminal observation led to the development of redox compartmentalization and cell stress. Hence, the two central thiol/disulfide couples reduced to oxidized glutathione (GSH/GSSG), and cysteine/cystine (Cys/CysSS) couples varied little among healthy individuals but were maintained in disequilibrium relative to each other. Thereby, it was shown that the concept of oxidative stress as a global imbalance affecting the entire economy was incorrect because two central antioxidant systems were not in balance [49]. Central couples, which include GSH/GSSG, Cys/CysSS, and thioredoxins (Trx/TrsSS), function as reducing counterparts of H_2O_2 and other oxidants in controlling redox state of oxidizable thiols in proteins. These sulfur switches are used for cell signaling, protein structure, protein trafficking, and regulation of enzyme, transporter, receptor, and transcription factor activity. The rates of electron transfer implicated in redox signaling and control are extremely low as compared to those of energy production in the mitochondria and in the range of approximately 1 % of the overall oxygen consumption by cells. H_2O_2 is the presumed oxidant in most of these reactions although other oxidants could be also involved. Of remarkable interest is the fact that the rates of peroxide generation and use for signaling and control appear to be considerably greater than the rates of free radical reactions contributing to macromolecular damage [118]. Alteration of thiol-dependent signaling seems to be the most sensitive and quantitatively important processes in oxidative stress. Since these redox mechanisms control proinflammatory and pro-fibrotic signaling, cell proliferation, apoptosis, and a range of other biologic processes without a requirement for macromolecular damage, the results suggest that failure of free radical scavenger trials may have occurred because oxidative stress research overemphasized the importance of free radical mechanisms and macromolecular damage as an underlying mechanism [47].

The redox hypothesis postulates that oxidizable thiols are common control elements for biologic processes. These control elements are functionally organized in redox circuits, which are controlled by GSH/GSSG, thioredoxins, and other control nodes. These circuits are isolated from each other and are highly responsive for redox conditions and can function independently in signaling and regulation of different biologic processes. Hence, oxidative stress would be the consequence of the disruption of these circuits by different mechanisms [49].

Antioxidant Defenses (Fig. 11.2)

The balance between generation of ROS and adequate maintenance of the cellular redox status is highly dependent on the antioxidant defenses (AOD). AOD can include enzymatic and nonenzymatic mechanisms. Antioxidant enzymes catalytically remove ROS, thereby decreasing ROS reactivity, and protect proteins through the use of chaperones, transition metal-containing proteins (transferrin, ferritin, ceruloplasmin), and low-molecular-weight compounds that purposely function as oxidizing or reducing agents to maintain intracellular redox stability [130].

Superoxide dismutases (SODs) are a ubiquitous group of enzymes that catalyze dismutation of superoxide anions to H_2O_2. Three different SODs have been characterized, cytoplasmic SOD1 (Cu/Zn), mitochondrial SOD2 (Mn/Cu), and extracellular SOD3 (EC-SOD) (Zn/Cu). H_2O_2 is mostly degraded by peroxidases; however, the remaining will function as a relevant signaling molecule. Catalases (CAT) and glutathione peroxidases (GPx) convert H_2O_2 into H_2O and O_2. GPx couples H_2O_2 reduction to water with the oxidation of GSH to GSSG. GSSG is again reduced to GSH by the activity of the pentose shunt. Other systems to detoxify hydrogen peroxide in mitochondria and other organelles include glutaredoxin, thioredoxin, thioredoxin reductase, and the peroxiredoxins. Other enzymes with antioxidant and signaling functions are heme oxygenases (HO-1 and HO-2). HO-1 removes heme, a pro-oxidant, and generates biliverdin, an antioxidant-releasing iron, and carbon monoxide. Finally, nonenzymatic antioxidants such as reduced glutathione, vitamin C, vitamin E, and β-carotene also function to protect cells from damaging effects of ROS [4, 5, 66]. Enzymatic antioxidants' expression is post-conceptionally programmed, and, therefore preterm infants are more susceptible to ROS-mediated damage especially under pro-oxidant circumstances such as hypoxia-reoxygenation, inflammation, or infection. In addition, transplacental passage of antioxidants also occurs in the later part of gestation. In assays performed in human *abortus* materials, it has been shown that SOD, catalase, GPx, and glutathione reductase activities increased with advancing of gestation. Therefore, specific conditions occurring during pregnancy such as preeclampsia significantly alter placental antioxidant enzyme expression causing a pro-oxidant burden for the fetus [15]. Hypoxia during fetal to neonatal transition may stimulate the expression of certain antioxidant enzymes as HO-1 or thioredoxin. However, hypoxia-reoxygenation in term and preterm infants causes a significant increase in antioxidant enzyme activities and a reduction in the GSH/GSSG ratio [130].

Biomarkers of Oxidative Stress (Summarized in Table 11.1)

Reduced and Oxidized Glutathione Ratio (GSH:GSSG)

Analysis of changes in glutathione status, including GSH oxidation to GSSG and protein glutathionylation reactions (P-SH → P-SSG), provides a measure of the cell redox status [48, 114]. Thus, levels of GSH and GSSG are used to evaluate oxidative stress, and the ratio of GSH to GSSG (GSH:GSSG) serves as a representative marker of the antioxidant capacity of the cell [27]. In humans, the overall glutathione status may not only be determined as plasma GSH:GSSG but also as whole blood GSH:GSSG, where 0.5 % of glutathione derives from plasma and 99.5 % from erythrocytes. Both are simple and straightforward ways to evaluate the oxidative stress associated to pathologies or clinical interventions as well as to test the efficacy of clinical therapies [27, 48, 105, 114]. At present, liquid chromatography coupled to tandem mass spectrometry (LC-MS/MS) constitutes a sensitive and accurate methodology for simultaneous determination of both GSH and GSSG in complex samples [23, 42]. For proper GSH and GSSG, sample processing is extremely important to prevent artifacts. For this reason, the use of thiol-blocking agents such as maleimides or iodoacetamides preventing GSH oxidation is highly recommendable [23, 81], and enzymes such as γ-glutamyl-transpeptidase (γGT) or glutathione reductase (GR) should be inactivated, which is usually achieved by sample acidification [114]. In practice, it is recommended to first add N-ethylmaleimide (NEM) to blood samples as soon as possible because the reaction of *N*-ethylmaleimide with –SH is fast at neutral pH and hence only requires short incubation times followed by acidification with perchloric acid (PCA) [80].

Table 11.1 Main oxidative biomarkers used in experimental and clinical studies (references in text)

Oxidative biomarkers	Target biomolecule	Modification	Biological sampling	Analytical method
Glutathione (GSH/GSSG ratio)	Antioxidants	General redox status	Whole blood	LC-MS/MS
MDA	Lipids	PUFA peroxidation	Plasma	HPLC (UV detection)
HNE	Lipids	PUFA peroxidation	Plasma	HPLC
o-Tyrosine (o-Tyr/Phe ratio)	Proteins	Tyrosine hydroxylation	Urine	LC-MS/MS
m-Tyrosine (m-Tyr/Phe ratio)	Proteins	Tyrosine hydroxylation	Urine	LC-MS/MS
3N2-Tyrosine	Proteins	Tyrosine nitration	Urine	LC-MS/MS
8OHdG (8OHdG/2dG ratio)	Lipids	AA peroxidation	Urine/plasma	LC-MS/MS
F2-IsoPs	Lipids	AA peroxidation	Urine/plasma	GC-MS/MS; LC-MS/MS
D2/F2-ISoPs	Lipids	AA peroxidation	Urine/plasma	GC-MS/MS; LC-MS/MS
IsoFs	Lipids	AA peroxidation	Urine/plasma	GC-MS/MS; LC-MS/MS
NeuPs	Lipids	DHA peroxidation	Urine/plasma	GC-MS/MS; LC-MS/MS
NeuFs	Lipids	DHA peroxidation	Urine/plasma	GC-MS/MS; LC-MS/MS

Abbreviations: GSH: reduced glutathione; GSSG: oxidized glutathione; MDA: malondialdehyde; HNE: 4-hydrox-2-nonenal; o-Tyr: ortho-tyrosine; m-Tyr: meta-tyrosine; 3N2-Tyrosine: 3-nitrotyrosine; 8OHdG: 8-hydroxi-2'-deoxiguanosine; 2dG: 2'-deoxiguanosine; IsoPs: isoprostanes; IsoFs: isofurans; NeuPs: Neuroprostanes; NeuFs: neurofurans; AA: arachidonic acid; DHA: docosa-hexanoic acid; LC: liquid chromatography: GC: gas chromatography; MS/MS: tandem mass spectrometry

In the clinical setting, this time-critical derivatization step might be a practical limitation. However, GSH/GSSG quotient has been very extensively used as a reliable marker of oxidative stress in neonatology, especially in studies related to asphyxia and reoxygenation with different oxygen concentrations [125, 127, 131, 132].

The evaluation of GSH/GSSG in complex whole blood samples using LC-MS/MS requires appropriate sample preparation and dilution to ensure accurate and reproducible determinations. This complex and time-consuming process may be difficult to employ use under stressful and highly dynamic situations such as post-natal resuscitation.

Biomarkers of Protein Oxidation

ROS may inactivate enzymes such as DNA repairing or polymerases enzymes, thus causing secondary damage to other biomolecules [40]. Protein oxidation may be reversible or irreversible depending on the ROS molecule and the severity of oxidative stress. Among oxygen free radicals, hydroxyl radicals (HO·) are the most highly reactive and may attack indiscriminately and irreversibly many biological molecules including amino acids [11]. The most common protein modifications used as oxidative biomarkers are protein carbonylation and tyrosine oxidation and nitration.

Protein Carbonyls. Hydroxyl radical attack of proteins may lead to protein carbonylation with the formation of aldehydes (RHC=O) and ketones (RR'C=O) in the presence of oxygen (O_2) [72, 136]. In addition, carbonyls may be formed by reaction with aldehydes or reactive carbonyls, which may arise from lipid peroxidation or glycation, respectively. Side-chain lysine, arginine, proline, and threonine are the main amino acid residues in proteins susceptible to yield carbonyl derivatives [72, 114]. Due to their chemical stability, protein carbonyls have become the most general and widely used marker of protein oxidation damage in experimental and clinical studies [6]. Protein carbonyls may be detected in tissue or biofluids (plasma, serum) by several techniques such as spectrophotometry, enzyme-linked immunosorbent assay (ELISA), or immune-blotting after derivatization to 2,4-dinitrophenylhydrazone (DNP) by reaction with 2,4-dinitrophenylhydrazine (DNPH) [12, 40, 114]. The availability of methodologically simple and reliable commercial kits for DNPH derivatization and anti-DNP antibodies has rendered protein carbonylation a useful diagnostic biomarker [6, 12].

On the other hand, carbonyls cannot be considered specific markers of oxidative damage to proteins since bound aldehydes and glycated protein are also measured. In addition, only few proteins are prone to be carbonylated, and carbonylation of abundant proteins as, e.g., albumin in plasma, might induce overestimation of the global protein carbonylation in the sample. Therefore, special care must be taken in this regard. Nevertheless, carbonyl assay may be a very sensitive and accurate technique if it is combined with proteomics to identify specific proteins regulated by carbonyl oxidation [40].

Oxidative and Nitrosative Tyrosine Derivatives. Under physiological conditions, oxidation of L-phenylalanine (L-phe) to L-para-tyrosine (L-tyr) is carried out enzymatically by the action of phenylalanine hydroxylase (PheH) [138]. However, in pathophysiologic situations, the spontaneous oxidation of L-phe to L-tyr isomers may be given in the presence of OH· leading to the production of the L-ortho-tyrosine (o-tyr) [64, 68] and L-meta-tyrosine (m-tyr) [68]. While o-tyr has been widely accepted, nonenzymatic oxidation of L-phe to m-tyr is more controversial since it could be produced by the action of phenylalanine 3-hydroxylase (Phe3H), a member of the PheH protein family [64]. In addition, the attack of ROS such as peroxynitrite ($ONOO^-$) or hypochlorous acid (HClO) may directly attack L-tyr producing oxidized L-tyr derivatives such as 3-nitrotyrosine (3Nitro-tyr) or 3-chlorotyrosine (3Cl-tyr), respectively [98]. L-phe and L-tyr derivatives have been described to be reliable biomarkers of oxidative damage to proteins. Furthermore, 3nitro-tyr is also used as an index of nitrosative stress and 3Cl-tyr as a marker of inflammation because it indicates myeloperoxidases (MPO) activation (MPO converts H_2O_2 to HClO) [14, 78, 98]. Analytical determination based on mass spectrometry (MS) of these L-phe and L-tyr oxidative derivatives has been widely probed in experimental and human urine samples in the newborn period as well as in amniotic fluid of pregnant women [25, 62, 108, 123, 131].

For these studies, a method employing liquid chromatography-tandem mass spectrometry (LC-MS/MS) was developed enabling the simultaneous quantification of L-phe, L-tyr, o-tyr, m-tyr, 3nitro-tyr, and 3Cl-tyr in urine samples establishing typical ranges of these biomarkers in newborns. Other LC and gas chromatography (GC)-based approaches for these analytes in human urine and plasma have been reported [14, 68, 78, 98]. MS approaches are useful tools for research and clinical applications allowing an accurate measurement of a set of metabolites in relatively small sample volumes. However, drawbacks are the need for valuable equipment and qualified staff. Furthermore, method development and optimization to ensure reliable results and control possible matrix effects can be time-consuming.

Biomarkers of DNA Oxidation

Hydroxyl radicals may attack deoxyribose phosphate backbone as well as the nitrogenous bases (nucleobases) of DNA nucleotides, generating a broad variety of base and sugar modification products [12]. Physiologically, oxidative DNA modifications are being produced continuously, and enzymatic DNA repair mechanisms are crucial to maintain a low steady state of oxidative DNA damage [117]. Therefore, DNA oxidation may be secondarily aggravated by oxidative protein damage to repair mechanisms, and urine determination of oxidative DNA derivatives may reflect the balance between DNA damage and repair [114]. Among the oxidative DNA derivatives, interest is focused specially on 8-oxo-2′-deoxyguanosine (8-oxodG) because it is an abundant lesion formed in vivo, which can be quantitatively measured [18, 117]. Usually, 8-OHdG and 8-oxodG are considered as the same compound [56]. 8-OHdG is by far the most extensively used biomarker of oxidative damage to DNA, and it can be measured in urine or directly from tissue samples, prior DNA isolation, and in vitro degradation [12, 40, 95, 114, 117].

8-OHdG detection and quantification can be performed by instrumental methods (HPLC, LC-MS/MS) and immunological techniques (ELISA, immunostaining). It has to be underlined that the determination of cellular 8-OHdG must be carried out carefully since 8-OHdG artifacts may be formed during DNA processing [40]. The use of urinary 8-OHdG is limited as it is a biased measure of damage to guanosine that cannot rule out other origins rather than disease (for instance, from diet), but to date 8-OHdG is the most accurate biomarker for evaluating DNA oxidation. The advantage of measurements in body fluids such as urine or amniotic fluid by LC-MS/MS is that it allows the simultaneous determination of both 8-OHdG and 2-dG as well as other oxidative biomarkers like L-phe and L-tyr derivatives in complex biological samples without tedious sample preparation steps. Moreover, it is easily performed in the newborn animals or human neonatal patients without being invasive and has therefore been widely used in experimental and clinical settings [25, 62, 108, 123, 131].

Biomarkers of Lipid Oxidation

The oxidative damage to lipids induced by ROS and RNS (not including H_2O_2, NO·, or O2·⁻ which poorly react with lipids) is known as lipid peroxidation (LPO) [38, 40]. Actually, LPO can originate from different sources, and it may be classified as *enzymatic*, *nonenzymatic non-radical peroxidation*, and *nonenzymatic free-radical-mediated peroxidation* [38]. Complexity of LPO originates from the large number of products that can be produced. Among PUFAs, peroxidation of the omega-6 (ω-6) fatty acid arachidonic leads to the formation of leukotrienes (LTs), hydroxytetronic acid derivatives (HETEs), prostaglandins (PGs), and thromboxanes (TXs), respectively; all of them are known to be inflammatory mediators and referred to as eicosanoids [22]. On the other hand, nonenzymatic PUFA peroxidation mediated by free radicals has also been studied in depth, and it occurs in three steps: (1) initiation, OH· attacks a PUFA generating a lipid radical (L·) ($LH + OH^{\cdot} \rightarrow L^{\cdot} + H_2O$); (2) propagation, $L^{\cdot}$ reacts with oxygen to form a lipo-peroxyl radical ($LOO^{\cdot}$) ($L\cdot + O^2 \rightarrow LOO\cdot$) that may react with another PUFA to yield a new L· and a lipid hydroperoxide (LOOH) ($LOO\cdot + LH \rightarrow LOOH + L\cdot$); and (3) termination, a series of radical reactions ending by the generation of non-radical products (NRPs) ($LOO^{\cdot} + LOO^{\cdot} \rightarrow NRP$) [38]. Free radical-mediated LPO generates stable products, which can be divided into primary and secondary lipid hydroperoxide products the latter being metabolites such as aldehydes or isoprostanes [114]. Malondialdehyde (MDA), 4-hydroxy-2-nonenal (HNE), and F_2-isoprostanes (F_2-IsoPs) are LPO products frequently used as lipid peroxidation biomarkers.

4-Hydroxynonenal (*HNE*). HNE is an α,β-unsaturated hydroxyalkenal originating from lipid peroxidation which has three reactive groups, an aldehyde, a double-bond at carbon 2, and a hydroxy group at carbon 4, and therefore it is a highly reactive molecule. It may form adducts with other biomolecules such as proteins, peptides, nucleic acids, or lipids. Under physiological conditions, HNE may act as signaling molecule by inactivating phosphatases [106] or by activating the NF-E2-related factor 2 (NRF2) that leads the antioxidant response [115]. In addition, a strong electrophile, HNE, can also significantly alter the cellular redox status by depleting thiol compounds such as GSH. An increase of HNE has been detected after perinatal asphyxia [103]. HNE determination can be performed by liquid chromatography with ultraviolet detection (LC-UV). However, this method may underestimate HNE due to its high reactivity with other molecules [112]. To avoid this inconvenience, ELISA kits that use anti-HNE-histidine conjugate antibodies have been developed [7]. In addition, analytical methods for the detection of HNE-protein adducts or HNE derivatives based on proteomics and lipidomics using LC-MS and GC-MS techniques have been developed. However, they are expensive and rather tedious and therefore of limited routine use [112].

Malondialdehyde (*MDA*). MDA is a three-carbon di- or keto-aldehyde containing a carbonyl group produced by cleavage of oxidized PUFAs as arachidonic acid (AA) or docosahexaenoic acid (DHA) [38]. MDA is highly reactive and able to form adducts with biomolecules. Elevated levels of MDA have also been reported in asphyxiated neonates and in preterm babies subjected to painful procedures [107]. The most common method to measure LPO based on determining thiobarbituric acid (TBA) reactive substances (TBARS) assay is questionable since many other molecules rather than MDA may react with TBA, and most TBA derivatives in human body fluids are not related to lipid peroxidation [65]. Nevertheless, the use of analytical methods like LC-UV has improved the accuracy in the detection of MDA-TBA adducts. Furthermore, due to its carbonyl group, MDA-DNP adducts (prior DNPH derivatization) may be also determined by LC-UV [7]. Other approaches to detect MDA are based on the use of antibodies against MDA protein adducts, such as the dihydropyridine-lysine (DHP-lysine) antibody [12]. Urinary levels of MDA may arise due to artifacts or derive from the diet. Therefore, for clinical purposes, plasma or serum MDA are more suitable [40].

Isoprostanes (*IsoPs*). During nonenzymatic LPO mediated by free radicals, peroxyl radicals may suffer cyclization to generate endoperoxides products like isoprostanes (IsoPs) by further oxidation [112]. IsoPs are a family of eicosanoids of nonenzymatic origin with a broad variety of isomers. F_2-isoprostanes (F_2-IsoPs) are prostaglandin F_2 (PGF_2)-like compounds that contain the F-type prostane ring of PGF_2. Currently, IsoPs are considered as the best available biomarker of LPO, being F_2-IsoPs the most studied class of IsoPs, due to their great stability [71]. Of note is that despite IsoPs may be detected in foods they are not absorbed in the gut in sufficient quantities to produce artifacts in the urinary or plasma IsoPs levels, and hence it is assumed that urinary and circulating F_2-IsoPs in humans arise from free radical-mediated LPO [40]. Increased levels 8-iso-PGF_2 have been found in plasma, urine, or bronchoalveolar lavage fluid (BAL) in neonates with bronchopulmonary dysplasia (BPD) or acute respiratory distress syndrome [12, 131]. Furthermore, the ratio of D_2/E_2-IsoPs to F_2-IsoPs provides information about the LPO and about the redox environment [71]. Several methodologies have been developed for detection of IsoPs. Immunoassays based on immunological reactions of antibodies against specific IsoPs such as 8-iso-$PGF_2\alpha$ have been frequently employed. However, cross-sensitivity between IsoPs isomers and homologous prostaglandins constitute the main drawback of this technique. Currently, new LC-MS/MS approaches are emerging which provide straightforward measurement and help to partly avoid time-consuming sample preparation steps [8].

Neuroprostanes (*NeuPs*). DHA is an important ω-3-PUFA that is the predominant structural component of the central nervous system tissues. DHA is more prone to oxidation than AA because of the higher number of double bonds, and neuroprostanes (NeuPs) arise from its oxidation in a similar manner to IsoPs [71, 93]. NeuPs have been suggested as neuronal oxidative damage markers more adequate than IsoPs that may be measured in neuronal tissues as well as in urine, plasma, or cerebrospinal fluid [71, 93]. The determination of NeuPs may be performed as the sum of total NeuPs by GC-MS/MS or LC-MS/MS. [61, 109, 110] However, both approaches are limited by the lack of commercially available, pure NeuP standards necessary for a reliable quantification of NeuPs.

Isofurans (IsoFs) and Neurofurans (NeuFs). IsoFs are mainly formed under conditions of elevated oxygen tension. Oxygen interacts with arachidonic acid (AA) inducing the generation of a tetrahydrofuran ring instead of a prostane ring. Therefore, IsoFs have been suggested as reliable biomarkers of LPO produced under hyperoxic conditions that may be detected in fluids and tissues [71, 110]. IsoFs are increased with the use of higher oxygen concentrations in the resuscitation of asphyxiated newborns in experimental [109] and clinical studies [131]. NeuFs formation mechanisms are not entirely clear. Hence, a direct action of oxygen upon DHA as similar to AA has not been proven in vitro studies. Another disadvantage for the clinical application of NeuFs is that urine and plasma levels may be below the achievable limits of detection employing state-of-the-art instruments. Nevertheless, NeuFs may be determined in brain tissue or cerebrospinal fluid [2]. Similarly to NeuPs, determination of IsoFs and NeuFs may be performed by GC-MS/MS or LC-MS/MS [61, 109] suffering from the above described limitations.

The Role of Oxygen in the Transition from Fetal to Extrauterine Life

Oxygen in the Fetal to Neonatal Transition

Fetal life develops in a relatively hypoxic environment. Hence, arterial partial pressure of oxygen (p_aO_2) in utero is approximately 4 kPa (25–30 mmHg) as compared to 9 kPa (80–90 mmHg) in the mother. Interestingly, this situation is partially compensated by the presence of fetal hemoglobin with a greater affinity for oxygen thus favoring placental oxygen uptake by fetal blood and increased oxygen saturation for a given p_aO_2. Immediately after birth and with the initiation of spontaneous respiration alveolar-capillary gas exchange, p_aO_2 rises to 80–90 mmHg in the first 5–10 min after birth [21, 66, 123, 139]. This abrupt change results in a physiological oxidative stress necessary to trigger the expression of a number of significant genes necessary for postnatal adaptation. Of note is that in the last trimester of gestation, both transplacental passage of antioxidant enzymes and increased activity of proper fetal antioxidant enzymes prepare the fetus for lung respiration [15, 31]. In the fetal to neonatal transition, both blood oxygen content and oxygen availability to tissue increase in a few minutes to almost adult values contributing to the generation of a burst of ROS which may act as signaling molecules modulating maturation of specific metabolic pathways [24, 32, 84]. Interestingly, the use of antenatal steroids, which has dramatically improved morbidity free survival of preterm infants, is associated with enhanced activity of the antioxidant defense system thus favoring postnatal adaptation to a relative hyperoxic situation [123].

Pulse Oximetry in the Delivery Room

Pulse oximetry is now commonly used to monitor peripheral oxygenation saturation (oxyhemoglobin saturation or SpO_2) in neonatal intensive care units (NICU) where it has been described as the "fifth vital sign" [73] but also in the delivery room (DR) where it has the benefit of measuring SpO_2 and heart rate (HR) continuously and noninvasively, without the need for calibration and correlates closely with arterial oxygen saturation except for the higher (>95 %) and lower (<50 %) ranges [44, 53, 55, 57]. National and international [3, 58, 74] resuscitation guidelines now recommend using pulse oximetry in the delivery room when managing preterm infants and when infants are receiving supplemental oxygen or advanced respiratory support. In the delivery room, assessment of an infant's color, as a surrogate measure of oxygenation, has constituted one-fifth of the Apgar score. However, O'Donnell showed that there is substantial inter- and intra-observer variability when assessing color in the minutes after birth and therefore is a poor proxy for tissue oxygenation during the first minutes of life [76]. In contrast, pulse oximetry provides an objective measure of oxygenation. The important, emerging role of pulse oximetry in the DR was recognized by the American Academy of Pediatrics (AAP) [58] in their 2010 guidelines for neonatal resuscitation: "It is recommended that oximetry be used when resuscitation can be anticipate, when positive pressure is administered for more than a few breaths, when cyanosis is persistent, or when supplementary oxygen is administered" [58].

Effect of Mode of Delivery

In the first 5 min after birth, infants born via caesarean have significantly lower SpO_2 measurements than those born vaginally [1, 16, 41, 54, 90]. Harris [41] postulated that this difference was due to the increased amount of lung fluid after caesarean section which may interfere with transition to air breathing. In contrast, other researchers found no significant difference in SpO_2 measurements in infants born vaginally versus caesarean births [20, 45, 85]. The latter group of studies had smaller samples and used older-generation pulse oximeter that may have contributed to their findings.

Effect of Gestational Age

While there are several reports on SpO_2 measurements in term infants in the minutes after birth, there are fewer reports on the change in SpO_2 in preterm infants. Dawson [16] showed that in 160 preterm infants, the median SpO_2 at 5 min was 86 %, versus 92 % in term infants ($p<0.001$). Kamlin [54] reported in 54 preterm infants the median SpO_2 at 5 min after birth was 87 %, significantly lower than term infants where the median SpO_2 at 5 min was 90 % ($p<0.001$). Nuntnarumit [75] studied 75 preterm infants <35 weeks who did not receive supplemental oxygen in the DR and reported median (IQR) SpO_2 at 2, 3, 4, 5, and 6 min was 77 % (72–92), 84 % (75–94), 88 % (80–94), 90 % (79–95), and 95 % (85–97), respectively. In Kopotic's observational study of 15 infants born at 24–29 weeks' gestation, the SpO_2 was ≥80 % by 4.4 (1.9–40) min, median (range). However, infants in the Kopotic study may have received oxygen and other interventions which might account for the shorter time to achieve an $SpO_2 \geq 80$ % [59].

Sensor Location

In the DR, SpO_2 is changing rapidly; therefore, it is important to use the best technique to place the sensor that obtains a reliable signal in the shortest possible time. The Masimo [77] and Nellcor oximeters [97] provide a measurement fastest when the oximeter is turned on; the sensor is applied to the infant before connecting the sensor to the oximeter cable. Pulse oximeter sensors can be placed on an infant's hand or wrist, anterior, dorsal aspect of the foot, and other locations. Sensors placed on the right hand or wrist inform on preductal oxygen saturation. Preductal measurements have been shown to be significantly higher than postductal measurements soon after birth and remain so for the first 15 min after birth [67]. In the DR where the presence of intra- and extracardiac shunts persist, the right hand or wrist is the best site for monitoring SpO_2 measurements and titrating administration of oxygen since these readings are more representative of SpO_2 values for the brain [70].

Motion Artifact

One of the concerns when using oximetry to determine oxygen saturation is motion artifact. Motion artifact can cause false readings when using oximetry. Motion artifact can cause both false-positive (false alarm) and false-negative (missed hypoxemia). New generation oximeters have electronic signal processing techniques and sensor features that decrease or eliminate motion artifact [43]. Premature infants may be less susceptible to motion artifact than more active term infants. When the sensor is attached correctly and firmly to the limb, motion artifact is reduced.

The Oxygen Saturation Centile Charts (Figs. 11.3, 11.4, and 11.5)

Previous researchers have presented SpO_2 measurements from infants not requiring interventions in the DR as box plots or as line graphs illustrating the interquartile range for pre-preductal SpO_2, in the first minutes after birth [20, 29, 45, 54, 70, 90]. The Dawson SpO_2 centile charts provide a reference range for SpO_2 in the first ten minutes after birth. These charts are unique due to the large number of data points (61,650) and the large number of infants studied ($n=468$) who did not receive any medical interventions other than warmth and stimulation in the first 10 min after birth.

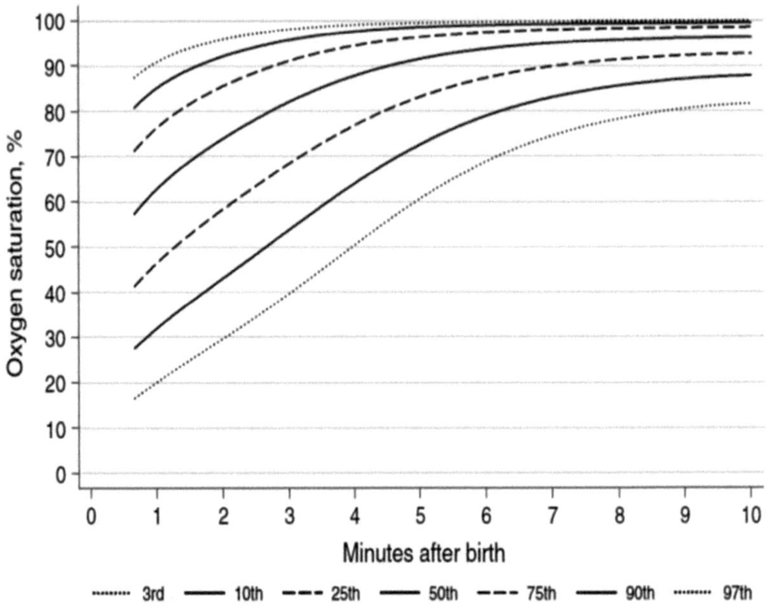

Fig. 11.3 Third, 10th, 25th, 50th, 75th, 90th, and 97th SpO_2 percentiles for newborn infants of all gestational ages with no medical intervention after birth (Reprinted from Dawson et al. [16]. With permission from *Pediatrics*)

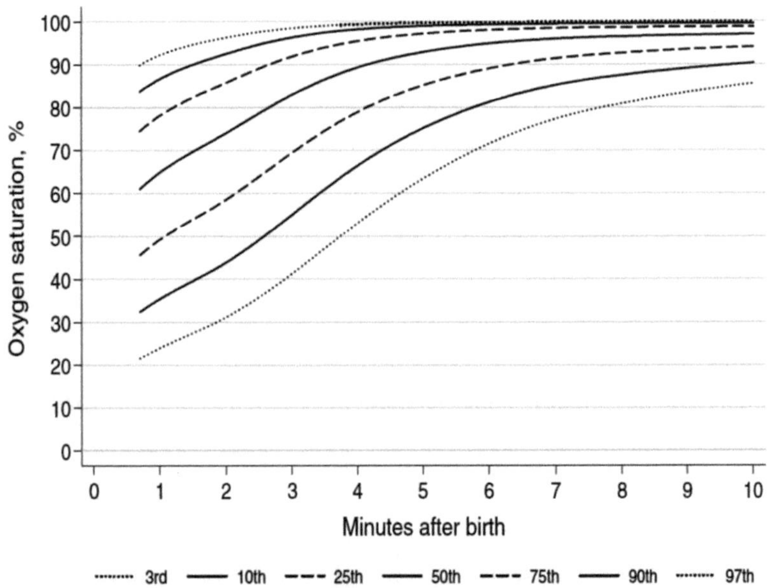

Fig. 11.4 Third, 10th, 25th, 50th, 75th, 90th, and 97th SpO_2 percentiles for all infants ≥ 37 weeks' gestation with no medical intervention after birth (Reprinted from Dawson et al. [16]. With permission from *Pediatrics*)

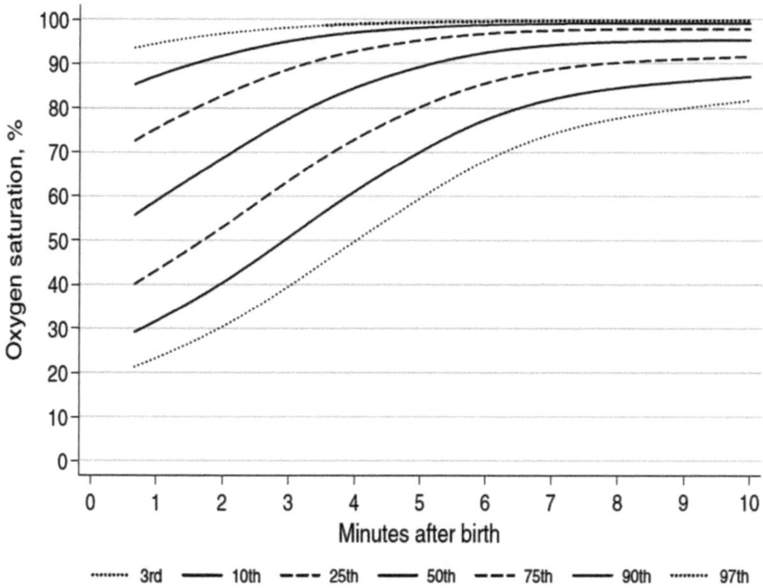

Fig. 11.5 Third, 10th, 25th, 50th, 75th, 90th, and 97th SpO_2 percentiles for all infants 32–36 weeks' gestation with no medical intervention after birth (Reprinted from Dawson et al. [16]. With permission from *Pediatrics*)

These data points were used to construct smoothed centile curves to illustrate the changing distribution of SpO_2 measurements after birth rather than "snapshots" of the data at each minute after birth. These data were used to develop centile charts for <32 weeks' gestation, 32–36 weeks' gestation, and term infants ≥37 weeks' gestation. For all 468 infants at 1 min, the 3rd, 10th, 50th, 90th, and 97th percentiles were 29, 39, 66, 87, and 92 %, respectively; at 2 min, 34, 46, 73, 91, and 95 %; and at 5 min, 59, 73, 89, 97, and 98 %. It took a median (IQR) of 7.9 (5.0–10.0) min to reach an SpO_2>90. At all time points, the median SpO_2 was significantly lower for preterm infants than for term infants. Until there is a consensus regarding the definition of "normoxia," the Dawson centile charts provide our "best guess" at the appropriate targets for oxygen saturation during resuscitation.

Oxygen Saturation in Healthy Preterm Using Continuous Positive Airway Pressure (CPAP) and Air (Fig. 11.6)

The use of CPAP in the DR has been recommended to facilitate an early achievement of a functional residual capacity (FRC) in preterm babies immediately after birth [17]. It has recently been shown that spontaneously breathing preterm babies using face mask CPAP and air achieved higher SpO_2s significantly earlier than that

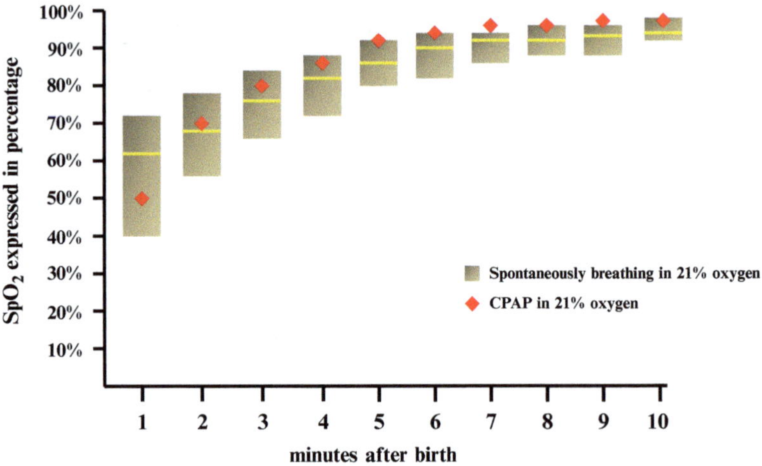

Fig. 11.6 SpO_2s in the first 10 min after birth in preterm babies of <32 weeks' gestation who received continuous positive pressure ventilation (PPV) with face and mask and air. Comparison established with Dawson's nomogram [16] for preterm infants using χ^2 test

reported in the reference nomogram [16, 129]. This seems to support the speculation that achieving stable lung aeration and functional residual capacity (FRC) might reduce the need for supplementary oxygen in preterm infants in the delivery room [88].

Brain Oxygenation and Cerebral Blood Flow

Cerebral Blood Flow: Physiological Considerations

Cerebral metabolic rate and the related rate of oxygen consumption is dependent on the species studied and on brain maturation, which may influence baseline cerebral blood flow and its relative changes related to hypoxic and hypercapnic challenge [94]. During conditions of hypoxia, cerebral blood flow increases to compensate for decreased arterial oxygen concentration [50]. Although a clear autoregulatory plateau for cerebral blood flow in response to changes in blood pressure has been demonstrated in fetal sheep [79], this may not be present consistently in premature infants after birth, where blood pressure fluctuations, arterial hypotension, or hypercapnia may result in significant changes of cerebral blood flow [52]. Doppler studies in human infants suggest that autoregulation of cerebral blood flow in response to blood pressure changes increases with a higher gestational age [91], and the range of autoregulation seems to be narrowed in very preterm infants [120]. Studies of cerebral blood flow in preterm infants in relation to changes in blood pressure or PCO_2 suggest a relationship between low cerebral blood flow and the loss of vasoreactivity, which may lead to the development of intraventricular hemorrhage [86]. Furthermore, the loss of vasoreactivity and persistently high cerebral blood flow velocity seems to be associated with poor outcome in asphyxiated full-term newborns [87]. Doppler measurements of cerebral blood flow (velocity) are, however, difficult to obtain for longitudinal observations and even more difficult if not impossible during resuscitation of sick newborn infants immediately after birth. Therefore, a simple system to monitor cerebral blood flow and/or cerebral tissue oxygenation, similar to pulse oximetry may be more helpful in these circumstances.

Measurements of Cerebral Oxygenation with Near-Infrared Spectroscopy (NIRS) in Different Clinical Settings

Near-Infrared Spectroscopy (NIRS) is a noninvasive technique to measure oxyhemoglobin, deoxyhemoglobin, and oxidized cytochrome aa_3 [46, 92, 134]. This technology is based on the absorption of near-infrared light by iron and copper atoms in hemoglobin and cytochrome aa_3. Infrared light penetrates several centimeters into the brain tissue and provides information on oxygenation of the brain. It measures tissue oxygenation underneath the optical pathway and gives real-time information to the clinician. NIRS has been used clinically since the 1970s, is noninvasive, provides continuous information, does not require pulsatile blood flow, and reflects a balance between oxygen delivery and consumption [46]. NIRS has been suggested as a useful tool to guide treatment in critically ill infants [137].

Technical Considerations with NIRS

There are no uniform standards for NIRS technology and for the algorithms used by different manufacturers. Clinicians need to be aware that the displayed absolute values are to some degree device-specific. Therefore, clinicians using NIRS should be aware that when they are interpreting NIRS data, there are no uniform standard. Experts in using NIRS have recommended standardizing devices, thus increasing validity of measurements in neonatal patients and to improve measurement techniques. This would enable more accurate comparison between data obtained from different models for NIRS [83]. Measured values of cerebral oxygenation may be affected by extracranial contamination. In healthy volunteers, it has been shown that tissue ischemia of the skin underneath the sensor site affects the measured values, and the magnitude of the measured decrease in cerebral oxygenation caused by scalp ischemia seems to be dependent on the specific device used [13]. This has not been studied in neonates, but may be an important issue as the scalp may be exposed to birth trauma secondary to local forces especially during vaginal delivery. There are a number of other pitfalls to consider when NIRS is used to monitor cerebral oxygen saturation in neonates, and these have been nicely reviewed by van Bel 2008 [119].

Measurements of Cerebral Oxygenation with Near-Infrared Spectroscopy (NIRS) in Neonates During Transition Shortly After Birth

The studies in adults and children undertaken in surgical settings suggest that monitoring cerebral oxygenation during surgery may help to identify those patients at risk for adverse outcomes. Cerebral oxygenation may be affected during delivery and by adverse conditions immediately after birth. Monitoring cerebral oxygenation may be helpful to identify those infants who need respiratory and/or cardiovascular support at this time. Mc Neill [69] has recently published data for preterm infants from the first day of life to postnatal day 21. If brain oxygenation, measured by NIRS, is used to guide interventions in the DR, there is an urgent need for reference standards for newly born premature and full-term infants.

Fauchere et al. measured cerebral oxygen saturation in 20 healthy term newborn infants during the first 15 min after elective caesarean section using the NIRO 300 device (Hamamatsu Photonics, Hamamatsu, Japan) [28]. All infants had an uneventful transition and did not need any resuscitative support or supplemental oxygen. They were able to obtain a signal within 2 (1–4) min (median [range]) after birth and found a median tissue oxygenation index of 52 % at 3 min of age, which increased to approximately 69 % at 6 min of age, with no further increase until 15 min of age. During the same time, the median SpO_2 increased from 71 to 88 %. Urlesberger et al. measured regional tissue oxygenation of the brain along with SpO_2 in 59 healthy term infants with no need for respiratory support or supplemental oxygen immediately after elective caesarean section in an observational study using the INVOS 5100 device (Somanetics. Troy, Michigan) [116]. The sensor was positioned at the left frontoparietal forehead. The authors were able to obtain regional cerebral saturation (rSO_2) values from 3 min after birth and could show that it increased from a mean of 44 % at 3 min to 76 % at 7 min, with no further increase during the 10 min observation period. The rate of increase was faster than the rise in preductal SpO_2, consistent with the observation of other researchers [28]. This may indicate preferential oxygen delivery to the brain immediately after birth. In another, study rSO_2 was measured in 46 healthy term newborn infants ($n=20$ after spontaneous vaginal delivery, $n=22$ after caesarean section, and $n=4$ after vaginal-assisted delivery) during the first 10 min of life using a laser light source oximeter (FORE-SIGHT, Casmed, Branford, CT) [33]. The median (interquartile range) cerebral oxygen saturation at 2 min after birth was 42 (39–46) % after spontaneous vaginal delivery, 42 (30–52) % after caesarean section, and 36 (20–53) % after assisted instrumental vaginal delivery (no significant difference between groups). rSO_2 increased continuously and reached a steady state approximately 8 min after birth. The differences in values measured in the three studies [28, 33, 116] may be related to the use of different devices or by differences in the amount of infant handling.

In another study in 51 VLBWI, it was shown that cerebral tissue oxygenation using NIRS can be measured in the delivery room within 52 (44–68) seconds of life [median (interquartile range)]. After delivery, cerebral tissue oxygen saturation rose continuously from 37 (31–49)% at 1 min reaching a steady state after approximately 7 min of life with a range of 61–84 % [33]. Regional cerebral oxygen saturation increased at least as fast as in the full-term cohort studied by the same group of researchers [33].

Oxygen Administration in the Delivery Room

Oxygen Supplementation in the Term Newborn Infant

Perinatal asphyxia is a relatively common condition especially in developing countries. Hence, approximately four million babies will suffer for intrapartum asphyxia, and out of these, one million will die and another million will have important neurocognitive impairment [133]. Interestingly, many so-called fresh stillbirths occurring in rural settings where supplemental oxygen was not available may have been successfully revived if positive pressure ventilation with air (21 % oxygen) had been initiated [36]. However, in the last two decades, a body of evidence has demonstrated in experimental [101] and clinical [99, 100] studies that in newly born asphyxiated term, infants' resuscitation with air may offer substantial advantages over the use of 100 % oxygen. Room air shortens the time needed to initiate spontaneous respiration and improves Apgar score [125]. Moreover, room air has been shown to reduce oxidative stress ([125–127, 132]; Martin 2008). In randomized studies where clinicians were blinded to the gas admixture, it has been shown that in infants receiving 21 % or 100 % oxygen for resuscitation, both groups demonstrated evidence of oxidative stress. However, infants receiving air had significantly higher reduced oxidized to glutathione ratios (index of cytoplasmic oxidative stress) at day 3 and day 28 after birth [125]. In addition, striking differences in the glutathione redox cycle enzymes in newborn infants resuscitated with room air versus 100 % oxygen confirmed the alteration of the cellular redox status caused by the use of higher oxygen loads during resuscitation [127, 132]. These findings suggest that excess of oxygen and OS in a "sensitive" period such as the fetal to neonatal transition could trigger a biologic response consisting of an oxidative stress mediated by gene activation. In addition to the alteration of biomarkers of oxidative stress, further studies have found that administration of 100 % oxygen in severely asphyxiated babies increased oxidative damage to myocardium and renal proximal tubule as compared to room air [132]. In a recent meta-analysis, it was shown that the use of room air for resuscitation significantly reduces mortality in the delivery room in term neonates [99]. In response to these findings, the 2010 International Liaison Committee on Resuscitation guidelines recommended the use of air as the initial gas admixture for the depressed term infant [82]. Thus, resuscitation of depressed term newborn infants should be initiated with positive pressure ventilation and 21 % oxygen. Lung expansion and early achievement of a functional residual capacity is essential for allowing a good gas exchange and pulmonary circulation. Under these circumstances, oxygenation and heart rate should be monitored with preductal pulse oximetry following centiles illustrated in the Dawson nomogram [16, 133]. FiO_2 should be individually adjusted, or titrated, according to an infant's response and SpO_2 measurements. In very severe clinical situations where an infant does not respond to PPV with 21 % oxygen, then endotracheal intubation and administration of a higher oxygen concentration are warranted [133].

Titrating Oxygen Needs in Preterm Infants

Pulse oximetry measurement of SpO_2 has replaced assessment of color as an indication for administering supplemental oxygen in several national and international guidelines, e.g., the AAP [58], European Resuscitation Council (ERC [74]) and the Australian Resuscitation Council [3]. The 2010 ILCOR guidelines suggest using PO to guide administration of supplemental oxygen in the DR to avoid hyperoxia and hypoxia particularly in extremely preterm infants. Suggested targets for SpO_2 at intervals following birth are provided [3, 74, 82]. The Dawson SpO_2 reference charts can be used in the delivery room as a guide to prevent hypo-/hyperoxia. This is especially important when managing extremely preterm infants at risk of developing hyperoxia [124, 128]. Kopotic and Lindner [59] studied 50 high-risk infants; 25 were managed with pulse oximetry, while the remaining 25 infants were not monitored. Infants managed with oximetry were less likely to be admitted to the special care nursery (32 % vs. 52 %; $p = 0.04$). Moreover, they observed the influence of oximetry during resuscitation in 15 infants <30 weeks' gestation (suppress in a second study!) [59]. Oxygen was started at 100 % and titrated to achieve an SpO_2 between 80 and 92 %. They demonstrated that they were able to reduce the fraction of inspired oxygen (FiO_2) from 1.0 to, on average, 0.40 [59]. Deckardt [19] used an $SpO_2 < 80$ % at 5 min to determine whether infants should receive continuous positive airway pressure (CPAP) with a mask and 100 % oxygen. The treatment stopped once the SpO_2 reached 90 %. The observational studies of Deckardt [19] and Kopotic [59] suggest that using SpO_2 in the DR is valuable in managing resuscitation. Finer and Leone have previously advocated a targeted oxygen saturation protocol in the DR [30]. Three randomized trials [26, 89, 135] have shown it is possible to titrate the administration of supplemental oxygen in the DR using SpO_2 target ranges. However, during resuscitation, an appropriate SpO_2 target below which oxygen therapy does more good than harm has not been determined. The highest safe level of SpO_2 is also unclear. Even short-term exposure to a high FiO_2 can generate damaging reactive oxygen (ROS) and nitrogen species (RNS) [128]. These free radicals are associated with short- and long-term morbidity, with preterm infants at most risk of harm from exposure to excess oxygen [102].

There are no guidelines recommending how fast or by how much oxygen should be increased if SpO_2 measurements are below the recommended target range. Nor are there guidelines for how quickly and by how much FiO_2 should be decreased when SpO_2 above the target range. We have recommended that once effective ventilation is confirmed, if SpO_2 is below the 10th centile, the FiO_2 should be increased until the SpO_2 reaches at least the 10th centile. Ongoing supplemental oxygen should be given targeting the 50th centile and avoiding saturations >90th centile. Since oxygen saturation >95 % may be associated with unpredictably high arterial oxygen concentration, supplemental oxygen should be reduced to maintain oxygen saturation <96 % at all times in the delivery room [17, 122]. Several NICU studies have shown how difficult it is to keep infants SpO_2 with a narrow target range [10, 39]. In the DR, it is likely to be more difficult to keep infants who are receiving supplemental oxygen within a target range when SpO_2 measurements are normally rising rapidly. Gandhi et al. [35] describes a novel system for improving the amount of time infants remain within a target range. At UCSD (San Diego, USA), the Transitional Oxygen Targeting System (TOTS) plots real-time SpO_2 values in relation to the 10th and 50th centiles oxygen saturation curves from Dawson [16]. Infants ≤36 weeks' gestation whose supplemental oxygen was managed with TOTS were kept within the specified target range significantly longer when compared to control infants managed without TOTS ($p=0.026$). Experts in neonatal resuscitation remind us that the cornerstone of resuscitation is to perform effective ventilation to establishing an adequate functional residual capacity before considering commencing or increasing supplemental oxygen [17, 63]

Interventions in the Delivery Room: Effects on Cerebral Oxygenation

There is very little data available on the effects of interventions undertaken during the first seconds or minutes of life on cerebral oxygenation. Out of the 51 infants studied by Fuchs et al., 10 received only CPAP or no respiratory support [33]. There was very little difference in cerebral oxygen saturation comparing those infants receiving respiratory support compared against those who did not receive respiratory support (Fig. 11.7). Furthermore, it is very interesting that the two infants who later developed intraventricular hemorrhage demonstrated cerebral oxygen saturation well below the 10th percentile compared to the other study infants, suggesting that adverse cerebral blood flow was responsible for this observation. We speculate that measuring cerebral oxygenation using NIRS immediately after birth may be useful to detect patients at high risk for adverse cerebral outcome (Fig. 11.7).

The immediate effects of sustained lung inflations on arterial and cerebral oxygenation were studied in 24 VLBWI with a mean gestational age of 28 weeks [34]. All study infants received one sustained inflation with 20 cmH_2O for 15 s duration and CPAP at 5 cmH_2O thereafter. 16/24 (67 %) received a second inflation with 25 cmH_2O for 15 s, and 2/24 (8 %) received a third inflation with 30 cmH_2O/15 s as per unit protocol if the infants did not respond to the previous inflation(s). Thereafter, all infants were supported with CPAP and eventually nasopharyngeal ventilation at the discretion of the neonatologist taking care. During sustained inflations, there was a rapid increase in heart rate, cerebral tissue, and SpO_2 [34]. In this observational study by Fuchs, there was no control group receiving an alternative mode of respiratory support. Therefore, it is presently unclear if sustained inflations are superior in stabilizing tissue oxygenation after birth. To evaluate the effect of sustained inflations on cerebral oxygenation, we need a properly powered randomized trial comparing infants receiving sustained inflations with a control group who do not receive sustained inflations.

Monitoring cerebral oxygen saturation in the delivery room and during the early hours of life might be useful to assess tissue oxygenation and may help to provide

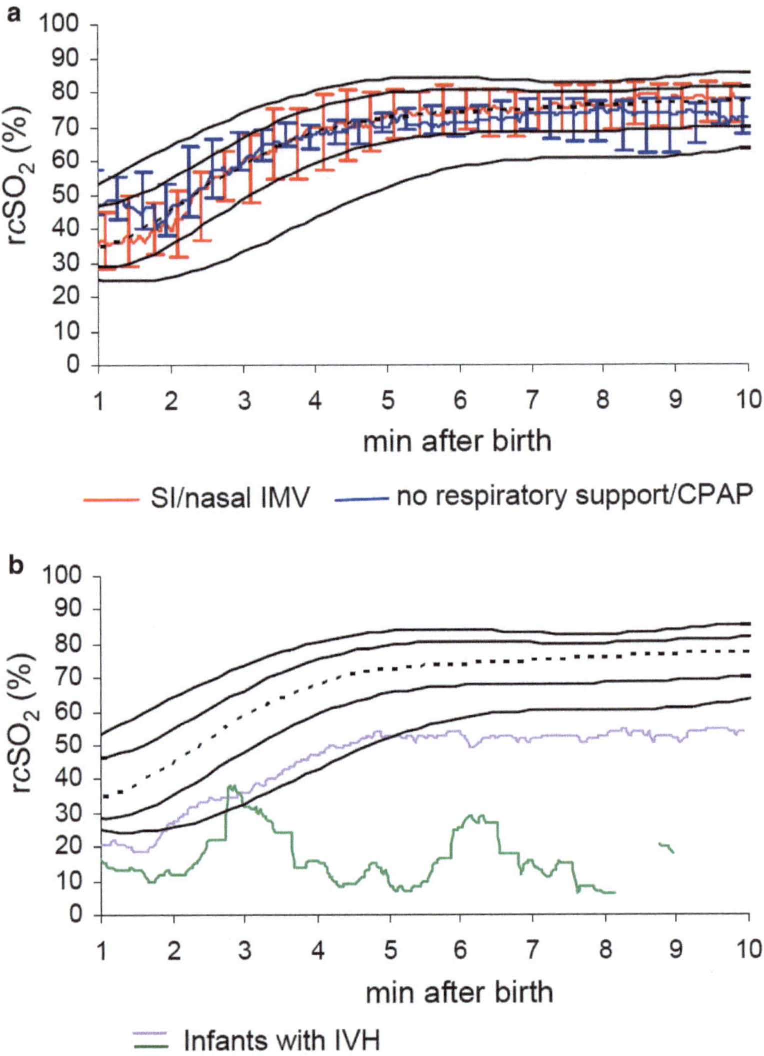

Fig. 11.7 Panel (**a**). Comparison between cerebral oxygen saturation (rSO_2) in infants receiving respiratory support as compared against those who did not receive respiratory support. As shown in the figure, there were not significant differences. Panel (**b**). The figures show the rSO_2 graph of two infants who later developed intraventricular hemorrhage and who demonstrated cerebral oxygen saturation well below the 10th percentile compared to the other study infants suggesting that adverse cerebral blood flow was responsible for this observation [33]

information to adjust FiO_2 and other interventions, such as ventilator settings in infants with lung disease. There is some evidence that the brain of stable VLBW infants may be hyperoxygenated during the first hours and days of life as shown in a cohort study of 46 preterm infants compared against 25 healthy term infants [111].

Technology is now available in the delivery room to measure the effect of interventions such as different modes of respiratory or cardiovascular support and their effect on cerebral oxygenation. We hope that further studies using this technology to monitor infants in the first minutes after birth may eventually help to improve survival and neurodevelopmental outcome of extremely low birth weight infants.

References

1. Altuncu E, Ozek E, Bilgen H, Topuzoglu A, Kavuncuoglu S. Percentiles of oxygen saturations in healthy term newborns in the first minutes of life. Eur J Pediatr. 2008;167:687–8.
2. Arneson KO, Roberts 2nd LJ. Measurement of products of docosahexaenoic acid peroxidation, neuroprostanes, and neurofurans. Methods Enzymol. 2007;433:127–43.
3. Australian Resuscitation Council. 2010. http://www.resus.org.au/
4. Auten RL, Davis JM. Oxygen toxicity and reactive oxygen species: the devil is in the details. Pediatr Res. 2009;66:121–7.
5. Bartz RR, Piantadosi CA. Clinical review: oxygen as a signaling molecule. Crit Care. 2010;14:234.
6. Breusing N, Grune T. Biomarkers of protein oxidation from a chemical, biological and medical point of view. Exp Gerontol. 2010;45:733–7.
7. Breusing N, Grune T, Andrisic L, Atalay M, Bartosz G, Biasi F, Borovic S, Bravo L, Casals I, Casillas R, Dinischiotu A, Drzewinska J, Faber H, Fauzi NM, Gajewska A, Gambini J, Gradinaru D, Kokkola T, Lojek A, Luczaj W, Margina D, Mascia C, Mateos R, Meinitzer A, Mitjavila MT, Mrakovcic L, Munteanu MC, Podborska M, Poli G, Sicinska P, Skrzydlewska E, Vina J, Wiswedel I, Zarkovic N, Zelzer S, Spickett CM. An inter-laboratory validation of methods of lipid peroxidation measurement in UVA-treated human plasma samples. Free Radic Res. 2010;44:1203–15.
8. Casetta B, Longini M, Proietti F, Perrone S, Buonocore G. Development of a fast and simple LC-MS/MS method for measuring the F2-isoprostanes in newborns. J Matern Fetal Neonatal Med. 2012;25 Suppl 1:114–8.
9. Circu ML, Aw TY. Reactive oxygen species, cellular redox systems and apoptosis. Free Radic Biol Med. 2010;48:749–62.
10. Clucas L, Doyle LW, Dawson J, Donath S, Davis PG. Compliance with alarm limits for pulse oximetry in very preterm infants. Pediatrics. 2007;119:1056–60.
11. D'Autréaux B, Toledano MB. ROS as signaling molecules: mechanisms that generate specificity in ROS homeostasis. Nat Rev Mol Cell Biol. 2007;8:813–24.
12. Dalle-Donne I, Rossi R, Colombo R, Giustarini D, Milzani A. Biomarkers of oxidative damage in human disease. Clin Chem. 2006;52:601–23.
13. Davie SN, Grocott HP. Impact of extracranial contamination on regional cerebral oxygen saturation: a comparison of three cerebral oximetry technologies. Anesthesiology. 2012; 116:834–40.
14. Davies MJ, Fu S, Wang H, Dean RT. Stable markers of oxidant damage to proteins and their application in the study of human disease. Free Radic Biol Med. 1999;27:1151–63.
15. Davis JM, Autent RL. Maturation of the antioxidant system and the effects on preterm birth. Semin Fetal Neonatal Med. 2010;15:191–5.
16. Dawson JA, Kamlin COF, Vento M, Wong C, Cole TJ, Donath SM, Davis PG, Morley CJ. Defining the reference range for oxygen saturation for infants after birth. Pediatrics. 2010;125:e1340–7.
17. Dawson JA, Vento M, Finer NN, Rich W, Saugstad OD, Morley CJ, Davis PG. Managing oxygen therapy during delivery room stabilization of preterm infants. J Pediatr. 2012; 160:158–61.

18. De Bont R, van Larebeke N. Endogenous DNA damage in humans: a review of quantitative data. Mutagenesis. 2004;19:169–85.
19. Deckardt R, Schneider KT, Graeff H. Monitoring arterial oxygen saturation in the neonate. J Perinat Med. 1987;15:357–60.
20. Dimich I, Singh PP, Adell A, Hendler M, Sonnenklar N, Jhaveri M. Evaluation of oxygen saturation monitoring by pulse oximetry in neonates in the delivery room. Can J Anaesth. 1991;38:985–8.
21. East CE. Fetal pulse oximetry during labor: theory, clinical use and the future. Pediatr Health. 2008;2:787–95.
22. Ecker J. Profiling eicosanoids and phospholipids using LC-MS/MS: principles and recent applications. J Sep Sci. 2012;35:1227–35.
23. Enomoto M, Gosal K, Cubells E, Escobar J, Vento M, Jankov RP, Belik J. Sex-dependent changes in the pulmonary vasoconstriction potential of newborn rats following short-term oxygen exposure. Pediatr Res. 2012. doi:10.1038/pr.2012.120.
24. Escobar J, Cernada M, Vento M. Oxygen and oxidative stress in the neonatal period. Neoreviews. 2011;12:e613–24.
25. Escobar J, Teramo K, Stefanovic V, Andersson S, Asensi MA, Arduini A, Cubells E, Sastre J, Vento M. Amniotic fluid oxidative and nitrosative stress biomarkers correlate with fetal chronic hypoxia in diabetic pregnancies. Neonatology. 2012. doi:10.1159/000345194.
26. Escrig R, Luis A, Izquierdo I, Villar G, Gimeno A, Sáenz P, Gimeno A, Moro M, Vento M. Achievement of target oxygen saturation in extremely low gestational neonates resuscitated with different oxygen concentrations: a prospective randomized clinical trial. Pediatrics. 2008;121:875–81.
27. Exner R, Wessner B, Manhart N, Roth E. Therapeutic potential of glutathione. Wien Klin Wochenschr. 2000;112:610–6.
28. Fauchère JC, Schulz G, Haensse D, Keller E, Ersch J, Bucher HU, Wolf M. Near-infrared spectroscopy measurements of cerebral oxygenation in newborns during immediate postnatal adaptation. J Pediatr. 2010;156:372–6.
29. Field JM, Hazinski MF, Sayre MR, Chameides L, Schexnayder SM, Hemphill R, Samson RA, Kattwinkel J, Berg RA, Bhanji F, Cave DM, Jauch EC, Kudenchuk PJ, Neumar RW, Peberdy MA, Perlman JM, Sinz E, Travers AH, Berg MD, Billi JE, Eigel B, Hickey RW, Kleinman ME, Link MS, Morrison LJ, O'Connor RE, Shuster M, Callaway CW, Cucchiara B, Ferguson JD, Rea TD, Vanden Hoek TL. Part 1. Executive summary: 2010 American Heart Association Guidelines for Cardiopulmonary Resuscitation and Emergency Cardiovascular Care. Circulation. 2010;122(18 Suppl 3):S640–56.
30. Finer N, Leone T. Oxygen saturation monitoring for the preterm infant: the evidence basis for current practice. Pediatr Res. 2009;65:375–80.
31. Forman HJ, Fukuto JM, Miller T, Zhang H, Rinna A, Levy S. The chemistry of cell signaling by reactive oxygen species and nitrogen species and 4-hydroxynonenal. Arch Biochem Biophys. 2008;477:183–95.
32. Friel JK, Friesen RW, Harding SV, Roberts LJ. Evidence of oxidative stress in full-term healthy infants. Pediatr Res. 2004;56:878–82.
33. Fuchs H, Lindner W, Buschko A, Almazam M, Hummler HD, Schmid MB. Brain oxygenation monitoring during neonatal resuscitation of very low birth weight infants. J Perinatol. 2012;32:356–62.
34. Fuchs H, Lindner W, Buschko A, Trischberger T, Schmid M, Hummler HD. Cerebral oxygenation in very low birth weight infants supported with sustained lung inflations after birth. Pediatr Res. 2011;70:176–80.
35. Gandhi BB, Rich WD, Finer NN. Improving oxygen saturation targeting during neonatal resuscitation. E-PAS. 2012;2855–6.
36. Garcés A, McClure EM, Chomba E, Patel A, Pasha O, Tshefu A, Esamai F, Goudar S, Lokangaka A, Hambidge KM, Wright LL, Koso-Thomas M, Bose C, Carlo WA, Liechty EA, Hibberd PL, Bucher S, Whitworth R, Goldenberg RL. Home birth attendants in low income

countries: who are they and what do they do? BMC Pregnancy Childbirth. 2012. doi:10.1186/1471-2393-12-34.
37. Giordano J. Oxygen, oxidative stress, hypoxia and heart failure. J Clin Invest. 2005; 115:500–8.
38. Guéraud F, Atalay M, Bresgen N, Cipak A, Eckl PM, Huc L, Jouanin I, Siems W, Uchida K. Chemistry and biochemistry of lipid peroxidation products. Free Radic Res. 2010; 44:1098–124.
39. Hagadorn JI, Furey AM, Nghiem TH, Schmid CH, Phelps DL, Pillers DA, Cole CH, AVIOx Study Group. Achieved versus intended pulse oximeter saturation in infants born less than 28 weeks' gestation: the AVIOx study. Pediatrics. 2006;118:1574–82.
40. Halliwell B, Whiteman M. Measuring reactive species and oxidative damage in vivo and in cell culture: how should you do it and what do the results mean? Br J Pharmacol. 2004;142:231–55.
41. Harris AP, Sendak MJ, Donham RT. Changes in arterial oxygen saturation immediately after birth in the human neonate. J Pediatr. 1986;109:117–9.
42. Harwood DT, Kettle AJ, Brennan S, Winterbourn CC. Simultaneous determination of reduced glutathione, glutathione disulphide and glutathione sulphonamide in cells and physiological fluids by isotope dilution liquid chromatography-tandem mass spectrometry. J Chromatogr B Analyt Technol Biomed Life Sci. 2009;877:3393–9.
43. Hay Jr WW. Pulse oximetry: as good as it gets? J Perinatol. 2000;20:181–3.
44. Hay Jr WW, Brockway JM, Eyzaguirre M. Neonatal pulse oximetry: accuracy and reliability. Pediatrics. 1989;83:717–22.
45. House JT, Schultetus RR, Gravenstein N. Continuous neonatal evaluation in the delivery room by pulse oximetry. J Clin Monit. 1987;3:96–100.
46. Jöbsis FF. Noninvasive, infrared monitoring of cerebral and myocardial oxygen sufficiency and circulatory parameters. Science. 1997;198:1264–7.
47. Jones DP. Radical-free biology of oxidative stress. Am J Physiol Cell Physiol. 2008; 295:C849–68.
48. Jones DP, Carlson JL, Mody VC, Cai J, Lynn MJ, Sternberg P. Redox state of glutathione in human plasma. Free Radic Biol Med. 2000;28:625–35.
49. Jones DP, Go MY. Redox compartmentalization and cellular stress. Diabetes Obes Metab. 2010;12 Suppl 2:116–25.
50. Jones MD, Traystman RJ. Cerebral oxygenation of the fetus, newborn, and adult. Semin Perinatol. 1984;8:205–16.
51. Kadenbach B, Ramzan R, Wen L, Vogt S. New extension of the Mitchell theory for oxidative phosphorylation in mitochondria in living organisms. Biochim Biophys Acta. 2010;1800:205–12.
52. Kaiser JR, Gauss CH, Williams DK. The effects of hypercapnia on cerebral autoregulation in ventilated very low birth weight infants. Pediatr Res. 2005;58:931–5.
53. Kamlin CO, Dawson JA, O'Donnell CP, Morley CJ, Donath SM, Sekhon J, Davis PG. Accuracy of pulse oximetry measurement of heart rate of newborn infants in the delivery room. J Pediatr. 2008;152:756–60.
54. Kamlin CO, O'Donnell CP, Davis PG, Morley CJ. Oxygen saturation in healthy infants immediately after birth. J Pediatr. 2006;148:585–9.
55. Kamlin COF, O'Donnell CPF, Everest N, Davis PG, Morley CJ. Accuracy of clinical assessment of infant heart rate in the delivery room. Resuscitation. 2006;71:319–21.
56. Kasai H. Analysis of a form of oxidative DNA damage, 8-hydroxy-2′-deoxyguanosine, as a marker of cellular oxidative stress during carcinogenesis. Mutat Res. 1997;387:147–63.
57. Katheria A, Rich W, Finer N. Electrocardiogram provides a continuous heart rate faster than oximetry during neonatal resuscitation. Pediatrics. 2012;130(5):e1177–81.
58. Kattwinkel J, Perlman JM, Aziz K, Colby C, Fairchild K, Gallagher J, Hazinski MF, Halamek LP, Kumar P, Little G, McGowan JE, Nightengale B, Ramirez MM, Ringer S, Simon WM, Weiner GM, Wyckoff M, Zaichkin J. Part 15: neonatal resuscitation: 2010 American Heart

Association Guidelines for Cardiopulmonary Resuscitation and Emergency Cardiovascular Care. Circulation. 2010;122(18 Suppl 3):S909–19.
59. Kopotic RJ, Lindner W. Assessing high-risk infants in the delivery room with pulse oximetry. Anaesth Analog. 2002;94:S31–6.
60. Lakshminrusimha S, Swartz DD, Gugino SF, Ma CX, Wynn KA, Ryan RM, Russell JA, Steinhorn RH. Oxygen concentration and pulmonary hemodynamics in newborn lambs with pulmonary hypertension. Pediatr Res. 2009;66:539–44.
61. Lawson JA, Kim S, Powell WS, FitzGerald GA, Rokach J. Oxidized derivatives of omega-3 fatty acids: identification of IPF3 alpha-VI in human urine. J Lipid Res. 2006;47: 2515–24.
62. Ledo A, Arduini A, Asensi MA, Sastre J, Escrig R, Brugada M, Aguar M, Saenz P, Vento M. Human milk enhances antioxidant defenses against hydroxyl radical aggression in preterm infants. Am J Clin Nutr. 2009;89:210–5.
63. Leone TA, Finer NN, Rich W. Delivery room respiratory management for the term and preterm infant. Clin Perinatol. 2012;39:431–40.
64. Lubec G, Widness JA, Hayde M, Menzel D, Pollak A. Hydroxyl radical generation in oxygen-treated infants. Pediatrics. 1997;100:700–4.
65. Lykkesfeldt J. Malondialdehyde as biomarker of oxidative damage to lipids caused by smoking. Clin Chim Acta. 2007;380:50–8.
66. Maltepe E, Saugstad OD. Oxygen in health and disease: regulation of oxygen homeostasis; clinical implications. Pediatr Res. 2009;65:261–8.
67. Mariani G, Dik PB, Ezquer A, Aguirre A, Esteban ML, Perez C, Fernandez Jonusas S, Fustiñana C. Pre-ductal and post-ductal O2 saturation in healthy term neonates after birth. J Pediatr. 2006;150:418–21.
68. Matayatsuk C, Poljak A, Bustamante S, Smythe GA, Kalpravidh RW, Sirankapracha P, Fucharoen S, Wilairat P. Quantitative determination of ortho- and meta-tyrosine as biomarkers of protein oxidative damage in beta-thalassemia. Redox Rep. 2007;12:219–28.
69. McNeill S, Gatenby JC, McElroy S, Engelhardt B. Normal cerebral, renal and abdominal regional oxygen saturations using near-infrared spectroscopy in preterm infants. J Perinatol. 2011;31:51–7.
70. Meier-Stauss P, Bucher HU, Hürlimann R, König V, Huch R. Pulse oximetry used for documenting oxygen saturation and right-to-left shunting immediately after birth. Eur J Pediatr. 1990;149:851–5.
71. Milne GL, Yin H, Hardy KD, Davies SS, Roberts 2nd LJ. Isoprostane generation and function. Chem Rev. 2011;111:5973–96.
72. Morgan PE, Pattison DI, Davies MJ. Quantification of hydroxyl radical-derived oxidation products in peptides containing glycine, alanine, valine, and proline. Free Radic Biol Med. 2012;52:328–39.
73. Neff TA. Routine oximetry. A fifth vital sign? Chest. 1988;94:227.
74. Nolan JP, Soar J, Zideman DA, Biarent D, Bossaert LL, Deakin C, Koster RW, Wyllie J, Böttiger B, on behalf of the European Resuscitation Council Guidelines for Resuscitation 2010. Section 1. Executive summary. Resuscitation. 2010;81:1219–76.
75. Nuntnarumit P, Rojnueangnit K, Tangnoo A. Oxygen saturation trend in preterm infants during the first 15 min after birth. J Perinatol. 2009;130:399–402.
76. O'Donnell CP, Kamlin CO, Davis PG, Carlin JB, Morley CJ. Clinical assessment of infant colour at delivery. Arch Dis Child Fetal Neonatal Ed. 2007;92:F465–7.
77. O'Donnell CP, Kamlin CO, Davis PG, Morley CJ. Feasibility of and delay in obtaining pulse oximetry during neonatal resuscitation. J Pediatr. 2005;147:698–9.
78. Orhan H, Vermeulen NP, Tump C, Zappey H, Meerman JH. Simultaneous determination of tyrosine, phenylalanine and deoxyguanosine oxidation products by liquid chromatography-tandem mass spectrometry as non-invasive biomarkers for oxidative damage. J Chromatogr B Analyt Technol Biomed Life Sci. 2004;799:245–54.
79. Papile LA, Rudolph AM, Heymann MA. Autoregulation of cerebral blood flow in the preterm fetal lamb. Pediatr Res. 1985;19:159–61.

80. Paulech J, Solis N, Cordwell SJ. Characterization of reaction conditions providing rapid and specific cysteine alkylation for peptide-based mass spectrometry. Biochim Biophys Acta. 2012. pii: S1570-9639(12)00175-6. doi:10.1016/j.bbapap.2012.08.002.
81. Pereda J, Pérez S, Escobar J, Arduini A, Asensi M, Serviddio G, Sabater L, Aparisi L, Sastre J. Obese rats exhibit high levels of fat necrosis and isoprostanes in taurocholate-induced acute pancreatitis. PLoS One. 2012;7:e44383.
82. Perlman JM, Wyllie J, Kattwinkel J, Atkins DL, Chameides L, Goldsmith JP, Guinsburg R, Hazinski MF, Morley C, Richmond S, Simon WM, Singhal N, Szyld E, Tamura M, Velaphi S. Neonatal Resuscitation Chapter Collaborators. Part 11: neonatal resuscitation 2010 International Consensus on Cardiopulmonary resuscitation and Emergency Cardiovascular Care Science with treatment recommendation. Circulation. 2010;112(16 Suppl 2):S516–38.
83. Pichler G, Wolf M, Roll C, Weindling MA, Greisen G, Wardle SP, Zaramella P, Naulaers G, Pellicer A, Austin T, Bartocci M, Urlesberger B. Recommendations to increase the validity and comparability of peripheral measurements by near infrared spectroscopy in neonates. "Round table", section of haematology, oxygen transport and microcirculation, 48th annual meeting of ESPR, Prague 2007. Neonatology. 2008;94:320–2.
84. Plank MS, Calderon TC, Asmerom Y, Boskovic DS, Angeles DM. Biochemical measurement of neonatal hypoxia. J Vis Exp. 2011. pii: 2948. doi:10.3791/2948.
85. Porter KB, Golhamer R, Mankad A, Peevy K, Gaddy J, Spinnato JA. Evaluation of arterial oxygen saturation in pregnant patients and their newborns. Obstet Gynecol. 1988;71:354–7.
86. Pryds O, Greisen G, Lou H, Friis-Hansen B. Heterogeneity of cerebral vasoreactivity in preterm infants supported by mechanical ventilation. J Pediatr. 1989;115:638–45.
87. Pryds O, Greisen G, Lou H, Friis-Hansen B. Vasoparalysis associated with brain damage in asphyxiated term infants. J Pediatr. 1990;117(1 Pt 1):119–25.
88. Rabi Y. Oxygen in the delivery room. Neoreviews. 2010;11:e130–8.
89. Rabi Y, Singhal N, Nettel-Aguirre A. Room-air versus oxygen administration during resuscitation of preterm infants: the ROAR Study. Pediatrics. 2011;128:e374–81.
90. Rabi Y, Yee W, Chen SY, Singhal N. Oxygen saturation trends immediately after birth. J Pediatr. 2006;148:590–4.
91. Ramaekers VT, Casaer P, Daniels H, Marchal G. Upper limits of brain blood flow autoregulation in stable infants of various conceptional age. Early Hum Dev. 1990;24(3):249–58.
92. Rea PA, Crowe J, Wickramasinghe Y, Rolfe P. Non-invasive optical methods for the study of cerebral metabolism in the human newborn: a technique for the future? J Med Eng Technol. 1985;9:160–6.
93. Reed TT. Lipid peroxidation and neurodegenerative disease. Free Radic Biol Med. 2011;51:1302–19.
94. Rosenberg AA, Jones MD, Traystman RJ, Simmons MA, Molteni RA. Response of cerebral blood flow to changes in PCO2 in fetal, newborn, and adult sheep. Am J Physiol. 1982;242:H862–6.
95. Rozalski R, Winkler P, Gackowski D, Paciorek T, Kasprzak H, Olinski R. High concentrations of excised oxidative DNA lesions in human cerebrospinal fluid. Clin Chem. 2003;49:1218–21.
96. Sanderson TH, Reynolds CA, Kumar R, Przyklenk K, Hütteman M. Molecular mechanism of ischemia-reperfusion injury in brain: pivotal role of mitochondrial membrane potential in reactive oxygen species generation. Mol Neurobiol. 2012. doi:10.1007/s12035-012-8344-z.
97. Saraswat A, Simionato L, Dawson JA, Thio M, Kamlin COF, Owen L, Schmölzer GB, Davis PG. Determining the best method of Nellcor pulse oximeter sensor application in neonates. Acta Paediatr. 2012;101:484–7.
98. Saravanabhavan G, Blais E, Vincent R, Kumarathasan P. A high performance liquid chromatography-electrochemical array method for the measurement of oxidative/nitrative changes in human urine. J Chromatogr A. 2010;1217:3269–74.
99. Saugstad OD, Ramji S, Soll RF, Vento M. Resuscitation of newborn infants with 21 % or 100 % oxygen: an updated systematic meta-analysis. Neonatology. 2008;94:176–82.

100. Saugstad OD, Rootwelt T, Aalen O. Resuscitation of asphyxiated newborn infants with room air or oxygen: an international controlled trial: the Resair 2 study. Pediatrics. 1998;102:e1.
101. Saugstad OD, Sejersted Y, Solberg R, Wollen EJ, Bjørås M. Oxygenation in the newborn: a molecular approach. Neonatology. 2012;101:315–25.
102. Saugstad OD, Speer CP, Halliday HL. Oxygen saturation in immature babies: revisited and updated information. Neonatology. 2011;100:217–8.
103. Schmidt H, Grune T, Müller R, Siems WG, Wauer RR. Increased levels of lipid peroxidation products malondialdehyde and 4-hydroxynonenal after perinatal hypoxia. Pediatr Res. 1996;40:15–20.
104. Semenza GL. Regulation of oxygen homeostasis by hypoxia-inducible factor 1. Physiology (Bethesda). 2008;24:97–106.
105. Serru V, Baudin B, Ziegler F, David JP, Cals MJ, Vaubourdolle M, Mario N. Quantification of reduced and oxidized glutathione in whole blood samples by capillary electrophoresis. Clin Chem. 2001;47:1321–4.
106. Shearn CT, Smathers RL, Stewart BJ, Fritz KS, Galligan JJ, Hail Jr N, Petersen DR. Phosphatase and tensin homolog deleted on chromosome 10 (PTEN) inhibition by 4-hydroxynonenal leads to increased Akt activation in hepatocytes. Mol Pharmacol. 2011;79:941–52.
107. Slater L, Asmerom Y, Boskovic DS, Bahjri K, Plank MS, Angeles KR, Phillips R, Deming D, Ashwal S, Hougland K, Fayard E, Angeles DM. Procedural pain and oxidative stress in premature neonates. J Pain. 2012;13:590–7.
108. Solberg R, Andresen JH, Escrig R, Vento M, Saugstad OD. Resuscitation of hypoxic newborn piglets with oxygen induces a dose-dependent increase in markers of oxidation. Pediatr Res. 2007;62:559–63.
109. Solberg R, Longini M, Proietti F, Vezzosi P, Saugstad OD, Buonocore G. Resuscitation with supplementary oxygen induces oxidative injury in the cerebral cortex. Free Radic Biol Med. 2012;53:1061–7.
110. Song WL, Lawson JA, Reilly D, Rokach J, Chang CT, Giasson B, FitzGerald GA. Neurofurans, novel indices of oxidant stress derived from docosahexaenoic acid. J Biol Chem. 2008;283:6–16.
111. Sorensen LC, Greisen G. The brains of very preterm newborns in clinically stable condition may be hyperoxygenated. Pediatrics. 2009;124:e958–63.
112. Spickett CM, Wiswedel I, Siems W, Zarkovic K, Zarkovic N. Advances in methods for the determination of biologically relevant lipid peroxidation products. Free Radic Res. 2010;44:1172–202.
113. Stanley WC, Recchia FA, Lopaschuck GD. Myocardial substrate metabolism in the normal and failing heart. Physiol Rev. 2005;85:1093–129.
114. Thérond P, Bonnefont-Rousselot D, Davit-Spraul A, Conti M, Legrand A. Biomarkers of oxidative stress: an analytical approach. Curr Opin Clin Nutr Metab Care. 2000;3:373–84.
115. Ueda K, Ueyama T, Yoshida K, Kimura H, Ito T, Shimizu Y, Oka M, Tsuruo Y, Ichinose M. Adaptive HNE-Nrf2-HO-1 pathway against oxidative stress is associated with acute gastric mucosal lesions. Am J Physiol Gastrointest Liver Physiol. 2008;295:G460–9.
116. Urlesberger B, Grossauer K, Pocivalnik M, Avian A, Müller W, Pichler G. Regional oxygen saturation of the brain and peripheral tissue during birth transition of term infants. J Pediatr. 2010;157:740–4.
117. Valavanidis A, Vlachogianni T, Fiotakis C. 8-hydroxy-2′-deoxyguanosine(8-OHdG): a critical biomarker of oxidative stress and carcinogenesis. J Environ Sci Health C Environ Carcinog Ecotoxicol Rev. 2009;27:120–39.
118. Valko M, Leibfritz D, Moncol J, Cronin MTD, Mazur M, Telser J. Free radicals and antioxidants in normal physiological functions and human disease. Int J Biochem Cell Biol. 2007;39:44–84.
119. van Bel F, Lemmers P, Naulaers G. Monitoring neonatal regional cerebral oxygen saturation in clinical practice: value and pitfalls. Neonatology. 2008;94:237–44.

120. van de Bor M, Walther FJ. Cerebral blood flow velocity regulation in preterm infants. Biol Neonate. 1991;59:329–35.
121. Vander Heiden MG, Cantley LC, Thompson CB. Understanding the Warburg effect: the metabolic requirements of cell proliferation. Science. 2009;324:1029–33.
122. Vento M. Tailoring oxygen needs of extremely low birth weight infants in the delivery room. Neonatology. 2011;99:342–8.
123. Vento M, Aguar M, Escobar J, Arduini A, Escrig R, Brugada M, Izquierdo I, Asensi MA, Sastre J, Saenz P, Gimeno A. Antenatal steroids and antioxidant enzyme activity in preterm infants: influence of gender and timing. Antioxid Redox Signal. 2009;11:2945–55.
124. Vento M, Aguar M, Leone TA, Finer NN, Gimeno A, Rich W, Saenz P, Escrig R, Brugada M. Using intensive care technology in the delivery room: a new concept for the resuscitation of extremely preterm neonates. Pediatrics. 2009;122:1113–6.
125. Vento M, Asensi M, Sastre J, García-Sala F, Viña J. Resuscitation with room air instead of 100 % oxygen prevents oxidative stress in moderately asphyxiated term neonates. Pediatrics. 2001;107:642–7.
126. Vento M, Asensi M, Sastre J, Lloret A, García-Sala F, Miñana JB, Viña J. Hyperoxemia caused by resuscitation with pure oxygen may alter intracellular redox status by increasing oxidized glutathione in asphyxiated newly born infants. Semin Perinatol. 2002;26:4–10.
127. Vento M, Asensi M, Sastre J, Lloret A, García-Sala F, Viña J. Oxidative stress in asphyxiated term infants resuscitated with 100 % oxygen. J Pediatr. 2003;142:240–6.
128. Vento M, Cheung PY, Aguar M. The first golden minutes of the extremely low-gestational-age neonate: a gentle approach. Neonatology. 2009;95:286–98.
129. Vento M, Cubells E, Escobar JJ, Escrig R, Aguar M, Brugada M, Cernada M, Saénz P, Izquierdo I. Oxygen saturation after birth in preterm infants treated with continuous positive airway pressure and air: assessment of gender differences and comparison with a published nomogram. Arch Dis Child Fetal Neonatal Ed. 2012. doi:10.1136/archdischild-2012-302369.
130. Vento M, Escobar J, Cernada M, Escrig R, Aguar M. The use and misuse of oxygen during the neonatal period. Clin Perinatol. 2012;39:165–76.
131. Vento M, Moro M, Escrig R, Arruza L, Villar G, Izquierdo I, Roberts 2nd LJ, Arduini A, Escobar JJ, Sastre J, Asensi MA. Preterm resuscitation with low oxygen causes less oxidative stress, inflammation, and chronic lung disease. Pediatrics. 2009;124:e439–49.
132. Vento M, Sastre J, Asensi MA, Viña J. Room-air resuscitation causes less damage to heart and kidney than 100 % oxygen. Am J Respir Crit Care Med. 2005;172:1393–8.
133. Vento M, Saugstad OD. Oxygen supplementation in the delivery room: updated information. J Pediatr. 2011;158:S5–8.
134. von Siebenthal K, Bernert G, Casaer P. Near-infrared spectroscopy in newborn infants. Brain Dev. 1992;14:135–43.
135. Wang CL, Anderson C, Leone TA, Rich W, Govindaswami B, Finer NN. Resuscitation of preterm neonates by using room air or 100 % oxygen. Pediatrics. 2008;121:1083–9.
136. Wong CM, Marcocci L, Liu L, Suzuki YJ. Cell signaling by protein carbonylation and decarbonylation. Antioxid Redox Signal. 2010;12:393–404.
137. Wyatt JS, Edwards AD, Azzopardi D, Reynolds EO. Magnetic resonance and near infrared spectroscopy for investigation of perinatal hypoxic-ischaemic brain injury. Arch Dis Child. 1989;64(7 Spec No):953–63.
138. Zhang W, Ames BD, Walsh CT. Identification of phenylalanine 3-hydroxylase for meta-tyrosine biosynthesis. Biochemistry. 2011;50:5401–3.
139. Zubarioglu U, Uslu S, Can E, Bülbül A, Nuhoglu A. Oxygen saturation levels during the first minutes of life in healthy term neonates. Tohoku J Exp Med. 2011;224:273–9.

Chapter 12
Antioxidant Properties of Surfactant

Carlo Dani and Chiara Poggi

List of Abbreviations

BPD	Bronchopulmonary dysplasia
CAT	Catalase
DPPC	Dipalmitoylphosphatidylcholine
FRs	Free radicals
PUFAs	Polyunsaturated fatty acids
PUPLs	Polyunsaturated phospholipids
RDS	Respiratory distress syndrome
ROS	Reactive oxygen species
SOD	Superoxide dismutase
SP-A	Surfactant protein A
SP-D	Surfactant protein D

Introduction

Preterm birth affects about 5–9 % of gestations in European countries and 12–13 % in the USA [1] and is at present the leading cause of overall infant death in developed countries [2]. Despite general improvement in perinatal care, preterm birth rates are still increasing [3]. Respiratory distress syndrome (RDS), the major acute complication of prematurity, occurs in about 50 % of preterm neonates born before 30 weeks of gestation [4] ranging from 100 % at 23 weeks to 50 % at 29 weeks [5].

C. Dani, M.D. (✉) • C. Poggi, M.D.
Department of Critical Care Medicine and Surgery, Section of Neonatology,
Careggi University Hospital, Florence, Italy
e-mail: carlo.dani@unifi.it

P.A. Dennery et al. (eds.), *Perinatal and Prenatal Disorders*,
Oxidative Stress in Applied Basic Research and Clinical Practice,
DOI 10.1007/978-1-4939-1405-0_12

RDS remains at present the leading cause of death in extremely preterm infants and often requires surfactant replacement therapy and respiratory support [5].

Endotracheal surfactant administration has affirmed as one of the milestones of RDS treatment, as it is associated with reduced mortality, air leaks, and need for mechanical ventilation, although it has been proved ineffective in preventing bronchopulmonary dysplasia (BPD) [6].

Surfactant is produced by mammalians type II alveolar cells and is a complex mixture of 90 % lipids and 10 % proteins. The most abundant phospholipid in surfactant is dipalmitoylphosphatidylcholine (DPPC), which is the main agent responsible for surface tension lowering at the air-liquid interface, probably with the contribution of other lipid compounds, as phosphatidylglycerol [7]. Surfactant proteins (SPs) constitute a mixture of hydrophilic and hydrophobic components of the surfactant, which contribute to surface tension lowering through specific interactions with DPPC and involvement in surfactant turnover process. Because of its well-known tensioactive properties, exogenous surfactant administration in RDS results in reduction of alveolar opening pressure, increase of lung volume at a certain distending pressure, and stabilization of lung volume during the deflation phases, thus reducing the work of breathing [8, 9].

However, surfactant is a pleiotropic compound. In fact, besides its chemicophysical properties, surfactant was also demonstrated to exert consistent anti-inflammatory and antioxidant activities, for which nonenzymatic and enzymatic proteins and some minor lipid components appear mainly responsible [10–12]. Although less extensively investigated, these functions may contribute to the efficacy of exogenous surfactant administration in preterm neonates with RDS.

Oxidative Stress in Preterm Newborn Lungs

Oxidative stress occurs when a relative unbalance exists between pro-oxidant agents and antioxidant pathways, leading to accumulation of reactive oxygen species (ROS) and other reactive components within different tissues, resulting in cascade damages of lipids, proteins, polysaccharides, and DNA strains [13, 14]. Antioxidant defense consists of a complex web of interacting molecules which exert scavenging activity both by enzymatic pathways, such as catalase (CAT), superoxide dismutase (SOD), glutathione peroxidase, and nitric oxide synthase, and nonenzymatic pathways, such as vitamins C and E and uric acid [15, 16].

Oxidative stress was demonstrated to play a significant role in the development of several complications of preterm birth, as RDS and BPD, but also retinopathy of prematurity and intraventricular hemorrhage, which are all grouped together as "free radical-related diseases" [13, 14].

Preterm newborns have to deal with a consistent load of ROS and other reactive species, as a consequence of several perinatal conditions related to preterm birth determining activation of pro-oxidant pathways, as hyperoxia, tissue perfusion impairment and reperfusion, infections, and rapidly growing energy demand.

However, antioxidant defenses show impaired activity because of developmental arrest of both placental transfer and endogenous production of enzymatic and nonenzymatic compounds, thus leading to the occurrence of a pro-oxidant status [13–16]. In fact, when compared to term newborns, preterm newborns present significantly reduced activity of glutathione peroxidase and SOD for several weeks after birth [17] and lower circulating levels of vitamin E and ascorbic acid [13, 14]. Moreover, susceptibility to oxidative stress may be further increased by high concentration of polyunsaturated fatty acids (PUFAs) which are highly sensitive to peroxidation injuries, particularly in the neuronal membranes, and relatively increased free iron release favoring free radicals (FRs) formation [13, 14]. It was also demonstrated that preterm newborns lack the capacity to induce a proper antioxidant enzymatic activity in response to oxidative challenges [18].

Lung tissues of preterm newborns are particularly exposed to ROS activity, as the main treatment options for RDS, oxygen supplementation, and mechanical ventilation are known to promote oxidative stress and local proinflammatory responses [19, 20]. Such detrimental effects of RDS therapy lead in turn to endogenous surfactant inactivation, lung tissue damage, and further aggravation of respiratory failure [19, 20]. Moreover, although BPD is definitely a multifactorial disease, a close relationship has been established between BPD and oxidative and inflammatory stress, as ROS and inflammatory cytokines production as a consequence of hyperoxia, volutrauma, chorionamnionitis, and postnatal infections are clearly related to the risk of BPD [21]. Moreover, several single nucleotide polymorphisms of genes involved in the inflammatory response, oxidative stress, and antioxidant enzymes show a definite relation with the risk of BPD [22–24], thus reinforcing the importance of FRs and their clearance in determining such complications of prematurity.

Antioxidant Effects of Surfactant

Endotracheal surfactant administration prevents oxidative alveolar injury firstly by the replacement of preexisting surfactant partially damaged by ROS [11]. However, the main antioxidant effects of surfactant probably relay on different mechanisms, since surfactant itself shows a consistent antioxidant activity attributable both to enzymatic and nonenzymatic scavenger molecules naturally contained in the mixture. Consequently, exogenous surfactant would be able to abate the local ROS concentration in the alveolar areas, thus directly preventing both local damage and ROS diffusion to the surrounding tissues and vessels [11, 12, 25, 26].

It is well known that antioxidant enzymes and other antioxidant molecules are commonly found in the epithelial lining fluid of the normal human lower airways [27]. Basing on the observation that bronchoalveolar lavage of preterm neonates treated with surfactant presented lower levels of pro-oxidant markers than untreated neonates [28], it was demonstrated in the animal model that natural calf lung surfactant contains a measurable amount of SOD and CAT, which have been also demonstrated to exert consistent scavenging activity when incubated with a definite amount of H_2O_2 [11].

Moreover, endotracheal administration of surfactant also induced a significant increase of SOD content of alveolar type II cells, demonstrating the occurrence of enzyme uptake via liposome during the surfactant recycle process [11]. Therefore, the positive effects of surfactant administration in preterm neonates may be related not only to direct improvement of lung mechanics but also to local antioxidant activity of surfactant and to increased antioxidant potential of alveolar cells. Interestingly, the comparison between natural lung surfactant and a lung surfactant extract containing only 1 % of proteins showed that the latter contains no significant amount of antioxidant enzymes and does not exert any scavenging activity against H_2O_2 [11]. These data reinforce the concept that surfactant mixtures not containing proper amount of protein compounds, as synthetic surfactants, are inadequate for clinical purpose at present, probably not only because of suboptimal tensoactive properties but also for the lack of enzymes naturally present in lung surfactant.

A recent study confirmed that natural porcine and bovine surfactants contain definite amounts of SOD and CAT, which have been measured in four different natural surfactants, namely, Poractant (Curosurf), Beractant (Survanata), Calfactant (infasurf), and Bovactant (Alveofact) [12]. Comparing these mixtures, Beractant roughly showed the highest content of SOD per mg of phosholipids, while Calfactant showed the highest content of CAT, but when enzymatic activities were expressed per mL of surfactant, Poractant provided the highest content of SOD while Calfactant was confirmed to contain the highest amount of CAT. Moreover, considering the recommended doses for clinical use, Poractant was shown to provide the major amount of SOD, while Calfactant was proved to contain the major amount of CAT (Table 12.1). After incubation with different amounts of H_2O_2, Curosurf showed the

Table 12.1 Antioxidants content in natural surfactants

	Poractant	Beractant	Bovactant	Calfactant
Doses				
mg of PLs/Kg	200	100	100	100
mL of surfactant/Kg	2.5	4	2.2	2.86
SOD				
U/mg of PLs	0.396*	0.474	0.027#	0.383§
U/mL of surfactant	31.7	11.9	1.21	13.4
U/dose per Kg	73.3	47.6	2.6	38.3
CAT				
nmol/min/mL of PLs	0.81°	2.60	1.58	3.23
nmol/min/mL of surfactant	64.80	65.00	71.10	113.10
U/dose per Kg	149.80	260.00	157.80	323.50
Plasmalogens				
mol % of total PLs	3.8±0.1	1.5±0.2	0.9±0.3	n.a.
PUPLs				
mol % of total PLs	26±1	6±1	11±1	n.a.

Modified from Refs. [12, 25]

PLs phospholipids, *PUPLs* polyunsaturated phospholipids, *n.a.* not available

*$p=0.019$ vs. Survanta; #$p<0.0001$ vs. Survanta; §$p=0.003$ vs. Survanta; °$p<0.0001$ vs. Infasurf

Table 12.2 Scavenger activity of natural surfactants

	Poractant	Beractant	Bovactant	Calfactant
25 μM	29±8*	17±4	47±6 * # §	36±5 * #
50 μM	30±6	32±4	75±8 * # §	56±9 * #
100 μM	66±8	64±9	113±11 * #	117±10 * #
250 μM	201±15	214±20	278±23 * #	332±18 * #

H_2O_2 concentration[a] after incubation with 25, 50,100 and 250 μM of H_2O_2 (mean ± SD) (Modified from Dani et al. [12])

*$p<0.05$ vs. Survanta; #$p<0.05$ vs. Curosurf; $p<0.05$ vs. Infasurf

[a]Data report the difference between H_2O_2 concentration at time 0 and at the end of the experiment

highest scavenger activity, except that at the lowest studied concentrations of H_2O_2 when Beractant exerted the highest scavenger activity. For any concentration of H_2O_2, Calfactant and Bovactant showed significantly lower scavenger activity vs. Poractant and Beractant (Table 12.2) [12].

However, not only SOD and CAT but also different antioxidant enzymes, as glutathione peroxidase and reductase, may be involved in determining the net antioxidant effect of surfactant, but unfortunately their activity in natural surfactants have not been detailed yet.

Moreover, the complex mixture of natural surfactants also contains nonenzymatic antioxidant molecules which might further contribute to the overall antioxidant activity of the mixture. Even if these aspects have been poorly investigated as far as now, plasmalogens and polyunsaturated phospholipids (PUPLs) are putatively the main responsible molecules for nonenzymatic antioxidant activity of natural surfactants.

Plasmalogens, a subgroup of phospholipids normally contained in cellular membranes, present definite scavenger activity against FRs due to the presence of a specific reductant group and are also minor components of natural surfactants [25]. Plasmalogens were recently considered of great importance to the proper biophysical function of the surfactant. According to the "surface-associated reservoir" model, large areas of surfactant layer are folded during the expiration phase, but remain adherent to the monolayer present at the air-liquid interface and extend during inspiration. Plasmalogens, together with cholesterol, were proved to work synergistically with the hydrophobic SP-B and SP-C for the proper formation of the reservoir and the spreading of DPPC and, in the experimental model, plasmalogens addition to surfactant achieved further reduction of surface tension and viscosity in comparison to surfactant alone [8, 25, 29]. Besides surface-active properties, plasmalogens were recently demonstrated to confer consistent antioxidant protection against ultraviolet light-induced lipid peroxidation within cell membranes [30] and also to exert antioxidant functions in low density lipoproteins [31]. The content of plasmalogens has been detailed for Poractant, Beractant, and Bovactant, but not for Calfactant [8, 25], and resulted consistently higher for Poractant than Beractant and Bovactant (Table 12.1). These results could partially explain the different scavenger performances against H_2O_2 observed for the natural surfactants, together with the differences in the content of antioxidant enzymes [12].

Moreover, along with plasmalogens, PUFAs are considered one of the main substrates for lipid peroxidation in lung surfactant and they were also recently demonstrated to exert consistent scavenger activity against ROS in cellular models [32]. Peroxidation of PUPLs induced by ROS at the air-liquid interface would then initiate a chain reaction, increasing in turn the availability of ROS in the alveolar spaces [25]. Interestingly, Poractant presents the highest concentration of PUPLs among natural surfactants [25], and, because of the highest concentration in the recommended dose, Poractant putatively has also the highest scavenger activity due to PULPs per treatment dose [12].

Plasmalogens and PUFAs content were detailed in tracheal aspirates collected at birth in a cohort of preterm neonates. Plasmalogens and PUFAs content resulted significantly higher in patients who developed BPD in comparison to controls of the same gestational age who did not develop BPD, suggesting a general protective role toward lung parenchyma, probably due to an overall antioxidant effects and not only to the surface tension-lowering properties [26].

Therefore, the importance of some protein and lipid components of surfactant, which are usually believed to be minor elements, should be properly considered in the setting of new synthetic surfactant preparations as beneficial effects of surfactant probably relay not only on its biophysical properties at the air-liquid interface but also on its complex antioxidant effects in the alveolar environment, which is burdened with high load of ROS in case of preterm birth.

Antioxidants Addition to Surfactant

SOD, CAT, and other antioxidants are physiologically present in measurable concentrations in natural surfactants and lung epithelial lining fluid and take part in the regulation of postnatal lung vascular development [33] and in the protection of microvasculature from ROS-induced injury [34].

Since exogenous surfactant is rapidly taken up by type II alveolar cells, it appears reasonable to supplement surfactant with antioxidant molecules in order to enhance its antioxidant potential, as both surfactant and lung antioxidant enzymes are deficient until the final 10–15 % of gestation [35].

Beractant was firstly studied as a vehicle for antioxidant enzymes to be delivered to the alveolar epithelium [36]. Incubation of lung epithelial cells with an emulsion of Beractant plus SOD and CAT resulted in significantly higher SOD and CAT activity in comparison with incubation with CAT and SOD alone or surfactant alone. These results were confirmed in lung homogenates obtained by animal models after endotracheal administration of Beractant plus SOD and CAT, suggesting that surfactant supplementation with antioxidant enzymes is an efficacious way to increase antioxidant defenses in the alveolar environment. Moreover, since intracellular localization of antioxidant enzymes may be a crucial point for their functionality, liposome encapsulation of SOD and CAT was studied, in order to enhance cellular delivery of the enzymes. Such method was proved effective in increasing the antioxidant enzymes activity in the alveolar cells in the animal model of hyperoxia-induced lung injury of prematurity [37].

The addition of SOD and CAT to four different natural surfactants was recently studied in vitro [12]. As expected, the addition of SOD to Poractant, Beractant, and Bovactant resulted in increased scavenger activities in comparison to the surfactants alone, while the addition of SOD to Calfactant resulted in a paradoxical reduction of its scavenger activity, which was putatively attributed to other mechanisms induced by SOD overexpression, as increased H_2O_2 production and hydroxyl radical formation. CAT addition to the four surfactants resulted in increased scavenger activity of the surfactants even if it did not reach significance in the case of Beractant. The addition of both SOD and CAT induced a further increase in comparison with SOD or CAT addition alone, suggesting a synergic activity of the two enzymes.

On the other hand, previous studies by Davies et al. demonstrated the safety of intratracheal administration of rhSOD in preterm infants [38] and its effectiveness in decreasing the need of asthma medications and emergency department visits in preterm infants during the first year of life [39].

Basing on the available data, the addition of antioxidant enzymes to surfactant appears as a possible future strategy in order to improve the lung antioxidant defenses during the phase of RDS treatment. However, clinical studies are mandatory in order to confirm these preliminary experimental results.

Anti-inflammatory and Antibacterial Effects of Surfactant

Inflammation and oxidative stress are strictly related processes, since infectious and inflammatory stimuli are the main triggers of pro-oxidant pathways in biological systems. In addition to specific antioxidant properties, surfactant has been shown to posses anti-inflammatory and immune-modulating activities, which indirectly contribute to the antioxidant effects, because of the interactions between inflammatory and oxidative stress.

The modulation of inflammatory processes by surfactant is mainly ruled by protein compounds, specifically SP-A and SP-D [10, 40]. SP-A and SP-D are two hydrophilic proteins which belong to the family of collectins and were shown to affect several steps of the immune response to several pathogens both in vivo and in vitro [41, 42]. These proteins have to property to bind and aggregate to virus, bacteria, fungi, and endotoxins present in the alveolar spaces, favoring their phagocytosis and killing by cells of the innate immune response, namely, the alveolar macrophages and neutrophil granulocytes. [10, 40]. As a consequence, the transgenic mice model not expressing SP-A and SP-D presents increased susceptibility to lung infections and to the development of severe lung lesions from several pathogens [41, 42]. Moreover, it was recently demonstrated that SP-A downregulate inflammation in presence of LPS suppressing the proinflammatory pathways mediated by NF-kB [43]. On the other hand, SP-A and SP-D downregulate the specific immune response to pathogens, through the inhibition of proinflammatory cytokines secretion, as TNF-alpha, and the modulation of lymphocytes proliferation [40].

The combined effects of enhanced pathogens clearance from the alveolar spaces and reduced specific immune response activation, together with the maintenance of a proper surfactant layer, result in an overall reduction of oxidative stress in the alveoli and the surrounding tissues.

It is noteworthy that oxidative modifications of SP-A by reactive oxygen/nitrogen intermediates interfere with their ability to enhance killing of pathogens [44]. Furthermore, nitration of SP-A that has been described in vivo in the lungs of patients with ARDS [45] decreased its ability to act synergistically with SP-B and SP-C to lower surface tension [46].

However, SP-A and SP-D probably play a more complex role in lung immunemodulation. In fact, these proteins present different behavior in the presence of different stimuli, exerting proinflammatory or anti-inflammatory activity according to different conditions. Particularly, SP-A was proved to suppress NO production by LPS-activated macrophages [47], but increases NO production in the presence of other pathogens such as Mycoplasma [48], thus suggesting differential activities in response to different triggers.

References

1. Slattery MM, et al. Preterm delivery. Lancet. 2002;360:1489–97.
2. Callaghan WM, et al. The contribution of preterm birth to infant mortality in the United States. Pediatrics. 2006;118:1566–73.
3. Martin JA, et al. Annual summary of vital statistics: 2006. Pediatrics. 2008;121:788–801.
4. Horbar JD, et al. A multicenter randomized trial comparing two surfactants for the treatment of neonatal respiratory distress syndrome. J Pediatr. 1993;123:757–66.
5. Rubaltelli FF, Dani C, Reali MF, Bertini G, Wiechmann L, Tangucci M, Spagnolo A. And the Italian Group of Neonatal Pneumology. Acute neonatal respiratory distress in Italy: a one year prospective study. Acta Paediatr. 1988;87:1261–8.
6. Soll RF. Prophylactic natural surfactant extract for preventing morbidity and mortality in preterm infants. Cochrane Database Syst Rev. 2000;2:CD000511.
7. Ramanathan R. Choosing a right surfactant for respiratory distress syndrome treatment. Neonatology. 2009;95:1–5.
8. Ramanathan R. Animal-derived surfactants: where are we? The evidence from randomized, controlled clinical trials. J Perinatol. 2009;29:S38–43.
9. Robertson B, et al. Experimental evaluation of surfactant for replacement therapy. Exp Lung Res. 1988;14:279.
10. Matalon S, et al. Surfactant proteins and inflammation: the Yin and the Yang. Am J Respir Cell Mol Biol. 2004;31:585–6.
11. Matalon S, et al. Characterization of antioxidant activities of pulmonary surfactant mixtures. Biochim Biophys Acta. 1990;1035:121–7.
12. Dani C, et al. Superoxide dismutase and catalase activity in naturally derived commercial surfactants. Pediatr Pulmonol. 2009;44:1125–31.
13. Perrone S. Early identification of the risk for free radical-related diseases in preterm newborns. Early Hum Dev. 2010;86:241–4.
14. Perrone S. Oxidative stress and antioxidant strategies in newborns. J Matern Fetal Neonatal Med. 2010;23:63–5.
15. Gitto E, et al. Oxidative stress in resuscitation and ventilation of newborns. Eur Respir J. 2009;34:1461–9.

16. Davis JM, et al. Maturation of the antioxidant system and effects on preterm birth. Semin Fetal Neonatal Med. 2010;1:191–5.
17. Nassi N, et al. Antioxidant enzymes and related elements in term and preterm newborns. Pediatr Int. 2009;51:183–7.
18. Asikainen TM, et al. Cell specific expression of manganese superoxide-dismutase protein in the lung of patients with respiratory distress syndrome, chronic lung disease or persistent pulmonary hypertension. Pediatr Pulmonol. 2001;32:193–200.
19. Pierce M, et al. The role of inflammation in the pathogenesis of bronchopulmonary dysplasia. Pediatr Pulmonol. 1995;19:371–8.
20. Worthen GS, et al. Neutrophil-mediated pulmonary vascular injury. Am Rev Respir Dis. 1987;136:19–28.
21. Viscardi RM, et al. Perinatal inflammation and lung injury. Semin Fetal Neonatal Med. 2012;17:30–5.
22. Bokodi G, et al. Dysplasia: a review. Pediatr Pulmonol. 2007;42:952–61.
23. Poggi C, et al. Genetic polymorphism of antioxidant enzymes in preterm infants. J Matern Fetal Neonatal Med. 2012;25 Suppl 4:131–4.
24. Giusti B, et al. Genetic polymorphisms of antioxidant enzymes as risk factors for oxidative stress-associated complications in preterm infants. Free Radic Res. 2012;46:1130–9.
25. Rüdiger M, et al. Naturally derived commercial surfactants differ in composition of surfactant lipids and in surface viscosity. Am J Physiol Lung Cell Mol Physiol. 2005;288:L379–83.
26. Rüdiger M, et al. Preterm infants with high polyunsaturated fatty acid and plasmalogens content in tracheal aspirates develop bronchopulmonary dysplasia less often. Pediatr Crit Care. 2000;28:1572–7.
27. Cantin AM, et al. Antioxidant macromolecules in the epithelial lining fluid of the normal human lower respiratory tract. J Clin Invest. 1990;86:962–71.
28. Merritt TA, et al. Human surfactant treatment of severe respiratory distress syndrome: pulmonary effluent indicators of lung inflammation. J Pediatr. 1986;108:741–8.
29. Tolle A, et al. Plasmalogens reduce the viscosity of surfactant-like phospholipid monolayer. Chem Phys Lipids. 1999;100:81–7.
30. Zoeller RA, et al. A possible role for plasmalogens in protecting animal cells against photosensitized killing. J Biol Chem. 1988;11590–6.
31. Jurgens G, et al. Delay of copper-catalyzed oxidation of low density lipoprotein by in vitro enrichment with choline or ethanolamine plasmalogens. Chem Phys Lipids. 1995;77:25–31.
32. Richard D, et al. Polyunsaturated fatty acids as antioxidants. Pharmacol Res. 2008;57:451–5.
33. Sharma S, et al. Lung antioxidant enzymes are regulated by development and increased pulmonary blood flow. Am J Physiol Lung Cell Mol Physiol. 2007;293:L960–71.
34. Barnard ML, et al. Mitigation of oxidant injury to lung microvasculature by intratracheal instillation of antioxidant enzymes. Am J Physiol. 1993;265:L340–5.
35. Frank L, et al. Prenatal development of lung antioxidant enzymes in four species. J Pediatr. 1987;110:106–10.
36. Nieves-Cruz B, et al. Clinical surfactant preparations mediate SOD and catalase uptake by type II cells and lung tissue. Am J Physiol. 1996;270:L659–67.
37. Walther FJ, et al. Antioxidant-surfactant liposomes mitigate hyperoxic lung injury in premature rabbits. Am J Physiol. 1995;269:L613–7.
38. Davis JM, et al. Safety and pharmacokinetics of multiple doses of recombinant human CuZn superoxide dismutase administered intratracheally to premature neonates with respiratory distress syndrome. Pediatrics. 1997;100:24–30.
39. Davis JM, et al. Pulmonary outcome at 1 year corrected age in premature infants treated at birth with recombinant human CuZn superoxide dismutase. Pediatrics. 2003;111:469–76.
40. Vaandrager AB, et al. Lung surfactant protein A and D and innate immune defense. Biol Neonate. 2000;77:9–13.
41. Crouch E, et al. Surfactant proteins A and D and pulmonary host defense. Annu Rev Physiol. 2001;63:521–54.

42. Shepherd V, et al. Distinct roles for lung collectins in pulmonary host defense. Am J Respir Cell Mol Biol. 2002;26:257–60.
43. Wu Y, et al. Accumulation of inhibitory of kB-α as a mechanism contributing to the anti-inflammatory effects of surfactant protein A. Am J Respir Cell Mol Biol. 2004;31:587–94.
44. Matalon S, et al. Modification of surfactant protein D by reactive oxygen-nitrogen intermediates is accompanied by loss of aggregating activity, in vitro and in vivo. FASEB J. 2009;23:1415–30.
45. Haddad IY, et al. Mechanisms of peroxynitrite-induced injury to pulmonary surfactants. Am J Physiol. 1993;265:L555–64.
46. Zhu S, et al. Increased levels of nitrate and surfactant protein a nitration in the pulmonary edema fluid of patients with acute lung injury. Am J Respir Crit Care Med. 2001;163:166–72.
47. Stamme C, et al. Surfqactant protein A differentially regulates IFN-γ and LPS-induced nitrite production by rat alveolar macrophages. Am J Respir Cell Mol Biol. 2000;23:772–9.
48. Hickman-Davis JM, et al. Surfactant protein A mediates mycoplasmacidal activity of alveolar macrophages. Am J Physiol. 1998;274:L270–7.

Chapter 13
Oxidative Stress and Glutathione Synthesis Rates in Early Postnatal Life

Denise Rook and Johannes B. van Goudoever

Introduction

Oxidative stress is the resultant of an imbalance between oxidants and antioxidants. An oxidant is a substance that removes electrons from another substance in a redox reaction, thereby reducing itself and oxidizing the other substance. Reactive oxygen species (ROS) are oxidants produced during, among others, normal oxygen metabolism and include the superoxide anion (O_2^-), the hydrogen peroxide (H_2O_2), and the hydroxyl radical (OH^-). ROS are highly reactive molecules which are able to cause oxidative damage to proteins, lipids, and DNA resulting in altered enzyme activities, damage of cellular and mitochondrial membrane, altered signal transduction, and even apoptosis. Besides being toxic, ROS also have crucial beneficial actions, e.g., regulating vascular tone, as part of physiological response to kill pathogens, inter- and intracellular signaling, activating protein cascades, and control gene expression. Thus, ROS are essential on the one hand but are dangerous if in excess.

To counteract the detrimental actions of ROS, the body has numerous antioxidant defense mechanisms (antioxidants), in the form of enzymes, vitamins, and other agents (Table 13.1). An antioxidant is defined as "any substance that delays, prevents, or removes oxidative damage," thereby counteracting the deleterious effects of oxidants. Glutathione (GSH), a tripeptide, is the most important intracellular antioxidant and its role in early postnatal life will be discussed in this chapter.

D. Rook
Department of Pediatrics, Erasmus Medical Centre - Sophia Children's Hospital, Dr. Molewaterplein 60, Rotterdam, The Netherlands

J.B. van Goudoever (✉)
Department of Pediatrics, AMC-Emma Children's Hospital
c/o Room H7-276, P.O. Box 22660, 1100 DD, Amsterdam, The Netherlands
e-mail: h.vangoudoever@amc.uva.nl

P.A. Dennery et al. (eds.), *Perinatal and Prenatal Disorders*,
Oxidative Stress in Applied Basic Research and Clinical Practice,
DOI 10.1007/978-1-4939-1405-0_13

Table 13.1 Antioxidants

Class	Antioxidant
Enzymes	Superoxide dismutase (SOD)
	Catalase
	Glutathione peroxidase (GPx)
	Glutathione reductase (GR)
Vitamins	Vitamin E
	Vitamin A
	Vitamin C
	Coenzyme Q
	β-carotene
Reducing agents	Glutathione (GSH)
	Cysteine
	Thioredoxin
Binding agents	Albumin
	Ceruloplasmin
	Lactoferrin
	Transferrin
Constituent of enzymes	Coper, zinc, selenium
Others	Uric acid
	Bilirubin
	Erythropoietin

Oxidative Stress in Newborns

Fetal development takes place in a relative hypoxic environment, with the partial pressure of oxygen (pO_2) in the fetus rarely above 4 kPa [1]. As soon as infants start breathing, the pO_2 rises sharply. This sharp increase in oxygen pressure is believed to generate a burst of ROS [2, 3] and thus results in physiologic oxidative stress immediately after birth in newborn infants [4]. This sudden increase in pO_2 at birth is essential for cardiopulmonary adaptation at birth, which includes a decrease of pulmonary vascular resistance and vascular remodeling, like the closure of the ductus arteriosus [5, 6]. Besides this hyperoxic challenge, the increase in metabolic rate after birth adds to this physiological oxidative stress [7].

In term infants, antioxidant defenses are present at birth to counteract the hyperoxic challenge at birth. Animal studies demonstrated that the antioxidant enzymes superoxide dismutase (SOD) and glutathione peroxidase (GPx) mature during the last trimester of pregnancy [8, 9]. Furthermore, the lungs of term rabbits have the capability to increase the normal antioxidant enzymes in reaction to the sudden O_2 exposure at birth as compared to preterm rabbits [10].

Oxidative Stress in Preterm Infants

Preterm infants are highly susceptible to oxidative stress, because they have both increased ROS formation and compromised antioxidant defenses. Besides the physiological oxidative challenge at birth, premature infants often require additional oxygen during resuscitation in order to achieve an appropriate rise in SpO_2 levels. Oxygen supplementation during resuscitation after birth is associated with increased oxidative stress [11–13], and this seems to be a dose-dependent association as demonstrated in animal studies [14]. Furthermore, preterm birth is often associated with fetal or maternal morbidities, such as preeclampsia or chorioamnionitis, which are linked to increased oxidative stress [15, 16]. Also in the neonatal period, preterm infants are at risk for increased ROS formation. Oxygen dependency due to immature lungs, infections, intravenous lipids, and increased free iron due to multiple blood transfusions are thought to contribute to increased oxygen radical formation [17–20]. In the first days and weeks after birth, levels of oxidative stress are increased in preterm infants [21, 22].

While ROS formation is increased, the antioxidant defenses are not fully matured or present after preterm birth, and the increased antioxidant consumption during the neonatal period further reduces antioxidant capacity of preterm infants [21, 23, 24]. The nutritional state of the newborn modulates the antioxidant defenses [25, 26]. Breast milk is a great source of antioxidant [27, 28]. However, preterm infants do not receive high amounts of breast milk in the first days of life [29]. Furthermore, breast milk of mothers after term delivery has higher antioxidant power than that of mothers after preterm delivery [28].

Glutathione

Glutathione (GSH) is one of the most abundant intracellular nonenzymatic antioxidants. With cellular concentrations in the millimolar range, GSH probably subserves the greatest intracellular antioxidant properties.

Glutathione Synthesis

Dietary GSH is only minimally released in the bloodstream and the body is thus dependent on de novo synthesis of GSH [30]. Although synthesized in all tissues, GSH is mainly synthesized in liver and erythrocytes. GSH is a small tripeptide synthesized from the amino acids glutamate, cysteine, and glycine (Fig. 13.1). Synthesis of GSH consists of two steps. The first, rate-limiting, step is the formation of γ-glutamylcysteine, catalyzed by glutamate-cysteine ligase, which is feedback

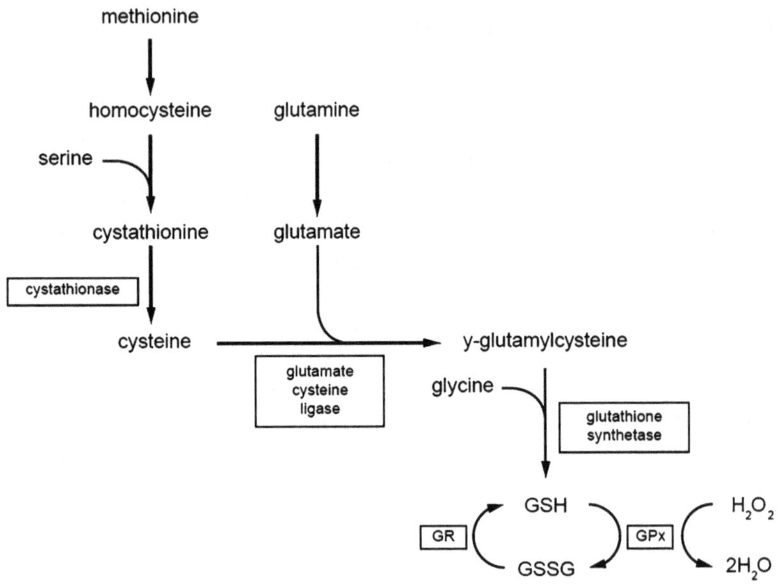

Fig. 13.1 Synthesis of glutathione. *GR* glutathione reductase, *GPx* glutathione peroxidase, *GSH* reduced glutathione, *GSSG* oxidized glutathione

inhibited by GSH. Availability of cysteine has been considered to be the rate-limiting factor in synthesis of GSH [31]. In the second step, catalyzed by glutathione synthase, a glycine residue is added to complete the formation.

Antioxidant Properties of Glutathione

The main function of GSH is maintaining the redox balance [23, 31–34]. This is achieved directly by oxidation of GSH to its dimeric form (GSSG) by GSH peroxidases (GPx). For this, the cysteine residue with its reducing sulfydryl group (electron donating group) is responsible. GPx reduces hydrogen peroxide by transferring the energy of the reactive peroxides to the sulfur-containing GSH. GSSG can be reconverted to GSH by the enzyme glutathione reductase (GR), using NADPH as source of electrons. The NADPH required in this reaction as a reducing equivalent is mainly provided by the oxidative pentose phosphate pathway.

Besides owning intrinsic antioxidant properties, GSH is also an essential cofactor for enzymes such as GSH S-transferases (GST). Glutathione S-transferase (GST) catalyzes binding with lipid oxidation products and thereby protects against these potentially cytotoxic compounds. Furthermore, by binding of toxic metals to its sulfhydryl group, GSH prevents the formation of ROS through the Fenton chemistry.

Other Functions of Glutathione

Other than its antioxidant properties, GSH exerts many other functions. By affecting cellular redox status, GSH influences the regulation of signal transduction pathways and gene transcription. The redox status is also important for modulation of several enzymes. GSH maintains the reduced state of the sulfhydryl groups of many proteins, which is required for their normal function. Furthermore, GSH functions as storage and transport of cysteine. In this manner, GSH prevents auto-oxidation of circulating cysteine (thereby forming the highly reactive hydroxyl radical [35]), serves as a cysteine reserve during food deprivation, and is source of cysteine for lymphocytes. GSH also plays an important role in the response of the immune system to infections, such as in the synthesis of leukotrienes and in lymphocyte proliferation.

Glutathione as Marker of Oxidative Stress

Since GSH is one of the most important antioxidants, its dynamics are an excellent marker of oxidative stress. In human studies, GSH can be measured in erythrocytes as marker of oxidative stress. Besides being readily accessible as opposed to other tissues, erythrocytes are also suggested to function as antioxidant defense by being a physiological source of GSH and by taking up ROS [23, 36]. Giustarini et al. [36] provided strong evidence for a role of erythrocytes as GSH donor for other tissues [36]. In this manner, erythrocytes can provide protection against oxidative injury to other tissues, such as the lungs, by supplying them with intracellular antioxidant [23, 34, 37–39]. Furthermore, membranes of erythrocytes are permeable to superoxide and hydrogen peroxide. By taking up these ROS, erythrocytes are also able to protect other tissues against oxidative damage [40, 41]. Taken together, erythrocytes seem to be a good representative of systemic oxidative stress.

Glutathione Synthesis Rates

Quantification of the utilization and synthesis rates provides a dynamic insight into its metabolism under pathological conditions, such as oxidative stress, or in response to interventions. By using stable isotopes, fractional synthesis rates (FSR) and absolute synthesis rates (ASR) can be measured in erythrocytes. FSR determination of GSH in erythrocytes requires suitable methods to measure both low level of isotopic enrichment in GSH and in its precursor. Moreover, determination in neonates is complicated because only small amounts of blood can be sampled. Recently, a new method was developed for simultaneous measurement of ^{13}C-glutathione as its dimeric form (GSSG) and its precursor [1-^{13}C] glycine in erythrocytes using liquid chromatography/isotope ratio mass spectrometry (LC-IRMS) [42]. Glutathione has a rapid turnover. The fractional synthesis rate (FSR) reported in healthy adult volunteers varied from 63 to 83 %/day in the various studies [43–44]. In neonates, it was found to be much lower, from 35 to 55 %/day [45]. Nevertheless, in all cases, the total pool is renewed within 3 days.

GSH – GSSG Redox Ratio

As mentioned previously, GSH can be oxidized to its dimeric form GSSG; thereby, GPx reduces transferring the energy of the reactive peroxides to the sulfur containing GSH. GSH is mostly kept in reduced form by GR (99 %) [46], with concentrations of GSSG being 1/1,000 of total glutathione. Plasma is lacking reductases and NADPH, so all GSSG is reduced intracellular [36].

The ratio between GSH and its oxidized form, i.e., the GSH – GSSG ratio, is widely used as a marker for oxidative stress in (preterm) infants [12, 47, 48]. A higher recycling of GSSG (dimeric oxidized form of GSH) into GSH in erythrocytes of preterm infants was observed compared to adults [49]. This more efficient recycling could decrease the need for de novo GSH synthesis, since there is relatively more effective GSH. It can, therefore, be questioned whether GSH synthesis rate alone is a suitable marker to study oxidative stress in preterm infants. Simultaneous measurement of all players of the GSH antioxidant defense system, namely, GSH synthesis, the ratio of GSH and its oxidized GSSG, and the enzymes involved in GSH synthesis and recycling, should provide a better inside into GSH in preterm infants.

Glutathione Synthesis in Early Postnatal Life

Apart from an initial high GSH concentration found in erythrocytes and plasma of preterm infants immediately after birth, concentrations drop rapidly after birth [50], and these concentrations were significantly lower than in term neonates [23, 51, 52]. The decreased concentrations of glutathione in preterm infants are most likely caused by increased usage due to increased oxidative stress. It has also been suggested that GSH synthesis might be hampered by immaturity of the enzymatic apparatus or by lack of substrate.

Glutamate-cysteine ligase – the rate-limiting enzyme in GSH synthesis – is present and active in preterm infants [53]. However, literature about cystathionase, which converts homocysteine to L-cysteine, is conflicting [4, 54–56]. It has been suggested that cystathionase is not fully matured in preterm infants, making cysteine a conditionally essential AA in preterm infants. On the other hand, stable isotope studies have shown that conversion of homocysteine to cysteine is not hampered in preterm infants [57, 58]. Moreover, biosynthesis of glutathione has proven to be active in leukocytes from preterm infants [59], and GSH synthesis is not dependent on gestational age (Rook et al. submitted).

Nutritional Modulation of Glutathione Kinetics

The nutritional state of the newborn modulates the antioxidant defenses [60]. Protein synthesis, like the synthesis of GSH, is dependent on adequate substrate, i.e., amino acids (AA). In addition, a positive energy balance is important for protein synthesis, since protein synthesis is an energy demanding process. Nonprotein energy deficits are, however, still common in premature infants during the first week of life [61]. Optimizing early nutrition by increasing AA intake and early administration of lipids might improve antioxidant defenses by increased GHS synthesis.

Amino Acids

Early administration of 2.4 g/(kg · d) amino acids (AA) to preterm infants resulted in increased concentration of glutathione on the second day of life, without a concomitant rise in the fractional synthesis rate of GSH (FSR_{GSH}) [45]. In follow-up of this study, significantly lower GSH concentrations within a few hours after birth were demonstrated when compared to the second day of life in infants receiving 2.4 g/(kg · d) AAs [50]. Also in this study, FSR_{GSH} was not increased. These data suggest that the increased GSH concentration after initial low concentrations after birth could be the result of decreased GSH consumption upon AA administration. Extracellular amino acid substrate availability suppresses the cellular glutathione loss through oxidation, thereby facilitating the preservation of cellular glutathione content [26]. Consequently, increased AA availability limits GSH recycling and the need for de novo synthesis. Other possible explanations for this might be that AAs, including methionine and cysteine, can serve as antioxidants themselves [62] or that increased availability of AAs upregulates synthesis of other antioxidants like albumin synthesis [63]. Besides antioxidant properties, GSH also functions as a cysteine reservoir [64, 65], and it might be possible that increased availability of cysteine reduces breakdown of GSH to generate free cysteine.

To further improve the GSH availability of preterm infants, we studied the effect of increasing AA administration to 3.6 g/(kg · d) and demonstrate that a further increase of AA administration does not result in either an increased GSH availability or synthesis rate suggesting a limit to this sparing effect of AAs on GSH (Rook et al. submitted).

Lipids

Besides AA, lipids are an essential component of the parenteral nutrition of preterm infants. In addition to supply of essential fatty acids, lipids are a source of high energy density. Since lipids become the main source of energy within hours after administration [66], early lipid administration may result in protein sparing. However, we demonstrated that increasing energy balance via lipid administration with a starting dose of 2 g/(kg · d) directly after birth and 3 g/(kg · d) on the following days did not result in increased GSH synthesis (Rook et al. submitted).

On the other hand, lipids are vulnerable for lipid peroxidation and could contribute to increased oxidative stress in preterm infants. Total isofurans and neurofurans were increased on postnatal day 2 upon early lipid administration. However, these differences disappeared in the consecutive days when all groups received lipids and other oxidative stress markers were not different between groups. Furthermore, early initiation of lipids did not result in differences in short-term clinical outcome [67]. Therefore, it seems that lipid administration from birth onward does not increase oxidative stress compared to initiation of lipids later, but long-term follow-up will determine whether the transient increase of isofurans and neurofurans upon early initiation of lipids is indeed clinically irrelevant.

Besides the timing of parenteral lipid administration, we also studied the effect of the type of lipid emulsion administered. Traditional lipid emulsions are mainly manufactured from soybean oil and are very rich in ω-6 polyunsaturated fatty acids (PUFAs), which are highly susceptible for lipid peroxidation resulting in increased oxidative stress [20, 68, 69]. A multicomponent lipid emulsion, containing soya bean oil, medium-chain triglycerides, olive oil, and fish oil with increased added α-tocopherol, might positively influence oxidative stress due a more balanced ω-3:ω-6 fatty acid ratio and the added α-tocopherol. Also, the content is increased. However, literature comparing the effect of a multicomponent lipid emulsion with a pure soybean oil emulsion on oxidative stress is conflicting [70–74]. Also, parenteral lipids were supplied either with low doses form birth onward or lipids were not started directly from birth onward in most of these studies. Therefore, we performed a double-blind randomized study comparing a multicomponent lipid with a pure soybean oil emulsion (Rook et al. submitted). GSH concentration and synthesis rate were not altered following administration of different lipids emulsions from birth onward in preterm infants. Besides a decrease in isofurans (indicative of a reduced arachidonic acid peroxidation), all other markers of oxidative stress were not reduced in infants receiving a multicomponent lipid emulsion. Although the association of a reduction in isofurans seems to suggest an interplay between lipid peroxidation and oxygen rather than an effect of sole lipid peroxidation, the other markers were not affected and type of lipid does not affect neonatal morbidities [75], and the implication of differences in isofurans is unknown. Taken together, there does not seem to be a clear reduction of oxidative stress upon administration of the multicomponent lipid emulsion compared to a pure soybean oil emulsion.

Conclusion

Initiation of AAs from birth onward resulted in increased GSH concentration, probably caused by decreased GSH consumption. Extracellular amino acid substrate availability suppresses the cellular glutathione loss through oxidation, thereby facilitating the preservation of cellular glutathione content [26]. Consequently, increased AA availability limits GSH recycling and the need for de novo synthesis. However, there appears to be a limit to this effect on GSH, since high-dose AA administration did not have an additive effect.

Increasing energy intake via intravenous lipids also did not result in increased GSH synthesis. However and fortunately, timing and type of lipid emulsion supplied as part of parenteral nutrition to preterm infants do not seem to influence oxidative stress significantly. Probably, oxidative stress is already so highly increased in preterm infants that parenteral lipids do not have a detectable effect. However, long-term follow-up will reveal whether the differences in isofurans and neurofurans are of clinical relevance.

In conclusion, nutritional modulation of GSH synthesis in preterm infants seems to warrant only minimal effects. Increased AAs and/or lipids do not increase GSH synthesis rate. Maximum synthetic capacity caused by immaturity of the enzymatic apparatus in preterm infants is not the (sole) explanation for this, since preterm infants seem to be capable of synthesizing GSH (discussed elsewhere in this chapter). Shortage of substrate should have been intercepted by increased AA administration, but the composition of the AA might have been suboptimal. Protein synthesis is dependent on availability of all precursor AAs and the rate of protein synthesis is thus determined by the first limiting AA.

Other Modulators of GSH Synthesis in Preterm Infants

Neonatal Sepsis on Oxidative Stress

Many preterm infants will develop nosocomial sepsis during the first weeks of life, with reported incidences of 21–36 % [29, 76–78]. Despite major improvements in neonatal care, sepsis remains an important cause of morbidity and mortality in preterm infants. Among others, sepsis in preterm infants is associated with severe adverse neurodevelopmental outcome at 2 years of age [79, 80]. The relation between neonatal sepsis and the adverse outcome is likely to be multifactorial, but oxidative stress is thought to be one of the major contributors [79, 81]. During sepsis, ROS formation is increased as part of the defensive response against pathogens through the stimulation of proinflammatory cytokine release, the formation of chemotactic factors, and neutrophil recruitment [33, 82–84]. In both adults and children, overstimulation of ROS production or an insufficient antioxidant capacity is associated with increased morbidity and mortality [85–87]. A compromised antioxidant defense makes preterm infants more susceptible to oxidative stress [23, 88], and the increased ROS formation during sepsis might be an etiological factor to the detrimental long-term outcome in preterm infants after neonatal sepsis. However, we demonstrated that nosocomial sepsis in premature infants did not result in increased oxidative stress and therefore there was no upregulation of GSH synthesis (Rook et al. submitted).

Our study design might have been suboptimal. We designed the study to detect differences in GSH kinetics, and the study was therefore not powered to detect differences in oxidative stress markers. Nowadays, less invasive methods to measure oxidative stress markers, in, for example, urine, are available. This allows future studies to include more preterm infants and also to include "healthy" preterm infants as controls. Our data do not explain the association between nosocomial neonatal sepsis in preterm infants and adverse long-term outcome. It would be of great interest to further explore other mechanisms, such as the production of proinflammatory cytokines which are known to be neurotoxic and altered cerebral blood flow through circulatory insufficiency.

Determinants of GSH Synthesis

Shorter gestation, lower birth weight (BW), and male gender are associated with poorer outcome. This effect might be partly mediated by differences in antioxidant capacity. Since oxidative stress is associated with both morbidity and mortality in preterm infants, we hypothesized that gender, GA, and BW would affect GSH kinetics. However, differences in outcome related to GA, BW, and gender seem to be unexplained by differences in GSH synthesis rates (Rook et al. submitted).

A possible explanation for differences in GSH synthesis, and also GSH recycling, might be found in genetic differences in the GSH synthesis apparatus. Research in this field is still in its infancy and has been mainly focused on the risk of adult cancer. Polymorphisms in the glutathione S-transferase (GST) and SOD gene have been associated with the risk of developing infant respiratory distress syndrome, BPD, IVH, and ROP in preterm infants [89–91]. Polymorphisms in microsomal epoxide hydrolase (EPHX), involved in the detoxification of epoxides which may be extremely toxic by the induction of oxidative stress, have also been associated with perinatal mortality [92]. Polymorphisms in antioxidant enzymes have indeed been found to alter enzyme activity in response to oxidative stress and might therefore be associated with differences in the ability to counteract oxidative stress [93, 94]. It would be interesting to investigate whether polymorphisms in enzymes involved in GSH synthesis might alter GSH synthesis and its response to oxidative stress.

Conclusions

Increased oxidative stress can have detrimental consequences, especially in preterm infants who are experiencing equivalent intrauterine development stage in an oxygen-rich environment postnatally. However, it is important to remember that oxidative stress also has a physiological function as in cardiopulmonary adaptation at birth, the defense against microorganisms, and in inter- and intracellular signaling. So, there seems to be a delicate balance between pathologic and physiologic oxidative stress.

Although we could not demonstrate differences in GSH synthesis upon nutritional intervention, it is evident that oxidative stress plays an important role in neonatal morbidity and mortality. Clearly, the mechanisms of oxidative stress and antioxidant defense are complex and warrant further research. We suggest that future research on oxidative stress in preterm infants should focus on finding more sensitive biomarkers of oxidative stress, genetic polymorphisms, and the adaptive mechanisms.

References

1. Maltepe E, Saugstad OD. Oxygen in health and disease: regulation of oxygen homeostasis – clinical implications. Pediatr Res. 2008;65(3):261–8.
2. Vento M, Asensi M, Sastre J, Lloret A, Garcia-Sala F, Minana JB, et al. Hyperoxemia caused by resuscitation with pure oxygen may alter intracellular redox status by increasing oxidized glutathione in asphyxiated newly born infants. Semin Perinatol. 2002;26(6):406–10.
3. Comporti M, Signorini C, Leoncini S, Buonocore G, Rossi V, Ciccoli L. Plasma F2-isoprostanes are elevated in newborns and inversely correlated to gestational age. Free Radic Biol Med. 2004;37(5):724–32.
4. Pallardo FV, Sastre J, Asensi M, Rodrigo F, Estrela JM, Vina J. Physiological changes in glutathione metabolism in foetal and newborn rat liver. Biochem J. 1991;274(Pt 3):891–3.
5. Noori S, Wlodaver A, Gottipati V, McCoy M, Schultz D, Escobedo M. Transitional changes in cardiac and cerebral hemodynamics in term neonates at birth. J Pediatr. 2012;160(6):943–8.
6. Noori S, Stavroudis TA, Seri I. Systemic and cerebral hemodynamics during the transitional period after premature birth. Clin Perinatol. 2009;36(4):723–36, v.
7. Friel JK, Friesen RW, Harding SV, Roberts LJ. Evidence of oxidative stress in full-term healthy infants. Pediatr Res. 2004;56(6):878–82.
8. Frank L, Groseclose EE. Preparation for birth into an O2-rich environment: the antioxidant enzymes in the developing rabbit lung. Pediatr Res. 1984;18(3):240–4.
9. Frank L, Sosenko IR. Prenatal development of lung antioxidant enzymes in four species. J Pediatr. 1987;110(1):106–10.
10. Frank L, Sosenko IR. Failure of premature rabbits to increase antioxidant enzymes during hyperoxic exposure: increased susceptibility to pulmonary oxygen toxicity compared with term rabbits. Pediatr Res. 1991;29(3):292–6.
11. Ezaki S, Suzuki K, Kurishima C, Miura M, Weilin W, Hoshi R, et al. Resuscitation of preterm infants with reduced oxygen results in less oxidative stress than resuscitation with 100 % oxygen. J Clin Biochem Nutr. 2009;44(1):111–8.
12. Vento M, Moro M, Escrig R, Arruza L, Villar G, Izquierdo I, et al. Preterm resuscitation with low oxygen causes less oxidative stress, inflammation, and chronic lung disease. Pediatrics. 2009;124(3):e439–49.
13. Wang CL, Anderson C, Leone TA, Rich W, Govindaswami B, Finer NN. Resuscitation of preterm neonates by using room air or 100 % oxygen. Pediatrics. 2008;121(6):1083–9.
14. Solberg R, Andresen JH, Escrig R, Vento M, Saugstad OD. Resuscitation of hypoxic newborn piglets with oxygen induces a dose-dependent increase in markers of oxidation. Pediatr Res. 2007;62(5):559–63.
15. Cheah FC, Pillow JJ, Kramer BW, Polglase GR, Nitsos I, Newnham JP, et al. Airway inflammatory cell responses to intra-amniotic lipopolysaccharide in a sheep model of chorioamnionitis. Am J Physiol Lung Cell Mol Physiol. 2009;296(3):L384–93.

16. Raijmakers MT, Roes EM, Poston L, Steegers EA, Peters WH. The transient increase of oxidative stress during normal pregnancy is higher and persists after delivery in women with pre-eclampsia. Eur J Obstet Gynecol Reprod Biol. 2008;138(1):39–44.
17. Moison RM, van Hoof EJ, Clahsen PC, van Zoeren-Grobben D, Berger HM. Influence of plasma preparations and donor red blood cells on the antioxidant capacity of blood from newborn babies: an in vitro study. Acta Paediatr. 1996;85(2):220–4.
18. Hirano K, Morinobu T, Kim H, Hiroi M, Ban R, Ogawa S, et al. Blood transfusion increases radical promoting non-transferrin bound iron in preterm infants. Arch Dis Child Fetal Neonatal Ed. 2001;84(3):F188–93.
19. Kapoor K, Basu S, Das BK, Bhatia BD. Lipid peroxidation and antioxidants in neonatal septicemia. J Trop Pediatr. 2006;52(5):372–5.
20. Laborie S, Lavoie JC, Chessex P. Increased urinary peroxides in newborn infants receiving parenteral nutrition exposed to light. J Pediatr. 2000;136(5):628–32.
21. Buonocore G, Perrone S, Longini M, Vezzosi P, Marzocchi B, Paffetti P, et al. Oxidative stress in preterm neonates at birth and on the seventh day of life. Pediatr Res. 2002;52(1):46–9.
22. Perrone S, Mussap M, Longini M, Fanos V, Bellieni CV, Proietti F, et al. Oxidative kidney damage in preterm newborns during perinatal period. Clin Biochem. 2007;40(9–10):656–60.
23. Ahola T, Levonen AL, Fellman V, Lapatto R. Thiol metabolism in preterm infants during the first week of life. Scand J Clin Lab Invest. 2004;64(7):649–58.
24. Georgeson GD, Szony BJ, Streitman K, Varga IS, Kovacs A, Kovacs L, et al. Antioxidant enzyme activities are decreased in preterm infants and in neonates born via caesarean section. Eur J Obstet Gynecol Reprod Biol. 2002;103(2):136–9.
25. Perrone S, Salvi G, Bellieni CV, Buonocore G. Oxidative stress and nutrition in the preterm newborn. J Pediatr Gastroenterol Nutr. 2007;45:S178–82.
26. Yeung MY. Influence of early postnatal nutritional management on oxidative stress and antioxidant defence in extreme prematurity. Acta Paediatr. 2006;95(2):153–63.
27. Friel JK, Martin SM, Langdon M, Herzberg GR, Buettner GR. Milk from mothers of both premature and full-term infants provides better antioxidant protection than does infant formula. Pediatr Res. 2002;51(5):612–8.
28. L'Abbe MR, Friel JK. Superoxide dismutase and glutathione peroxidase content of human milk from mothers of premature and full-term infants during the first 3 months of lactation. J Pediatr Gastroenterol Nutr. 2000;31(3):270–4.
29. Corpeleijn WE, Kouwenhoven SM, Paap MC, van Vliet I, Scheerder I, Muizer Y, et al. Intake of own mother's milk during the first days of life is associated with decreased morbidity and mortality in very low birth weight infants during the first 60 days of life. Neonatology. 2012;102(4):276–81.
30. Witschi A, Reddy S, Stofer B, Lauterburg BH. The systemic availability of oral glutathione. Eur J Clin Pharmacol. 1992;43(6):667–9.
31. Lyons J, Rauh-Pfeiffer A, Ming-Yu Y, Lu XM, Zurakowski D, Curley M, et al. Cysteine metabolism and whole blood glutathione synthesis in septic pediatric patients. Crit Care Med. 2001;29(4):870–7.
32. Malmezat T, Breuille D, Capitan P, Mirand PP, Obled C. Glutathione turnover is increased during the acute phase of sepsis in rats. J Nutr. 2000;130(5):1239–46.
33. Hansson L, Seidegard J, Johansson L, Jeppsson B. Influence of glutathione metabolising enzymes in rats with gram-negative sepsis. Eur J Surg. 2000;166(9):728–33.
34. Biolo G, Antonione R, De Cicco M. Glutathione metabolism in sepsis. Crit Care Med. 2007;35(9 Suppl):S591–5.
35. Kidd PM. Glutathione: systemic protectant against oxidative and free radical damage. Altern Med Rev. 1997;2(3):155–76.
36. Giustarini D, Milzani A, Dalle-Donne I, Rossi R. Red blood cells as a physiological source of glutathione for extracellular fluids. Blood Cells Mol Dis. 2007;40(2):174–9.
37. Berggren M, Dawson J, Moldeus P. Glutathione biosynthesis in the isolated perfused rat lung: utilization of extracellular glutathione. FEBS Lett. 1984;176(1):189–92.
38. Clahsen PC, Moison RM, Holtzer CA, Berger HM. Recycling of glutathione during oxidative stress in erythrocytes of the newborn. Pediatr Res. 1992;32(4):399–402.

39. Griffith OW, Meister A. Glutathione: interorgan translocation, turnover, and metabolism. Proc Natl Acad Sci U S A. 1979;76(11):5606–10.
40. Heffner JE, Repine JE. Pulmonary strategies of antioxidant defense. Am Rev Respir Dis. 1989;140(2):531–54.
41. van Asbeck BS, Hoidal J, Vercellotti GM, Schwartz BA, Moldow CF, Jacob HS. Protection against lethal hyperoxia by tracheal insufflation of erythrocytes: role of red cell glutathione. Science. 1985;227(4688):756–9.
42. Schierbeek H, Rook D, te Braake FW, Dorst KY, Voortman G, Godin JP, et al. Simultaneous analysis of (13)C-glutathione as its dimeric form GSSG and its precursor [1-(13)C]glycine using liquid chromatography/isotope ratio mass spectrometry. Rapid Commun Mass Spectrom. 2009;23(18):2897–902.
43. Darmaun D, Smith SD, Sweeten S, Sager BK, Welch S, Mauras N. Evidence for accelerated rates of glutathione utilization and glutathione depletion in adolescents with poorly controlled type 1 diabetes. Diabetes. 2005;54(1):190–6.
44. Jahoor F, Jackson A, Gazzard B, Philips G, Sharpstone D, Frazer ME, et al. Erythrocyte glutathione deficiency in symptom-free HIV infection is associated with decreased synthesis rate. Am J Physiol. 1999;276(1 Pt 1):E205–11.
45. Te Braake FW, Schierbeek H, de Groof K, Vermes A, Longini M, Buonocore G, et al. Glutathione synthesis rates after amino acid administration directly after birth in preterm infants. Am J Clin Nutr. 2008;88(2):333–9.
46. Pastore A, Federici G, Bertini E, Piemonte F. Analysis of glutathione: implication in redox and detoxification. Clin Chim Acta. 2003;333(1):19–39.
47. Papp A, Nemeth I, Karg E, Papp E. Glutathione status in retinopathy of prematurity. Free Radic Biol Med. 1999;27(7–8):738–43.
48. Vento M, Aguar M, Escobar J, Arduini A, Escrig R, Brugada M, et al. Antenatal steroids and antioxidant enzyme activity in preterm infants: influence of gender and timing. Antioxid Redox Signal. 2009;11(12):2945–55.
49. Frosali S, Di Simplicio P, Perrone S, Di Giuseppe D, Longini M, Tanganelli D, et al. Glutathione recycling and antioxidant enzyme activities in erythrocytes of term and preterm newborns at birth. Biol Neonate. 2004;85(3):188–94.
50. Rook D, Te Braake FW, Schierbeek H, Longini M, Buonocore G, van Goudoever JB. Glutathione synthesis rates in early postnatal life. Pediatr Res. 2010;67(4):407–11.
51. Jain A, Mehta T, Auld PA, Rodrigues J, Ward RF, Schwartz MK, et al. Glutathione metabolism in newborns: evidence for glutathione deficiency in plasma, bronchoalveolar lavage fluid, and lymphocytes in prematures. Pediatr Pulmonol. 1995;20(3):160–6.
52. Jean-Baptiste D, Rudolph N. Sequential postnatal changes in erythrocyte glutathione and sulfhydryl content: a possible adaptational response to the extrauterine environment. Biol Neonate. 2003;84(2):142–6.
53. Levonen AL, Lapatto R, Saksela M, Raivio KO. Expression of gamma-glutamylcysteine synthetase during development. Pediatr Res. 2000;47(2):266–70.
54. Lestas AN, Rodeck CH. Normal glutathione content and some related enzyme activities in the fetal erythrocytes. Br J Haematol. 1984;57(4):695–702.
55. Martin JA, Pereda J, Martinez-Lopez I, Escrig R, Miralles V, Pallardo FV, et al. Oxidative stress as a signal to up-regulate gamma-cystathionase in the fetal-to-neonatal transition in rats. Cell Mol Biol (Noisy-le-grand). 2007;53(Suppl):OL1010–7.
56. Vina J, Vento M, Garcia-Sala F, Puertes IR, Gasco E, Sastre J, et al. L-cysteine and glutathione metabolism are impaired in premature infants due to cystathionase deficiency. Am J Clin Nutr. 1995;61(5):1067–9.
57. Riedijk MA, van Beek RH, Voortman G, de Bie HM, Dassel AC, van Goudoever JB. Cysteine: a conditionally essential amino acid in low-birth-weight preterm infants? Am J Clin Nutr. 2007;86(4):1120–5.
58. Thomas B, Gruca LL, Bennett C, Parimi PS, Hanson RW, Kalhan SC. Metabolism of methionine in the newborn infant: response to the parenteral and enteral administration of nutrients. Pediatr Res. 2008;64(4):381–6.

59. Lavoie JC, Chessex P. Development of glutathione synthesis and gamma-glutamyltranspeptidase activities in tissues from newborn infants. Free Radic Biol Med. 1998;24(6):994–1001.
60. Jobe AH, Kallapur SG. Long term consequences of oxygen therapy in the neonatal period. Semin Fetal Neonatal Med. 2010;15(4):230–5.
61. Hulst JM, Goudoever JB, Zimmermann LJI, Hop WCJ, Albers MJIJ, Tibboel D, et al. The effect of cumulative energy and protein deficiency on anthropometric parameters in a pediatric ICU population. Clin Nutr. 2004;23(6):1381–9.
62. Manzella D, Grella R, Esposito K, Cacciapuoti F, Arciello A, Giugliano D, et al. Oral amino acid administration decreases oxidative stress and improves brachial reactivity in elderly individuals. Am J Hypertens. 2005;18(6):858–63.
63. van den Akker CH, te Braake FW, Schierbeek H, Rietveld T, Wattimena DJ, Bunt JE, et al. Albumin synthesis in premature neonates is stimulated by parenterally administered amino acids during the first days of life. Am J Clin Nutr. 2007;86(4):1003–8.
64. Stipanuk MH, Coloso RM, Garcia RA, Banks MF. Cysteine concentration regulates cysteine metabolism to glutathione, sulfate and taurine in rat hepatocytes. Clin Nutr. 1992; 122(3):420–7.
65. Stipanuk MH, Dominy Jr JE, Lee JI, Coloso RM. Mammalian cysteine metabolism: new insights into regulation of cysteine metabolism. Clin Nutr. 2006;136(6 Suppl):1652S–959.
66. Vlaardingerbroek H, van Goudoever JB, van den Akker CH. Initial nutritional management of the preterm infant. Early Hum Dev. 2009;85(11):691–5.
67. Vlaardingerbroek H, Vermeulen MJ, Rook D, van den Akker CH, Dorst K, Wattimena JL, et al. Safety and efficacy of early parenteral lipid and high-dose amino acid administration to very low birth weight infants. J Pediatr. 2013;163(3):638–44.
68. Alexander-North LS, North JA, Kiminyo KP, Buettner GR, Spector AA. Polyunsaturated fatty acids increase lipid radical formation induced by oxidant stress in endothelial cells. J Lipid Res. 1994;35(10):1773–85.
69. Helbock HJ, Motchnik PA, Ames BN. Toxic hydroperoxides in intravenous lipid emulsions used in preterm infants. Pediatrics. 1993;91(1):83–7.
70. Skouroliakou M, Konstantinou D, Koutri K, Kakavelaki C, Stathopoulou M, Antoniadi M, et al. A double-blind, randomized clinical trial of the effect of omega-3 fatty acids on the oxidative stress of preterm neonates fed through parenteral nutrition. Eur J Clin Nutr. 2010;64(9):940–7.
71. Koksal N, Kavurt AV, Cetinkaya M, Ozarda Y, Ozkan H. Comparison of lipid emulsions on antioxidant capacity in preterm infants receiving parenteral nutrition. Pediatr Int. 2011; 53(4):562–6.
72. Roggero P, Mosca F, Gianni ML, Orsi A, Amato O, Migliorisi E, et al. F2-isoprostanes and total radical-trapping antioxidant potential in preterm infants receiving parenteral lipid emulsions. Nutrition. 2010;26(5):551–5.
73. Gobel Y, Koletzko B, Bohles HJ, Engelsberger I, Forget D, Le Brun A, et al. Parenteral fat emulsions based on olive and soybean oils: a randomized clinical trial in preterm infants. J Pediatr Gastroenterol Nutr. 2003;37(2):161–7.
74. Webb AN, Hardy P, Peterkin M, Lee O, Shalley H, Croft KD, et al. Tolerability and safety of olive oil-based lipid emulsion in critically ill neonates: a blinded randomized trial. Nutrition. 2008;24(11–12):1057–64.
75. Vlaardingerbroek H, Veldhorst MA, Spronk S, van den Akker CH, van Goudoever JB. Parenteral lipid administration to very-low-birth-weight infants – early introduction of lipids and use of new lipid emulsions: a systematic review and meta-analysis. Am J Clin Nutr. 2012;96(2):255–68.
76. Stoll BJ, Hansen N, Fanaroff AA, Wright LL, Carlo WA, Ehrenkranz RA, et al. Late-onset sepsis in very low birth weight neonates: the experience of the NICHD Neonatal Research Network. Pediatrics. 2002;110(2 Pt 1):285–91.
77. Stoll BJ, Hansen NI, Bell EF, Shankaran S, Laptook AR, Walsh MC, et al. Neonatal outcomes of extremely preterm infants from the NICHD Neonatal Research Network. Pediatrics. 2010;126(3):443–56.

78. van der Zwet WC, Kaiser AM, van Elburg RM, Berkhof J, Fetter WP, Parlevliet GA, et al. Nosocomial infections in a Dutch neonatal intensive care unit: surveillance study with definitions for infection specifically adapted for neonates. J Hosp Infect. 2005;61(4):300–11.
79. Adams-Chapman I, Stoll BJ. Neonatal infection and long-term neurodevelopmental outcome in the preterm infant. Curr Opin Infect Dis. 2006;19(3):290–7.
80. Stoll BJ, Hansen NI, Adams-Chapman I, Fanaroff AA, Hintz SR, Vohr B, et al. Neurodevelopmental and growth impairment among extremely low-birth-weight infants with neonatal infection. JAMA. 2004;292(19):2357–65.
81. Volpe JJ. Neurobiology of periventricular leukomalacia in the premature infant. Pediatr Res. 2001;50(5):553–62.
82. Martins PS, Kallas EG, Neto MC, Dalboni MA, Blecher S, Salomao R. Upregulation of reactive oxygen species generation and phagocytosis, and increased apoptosis in human neutrophils during severe sepsis and septic shock. Shock. 2003;20(3):208–12.
83. de Souza LF, Ritter C, Pens Gelain D, Andrades M, Bernard EA, Moreira JC, et al. Mitochondrial superoxide production is related to the control of cytokine release from peritoneal macrophage after antioxidant treatment in septic rats. J Surg Res. 2007;141(2):252–6.
84. Andrades M, Ritter C, Moreira JC, Dal-Pizzol F. Oxidative parameters differences during non-lethal and lethal sepsis development. J Surg Res. 2005;125(1):68–72.
85. Victor VM, Rocha M, De la Fuente M. Immune cells: free radicals and antioxidants in sepsis. Int Immunopharmacol. 2004;4(3):327–47.
86. Crimi E, Sica V, Slutsky AS, Zhang H, Williams-Ignarro S, Ignarro LJ, et al. Role of oxidative stress in experimental sepsis and multisystem organ dysfunction. Free Radic Res. 2006; 40(7):665–72.
87. von Dessauer B, Bongain J, Molina V, Quilodran J, Castillo R, Rodrigo R. Oxidative stress as a novel target in pediatric sepsis management. J Crit Care. 2011;26(1):103.e1–7.
88. Davis JM, Auten RL. Maturation of the antioxidant system and the effects on preterm birth. Semin Fetal Neonatal Med. 2010 Aug;15(4):191–5.
89. Poggi C, Giusti B, Vestri A, Pasquini E, Abbate R, Dani C. Genetic polymorphisms of antioxidant enzymes in preterm infants. J Matern Fetal Neonatal Med. 2012;25 Suppl 4:131–4.
90. Manar MH, Brown MR, Gauthier TW, Brown LA. Association of glutathione-S-transferase-P1 (GST-P1) polymorphisms with bronchopulmonary dysplasia. J Perinatol. 2004;24(1):30–5.
91. Oretti C, Marino S, Mosca F, Colnaghi MR, De Iudicibus S, Drigo I, et al. Glutathione-S-transferase-P1 I105V polymorphism and response to antenatal betamethasone in the prevention of respiratory distress syndrome. Eur J Clin Pharmacol. 2009;65(5):483–91.
92. Raijmakers MT, de Galan-Roosen TE, Schilders GW, Merkus JM, Steegers EA, Peters WH. The Tyr113His polymorphism in exon 3 of the microsomal epoxide hydrolase gene is a risk factor for perinatal mortality. Acta Obstet Gynecol Scand. 2004;83(11):1056–60.
93. Komina AV, Korostileva KA, Gyrylova SN, Belonogov RN, Ruksha TG. Interaction between single nucleotide polymorphism in catalase gene and catalase activity under the conditions of oxidative stress. Physiol Res. 2013;61(6):655–8.
94. Bresciani G, Gonzalez-Gallego J, da Cruz IB, de Paz JA, Cuevas MJ. The Ala16Val MnSOD gene polymorphism modulates oxidative response to exercise. Clin Biochem. 2013;46(4–5): 335–4.

Chapter 14
Oxygen Vulnerability in the Immature Brain

Ursula Felderhoff-Müser

Introduction

Major advances in obstetrics and neonatal intensive care have substantially improved survival of very preterm infants and critically ill term newborns. Perinatal mortality has decreased by 25 % over the last decade and has expanded the surviving population [1, 2].

However, despite all efforts, perinatal brain injury is still a leading cause of death and disability in children. Both asphyxia and premature birth represent major insults affecting brain development at an early age with lifelong individual, familial and socioeconomic consequences.

Birth asphyxia is mainly prevalent in the term infant population. In resource-rich countries 1–6 per 1,000 live-born infants experience a hypoxic–ischaemic (HI) insult during the perinatal period. In underdeveloped countries, the rate of affected children is significantly higher.

Injury can occur during the acute episode of hypoxia–ischaemia and involves many brain structures such as the basal ganglia and brain stem but also the subcortical white matter. In many cases, injury evolves for hours after resuscitation, during the recovery period triggered by inflammation, excitotoxicity and oxidative stress. Development of hypoxic–ischaemic encephalopathy is associated with cerebral palsy, epilepsy and visual impairment, as well as cognitive and motor deficits in later life [3, 4].

U. Felderhoff-Müser (✉)
Department of Pediatrics, Division of Neonatology, University Hospital Essen,
University Duisburg-Essen, Hufelandstrasse 55, 45122 Essen, Germany
e-mail: ursula.felderhoff@uk-essen.de

P.A. Dennery et al. (eds.), *Perinatal and Prenatal Disorders*,
Oxidative Stress in Applied Basic Research and Clinical Practice,
DOI 10.1007/978-1-4939-1405-0_14

The percentage of prematurely born infants has increased in industrialised countries during the last decades and accounts for 12–15 % of all live births [5]. This number is likely to rise further due to the increasing rate of infertility treatments, multiple pregnancies and older mothers [6]. Intact long-term survival for premature infants has become an almost expected outcome over the past two decades. Improvements in neonatal respiratory care have expanded the very low-birthweight population accounting for approximately 2.5 % of total annual births. With the improvement in survival rate, the focus of clinical efforts has shifted to the immediate and later consequences of prematurity. Unfortunately, still a substantial proportion of survivors have neurological deficits, which affect motor and cognitive function [7–9]. Approximately 10 % of survivors of very preterm birth later exhibit motor deficits characterising cerebral palsy (CP). Even in the absence of obvious intracranial pathology such as intraventricular haemorrhage, preterm infants are at high risk (~40 %) for neurodevelopmental impairment. Cognitive impairment, attention deficit disorder, behavioural problems, autism and development of psychiatric disease in later life are common findings in neurodevelopmental follow-up in children born preterm [10].

Perinatal brain injury largely affects the periventricular cerebral white matter and is the most common type of brain injury in the preterm population born between 23 and 32 weeks' gestation. It remains the leading cause of later neurological impairment. The spectrum of periventricular white matter injury is wide and includes focal cystic necrotic lesions, the so-called periventricular leukomalacia (PVL), and a diffuse form. Diffuse white matter injury is related to a large susceptibility of immature premyelinating oligodendrocytes to injury and leads to a global myelination delay in these patients and also to deep grey matter damage.

Preterm birth is associated with brain tissue volume alterations that become more pronounced in the presence of perinatal risk factors and white matter injury. For decades, white matter damage was the predominant type of injury in the preterm infant, and only recently, the neuronal loss that accompanies white matter damage has become into the focus of investigations. It has been shown that excitotoxic and inflammatory processes leading to white matter damage are also able to damage developing neurons migrating through these areas [11].

Thus, the aetiology and pathophysiology of perinatal brain injury is complex and involves grey and white matter structures to varying degrees, depending on gestational age and developmental stage. Reduction of grey and white matter volume, cortical folding and delayed maturation associated with adverse neurological development are common findings in ex-premature infants on magnetic resonance imaging at term. Neuroimaging at term further revealed a shift from cystic PVL to diffuse white matter injury and brain volume reduction possibly due to advances in neonatal care [12]. Several perinatal factors such as hypoxia–ischaemia, growth factor deficiency, maternal infection yielding excess inflammation, drug exposure, maternal stress, malnutrition and also genetic factors have been in part extensively studied and are likely to be important players in the pathophysiology of brain lesions associated with adverse neurological outcome (for review, [13, 14]).

It has been known for a long time that treatment of premature infants with high oxygen levels results in retinopathy of prematurity which can lead to severe visual impairment and blindness [15, 16]. However, a decade later, the induction of mechanical ventilation with high concentrations of oxygen improved survival of critically ill neonates with respiratory distress. Meanwhile, oxygen has been identified as one of the principal factors in the pathogenesis of chronic lung disease of prematurity (BPD) due to its detrimental effects on lung development. These findings have led to major changes in clinical recommendations towards the careful use of oxygen in infants. For perinatal brain injury, hypoxia–ischaemia and inflammation have been attributed as underlying cause of periventricular leukomalacia (PVL) and diffuse white matter injury disease (WMD). However, mounting evidence from various clinical and experimental observations suggests hyperoxia is, among other sources of oxidative stress, an important trigger of brain injury [17]. Although clinical practice regarding the use of supplementary oxygen has changed in recent years, chronic exposure to supraphysiological oxygen concentrations may also lead to malformation of neuronal circuits during development with dramatic deterioration of brain function in later life.

Pathogenesis of Neonatal Brain Injury

Development of the mammalian brain is a dynamic process involving structural and functional maturation processes. Structural brain evolution is characterised by neuronal cell development and proliferation, migration, glial cell proliferation, axonal and dendritic outgrowth, synaptogenesis and myelination of axons. At the border of viability of extremely preterm infants, neuronal migration processes are completed, but glial cell maturation, outgrowth and formation of connectivity are still in progress. Formation of neural electric activity strongly depends on mitochondrial development, cerebral vascular density and blood flow, maturation of glucose utilisation systems and cytochrome oxidase activity. Synaptic function is closely associated with total mitochondrial density, mitochondrial structure of cristae and volume, which is still in progress during this critical phase of brain development [18].

During normal brain development, initially formed supernumerary and unconnected neurons are deleted by physiological programmed cell death or apoptosis.

During a perinatal insult such as birth asphyxia or inflammation, accidental activation of the well-refined apoptotic cell death machinery may occur. Apoptotic cell death is executed via two different signalling routes. The extrinsic pathway involves extracellular signalling via cell death surface receptors (i.e. Fas, TNFα), and the intrinsic pathway is activated by cellular stress or DNA damage. Both pathways converge downstream at the mitochondrial level. Upon a strong injurious trigger, the mitochondrial inner membrane potential decreases and induces permeability with release of proapoptotic factors (i.e. apoptosis-inducing factor AIF, cytochrome c) in the cytosol, which in turn activates downstream death mechanisms including formation of the active apoptosome with Apaf-1. These mechanisms are modulated by several intrinsic factors such as members of the BcL-2 family. Depending on the type of insult, caspase-dependent and caspase-independent apoptosis signalling is induced [19, 20]. The majority of factors of the apoptotic machinery including caspases are highly expressed in the developing brain, resulting in increased susceptibility of the immature organism to injurious activation. Moreover, caspases seem to have important non-apoptotic functions in multiple cellular processes, such as synaptic plasticity, dendritic development, learning and memory [21].

Recently autophagy has also been described as a prominent feature of cell death during embryonic development. It displays a specific type of programmed cell death, suggesting that autophagic death may be distinct from apoptotic cell death. Additionally, crosstalking between autophagy and apoptosis has been described. In the early stages of programmed cell death, it is protective, but it can also promote apoptosis under particular circumstances such as hypoxia–ischaemia. Proteins well described in this process are Beclin-1 and microtubule-associated protein 1 light chain 3 (LC3). For the developing brain, data regarding autophagic pathways are sparse [22, 23].

Hypoxia–ischaemia, infection, inflammation, intrauterine growth restriction and excitotoxicity are important factors in evolution and also in modulation of brain injury. Furthermore, maturational age, sex, genetic background and developmental stage of the brain influence susceptibility to injurious stimuli. The aetiology of injury to the preterm brain is complex and multifactorial with sometimes several insults called the "multiple hit scenario", which makes it difficult to develop dose and timing of neuroprotective strategies [24].

The major proposed mechanisms are inflammation and hypoxia–ischaemia which itself can trigger long-lasting inflammatory processes [25]. Mounting evidence suggests that neuronal injury related to inflammation is crucial for the pathogenesis of developmental brain injury. However, inflammatory processes such as persistent expression of proinflammatory cytokines and the anti-inflammatory reparative formation of M2 microglia may also trigger repair [26, 27].

Brain injury evolves over time, and various mechanisms are crucial during the different phases of injury. Following a primary strong injurious insult such as hypoxia–ischaemia, a depletion of high-energy phosphates occurs. The second phase of injury is characterised by restoration of glucose utilisation, ATP and phosphocreatine and a delayed loss of high-energy phosphates implicating mitochondrial failure. In combination with calcium influx and accumulation of excessive formation of reactive oxygen species (ROS) and reactive nitrogen species (RNS), mostly during reperfusion, mitochondrial permeability leads to activation of the apoptotic machinery. Furthermore, ROS are essential for autophagy activation as an active response to degrade and recycle damaged cellular material [28].

Tertiary brain injury is defined as all events occurring following the acute phase of cell death in the second phase such as sensitisation to inflammation, susceptibility to seizures, impaired oligodendrocyte maturation and myelination, persistent inflammatory processes and astrogliosis. However, during this phase, intrinsic mechanisms facilitating repair and plasticity come in place.

Reactive Oxygen Species and Endogenous Free Radical-Scavenging Systems

Reactive oxygen species (ROS) include oxygen ions, free radicals and peroxides and are generated during normal mitochondrial respiration. They are generally very small molecules and are highly reactive due to the presence of unpaired valence shell electrons. ROS are formed as a natural by-product of the physiological metabolism of oxygen and have important roles in neuronal cell growth, function and signalling such as induction of host defence genes and mobilisation of ion transport systems. Oxidative stress on nervous tissue can produce damage by several interacting mechanisms, including increase in intracellular free Ca^{2+} and, possibly, release of excitatory amino acids [29].

Oxidative stress in the developing brain is characterised by an imbalance between ROS and RNS and the insufficient production of antioxidant defence mechanisms. Peroxidation of proteins, lipids and polysaccharides and RNA/DNA damage occur in this context. The immature brain is particularly vulnerable to these changes because of its high oxygen consumption and the high content of polyunsaturated fatty acids.

Rodent experiments revealed excessive amounts of ROS generation upon stimulation in comparison to the adult brain [30].

Because of their high numbers of mitochondria, synapses are highly susceptible to oxidative stress, which may either lead to synapse loss as a result of impaired mitochondrial function or synaptic overgrowth [31, 32].

The mammalian organism has evolved inducible responses to ROS that are generated as a consequence of physiological metabolism. ROS may also play a role in synaptic signalling during cognitive processes such as learning and memory. However, excessive generation of ROS results in activation of the same physiological transcription factors (i.e. JNK, autophagy) as mediators for learning and function. There is a great communality between pathways involved in synapse function and development and those who are involved in oxidative stress [33].

The most studied antioxidant enzymes in the developing brain are manganese-containing superoxide-dismutase (Mn-SOD), copper- and zinc-containing SOD (CuZn-SOD), glutathione peroxidase (GPx), glutathione, vitamins A, C and E and catalase. Mitochondria can also produce SOD, GPx and glutathione reductase.

The immature mammalian brain physiologically needs these protective antioxidant enzymes against oxidative stress that occurs at birth due to the relative hyperoxia compared to intrauterine conditions. Expression of SOD, catalase and GPx increase by 150 % in the third trimester of pregnancy. Furthermore, development of antioxidant capacities during the foetal period is associated with redox signalling for the maintenance of pregnancy.

Local nitric oxide (NO) generation as a relatively weak oxygen free radical in the placenta is important for vascular development. Mn-SOD seems crucial for the protection of immature oligodendrocytes, and production of CuZn-SOD increases significantly during myelination in postnatal rat brains [34, 35].

The influence of oxidative stress is also well investigated in the developing white matter: Baud and co-workers demonstrated that nitric oxide is more toxic to developing OLs than to mature OLs. A maturation-dependent vulnerability of pre-OLs to oxidative stress has been confirmed in several paradigms. Thus, oxidative stress produced by glutathione depletion results in marked cell death in pre-OLs, while mature OLs are resistant. This vulnerability appears to be related to accumulation of free radicals in pre-OLs but not in mature cells, suggesting a defect in the ability of immature OLs to remove ROS. This inability to remove ROS is related to relative deficiencies of antioxidant enzymes [36, 37].

Oxygen Vulnerability and Developmental Brain Injury

Fluctuating environmental oxygen conditions play a substantial role in cellular and organismal respiration and evolution. The increasing dependency of many organisms on a constant supply of oxygen in order to function is associated with the development of molecular hypoxia-dependent pathways (i.e. hypoxia-inducible factor HIF-1). Physiological hypoxia plays an important role in cellular development, differentiation and angiogenesis.

Foetal development occurs under relative hypoxic conditions in utero (PaO_2 about 25 mmHg) compared to extrauterine conditions (PaO_2 70 mmHg). The switch from placenta-mediated to lung-mediated oxygen supply during birth is associated with a sudden rise of tissue oxygen tension that amounts to relative hyperoxia in preterm infants (Fig. 14.1). Hyperoxia, during development in rats, results in hypoxic chemosensitivity ablation, carotid body hypoplasia, and reduced chemoafferents. Normoxia after hypoxia (8 % O_2) increased ROS levels by 19.2 % with a further increase of 54.8 % by hyperoxia (95 % O_2) compared to normoxia. Electron microscopy demonstrated that hyperoxia damages the ultrastructure within petrosal ganglion neurons [38].

A large number of studies have confirmed the role of hyperoxia in the pathogenesis of prematurity-related diseases such as ROP and BPD. In recent years, there is mounting evidence that hyperoxia might also have deleterious effects on the immature brain. In term infants, there is an increasing body of experimental and clinical evidence on oxygen toxicity following reoxygenation after birth asphyxia, whereas for the preterm brain, data indicate that oxygen may also be toxic by itself.

In term infants suffering from birth asphyxia, resuscitation with high oxygen concentrations significantly increases mortality and morbidity [39]. In a newborn piglet model of global hypoxia/reoxygenation levels of the neurotrophic factor, BDNF was reduced when animals received 100 % oxygen for resuscitation. In the same model, supplementary oxygen increased lipid peroxidation in cortical neurons [40, 41]. In a neonatal rat model of hypoxic–ischaemic brain injury, resuscitation with 100 % oxygen counteracted the therapeutic effect of hypothermia [42].

In preterm infants, chronic or fluctuating exposure to supraphysiological oxygen levels compared to intrauterine conditions may cause encephalopathy of prematurity

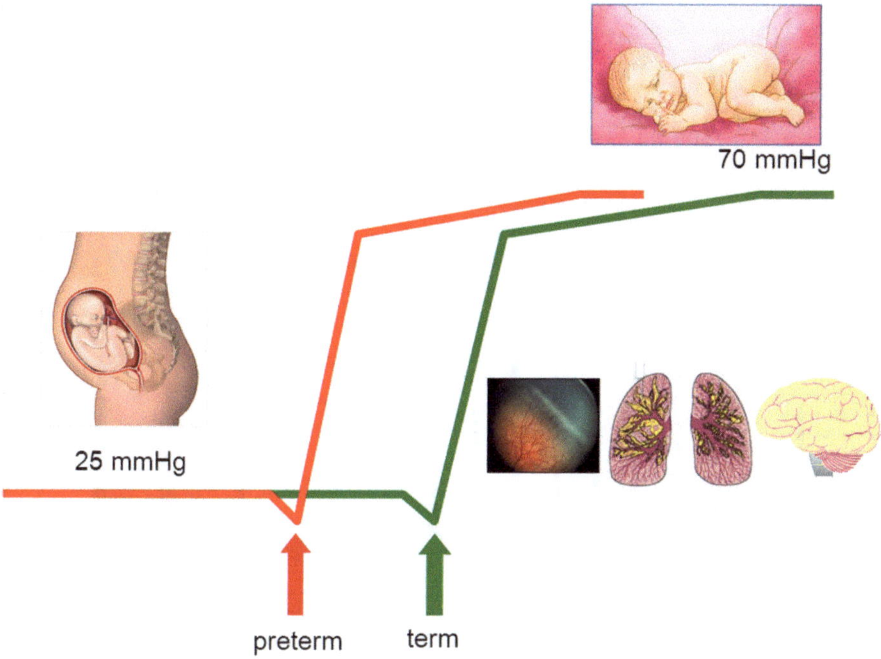

Fig. 14.1 Effect of the perinatal increase of oxygen on the developing infant

with cystic or diffuse periventricular leukomalacia [43, 44]. New insights may help refine the traditional view of how premature exposure to large amounts of oxygen in the neonatal period may disrupt brain maturation. Several animal models of perinatal brain damage have been used to unravel the underlying molecular and cellular mechanisms of oxygen-induced neurotoxicity.

In 2004, a preterm animal model of hyperoxia-induced developmental brain injury was established. Six-day-old rodents (P6) were used as a preterm experimental model since their maturational state of the brain corresponds well to the human infant in the last trimester of pregnancy. In rodents, physiological apoptosis is still ongoing at this age and myelination starts around P11 [45]. Experimental animals were subjected to increased oxygen levels of 85 % O_2 over a period of 6–24 h. This caused a widespread pattern of increased neuronal apoptosis when compared to physiological apoptosis typically seen at this developmental stage. Affected areas are the cortex, basal ganglia, hypothalamus, hippocampus and white matter tracts. Interestingly, oxygen sensitivity is maturation-dependent as 14-day-old rats are resistant to the effects of rapidly increasing oxygen supply; only analysis of the dentate gyrus of the hippocampus revealed areas of increased apoptotic cell death (Fig. 14.2).

Furthermore, autophagy as a self-degradation process that involves turnover and recycling of cytoplasmic components in healthy and diseased tissue seems to be

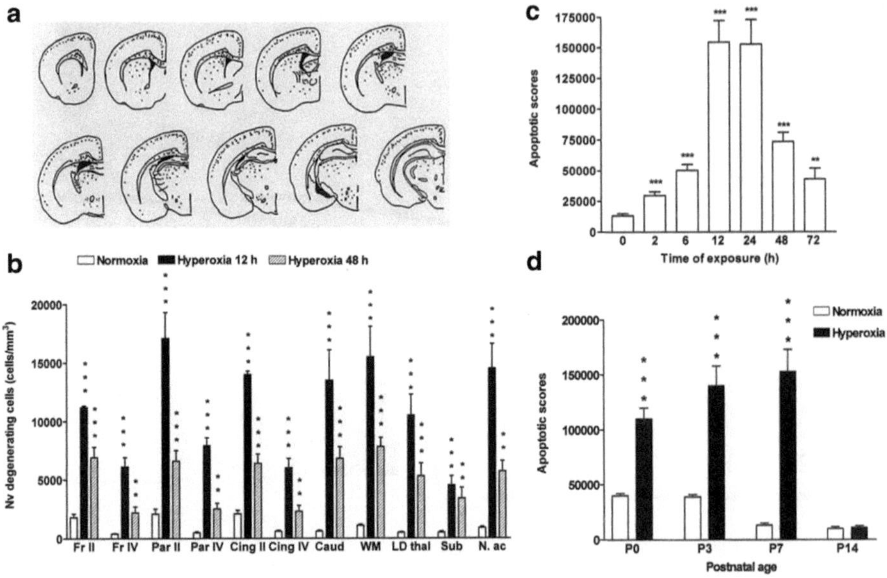

Fig. 14.2 *Panel A* Schematic illustration of the distribution pattern of hyperoxia-induced brain damage in infant rats. *Panel B* Numerical densities of degenerating cells in 11 brain regions in normoxic (*white columns*) rats and rats subjected to 12 h (*black columns*) or 48 h (*shaded columns*) of hyperoxia are shown. *Panel C* 2 h of exposure to 80 % O_2 triggered significant brain damage compared to normoxic littermates. The most severe damage was detected in the brains of rats subjected for 12 or 24 h to hyperoxia. No further increase could be detected by this method after longer exposure times, most likely because apoptotic cells had already been eliminated. *Panel D* shows the developmental vulnerability profile to hyperoxia. Vulnerability to hyperoxia subsided by P14 in the rat

involved in hyperoxia-induced developmental brain injury. Autophagy may be protective at the early stages of programmed cell death, but it can also promote apoptosis under certain conditions. An oxygen-exposure-dependent regulation of autophagy-related gene (Atg) proteins Atg3, Atg5 and Atg12, Beclin-1 and microtubule-associated protein 1 light chain 3 (LC3), LC3A-II and LC3B-II was observed in this experimental model [46].

In line with increased cell death observed in this model, treatment of immature rats with high doses of oxygen from birth during the first week of life results in a significant reduction of brain weight. Analysis of apoptotic pathways revealed receptor-mediated apoptosis as hyperoxia resulted in induction of Fas and its downstream signalling events, such as Fas-associated death-domain (FADD), the long and short form of FADD-like-IL-1β-converting enzyme (FLICE)-inhibitory protein (FLIP-L, FLIP-S) and cleavage of caspase-8 and caspase-3. Mice deficient in Fas (B6.MRL-*Tnfrsf6*^{lpr}) were protected against oxygen-mediated injury. Hyperoxia initiates intrinsic apoptotic pathways with upregulation of key proteins, namely, cytochrome c, apoptosis protease-activating factor-1 (Apaf-1) and the caspase-independent protein apoptosis-inducing factor (AIF), whereas members of the anti-apoptotic Bcl-2 family are downregulated. Cell death is also associated with decreased expression of neurotrophins and neurotrophin-regulated signalling pathways [47–50].

Even short exposures to non-physiological oxygen levels also change the balance of the ROS-dependent thioredoxin/peroxiredoxin system in the animal model described above. Oxygen toxicity significantly induced upregulation of peroxiredoxins 1 and 2, peroxiredoxin sulphonic form, thioredoxin 1 and sulfiredoxin 1 in the brains of immature rats. Hyperoxia reduced the level of DJ-1, a hydroperoxide-responsive protein in the developing rat brain [51].

Furthermore, hyperoxia exposure leads to nitrative stress, ensuing microvascular degeneration, diminished brain mass and neurophysiological function in immature rat pups. These effects are preceded by an upregulation of endothelial nitric oxide synthase (eNOS) in cerebral capillaries and a downregulation of Cu/Zn superoxide dismutase (SOD). A role for reactive nitrogen species in hyperoxic death is suggested by the observation that hyperoxia causes iNOS mRNA and protein upregulation in microglial cells and formation of nitrotyrosine in neurons of the immature rat brain [52].

In newborn piglets ventilated with 100 % oxygen for 4 h, levels of the neurotransmitters dopamine and 3,4-dihydroxyphenylacetic acid and serotonin were significantly lower after the end of hyperoxia in cortical areas. Several studies have shown that a small decrease in oxygen tension in the brain can cause a significant increase in extracellular dopamine concentrations, which may result from an increase in the releasable dopamine pool or from inhibition of reuptake. It has also been reported that hyperbaric oxygen therapy can dramatically inhibit cerebral monoamine oxidase B, which is responsible for the breakdown of dopamine [53]. Whether the observed alterations in the neurotransmitter production following a short time of oxygen exposure have long-term consequences remains unclear.

Since in the preterm brain diffuse periventricular leukomalacia (PVL) affecting the white matter is beside grey matter volume reduction the predominant type of injury, it can be hypothesised that oxygen exposure is also harmful to developing oligodendrocytes (OLs). Primary premyelinating oligodendrocytes (pre-OLs) are susceptible to hyperoxia-induced cell death via caspase-dependent pathways, whereas mature OLs are resistant.

Accumulation of superoxide and generation of reactive oxygen species (ROS) are detected as early as 2 h after oxygen exposure. Degeneration of pre-OLs is also seen in neonatal rats. P3 and P6 rat pups show bilateral reduction in MBP expression, whereas MBP expression in brains of P10 rats is similar in comparison to room-air-treated littermates [54, 55].

Despite apparent recovery in the glial population and in MBP levels 1 week after the hyperoxic insult, the disruption in oligodendroglia development and WM maturation during a critical period of vulnerability leads to long-term deficiencies in WM organisation and integrity as shown by a marked reduction in WM diffusivity on magnetic resonance imaging (MRI) in the adult mouse brain on P30 and P53. Decreased fractional anisotropy and increased radial diffusion coefficient are observed in the corpus callosum mice after neonatal hyperoxia on P6 compared to control mice. Furthermore, hippocampal and cerebellar volumes are reduced on MRI volumetry [56–58].

Hyperoxia does not affect survival or proliferation of astrocytes in vivo but modifies glial fibrillary acidic protein (GFAP) and glutamate–aspartate transporter (GLAST) expression indicating altered glutamate homeostasis. Furthermore, cultured astrocytes exposed to hyperoxia show a reduced capacity to protect oligodendrocyte progenitor cells (OPCs) against the toxic effects of exogenous glutamate. Neonatal hyperoxia also causes ultrastructural changes, including reduced myelin thickness, abnormal extramyelin loops and axonopathy. This disruption of axon–oligodendrocyte integrity results in the lasting impairment of conduction properties and might be responsible for neurodevelopmental disorders in prematurely born infants.

In a rodent model of intrauterine hypoxia, postnatal exposure to hyperoxia significantly potentiated the myelination delay and oligodendroglial dysmaturation induced by antenatal hypoxia. In a rat model (P7) of postnatal hypoxia–ischaemia followed by a 2 h exposure to 100 % hyperoxia magnetic resonance imaging showed deleterious effects for the developing grey and white matter. MRI including apparent diffusion coefficient (ADC) – maps, diffusion tensor imaging (DTI) and manganese enhancement indicating inflammation – showed lower fractional anisotropy (FA) values and increased diffusivity. These findings persisted from P21 until P42. On day 7 of life, increased manganese enhancement along with microglial activation was noted in the ipsilateral basal ganglia, thalamus and cortex, more pronounced in the hyperoxia group. These studies indicate that hyperoxia at birth should be avoided in preterm infants at risk of WMD [59–61].

Hyperoxia also generates inflammatory responses in the developing brain as demonstrated by a marked increase of mRNA and protein levels of caspase-1 and its downstream effectors Interleukin (IL)-1β, IL-18 and IL-18 receptor-α (IL-18Rα) in 6-day-old rodents. Intraperitoneal injection of recombinant human IL-18-binding protein (IL-18BP), a specific inhibitor of IL-18, attenuated hyperoxic brain injury. Mice deficient in IL-1 receptor-associated kinase-4 (IRAK-4), which is pivotal for both IL-1β and IL-18 signal transduction, are protected against oxygen-mediated neurotoxicity. These findings causally link inflammation through IL-1β and IL-18 to hyperoxia-induced cell death in the immature brain [62].

Recently, the hyperoxia model has been modified by addition of an inflammatory stimulus, which represents a clinically relevant problem in preterm brain injury. Intrauterine infection and inflammation are major reasons for preterm birth. Both infection/inflammation and hyperoxia have been shown to be involved in brain injury of preterm infants.

To explore the additive or synergistic effects of combined treatment with oxygen and inflammation, the influence of a systemic lipopolysaccharide (LPS) application on hyperoxia-induced white matter damage (WMD) in newborn rats was studied. Injection with lipopolysaccharide (LPS) as an inflammatory stimulus aggravates hyperoxia-induced damage due to microglial activation. Both noxious stimuli, hyperoxia and LPS caused hypomyelination and altered WM microstructure on diffusion tensor magnetic resonance imaging (DT-MRI). While hyperoxia induces cell death, LPS induces oligodendrocyte maturation arrest. Reduced expression of transcription factors controlling oligodendrocyte development and maturation further indicates oligodendrocyte maturation arrest [56, 63].

Analysis of molecular changes for disruption of myelination revealed altered CEACAM1 expression in the model of hyperoxia- and inflammation-induced encephalopathy of prematurity. Furthermore, primary oligodendrocytes stimulated with CEACAM1 show increased myelination. CEACAM1 is the founder molecule of the family of 'carcinoembryonic antigen-related cell adhesion molecules' and part of the immunoglobulin superfamily and is ontogenetically expressed in myelinating oligodendrocytes. Due to its role as a co-receptor to a variety of other receptors (e.g. Toll-like receptor 2, Toll-like receptor 4, T-cell receptor, B-cell receptor, epidermal growth factor receptor and vascular endothelial growth factor receptor) and its different isoforms, CEACAM1 is a multifunctional protein with an impact on proliferation and differentiation. Its function on other cell types in the context of hyperoxic brain injury remains to be investigated [64].

To identify further pathways engaged in a pathological modulation of maturation processes, analysis of acute (P7) and long-term (P14, P35) changes of the brain proteome in mice subjected to high oxygen levels for 12 h at postnatal day 6 revealed acute brain protein alterations. These indicate that vesicle trafficking (e.g. synapsin, pacsin), cell growth and differentiation (e.g. hnRNP, EF1), neuronal migration and axonal arborisation (e.g. TUC-2/4, GAP43, doublecortin) are affected. Late protein changes at P35 suggest long-term/chronic disruption of cytoskeletal organisation, intracellular transport, synaptic function and energy metabolism [65].

Clinical data confirm that higher oxygen levels in preterm infants may influence long-term neurological outcome. However, the aetiology for neurodevelopmental impairment in these patients has not yet been determined. Whether behavioural and cognitive deficits observed in ex-preterm infants at school age are related to hyperoxia remains unclear.

However, very few experimental studies have addressed this question: Behavioural testing in mice exposed to 48 h of hyperoxia revealed impaired motor activity in running wheels starting at adolescent age on postnatal day 30 (P30). Subsequently, from P44 to P53, adolescent mice had significantly higher values for maximum and mean velocity in regular wheels than controls. In complex running wheels, however, maximum velocity was decreased in animals after hyperoxia, as compared to controls. Hyperoxia causes hyperactivity and motor coordination deficits in adolescent and young adult mice. Testing of experimental animals exposed to 24 h of 85 % oxygen with the open field revealed increased anxiety and stress in hyperoxia-exposed animals.

Long-term exposure of newborn C57BL/6 mice to 85 % O_2 from postnatal days 1–14 revealed that hyperoxia-exposed mice performed poorly on the water maze and novel object recognition tests compared to air-exposed mice [57, 58], (Bendix I. pers. communication).

In conclusion, neonatal hyperoxia exposure in rodents leads to activation of various cell death cascades and impairment of maturation processes affecting all brain

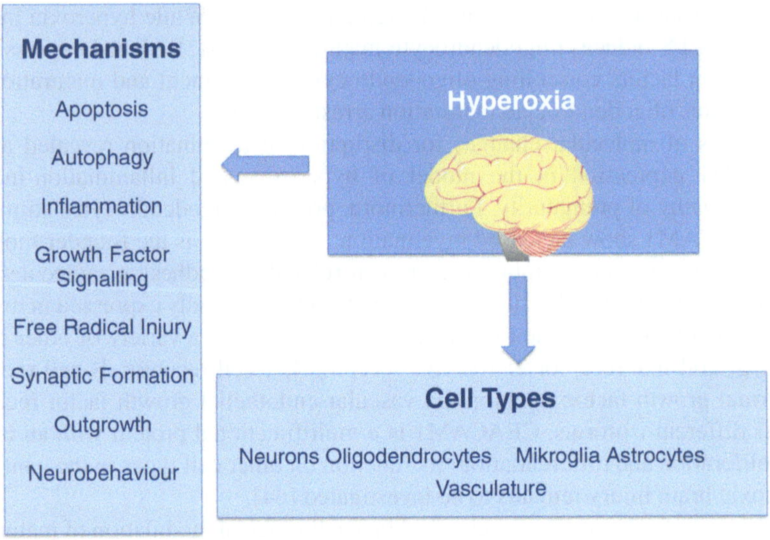

Fig. 14.3 Representative molecular pathways and cell types that may be affected by exposure to hyperoxia, as described in the book chapter

cell types. Abnormal neurobehaviour, primary deficits in spatial and recognition memory are also associated with smaller hippocampal sizes, similar to findings in ex-preterm infants at school age (Fig. 14.3).

Therapeutic Approaches

In the adult brain, ischaemic tolerance can be induced by hyperoxia and can protect against brain injury and neurodegenerative diseases. However, on the developing brain, oxygen has probably detrimental effects due to the immaturity of antioxidant capacities. Prevention and optimisation of the application of supplemental oxygen and the development of adjunctive therapies are highly warranted.

Despite its dangers and effects on the developing organisms, the use of supplemental oxygen cannot always be avoided in the care of preterm infants, sick neonates and young children. Thus, there is an urgent need to optimise oxygen therapy and to search for strategies that counteract oxygen toxicity.

In 2010, the new International Liaison Committee on Resuscitation and American Academy of Pediatrics guidelines for newborn resuscitation recommended starting a term or near-term newborn resuscitation on air rather than 100 % oxygen [66]. However, sparse clinical data exist regarding post-ischaemic oxygen supplementation and its effect on the developing central nervous system. Klinger et al. reported in 2005 an increased risk of poor outcome in asphyxiated babies with hyperoxaemia

($PaO_2 > 200$ mmHg) in the first 2 h of life [67]. However, the pathophysiology of asphyxia in combination with different levels of oxygenation may also be modulated by hypothermia, which is now recommended as the only proven neuroprotective therapy. In a clinical retrospective study, hyperoxaemia on admission was associated with a fourfold increased risk of moderate to severe HIE, and the incidence of HIE was directly related to the degree of hyperoxaemia. MRI abnormalities indicating HIE were present in 79 % of hyperoxaemic newborns and in 33 % of normoxic newborns, indicating that benefits of randomised studies may even underestimate the effect of cooling in normoxic infants [68, 69].

Furthermore, experimental findings support these clinical observations. In a neonatal rat model of hypoxic–ischaemic injury, reoxygenation with 100 % oxygen counteracted the neuroprotective effect of hypothermia. However, an early protective effect of hypothermia after hyperoxic, but not after normoxic reoxygenation, has been observed in a newborn piglet model of perinatal asphyxia [42, 70].

In search for adjuvant therapies in cases where oxygen cannot be avoided, it has to be taken into account that neonatal brain injury leads to damage which progresses of over a long period of time. For the term brain suffering from neonatal asphyxia, neuroprotective studies aiming at improving the effect the hypothermia are highly warranted. In an experimental study using a hypoxia–ischaemia piglet model, the intravenous application of 5 mg/kg/h melatonin, which has antioxidative properties, 10 min after the end of transient hypoxia–ischaemia over 6 h and repeated at 24 h augments hypothermic neuroprotection.

Findings were based on restoration of proton spectroscopy ^{1}H MRS (ratios deep grey matter lactate/N-acetyl aspartate and lactate/creatine) and also on phosphorus spectroscopy ^{31}P MRS biomarkers (whole ATP ^{31}P MRS NTP/exchangeable phosphate pool), known to change after hypoxia–ischaemia in the piglet.

Correlating with an improved cerebral energy metabolism, apoptotic TUNEL-positive nuclei were reduced in the melatonin-augmented hypothermia group in the thalamus, white matter, internal capsule, putamen and caudate. However, the effects of resuscitation with different oxygen levels were not investigated in this setting [71].

For preterm infants, new evidence is emerging to address the continued uncertainty regarding the optimal range of oxygen saturation levels. Between 2005 and 2007, five randomised trials were conducted in order to identify the optimum saturation targets for extremely preterm infants. All trials examined the efficacy and safety of supplemental oxygen to target arterial oxygen saturations of 85–89 % compared with 91–95 %. In the US Surfactant Positive Airway Pressure and Pulse Oximetry Trial (SUPPORT), severe retinopathy was reduced but mortality increased in the lower saturation target group. In the SUPPORT and also in the Canadian Oxygen Trial (COT), there was no difference between the two groups in the combined outcome of death or neurodevelopmental impairment at 18 months [72–74].

However, neurodevelopmental testing at the corrected age of 18 months may not be predictive for cognitive, behavioural and neuropsychiatric development frequently observed in later life of extremely preterm infants. These impairments are likely to be attributed to disturbance of maturational processes such as formation of synapses and myelination rather than large amounts of cell death.

Besides oxygen toxicity, there is a variety of consecutive insults affecting brain development at different time points responsible for injury and adverse neurodevelopmental outcome in preterm infants. Classically, therapies have targeted individual pathways during early phases of injury, but regenerative therapies such as growth factors may also enhance cell proliferation, differentiation and migration over time. Possibly, combined treatment following both strategies may reduce cell death and enhance repair mechanisms and/or formation of new cells.

For the preterm brain, there are no clinical data for neuroprotective strategies in the context of supplemental oxygen treatment available. The oxygen saturation target trials have not revealed any positive effects on neurodevelopmental outcome at the corrected age of 18 months for the low oxygen saturation group. However, data on the long-term consequences of restrictive oxygen supplementation until school age are still missing.

There is a small number of ongoing studies in preterm infants using compounds that may also be useful in the treatment of hyperoxia-induced brain injury such as recombinant erythropoietin (rEpo) and melatonin.

In the following, the current experimental evidence for therapeutic approaches in hyperoxia-induced brain injury will be reviewed. The major interest in recent years concentrated on the evaluation of neuroprotective strategies potentially applicable in young infants.

Studies from adult and also developmental brain injury models suggest similar molecular mechanisms involved in hyperoxia-induced injury to the immature brain, which potentially can be targeted by rEpo. This compound has a long track record of use in preterm infants to prevent anaemia of prematurity and has been approved by the US Food and Drug Administration for this clinical use in its recombinant form. Erythropoiesis was considered originally to be the sole physiological action of rEpo. This premise was changed through the knowledge that rEpo and its receptor are expressed in several organs including the central nervous system and the subsequent discovery of its neuroprotective properties in ischaemic stroke, traumatic brain injury, spinal cord injury and perinatal asphyxia. Recent research on the immature brain has identified numerous pathways influenced by this hormone, many of which are of significance for neuroprotective effects on developing brain tissue. However, larger doses of rEpo are required to obtain the desired neuroprotective effect [75].

In the rodent model of hyperoxia-induced brain damage at P6, a single treatment with 20,000 IE rEpo induced a significant reduction of the extent of apoptotic cell death and proapoptotic factors in the cortex, basal ganglia and white matter. In parallel, rEpo inhibited most brain proteome changes observed when hyperoxia was applied exclusively as demonstrated by two-dimensional electrophoresis and mass spectrometry. Analysis of its molecular mode of action suggests that rEpo generates its protective effect against oxygen toxicity through a reduction of oxidative stress, decreased expression of proinflammatory mediators (i.e. IL-1β and IL-18) and matrix metalloproteinases and also by limiting the stressor-inducible changes in both hemeoxygenase-1 and cholinergic functions. rEpo treatment results in a restoration of hyperoxia-induced increased levels of caspases and decreased levels of neurotrophins. However, proteome alterations detected following a single treatment also raise the question whether rEpo is merely beneficial to the immature brain [65].

The female hormone oestrogen (E2) has also neuroprotective properties in models of in vitro and in vivo neurodegeneration in the adult and also in the developing brain. These properties result from activation of oestrogen receptors and crosstalking with intracellular signalling pathways that are also activated by neurotrophins, i.e. the ERK1/2- and PI3-AKT-pathways. In addition, antioxidant properties and modulation of nitric oxide synthase have been assigned to the 17β-estradiol molecule. E2 also has anti-inflammatory properties by reducing microglial activation and the iNOS-mediated immune response as well as the production of several proinflammatory mediators including metalloproteinase, prostaglandin E2 (PGE2) and cyclooxygenase 2 (COX-2). Furthermore, profound effects on the function and plasticity of the brain and proliferation, differentiation and migration of neurons are controlled by E2 [76, 77].

During the last trimester of pregnancy, E2 plasma levels increase up to 15,000 pg/mL in the placenta. Both the mother and foetus are exposed to the same increasing levels. At birth, the levels of E2 decrease by a factor of 100 within 24 h in the mother (150 pg/mL) and by a factor of 1,000 in the neonate (15 pg/mL). Premature infants experience this hormone deprivation and simultaneous increase of the oxygen tissue tension much earlier than infants born at term [78].

In primary oligodendrocytes, the rat hyperoxia model, a single intraperitoneal injection of E2 provided significant dose-dependent protection against oxygen-induced apoptotic cell death. Treatment with E2 prevented hyperoxia-induced proapoptotic Fas upregulation and caspase-3 activation. Finally, E2 antagonises hyperoxia-induced inactivation of ERK1/2 and Akt, essential kinases of the MAPK and PI3-kinase pathways promoting cell survival, respectively.

E2 replacement therapy in extremely low-birthweight infants has been introduced in some centres with the goal to improve bone mineralisation, and no adverse side effects have been observed so far. Therefore, maintaining placental E2 plasma levels may be effective to protect neonates from brain injury [79, 80].

However, there was no effect on the prevention of death or development of bronchopulmonary dysplasia [81, 82]. More data are urgently needed concerning the safety and feasibility of oestrogen supplementation in the neonatal period.

Experimentally, inhibition of key players of the apoptotic cascade such as caspases appears a promising strategy for neuroprotection. Beside the receptor-mediated extrinsic apoptotic pathway, hyperoxia-mediated neurodegeneration in the developing brain is supported by intrinsic apoptosis, suggesting that the development of highly selective caspase inhibitors will represent a potential useful therapeutic strategy in prematurely born infants.

Injection of the selective caspase-8 inhibitor (TRP801), a downstream effector caspase in the receptor-mediated apoptotic pathway, at the beginning of hyperoxia, blocked subsequent caspase-3 cleavage and conferred neuroprotection in the hyperoxia model.

Elevated oxygen levels also trigger a marked increase in active caspase-2 expression, resulting in an initiation of the intrinsic apoptotic pathway involving the mitochondrial route with upregulation of key proteins, namely, cytochrome c, Apaf-1 and AIF. Single treatment with TRP601 at the beginning of hyperoxia reversed the detrimental effects in this model. Hyperoxia-mediated neurodegeneration is supported by intrinsic apoptosis, suggesting that the development of highly selective caspase inhibitors will represent a potential useful therapeutic strategy in prematurely born infants [47, 83].

Conclusions

Hyperoxia causes oxidative stress and contributes to the pathogenesis of injury in the preterm as well as in the term brain. During the critical time period of development, the immature central nervous system is particularly vulnerable to this type of stress. From current experimental evidence, it may be hypothesised that oxygen causes cell death and profoundly alters maturational processes. Multiple cell types such as neurons, oligodendrocytes, astrocytes, microglia and also the immature vasculature are affected. Behavioural studies in animals revealed changes such as motor hyperactivity but also cognitive impairment similar to those observed in ex-preterm infants at school age.

Furthermore, characteristic MRI findings in hyperoxia-exposed rodents such as reduced hippocampal size and white matter abnormalities resemble MR images of prematurely born infants at term.

Therapeutic efforts aiming at defining the optimum oxygen saturation and the development of adequate monitoring systems are highly warranted. Furthermore, in situations where oxygen supplementation cannot be avoided, the development of adjunctive therapies is a major challenge for current experimental research.

References

1. Kusuda S, et al. Trends in morbidity and mortality among very-low-birth-weight infants from 2003 to 2008 in Japan. Pediatr Res. 2012;72(5):531–8.
2. Stoll BJ, et al. Neonatal outcomes of extremely preterm infants from the NICHD Neonatal Research Network. Pediatrics. 2010;126(3):443–56.
3. Buchmann EJ, Pattinson RC, Nyathikazi N. Intrapartum-related birth asphyxia in South Africa – lessons from the first national perinatal care survey. S Afr Med J. 2002;92(11):897–901.
4. Thornberg E, et al. Birth asphyxia: incidence, clinical course and outcome in a Swedish population. Acta Paediatr. 1995;84(8):927–32.
5. Martin JA, et al. Births: final data for 2005. Natl Vital Stat Rep. 2007;56(6):1–103.
6. Ananth CV, et al. Trends in preterm birth and perinatal mortality among singletons: United States, 1989 through 2000. Obstet Gynecol. 2005;105(5 Pt 1):1084–91.
7. Marlow N, et al. Neurologic and developmental disability at six years of age after extremely preterm birth. N Engl J Med. 2005;352(1):9–19.

8. Hack M, et al. School-age outcomes in children with birth weights under 750 g. N Engl J Med. 1994;331(12):753–9.
9. Wood NS, et al. The EPICure study: associations and antecedents of neurological and developmental disability at 30 months of age following extremely preterm birth. Arch Dis Child Fetal Neonatal Ed. 2005;90(2):F134–40.
10. Doyle LW, Anderson PJ. Adult outcome of extremely preterm infants. Pediatrics. 2010; 126(2):342–51.
11. Kaindl AM, Favrais G, Gressens P. Molecular mechanisms involved in injury to the preterm brain. J Child Neurol. 2009;24(9):1112–18.
12. Brown NC, et al. Neurobehavior at term and white and gray matter abnormalities in very preterm infants. J Pediatr. 2009;155(1):32–8. 38.e1.
13. Ferriero DM. Neonatal brain injury. N Engl J Med. 2004;351(19):1985–95.
14. Girard S, et al. Role of perinatal inflammation in cerebral palsy. Pediatr Neurol. 2009;40(3): 168–74.
15. Ashton N, Ward B, Serpell G. Role of oxygen in the genesis of retrolental fibroplasia; a preliminary report. Br J Ophthalmol. 1953;37(9):513–20.
16. Campbell K. Intensive oxygen therapy as a possible cause of retrolental fibroplasia; a clinical approach. Med J Aust. 1951;2(2):48–50.
17. Ikonomidou C, Kaindl AM. Neuronal death and oxidative stress in the developing brain. Antioxid Redox Signal. 2011;14(8):1535–50.
18. Rousset CI, et al. Mitochondria and perinatal brain injury. J Matern Fetal Neonatal Med. 2012;25 Suppl 1:35–8.
19. Bratton SB, Salvesen GS. Regulation of the Apaf-1-caspase-9 apoptosome. J Cell Sci. 2010; 123(Pt 19):3209–14.
20. Reiss A, et al. Bacterial pore-forming cytolysins induce neuronal damage in a rat model of neonatal meningitis. J Infect Dis. 2011;203(3):393–400.
21. Li Z, Sheng M. Caspases in synaptic plasticity. Mol Brain. 2012;5:15.
22. Balduini W, Carloni S, Buonocore G. Autophagy in hypoxia-ischemia induced brain injury. J Matern Fetal Neonatal Med. 2012;25 Suppl 1:30–4.
23. Maiuri MC, et al. Self-eating and self-killing: crosstalk between autophagy and apoptosis. Nat Rev Mol Cell Biol. 2007;8(9):741–52.
24. Dammann O, et al. Lung and brain damage in preterm newborns. Are they related? How? Why? Biol Neonate. 2004;85(4):305–13.
25. Volpe JJ. Brain injury in premature infants: a complex amalgam of destructive and developmental disturbances. Lancet Neurol. 2009;8(1):110–24.
26. Czeh M, Gressens P, Kaindl AM. The yin and yang of microglia. Dev Neurosci. 2011; 33(3–4):199–209.
27. Kaur C, Ling EA. Periventricular white matter damage in the hypoxic neonatal brain: role of microglial cells. Prog Neurobiol. 2009;87(4):264–80.
28. Lee J, Giordano S, Zhang J. Autophagy, mitochondria and oxidative stress: cross-talk and redox signalling. Biochem J. 2012;441(2):523–40.
29. Halliwell B. Reactive oxygen species and the central nervous system. J Neurochem. 1992;59(5):1609–23.
30. Dreiem A, Gertz CC, Seegal RF. The effects of methylmercury on mitochondrial function and reactive oxygen species formation in rat striatal synaptosomes are age-dependent. Toxicol Sci. 2005;87(1):156–62.
31. Kim SM, et al. Fos and Jun potentiate individual release sites and mobilize the reserve synaptic vesicle pool at the Drosophila larval motor synapse. Proc Natl Acad Sci U S A. 2009;106(10): 4000–5.
32. Mast JD, et al. Reactive oxygen species act remotely to cause synapse loss in a Drosophila model of developmental mitochondrial encephalopathy. Development. 2008;135(15):2669–79.
33. Milton VJ, Sweeney ST. Oxidative stress in synapse development and function. Dev Neurobiol. 2012;72(1):100–10.

34. Davis JM, Auten RL. Maturation of the antioxidant system and the effects on preterm birth. Semin Fetal Neonatal Med. 2010;15(4):191–5.
35. Shim SY, Kim HS. Oxidative stress and the antioxidant enzyme system in the developing brain. Korean J Pediatr. 2013;56(3):107–11.
36. Baud O, et al. Glutathione peroxidase-catalase cooperativity is required for resistance to hydrogen peroxide by mature rat oligodendrocytes. J Neurosci. 2004;24(7):1531–40.
37. Baud O, et al. Developmental up-regulation of MnSOD in rat oligodendrocytes confers protection against oxidative injury. Eur J Neurosci. 2004;20(1):29–40.
38. Kwak DJ, Kwak SD, Gauda EB. The effect of hyperoxia on reactive oxygen species (ROS) in rat petrosal ganglion neurons during development using organotypic slices. Pediatr Res. 2006;60(4):371–6.
39. Vento M, Saugstad OD. Oxygen as a therapeutic agent in neonatology: a comprehensive approach. Semin Fetal Neonatal Med. 2010;15(4):185.
40. Solberg R, et al. Resuscitation of newborn piglets. short-term influence of FiO_2 on matrix metalloproteinases, caspase-3 and BDNF. PLoS One. 2010;5(12):e14261.
41. Solberg R, et al. Resuscitation with supplementary oxygen induces oxidative injury in the cerebral cortex. Free Radic Biol Med. 2012;53(5):1061–7.
42. Dalen ML, et al. Resuscitation with 100% oxygen increases injury and counteracts the neuroprotective effect of therapeutic hypothermia in the neonatal rat. Pediatr Res. 2012;71(3): 247–52.
43. Collins MP, et al. Hypocapnia and other ventilation-related risk factors for cerebral palsy in low birth weight infants. Pediatr Res. 2001;50(6):712–19.
44. Deulofeut R, Dudell G, Sola A. Treatment-by-gender effect when aiming to avoid hyperoxia in preterm infants in the NICU. Acta Paediatr. 2007;96(7):990–4.
45. Dobbing J, Sands J. Comparative aspects of the brain growth spurt. Early Hum Dev. 1979;3(1):79–83.
46. Bendix I, et al. Erythropoietin modulates autophagy signaling in the developing rat brain in an in vivo model of oxygen-toxicity. Int J Mol Sci. 2012;13(10):12939–51.
47. Dzietko M, et al. A critical role for Fas/CD-95 dependent signaling pathways in the pathogenesis of hyperoxia-induced brain injury. Ann Neurol. 2008;64(6):664–73.
48. Felderhoff-Mueser U, et al. Oxygen causes cell death in the developing brain. Neurobiol Dis. 2004;17(2):273–82.
49. Sifringer M, et al. Erythropoietin attenuates hyperoxia-induced oxidative stress in the developing rat brain. Brain Behav Immun. 2010;24(5):792–9.
50. Yis U, et al. Hyperoxic exposure leads to cell death in the developing brain. Brain Dev. 2008;30(9):556–62.
51. Bendix I, et al. Hyperoxia changes the balance of the thioredoxin/peroxiredoxin system in the neonatal rat brain. Brain Res. 2012;1484:68–75.
52. Hoehn T, et al. Hyperoxia causes inducible nitric oxide synthase-mediated cellular damage to the immature rat brain. Pediatr Res. 2003;54(2):179–84.
53. Tsao PC, et al. Cranial neurotransmitter alteration in newborn piglets exposed to oxygen. J Chin Med Assoc. 2012;75(9):449–53.
54. Gerstner B, et al. Maturation-dependent oligodendrocyte apoptosis caused by hyperoxia. J Neurosci Res. 2006;84(2):306–15.
55. Gerstner B, et al. Hyperoxia causes maturation-dependent cell death in the developing white matter. J Neurosci. 2008;28(5):1236–45.
56. Brehmer F, et al. Interaction of inflammation and hyperoxia in a rat model of neonatal white matter damage. PLoS One. 2012;7(11):e49023.
57. Ramani M, et al. Neurodevelopmental impairment following neonatal hyperoxia in the mouse. Neurobiol Dis. 2013;50:69–75.
58. Schmitz T, et al. Adolescent hyperactivity and impaired coordination after neonatal hyperoxia. Exp Neurol. 2012;235(1):374–9.
59. Morken TS, et al. Longitudinal diffusion tensor and manganese-enhanced MRI detect delayed cerebral gray and white matter injury after hypoxia-ischemia and hyperoxia. Pediatr Res. 2013;73(2):171–9.

60. Ritter J, et al. Neonatal hyperoxia exposure disrupts axon-oligodendrocyte integrity in the subcortical white matter. J Neurosci. 2013;33(21):8990–9002.
61. Vottier G, et al. Deleterious effect of hyperoxia at birth on white matter damage in the newborn rat. Dev Neurosci. 2011;33(3–4):261–9.
62. Felderhoff-Mueser U, et al. Caspase-1-processed interleukins in hyperoxia-induced cell death in the developing brain. Ann Neurol. 2005;57(1):50–9.
63. Favrais G, et al. Systemic inflammation disrupts the developmental program of white matter. Ann Neurol. 2011;70(4):550–65.
64. Prager S, et al. CEACAM1 expression in oligodendrocytes of the developing rat brain shows a spatiotemporal relation to myelination and is altered in a model of encephalopathy of prematurity. Dev Neurosci. 2013;35(2–3):226–40.
65. Kaindl AM, et al. Erythropoietin protects the developing brain from hyperoxia-induced cell death and proteome changes. Ann Neurol. 2008;64(5):523–34.
66. Perlman JM, et al. Part 11: neonatal resuscitation: 2010 international consensus on cardiopulmonary resuscitation and emergency cardiovascular care science with treatment recommendations. Circulation. 2010;122(16 Suppl 2):S516–38.
67. Klinger G, et al. Do hyperoxaemia and hypocapnia add to the risk of brain injury after intrapartum asphyxia? Arch Dis Child Fetal Neonatal Ed. 2005;90(1):F49–52.
68. Kapadia VS, et al. Perinatal asphyxia with hyperoxemia within the first hour of life is associated with moderate to severe hypoxic-ischemic encephalopathy. J Pediatr. 2013;163(4):949–54.
69. Sabir H, et al. Increased inspired oxygen in the first hours of life is associated with adverse outcome in newborns treated for perinatal asphyxia with therapeutic hypothermia. J Pediatr. 2012;161(3):409–16.
70. Dalen ML, et al. Early protective effect of hypothermia in newborn pigs after hyperoxic, but not after normoxic, reoxygenation. J Perinat Med. 2010;38(5):545–56.
71. Robertson NJ, et al. Melatonin augments hypothermic neuroprotection in a perinatal asphyxia model. Brain. 2013;136(Pt 1):90–105.
72. SUPPORT Study Group of the Eunice Kennedy Shriver NICHD Neonatal Research Network, et al. Target ranges of oxygen saturation in extremely preterm infants. N Engl J Med. 2010; 362(21):1959–69.
73. Saugstad OD, Aune D. In search of the optimal oxygen saturation for extremely low birth weight infants: a systematic review and meta-analysis. Neonatology. 2011;100(1):1–8.
74. Schmidt B, et al. Effects of targeting higher vs lower arterial oxygen saturations on death or disability in extremely preterm infants: a randomized clinical trial. JAMA. 2013;309(20):2111–20.
75. Juul S, Felderhoff-Mueser U. Epo and other hematopoietic factors. Semin Fetal Neonatal Med. 2007;12(4):250–8.
76. Asimiadou S, et al. Protection with estradiol in developmental models of apoptotic neurodegeneration. Ann Neurol. 2005;58(2):266–76.
77. Liu M, et al. Neuroprotection of sex steroids. Minerva Endocrinol. 2010;35(2):127–43.
78. Tulchinsky D, et al. Plasma estrone, estradiol, estriol, progesterone, and 17-hydroxyprogesterone in human pregnancy. I. Normal pregnancy. Am J Obstet Gynecol. 1972;112(8):1095–100.
79. Gerstner B, et al. 17beta-estradiol protects against hypoxic/ischemic white matter damage in the neonatal rat brain. J Neurosci Res. 2009;87(9):2078–86.
80. Gerstner B, et al. Estradiol attenuates hyperoxia-induced cell death in the developing white matter. Ann Neurol. 2007;61(6):562–73.
81. Trotter A, et al. Follow-up examination at the age of 15 months of extremely preterm infants after postnatal estradiol and progesterone replacement. J Clin Endocrinol Metab. 2001;86(2): 601–3.
82. Trotter A, et al. Effect of oestradiol and progesterone replacement on bronchopulmonary dysplasia in extremely preterm infants. Arch Dis Child Fetal Neonatal Ed. 2007;92(2):F94–8.
83. Sifringer M, et al. Prevention of neonatal oxygen-induced brain damage by reduction of intrinsic apoptosis. Cell Death Dis. 2012;3:e250.

Chapter 15
Hyperbilirubinemia and Antioxidant Defenses in the Neonate

Roland Stocker

This chapter focuses on the role of bilirubin as an antioxidant in the setting of neonatal hyperbilirubinemia. Originally, Dr. Antony F. McDonagh agreed to be the author. Unfortunately, however, Tony's unexpected early passing necessitated a change in authorship. More importantly, it also has left the bilirubin research community with a tremendous personal loss and an irreplaceable gap in wisdom about the "yellow pigment." I am humbled to have been asked to step in for a truly outstanding leader of the field, who introduced me to the world of bilirubin in the context of learning more about its antioxidant properties. In accepting this challenge, it is appropriate to dedicate this chapter to Tony McDonagh, as I try to encapsulate some of his views on the complexities and properties of bilirubin.

Formation and Metabolism of Bilirubin

Bilirubin exists in multiple structures, all of which are derived from the catabolism of heme catalyzed by the enzyme heme oxygenase (HO) [1]. Heme itself is derived predominantly from hemoglobin released by aged or damaged red blood cells (RBCs) that are removed by macrophages in the spleen, liver, and bone marrow.

R. Stocker, Ph.D. (✉)
Vascular Biology Division, Victor Chang Cardiac Research Institute,
Lowy Packer Building, 405 Liverpool Street, Darlinghurst, NSW 2010, Australia

School of Medical Sciences, Faculty of Medicine,
University of New South Wales, Kensington, Australia
e-mail: r.stocker@victorchang.edu.au

P.A. Dennery et al. (eds.), *Perinatal and Prenatal Disorders*,
Oxidative Stress in Applied Basic Research and Clinical Practice,
DOI 10.1007/978-1-4939-1405-0_15

HO cleaves the α-*meso* carbon of heme in three successive oxidation reactions during which stoichiometric amounts of carbon monoxide, ferrous iron, and biliverdin IXα are released sequentially [2]. The biliverdin formed is commonly considered a transient intermediate that does not accumulate to appreciable amounts due to its efficient reduction to 4*Z*,15*Z*-bilirubin IXα by cytosolic biliverdin reductase (see, however, Ref. [3]). The bilirubin formed is then released into the circulation where it exists as a reversible complex with albumin. While some uncertainties remain about the precise nature of the binding site of 4*Z*,15*Z*-bilirubin IXα on albumin [4–6], it is clear that most bilirubin in circulation is bound to albumin, with only a small fraction remaining unbound [7]. It is the concentration of this "free" bilirubin that is generally believed to be responsible for the neurotoxic effects of bilirubin in jaundiced neonates [8]. "Free" bilirubin is also a likely source of the pigment for cells exposed to circulating albumin, i.e., RBCs, white blood cells, and endothelial cells. Circulating bilirubin is transported to the liver where it becomes a substrate of uridine diphosphoglucuronosyl transferase 1A1 (UGT1A1). The resulting mono- and diglucuronides of bilirubin are then transported into bile by the apical ATP-dependent transporter ABCC2, also known as multidrug-resistant protein 2 [9].

In situations of deficient or decreased glucuronidation, as in neonates or in those with Gilbert or Crigler–Najjar syndrome, unconjugated bilirubin accumulates in the circulation and extravascular space. In conditions of impaired biliary excretion of bilirubin mono- and diglucuronides, conjugated forms of bilirubin accumulate in the circulation where they undergo a number of modifications. These chemical modifications include hydrolysis to unconjugated bilirubin and covalent binding to albumin, with the latter giving rise to the so-called δ-bilirubin [10]. Similarly, circulating unconjugated bilirubin undergoes a complex set of light-induced conversions when babies (and adults) are exposed to sunlight, ambient white room light, or phototherapy [11]. Photoisomers, including 4*Z*,15*E*-bilirubin IXα, are the major products resulting from such light-induced "rearrangements." These bilirubin photoisomers also form reversible complexes with albumin, but they no longer require glucuronidation for excretion due to their increased polarity and hence solubility in aqueous fluids. Considering their physicochemical properties, photoisomers of bilirubin likely have a decreased ability to cross the blood–brain barrier than the predominant 4*Z*,15*Z*-bilirubin IXα may be less toxic, although this requires unambiguous verification [12].

Neonatal Hyperbilirubinemia

In their first week of life, most infants develop jaundice or hyperbilirubinemia. In healthy term infants, this "physiological jaundice" usually peaks at around 100 μmol/L (5.85 mg/dL) total serum bilirubin between the third and fifth days of life [13, 14] and then declines to normal values of around 13 μmol/L (0.8 mg/dL).

Physiological jaundice in neonates arises from a combination of increased RBC turnover due to the decreased life span of infant compared with adult erythrocytes [15] and the low activity of UGT1A1 [16]. As a result, bilirubin is effectively the only species of bile pigment present in the blood of neonates. Depending on the light source and extent of light exposure to the neonate, up to 20–30 % of this bilirubin can be present as 4*Z*,15*E*-bilirubin IXα [17].

Protective Properties of Bilirubin

It has been puzzling why humans reduce the comparatively nontoxic biliverdin to the potentially neurotoxic bilirubin, which also requires glucuronidation for its elimination (see above). However, some 26 years ago, bilirubin was reported to have potent antioxidant activities. Specifically, at physiologically relevant low oxygen concentrations, the pigment was shown to protect membrane lipids from peroxyl radical-mediated oxidation more effectively than equimolar concentrations of α-tocopherol (α-TOH) [18]. α-Tocopherol is the most active form of vitamin E and generally considered to be the most important fat-soluble antioxidant in human blood plasma and RBCs [19]. Following this landmark publication, work from different laboratories has confirmed and supported the antioxidant activity of bilirubin (Table 15.1). In addition, potential beneficial roles other than antioxidant activities also have been reported for bilirubin (Table 15.1), although the mechanisms by which the pigment exerts such putative benefits remain largely unclear.

Table 15.1 Protective properties reported for bilirubin

Reported protective activities	Supporting references
Antioxidant activity	
Reaction with different reactive oxygen and nitrogen species	Reviewed in Ref. [20]
Inhibition of lipid oxidation	[18, 21–26]
Inhibition of lipoprotein oxidation	[27–33]
Inhibition of protein oxidation	[34–36]
Inhibition of NADPH oxidase	[37–39]
Protection of cells against oxidant-induced damage	[40–42]
Other protective activities	
Anti-inflammatory activities	[43]
Modulation of T-regulatory-cell differentiation	[44]
Protection against experimental endotoxemia	[45]
Protection against ischemia/reperfusion injury	[46, 47]
Inhibition of complement activation	[48]
Inhibition of inflammation and tissue damage in experimental pulmonary fibrosis	[49]

Antioxidants

Different reactive oxygen and nitrogen species (ROS and RNS, respectively) are required for normal cellular homeostasis and biological processes, including antimicrobial defenses by the immune system, vascular homeostasis, normal cell growth, DNA replication, and cell signaling [50]. However, when these reactive species are present in excess or are lacking, homeostasis is disturbed and physiological functions become impaired [50]. The steady-state concentration of different ROS and RNS is governed by both their rates of formation and removal. Antioxidants, commonly defined as a substance that, when present at a low concentration compared with that of an oxidizable substrate, inhibits oxidation of the substrate [51], are the primary cause for the removal of oxidants. A broader definition of antioxidants also includes enzymes that repair or facilitate the removal of oxidized macromolecules, to reestablish or maintain normal redox balance.

Humans possess a complex array of antioxidant defenses that may be divided into different categories of proteinaceous and non-proteinaceous compounds [51]. Proteinaceous compounds include: (1) enzymes that scavenge, degrade, or metabolize reactive species; (2) enzymes that repair oxidized macromolecules; and (3) proteins that prevent transition metals or species containing a redox-active metal from participating in inadvertent oxidation reactions. Non-proteinaceous antioxidants include low-molecular-weight compounds either derived from the diet (e.g., vitamins C and E) or formed during normal metabolism (e.g., glutathione, coenzyme Q, and bilirubin). It is increasingly recognized that antioxidant defenses are complex, vary in composition, and often act in a temporal and spatially restricted manner [52]. For example, cellular signaling by hydrogen peroxide (H_2O_2) induced by growth factors relies on a temporal inactivation of the H_2O_2-degrading enzyme peroxiredoxin-1 at the plasma membrane, allowing for a transient accumulation of H_2O_2 to participate in appropriate cell signaling [53]. The variation in composition can be exemplified by the differences in antioxidant defenses within and outside cells. Thus, while cellular defenses rely primarily on enzymatic antioxidants and glutathione, the defenses in human blood plasma rely more heavily on nonenzymatic proteinaceous antioxidants related to the sequestration of pro-oxidant transition metals as well as non-proteinaceous antioxidants (Table 15.2).

Circulating blood cells (and perhaps also endothelial cells) directly and indirectly contribute to the antioxidant defense of human plasma. For example, white blood cells effectively take up oxidized vitamin C and reduce it to the antioxidant active form, ascorbate [54], some of which may then be secreted back into the extracellular space. Also, erythrocytes can scavenge superoxide anion radical ($O_2^{-\bullet}$), and they effectively detoxify plasma H_2O_2 via catalase [55] and peroxiredoxin-2 [56, 57]. In addition, hemoglobin within intact RBCs binds and oxidizes nitric oxide ($^{\bullet}NO$), and it can convert nitrite (NO_2^-) to $^{\bullet}NO$ [58].

Table 15.2 Classification of major antioxidants defenses

	Human plasma (extracellular)	Cellular
Enzymes	Extracellular superoxide dismutase (SOD3)	Superoxide dismutases (SOD1, SOD2); peroxiredoxins; catalase; glutathione peroxidases, reductase, and transferases; thioredoxin and thioredoxin reductase; glutaredoxin and glutaredoxin reductase; methionine sulfoxide reductases; protein disulfide isomerase; heme oxygenases; phospholipase A_2
Proteins	Albumin; transferrin; lactoferrin; ceruloplasmin; haptoglobin; hemopexin	Ferritins; metallothioneins
Non-proteinaceous compounds	*Endogenous*: albumin-bound bilirubin; ubiquinol-10; urate; α-keto acids	*Endogenous*: reduced glutathione (GSH); ubiquinol-10; bilirubin
	Dietary: ascorbate (vitamin C); α- and γ-tocopherols (vitamin E)	*Dietary*: ascorbate (vitamin C); α- tocopherol (vitamin E)

Oxidants

It follows from the above that the antioxidant defenses in neonates are multifaceted and that the contribution and potential relevance of an individual component, such as bilirubin, to the overall endogenous defense requires consideration of multiple factors. These factors also include the tissue, cell type or body fluid of interest, the nature and concentrations of different antioxidants, as well as the target macromolecule or biological process concerned. Just as importantly, the efficacy of a particular antioxidant is also dependent on the chemical nature of the reactive species involved. This is because the reactivity of different ROS and RNS varies greatly, so that most antioxidants only provide effective defense against some but not other oxidants. For example, in the case of non-proteinaceous antioxidants, vitamin E rapidly scavenges certain 1-electron (or radical) oxidants such as peroxyl radicals ($ROO^{\bullet}$), whereas it is generally ineffective against 2-electron (or non-radical) oxidants such as H_2O_2.

Unfortunately, information on the above factors in neonates is scarce, limiting our ability to directly assess the importance of bilirubin as an antioxidant. What is known, however, is that the partial pressure of oxygen (pO_2) in blood increases sharply during early neonatal age, from 25 to 40 mmHg pre-birth to 50 to 70 mmHg 5 min after birth, and 70–95 mmHg at days 1–7. Such an increase in oxygen pressure is likely associated with a buildup of reactive oxygen species, because oxygen toxicity intensifies with rising oxygen pressure, and reactive oxygen species are principally responsible for oxygen toxicity [59]. The resulting increase in oxidative stress will be enhanced further if redox-active transition metals were also present. Indeed, using the bleomycin assay to assess the availability of non-transferrin iron, Evans and co-workers reported "redox-active" iron to be present in plasma of some full-term neonates but not adults [60]. Therefore, from the information currently available, it is reasonable to assume that neonates experience increased exposure to reactive oxygen species early in their life, although detailed information on the chemical nature of the oxidants involved remains largely unknown.

Bilirubin as an Antioxidant in Neonatal Hyperbilirubinemia

Given the limitations described above, the following will focus on our current knowledge of the *potential* antioxidant contribution of bilirubin in the setting of hyperbilirubinemia. Other chapters in this volume address the roles of proteinaceous and non-proteinaceous antioxidants other than bilirubin in neonatal hyperbilirubinemia.

Antioxidant Activity of Bilirubin in Neonatal Plasma

Currently available literature essentially limits an assessment of bilirubin as a biologically relevant antioxidant in neonates to blood and, more specifically, to blood plasma. Early studies have concluded that bilirubin significantly contributes to the antioxidant activity in neonatal plasma [13, 61] and that for 5-day-old term babies "the total plasma antioxidant activity and the bilirubin concentration show a direct, highly significant correlation" [13]. Importantly, the concentrations of non-proteinaceous antioxidants other than bilirubin, i.e., albumin, uric acid, ascorbate, α-TOH, and ubiquinol-10 (the reduced form of coenzyme Q_{10}), are comparable in plasma of neonates on the 5th day of life and adults [13, 62]. This suggests that the elevated concentrations of bilirubin in neonates provide "additional" rather than "compensatory" antioxidant defense in response to a deficiency of other antioxidants.

To assess the potential biological meaning of these early studies, it is important to understand the methodologies and tests employed. Typically, "total antioxidant activity" is measured in a "closed" system, such as a chamber of an oxygen electrode into which aerated plasma is placed together with a readily oxidizable lipid (typically linoleic acid) and an "initiator." The initiator is a thermolabile chemical that, upon heating to 37 °C, decomposes and generates $ROO^•$ at a constant rate. As long as effective antioxidants remain present, the lipid is protected from oxidation. At the time point when all of the effective antioxidants are consumed, lipid oxidation begins, and this is associated with a substantial increase in oxygen consumption. Total antioxidant activity is determined by the length of time prior to this onset of rapid oxygen consumption. As such, the method gives *quantitative* information about the "total antioxidant activity" contained in a biological sample such as plasma. In addition, the method can provide quantitative information on the relative contributions of individual antioxidants to the total antioxidant activity from the following additional parameters: (1) the concentration of individual antioxidants in the sample, (2) the rate at which ROO˙ are formed, and (3) the number of ROO˙ molecules each molecule of individual antioxidant can scavenge (commonly referred to as the "stoichiometric factor"). The stoichiometric factor of bilirubin is approximately 2. Based on this, bilirubin has been reported to contribute between 3 (at birth) and 23 % (at day 5) of the total antioxidant activity [13, 61].

However, this methodology has some limitations. Firstly, the information obtained is limited to the particular radical oxidant used, e.g., ROO˙. In addition, the information sought and gained is based on the requirement for all antioxidants to be consumed completely, a scenario unlikely to ever occur in vivo. This is because biological systems are "open" rather than "closed," i.e., compounds like bilirubin are formed continuously and hence replenished as they are consumed as a result of oxidant scavenging. Therefore, *qualitative* information about the respective contribution of individual antioxidants arguably is biologically more relevant. To obtain such information, plasma samples may be exposed to ROO˙ radicals or other any other oxidant, followed by the time-dependent determination of loss of individual antioxidants versus the accumulation of oxidized lipids, the latter being a measure of oxidative damage [63]. Although such experiments have not been reported for neonatal plasma samples, important information can nevertheless be gained from corresponding experiments performed with adult blood plasma samples. This is because it is the composition rather than the concentration of individual antioxidants in a particular biological sample that determines the relative importance of each antioxidant toward a particular oxidant species, and the antioxidant composition of neonatal and adult plasma are likely comparable.

The results of such oxidation experiments are summarized in Table 15.3. They show that plasma bilirubin provides a "second line" of antioxidant defense: ROO˙ causes disappearance of the pigment only after the "first-line" aqueous and lipid

Table 15.3 Qualitative contribution of bilirubin as an antioxidant in adult human plasma

Reactive species	Contribution by bilirubin	References
Aqueous ROO˙	"Second-line," behind ascorbate[a]	[21, 32, 63]
Lipophilic ROO˙	"Second-line," behind ubiquinol-10 and ascorbate	[27]
$O_2^{-\bullet}$+˙NO (peroxynitrite)[b]	"Second-line," behind ascorbate (prior to ubiquinol-10)	[28]
Activated neutrophils	Ineffective	[63]

Antioxidant efficacy was assessed by the extent of inhibition of oxidation of endogenous lipoprotein-associated lipids

[a]Potential contribution of ubiquinol-10 not assessed

[b]Generated by 3-morpholinosydnonimine-1 (SIN-1) which generates superoxide anion radical ($O_2^{-\bullet}$) and ˙NO in sequential reactions [64] and hence likely produces peroxynitrite at constant rates

phase antioxidant defenses, i.e., ascorbate and ubiquinol-10, respectively, are consumed [27, 63]. Similar results are obtained with 3-morpholinosydnonimine-1 (SIN-1) [28], a compound that decomposes at a constant rate and in doing so generates superoxide anion radical ($O_2^{-\bullet}$) and $^\bullet$NO in sequential reactions [64]. Superoxide and $^\bullet$NO readily combine to the strong oxidant peroxynitrite [65, 66]. In contrast to the situation with ROO$^\bullet$ and SIN-1, plasma bilirubin is not an efficient antioxidant against activated neutrophils [63] that generate $O_2^{-\bullet}$ and hypochlorite as major oxidants [67].

Antioxidant Activities of Albumin-Bound Bilirubin

It is particularly relevant to consider the antioxidant properties of albumin-bound bilirubin as it is the predominant form of the pigment present in blood of neonates (see above). Studies from several laboratories have established that bilirubin bound to human serum albumin (HSA–bilirubin) effectively protects the protein as well as bound fatty acids from oxidation induced by photooxidation, ROO$^\bullet$, hydroxyl radicals, and peroxynitrite [21, 34, 68, 69]. Although initially thought to directly react with ROO$^\bullet$ [27], recent studies indicate that consumption of albumin-bound bilirubin by aqueous peroxyl radicals is likely indirect via protein-derived radicals [70].

Interestingly, HSA–bilirubin also protects lipids carried in lipoprotein particles from oxidation [27]. This antioxidant activity of the pigment requires lipoproteins to contain α-TOH [27] and is due to bilirubin-mediated recycling of α-TOH via reduction of α-tocopheroxyl radical (α-TO$^\bullet$) (Fig. 15.1). Indeed, HSA–bilirubin effectively reduces α-TO$^\bullet$ in lipoproteins to α-TOH [27, 72–74] and thereby prevents α-TO$^\bullet$ from causing lipoprotein lipid peroxidation [72]. As a consequence of this reduction of α-TO$^\bullet$, ~30 % of HSA–bilirubin is oxidized to albumin-bound biliverdin (HSA–biliverdin) (Eq. 15.1): [27]

$$\text{HSA-bilirubin} + \alpha\text{-TO}^{\bullet}_{\text{lipoproteins}} \longrightarrow \text{HSA-biliverdin} + \alpha\text{-TOH}_{\text{Lipoprotein}} \qquad (15.1)$$

It follows from the above that the antioxidant activity of HSA–bilirubin in blood plasma is at least in part indirect, i.e., via interaction with α-TOH. A feature characteristic of such “co-antioxidant activity” [72] is that the pigment is consumed before α-TOH when plasma undergoes oxidation (see above). Unlike the situation with α-TOH, however, the co-antioxidant activity of HSA–bilirubin is independent of ascorbate and ubiquinol-10 [27]. Although plausible, it remains to be tested whether HSA–bilirubin similarly reduces α-TO$^\bullet$ in plasma membranes (Fig. 15.1). If so, it could explain how the pigment protects blood and endothelial cells from oxidative damage without the need for its physical incorporation into membranes, itself a process known to cause toxicity.

While in circulation, the vast majority of bilirubin is bound to albumin, a small fraction of the pigment will remain unbound. This raises the question of whether the co-antioxidant activity of HSA–bilirubin could in fact be ascribed to unbound pigment, according to Eqs. 15.2 and 15.3:

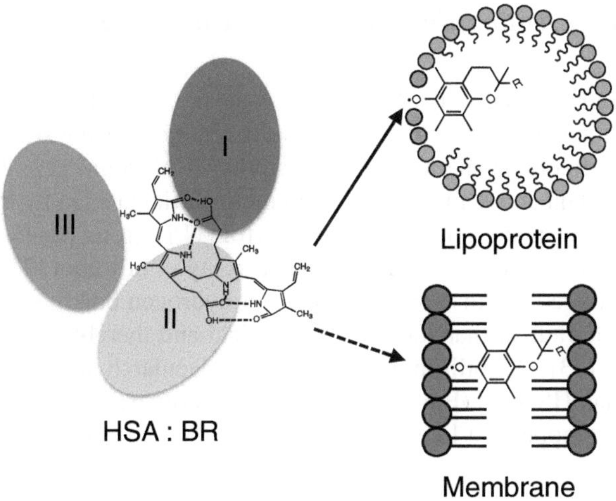

Fig. 15.1 Proposed model for co-antioxidant activity of albumin-bound bilirubin. 4*Z*,15*Z*-Bilirubin IX**α**, bound as the P-conformer to domain IIA of human serum albumin [6], protects lipoprotein lipids from oxidation by reducing **α**-tocopheroxyl radical within the lipoprotein particle (*arrow*, *top right*) [27]. By analogy, it is proposed that albumin-bound bilirubin may also be able to reduce **α**-tocopheroxyl radical within cell membranes (*broken arrow*, *bottom right*), thereby inhibiting membrane lipid peroxidation. It is assumed that the phenoxyl moiety of **α**-tocopheroxyl radical in lipoproteins and cell membranes undergoing oxidation is located at the lipid–water interface [71] and that bilirubin is sufficiently "accessible" to chemically reduce **α**-tocopheroxyl radical to **α**-tocopherol. $R = C_{15}H_{31}$

$$\text{HSA-bilirubin} \rightleftarrows \text{HSA} + \text{Bilirubin} \tag{15.2}$$

$$\text{Bilirubin} + \alpha\text{-TO}^{"}_{\text{lipoproteins}} \longrightarrow \text{Bilirubin}_{\text{ox}} + \alpha\text{-TOH}_{\text{lipoproteins}} \tag{15.3}$$

Indeed, compared with the albumin-bound pigment, unbound bilirubin is a superior co-antioxidant, and it protects lipoprotein lipids more effectively from oxidation than HSA–bilirubin [27]. However, inhibition of lipoprotein oxidation by HSA–bilirubin and "free" bilirubin differs in two distinct ways. Firstly, biliverdin is formed as an oxidation product only in the case of HSA–bilirubin. Secondly, inclusion of the enzyme bilirubin oxidase inhibits the antioxidant activity of the unbound pigment, but not that of HSA–bilirubin [27]. Therefore, current evidence suggests that it is albumin-bound rather than unbound bilirubin that provides co-antioxidant activity in the circulation. The importance of this to the overall antioxidant defense is expected to be higher in neonatal hyperbilirubinemia than in adults. This is because the concentrations of bilirubin are about 6 times higher in neonates than adults, whereas the concentrations of the "first-line" antioxidants ascorbate [13] and ubiquinol-10 [62] are comparable in neonates and adults.

Mechanism of Antioxidant Activity of Albumin-Bound Bilirubin

It was proposed originally that bilirubin acts as an antioxidant by hydrogen atom transfer from the methylene at C-10 [18]. However, more recent studies have shown strong media effects on the antioxidant activity of bilirubin [22, 23], suggesting that the antioxidant active moiety of the pigment participates in hydrogen bonding. Therefore, hydrogen atom transfer from a pyrrole N-H group has been proposed as an alternative to account for the antioxidant activity of bilirubin [23]. Such mechanism is also favored because of intramolecular hydrogen bonding within the intermediate pyrrole radical that stabilizes the radical and thereby promotes hydrogen atom transfer [23]. As bilirubin retains intramolecular hydrogen bonding when bound to human HSA [6] (Fig. 15.1), a hydrogen atom transfer involving a pyrrole N-H group also appears a likely mechanism for the antioxidant activity of HSA–bilirubin [24].

Cellular Antioxidant Activity of Bilirubin

In comparison with macrophages engaged in RBC removal, most other cells form only small amounts of bilirubin as a result of cellular heme turnover. Cells exposed to blood may also "acquire" additional small amounts of bilirubin after its dissociation from circulating albumin, a path enhanced in neonatal hyperbilirubinemia. Despite this, however, in most tissues and cells, the intracellular concentrations of bilirubin are likely orders of magnitudes lower than the micromolar concentrations present in blood plasma, raising the question of physiological relevance of cellular antioxidant protection by bilirubin.

Remarkably, Snyder and colleagues reported that even nanomolar concentrations of bilirubin can protect cells against the toxic effects of H_2O_2 [40, 75]. Specifically, it was suggested that 10 nM bilirubin is sufficient to protect cells against a 10,000-fold excess of H_2O_2 and that this is achieved by bilirubin being converted to biliverdin and then "recycled" by biliverdin reductase back to bilirubin. For such recycling to be efficient, bilirubin oxidation to biliverdin would have to be quantitative, a notion inconsistent with a large body of older as well as more recent literature [70, 76, 77]. Also, increasing or decreasing biliverdin reductase activity by gene overexpression or silencing, respectively, has subsequently been reported to have no effect on H_2O_2-mediated death of HeLa cells, irrespective of whether bilirubin or biliverdin was added to the cells as substrate for the putative redox cycle [76]. However, silencing biliverdin reductase (or HO-1) attenuates the protective effect of inducers of HO-1 on markers of oxidative stress in human endothelial cells [77]. The overall simplest interpretation of these results is that the bilirubin–biliverdin recycling theory of cellular protection [40, 75] should be considered highly improbable. In situations where HO-1 is induced, biliverdin reductase can contribute to the cellular antioxidant defense. However, this is due to formation of the antioxidant bilirubin from biliverdin derived from heme rather than bilirubin oxidation.

Bilirubin as an Antioxidant In Vivo

Although bilirubin is an effective reducing agent and possesses antioxidant activity in a number of different in vitro systems and against various reactive oxygen and nitrogen species [20], there is only indirect evidence in support of bilirubin as an in vivo antioxidant. For example, cells deficient in HO-1 have an increased susceptibility to H_2O_2 [78]. Conversely, HO-1 is induced by oxidants and conditions that exert an oxidative stress [79]. However, changes in HO activity also alter cellular heme and iron status [80], so that it is not possible from these types of experiments to assign any beneficial effect solely to bilirubin. A somewhat more direct approach to the in vivo antioxidant activity of bilirubin comes from studies by Dennery and co-workers. They reported markers of lipid oxidation to be decreased in hyperbilirubinemic Gunn rats after exposure to ≥95 % oxygen for 3 days compared with non-jaundiced littermates [81]. Unfortunately, the lipid oxidation markers used in that study were nonspecific, so that it will be interesting to examine Gunn rats without and with oxidant exposure for more specific markers of oxidative damage in plasma.

Bilirubin as a Pro-oxidant

When bilirubin accumulates beyond a certain threshold, the pigment undoubtedly becomes toxic. It is therefore prudent to also consider the possibility that bilirubin may act as a pro-oxidant and that such activity may contribute to its toxicity. Indeed, bilirubin has long been known to act as in vitro sensitizer through generation of singlet oxygen [82], including the photooxidation of RBC membranes [83] and cross-linking of proteins [84]. Also, in vitro albumin–bilirubin complexes with a molar pigment-to-albumin ratio of ≥3 have been reported to lose their antioxidant activity, assessed by pretreating erythrocytes with bilirubin followed by exposure to H_2O_2 plus Cu^{2+} [85]. However, the in vivo relevance of the conditions associated with a pro-oxidant activity of bilirubin is likely low. Also, bilirubin overall is a weak photosensitizer, and light-induced photooxidation reactions represent a minor pathway, with photoisomerization being more important. In addition, as far as tested, photoisomerization does not appear to decrease the antioxidant activity of bilirubin [18]. Therefore, a pro-oxidant activity of the pigment in vivo appears unprobable.

Summary

A biologically relevant antioxidant activity of bilirubin in hyperbilirubinemia is most likely for the HSA–bilirubin complex. In addition to direct antioxidant activities by scavenging reactive oxygen and nitrogen species, the pigment also has indirect, or co-antioxidant activities. These have been best studied in the case of interaction with α-tocopherol in circulating lipoproteins but could conceivably extend to α-tocopherol in plasma membranes of cells exposed to blood. Compared with blood plasma, the bilirubin concentrations in most cells exposed to blood are substantially smaller, and it remains to be verified whether the unbound pigment possesses relevant cellular antioxidant activity. A previously described bilirubin–biliverdin redox cycle is unlikely to provide major cellular antioxidant protection. Future work should focus on evaluating whether oxidative damage to macromolecules is attenuated in conditions of neonatal hyperbilirubinemia, using state-of-the-art technology and more specific methods than employed previously.

Acknowledgments R.S. is supported by a Senior Principal Research Fellowship from the Australian National Health and Medical Research Council (NHMRC) and the Victor Chang Cardiac Research Institute. His current research is supported by NHMRC Project Grants 1020400, 1020776, 1049381, and 1050776 and by the Australian Research Council Discovery Project 120103170.

References

1. Tenhunen R, Marver HS, Schmid R. The enzymatic conversion of heme to bilirubin by microsomal heme oxygenase. Proc Natl Acad Sci U S A. 1968;61:748–55.
2. Unno M, Matsui T, Ikeda-Saito M. Crystallographic studies of heme oxygenase complexed with an unstable reaction intermediate, verdoheme. J Inorg Biochem. 2012;113:102–9.
3. Gåfvels M, Holmström P, Somell A, Sjövall F, Svensson JO, Ståhle L, Broomé U, Stål P. A novel mutation in the biliverdin reductase-A gene combined with liver cirrhosis results in hyperbiliverdinaemia (green jaundice). Liver Int. 2009;29:1116–24.
4. Zunszain PA, Ghuman J, McDonagh AF, Curry S. Crystallographic analysis of human serum albumin complexed with 4*Z*,15*E*-bilirubin-IXα. J Mol Biol. 2008;381:394–406.
5. Minomo A, Ishima Y, Kragh-Hansen U, Chuang VT, Uchida M, Taguchi K, Watanabe H, Maruyama T, Morioka H, Otagiri M. Biological characteristics of two lysines on human serum albumin in the high-affinity binding of 4Z,15Z-bilirubin-IXα revealed by phage display. FEBS J. 2011;278:4100–11.
6. Goncharova I, Orlov S, Urbanová M. The location of the high- and low-affinity bilirubin-binding sites on serum albumin: ligand-competition analysis investigated by circular dichroism. Biophys Chem. 2013;180–181:55–65.
7. Daood MJ, McDonagh AF, Watchko JF. Calculated free bilirubin levels and neurotoxicity. J Perinatol. 2009;29:S14–9.
8. Wennberg RP, Ahlfors CE, Bhutani VK, Johnson LH, Shapiro SM. Toward understanding kernicterus: a challenge to improve the management of jaundiced newborns. Pediatrics. 2006;117:474–85.
9. Nies AT, Keppler D. The apical conjugate efflux pump ABCC2 (MRP2). Pflugers Arch Eur J Physiol. 2007;453:643–59.
10. McDonagh AF, Palma LA, Lauff JJ, Wu TW. Origin of mammalian biliprotein and rearrangement of bilirubin glucuronides in vivo in the rat. J Clin Invest. 1984;74:763–70.
11. McDonagh AF, Lightner DA. 'Like a shrivelled blood orange' – bilirubin, jaundice, and phototherapy. Pediatrics. 1985;75:443–55.
12. Ruud Hansen TW. Phototherapy for neonatal jaundice – therapeutic effects on more than one level? Semin Perinatol. 2010;34:231–4.
13. Gopinathan V, Miller NJ, Milner AD, Rice-Evans CA. Bilirubin and ascorbate antioxidant activity in neonatal plasma. FEBS Lett. 1994;349:197–200.

14. Dennery PA, Seidman DS, Stevenson DK. Neonatal hyperbilirubinemia. N Engl J Med. 2001;344:581–90.
15. Pearson HA. Life-span of the fetal red blood cell. J Pediatr. 1967;70:166–71.
16. Kawade N, Onishi S. The prenatal and postnatal development of UDP-glucuronyltransferase activity towards bilirubin and the effect of premature birth on this activity in the human liver. Biochem J. 1981;196:257–60.
17. Myara A, Sender A, Valette V, Rostoker C, Paumier D, Capoulade C, Loridon F, Bouillie J, Milliez J, Brossard Y, Trivin F. Early changes in cutaneous bilirubin and serum bilirubin isomers during intensive phototherapy of jaundiced neonates with blue and green light. Biol Neonate. 1997;71:75–82.
18. Stocker R, Yamamoto Y, McDonagh AF, Glazer AN, Ames BN. Bilirubin is an antioxidant of possible physiological importance. Science. 1987;235:1043–6.
19. Burton GW, Joyce A, Ingold KU. Is vitamin E the only lipid-soluble, chain-breaking antioxidant in human blood plasma and erythrocyte membranes? Arch Biochem Biophys. 1983;221:281–90.
20. Stocker R. Antioxidant activities of bile pigments. Antioxid Redox Signal. 2004;6:841–9.
21. Stocker R, Glazer AN, Ames BN. Antioxidant activity of albumin bound bilirubin. Proc Natl Acad Sci U S A. 1987;84:5918–22.
22. Hatfield GL, Barclay LR. Bilirubin as an antioxidant: kinetic studies of the reaction of bilirubin with peroxyl radicals in solution, micelles, and lipid bilayers. Org Lett. 2004;6:1539–42.
23. Chepelev LL, Beshara CS, MacLean PD, Hatfield GL, Rand AA, Thompson A, Wright JS, Barclay LR. Polypyrroles as antioxidants: kinetic studies on reactions of bilirubin and biliverdin dimethyl esters and synthetic model compounds with peroxyl radicals in solution. Chemical calculations on selected typical structures. J Org Chem. 2006;71:22–30.
24. MacLean PD, Drake EC, Ross L, Barclay C. Bilirubin as an antioxidant in micelles and lipid bilayers: its contribution to the total antioxidant capacity of human blood plasma. Free Radic Biol Med. 2007;43:600–9.
25. Stocker R, Lai A, Peterhans E, Ames BN. Medical, biochemical and chemical aspects of free radicals. In: Hayaishi O, Niki E, Kondo M, Yoshikawa Y, editors. Antioxidant properties of bilirubin and biliverdin. Amsterdam: Elsevier; 1988. p. 465–8.
26. Stocker R, Peterhans E. Synergistic interaction between vitamin E and the bile pigments bilirubin and biliverdin. Biochim Biophys Acta. 1989;1002:238–44.
27. Neuzil J, Stocker R. Free and albumin-bound bilirubin is an efficient co-antioxidant for α-tocopherol, inhibiting plasma and low density lipoprotein lipid peroxidation. J Biol Chem. 1994;269:16712–9.
28. Thomas SR, Davies MJ, Stocker R. Oxidation and antioxidation of human low-density lipoprotein and plasma exposed to 3-morpholinosydnonimine and reagent peroxynitrite. Chem Res Toxicol. 1998;11:484–94.
29. Wu TW, Fung KP, Wu J, Yang CC, Weisel RD. Antioxidation of human low density lipoprotein by unconjugated and conjugated bilirubins. Biochem Pharmacol. 1996;51:859–62.
30. Wu TW, Fung KP, Yang CC. Unconjugated bilirubin inhibits the oxidation of human low density lipoprotein better than Trolox. Life Sci. 1994;54:PL477–81.
31. Hulea SA, Wasowicz E, Kummerow FA. Inhibition of metal-catalyzed oxidation of low-density lipoprotein by free and albumin-bound bilirubin. Biochim Biophys Acta. 1995; 1259:29–38.
32. Thomas SR, Witting PK, Stocker R. 3-Hydroxyanthranilic acid is an efficient, cell-derived co-antioxidant for α-tocopherol, inhibiting human low density lipoprotein and plasma lipid peroxidation. J Biol Chem. 1996;271:32714–21.
33. Neuzil J, Thomas SR, Stocker R. Requirement for, promotion, or inhibition by α-tocopherol of radical-induced initiation of plasma lipoprotein lipid peroxidation. Free Radic Biol Med. 1997;22:57–71.
34. Minetti M, Mallozzi C, Di Stasi AM, Pietraforte D. Bilirubin is an effective antioxidant of peroxynitrite-mediated protein oxidation in human blood plasma. Arch Biochem Biophys. 1998;352:165–74.

35. Neuzil J, Stocker R. Bilirubin attenuates radical-mediated damage to serum albumin. FEBS Lett. 1993;331:281–4.
36. Adhikari S, Gopinathan C. Oxidation reactions of a bovine serum albumin-bilirubin complex. A pulse radiolysis study. Int J Radiat Biol. 1996;69:89–98.
37. Kwak JY, Takeshige K, Cheung BS, Minakami S. Bilirubin inhibits the activation of superoxide-producing NADPH oxidase in a neutrophil cell-free system. Biochim Biophys Acta. 1991;1076:369–73.
38. Lanone S, Bloc S, Foresti R, Almolki A, Taillé C, Callebert J, Conti M, Goven D, Aubier M, Dureuil B, El-Benna J, Motterlini R, Boczkowski J. Bilirubin decreases nos2 expression via inhibition of NAD(P)H oxidase: implications for protection against endotoxic shock in rats. FASEB J. 2005;19:1890–2.
39. Weinberger B, Archer FE, Kathiravan S, Hirsch DS, Kleinfeld AM, Vetrano AM, Hegyi T. Effects of bilirubin on neutrophil responses in newborn infants. Neonatology. 2013;103:105–11.
40. Doré S, Takahashi M, Ferris CD, Hester LD, Guastella D, Snyder SH. Bilirubin, formed by activation of heme oxygenase-2, protects neurons against oxidative stress injury. Proc Natl Acad Sci U S A. 1999;96:2445–50.
41. Wu TW, Wu J, Li RK, Mickle D, Carey D. Albumin-bound bilirubins protect human ventricular myocytes against oxyradical damage. Biochem Cell Biol. 1991;69:683–8.
42. Clark JE, Foresti R, Green CJ, Motterlini R. Dynamics of haem oxygenase-1 expression and bilirubin production in cellular protection against oxidative stress. Biochem J. 2000; 348:615–9.
43. Keshavan P, Deem TL, Schwemberger SJ, Babcock GF, Cook-Mills JM, Zucker SD. Unconjugated bilirubin inhibits VCAM-1-mediated transendothelial leukocyte migration. J Immunol. 2005;174:3709–18.
44. Rocuts F, Zhang X, Yan J, Yue Y, Thomas M, Bach FH, Czismadia E, Wang H. Bilirubin promotes de novo generation of T regulatory cells. Cell Transplant. 2010;19:443–51.
45. Wang WW, Smith DL, Zucker SD. Bilirubin inhibits iNOS expression and NO production in response to endotoxin in rats. Hepatology. 2004;40:424–33.
46. Clark JE, Foresti R, Sarathchandra P, Kaur H, Green CJ, Motterlini R. Heme oxygenase-1-derived bilirubin ameliorates postischemic myocardial dysfunction. Am J Physiol Heart Circ Physiol. 2000;278:H643–51.
47. Adin CA, Croker BP, Agarwal A. Protective effects of exogenous bilirubin on ischemia-reperfusion injury in the isolated, perfused rat kidney. Am J Physiol Renal Physiol. 2005;288:F778–84.
48. Arriaga SM, Mottino AD, Almara AM. Inhibitory effect of bilirubin on complement-mediated hemolysis. Biochim Biophys Acta. 1999;1473:329–36.
49. Wang HD, Yamaya M, Okinaga S, Jia YX, Kamanaka M, Takahashi H, Guo LY, Ohrui T, Sasaki H. Bilirubin ameliorates bleomycin-induced pulmonary fibrosis in rats. Am J Respir Crit Care Med. 2002;165:406–11.
50. Finkel T, Holbrook NJ. Oxidants, oxidative stress and the biology of ageing. Nature. 2000;408:239–47.
51. Halliwell B, Gutteridge JMC. Free radicals in biology and medicine. 4th ed. Oxford: Oxford University Press; 2007. p. 1–851.
52. Lönn ME, Dennis JM, Stocker R. Actions of "antioxidants" in the protection against atherosclerosis. Free Radic Biol Med. 2012;53:863–84.
53. Woo HA, Yim SH, Shin DH, Kang D, Yu DY, Rhee SG. Inactivation of peroxiredoxin I by phosphorylation allows localized H_2O_2 accumulation for cell signaling. Cell. 2010;140:517–28.
54. Washko PW, Wang Y, Levine M. Ascorbic acid recycling in human neutrophils. J Biol Chem. 1993;268:15531–5.
55. Agar NS, Sadrzadeh SMC, Hallaway PE, Eaton JW. Erythrocyte catalase. A somatic oxidant defense? J Clin Invest. 1986;77:319–21.

56. Low FM, Hampton MB, Peskin AV, Winterbourn CC. Peroxiredoxin 2 functions as a noncatalytic scavenger of low-level hydrogen peroxide in the erythrocyte. Blood. 2007;109:2611–7.
57. Bayer SB, Maghzal G, Stocker R, Hampton MB, Winterbourn CC. Neutrophil-mediated oxidation of erythrocyte peroxiredoxin 2 as a potential marker of oxidative stress in inflammation. FASEB J. 2013;27:3315–22.
58. Rong Z, Wilson MT, Cooper CE. A model for the nitric oxide producing nitrite reductase activity of hemoglobin as a function of oxygen saturation. Nitric Oxide. 2013;33C:74–80.
59. Gershman R, Gilbert DL, Nye SW, Dwyer P, Fenn WO. Oxygen poisoning and X-irradiation: a mechanism in common. Science. 1954;119:623–6.
60. Evans PJ, Evans R, Kovar IZ, Holton AF, Halliwell B. Bleomycin-detectable iron in the plasma of premature and full-term neonates. FEBS Lett. 1992;303:210–2.
61. Lindeman JH, van Zoeren-Grobben D, Schrijver J, Speek AJ, Poorthuis BJHM, Berger HM. The total free radical trapping ability of cord blood plasma in preterm and term babies. Pediatr Res. 1989;26:20–4.
62. Finckh B, Kontush A, Commentz J, Hübner C, Burdelski M, Kohlschütter A. Monitoring of ubiquinol-10, ubiquinone-10, carotenoids, and tocopherols in neonatal plasma microsamples using high-performance liquid chromatography with coulometric electrochemical detection. Anal Biochem. 1995;232:210–6.
63. Frei B, Stocker R, Ames BN. Antioxidant defenses and lipid peroxidation in human blood plasma. Proc Natl Acad Sci U S A. 1988;85:9748–52.
64. Bohn H, Schönafinger K. Oxygen and oxidation promote the release of nitric oxide from sydnonimines. J Cardiovasc Pharmacol. 1989;14 Suppl 11:S6–12.
65. Blough NV, Zafiriou OC. Reaction of superoxide with nitric oxide to form peroxynitrite in alkaline aqueous solution. Inorg Chem. 1985;24:3502–4.
66. Nauser T, Koppenol WH. The rate constant of the reaction of superoxide with nitrogen monoxide: approaching the diffusion limit. J Phys Chem A. 2002;106:4084–6.
67. Hampton MB, Kettle AJ, Winterbourn CC. Inside the neutrophil phagosome: oxidants, myeloperoxidase, and bacterial killing. Blood. 1998;92:3007–17.
68. Pedersen AO, Schønheyder F, Brodersen R. Photooxidation of human serum albumin and its complex with bilirubin. Eur J Biochem. 1977;72:213–21.
69. Neuzil J, Gebicki JM, Stocker R. Radical-induced chain oxidation of proteins and its inhibition by chain-breaking antioxidants. Biochem J. 1993;293:601–6.
70. McDonagh AF. The biliverdin-bilirubin antioxidant cycle of cellular protection: missing a wheel? Free Radic Biol Med. 2010;49:814–20.
71. Waldeck AR, Stocker R. Radical-initiated lipid peroxidation in low density lipoproteins: insights obtained from kinetic modeling. Chem Res Toxicol. 1996;9:954–64.
72. Bowry VW, Stocker R. Tocopherol-mediated peroxidation. The pro-oxidant effect of vitamin E on the radical-initiated oxidation of human low-density lipoprotein. J Am Chem Soc. 1993;115:6029–44.
73. Bowry VW, Mohr D, Cleary J, Stocker R. Prevention of tocopherol-mediated peroxidation of ubiquinol-10-free human low density lipoprotein. J Biol Chem. 1995;270:5756–63.
74. Witting PK, Westerlund C, Stocker R. A rapid and simple screening test for potential inhibitors of tocopherol-mediated peroxidation of LDL lipids. J Lipid Res. 1996;37:853–67.
75. Barañano DE, Rao M, Ferris CD, Snyder SH. Biliverdin reductase: a major physiologic cytoprotectant. Proc Natl Acad Sci U S A. 2002;99:16093–8.
76. Maghzal GJ, Leck MC, Collinson E, Li C, Stocker R. Limited role for the bilirubin-biliverdin redox amplification cycle in the cellular antioxidant protection by biliverdin reductase. J Biol Chem. 2009;284:29251–9.
77. Jansen T, Hortmann M, Oelze M, Opitz B, Steven S, Schell R, Knorr M, Karbach S, Schuhmacher S, Wenzel P, Münzel T, Daiber A. Conversion of biliverdin to bilirubin by biliverdin reductase contributes to endothelial cell protection by heme oxygenase-1-evidence for direct and indirect antioxidant actions of bilirubin. J Mol Cell Cardiol. 2010;49:186–95.

78. Poss KD, Tonegawa S. Reduced stress defense in heme oxygenase 1-deficient cells. Proc Natl Acad Sci U S A. 1997;94:10925–30.
79. Ryter SW, Alam J, Choi AM. Heme oxygenase-1/carbon monoxide: from basic science to therapeutic applications. Physiol Rev. 2006;86:583–650.
80. Li C, Lönn M, Xu X, Maghzal GJ, Frazer DM, Thomas SR, Halliwell B, Richardson DR, Anderson GJ, Stocker R. Sustained expression of heme oxygenase-1 alters iron homeostasis in nonerythroid cells. Free Radic Biol Med. 2012;53:366–74.
81. Dennery PA, McDonagh AF, Spitz DR, Rodgers PA, Stevenson DK. Hyperbilirubinemia results in reduced oxidative injury in neonatal Gunn rats exposed to hyperoxia. Free Radic Biol Med. 1995;19:395–404.
82. McDonagh AF. The role of singlet oxygen in bilirubin photo-oxidation. Biochem Biophys Res Commun. 1971;44:1306–11.
83. Deziel MR, Girotti AW. Photodynamic action of bilirubin on liposomes and erythrocyte membranes. J Biol Chem. 1980;255:8192–8.
84. Girotti AW. Bilirubin-photosensitized cross-linking of polypeptides in the isolated membrane of the human erythrocyte. J Biol Chem. 1978;253:7186–93.
85. Mireles LC, Lum MA, Dennery PA. Antioxidant and cytotoxic effects of bilirubin on neonatal erythrocytes. Pediatr Res. 1999;45:355–62.

Chapter 16
Pain and Oxidative Stress in Newborns

Carlo V. Bellieni and Giuseppe Buonocore

Pain and Brain Damage in Newborns

Neonates undergo many painful procedures daily, which can have long-term negative effects [1, 2]. Neonatal pain treatment is still far from being satisfactory, as reported by many recent reviews [3–5]. The concept that neonates feel pain, at least in most of North America and Europe, is now well accepted and pain has been associated with worse neurodevelopmental outcomes in preterm neonates [6, 7]. Newborns feel pain and even common routine procedures such as a heel prick or the suction of an intratracheal tube are painful [8]. Painful procedures may be a risk for brain damage [9, 10] through an increase in arterial and intracranial pressure and oxygen desaturation [11, 12] but also through the use of sedatives [13] or of oral sucrose [14]. The analgesic use of sucrose, given orally before heel lance, increases adenosine triphosphate production and oxidative stress (OS) in preterm neonates. OS has widely been described both in animals and in newborn infants [15, 16]. Reactions involving free radical (FR) toxicity in neonates have a high potential for damaging the fast growing tissues, leading to conditions such as chronic lung disease, retinopathy of prematurity, necrotizing enterocolitis, and periventricular leukomalacia [17–20]. The developing brain is particularly susceptible to FR toxicity, due to the high cellular growth rate, the peculiar lipid composition, the presence of low antioxidant defenses, and the high prenatal rate of FR generation. All these factors induce an increased lipid peroxidation [21–26].

C.V. Bellieni (✉)
Department of Pediatrics, Obstetrics and Reproductive Medicine,
University of Siena, Siena, Italy
e-mail: cvbellieni@gmail.com

G. Buonocore, M.D.
Departments of Pediatrics, Obstetrics and Reproductive Medicine and Molecular and Developmental Medicine, University of Siena, Siena, Italy
e-mail: giuseppe.buonocore@unisi.it

P.A. Dennery et al. (eds.), *Perinatal and Prenatal Disorders*,
Oxidative Stress in Applied Basic Research and Clinical Practice,
DOI 10.1007/978-1-4939-1405-0_16

Pain Is Mediated by OS

The correlation between FR production and pain has been elucidated in animal models and in adult human patients in the last few years. The main studies deal with the effectiveness of FR in mediating pain as shown by the action of FR chemical scavengers that blunt this mediation. Several studies show this phenomenon, since FR involvement in inflammation, fibrosis control, and pain nociception has been proven [27].

Painful stimulation increases lipoperoxidation; this has been shown to persist for up to 15 days after the painful stimulus has been discontinued [28]. A simultaneous injection of antioxidants decreases the levels of thiobarbituric acid reactive substances – TBARS – formed as a byproduct of lipid peroxidation and superoxide dismutase; however, antioxidants applied 1 week prior to the painful stimulation were ineffective [24]. Phenyl N-tert-butylnitrone, a free radical scavenger, reduces mechanical allodynia in chemotherapy-induced neuropathic pain in rats [29]. Free radical production is involved in the hyperalgesia induced by immobilization in rats and is reversed by injection of free radical scavengers: intraperitoneal injection of the free radical scavengers 4-hydroxy-2,2,6,6-tetramethylpiperydine-1-oxyl or N-acetylcysteine clearly inhibited mechanical hyperalgesia in bilateral calves and hindpaws in rats [30]. These results suggest that cast immobilization induces ischemia/reperfusion injury in the hindlimb and consequent production of oxygen free radicals, which may be involved in the mechanism of widespread hyperalgesia in rats [30]. Epigallocatechin-3-gallate (EGCG), the major catechin in green tea, is known to have antioxidant activity against nitric oxide (NO) by scavenging free radicals, chelating metal ions, and inducing endogenous antioxidant enzymes. Intrathecal EGCG attenuated mechanical allodynia in spinal nerve-ligated rats, compared to sham-operated rats [31].

Evidence has accumulated in recent years indicating structural, physiologic, and biochemical alterations in the brain of patients with chronic migraine (CM). Altered pharmacologic responses to opioids and other analgesics have also been reported [32]. Structural or morphologic changes include reduced cortical gray matter of the pain processing areas of the brain and iron accumulation in the periaqueductal gray matter (PAG), red nucleus, and basal ganglia structures. These changes correlate with the duration of migraine disorder.

The biochemical markers of lipid, protein and saccharide metabolisms, and free radicals as well as singlet oxygen can serve as very good indicators of the intensity of pain and nociception [33]. FR were measured in samples of the sensorimotor brain cortex and in the rat blood after noxious stimuli. Their concentration was measured under physiological conditions, after the nociceptive stimulation, and after the application of melatonin (a free radical scavenger), both in normal and stimulated animals. In the brain cortex only the singleton oxygen decreased after the nociceptive stimulation, whereas the nitroxide radicals and six lines spectra increased. The free hydroxyl radicals did not change significantly. In the blood serum, while the six lines spectra and concentration of nitroxide radical increased, the concentration of the free hydroxyl radicals did not change. Melatonin increased both the hydroxyl and nitroxide radicals. There was a nonsignificant decrease in the six lines spectra. Therefore, estimation of FR can be used as a tool for detecting metabolic changes and the consequences of different environmental influences, in our case the influence of nociception [34].

Lastly, after the application of contention stress, mimicking pain, GSH level increased insignificantly, while MDA level increased significantly compared with the normal animals, and nociceptive stress actioned the mechanisms of membrane lipid peroxidation [35]. The activity of the antioxidant enzymes GSH, GPX, and SOD decreased in the stressed animals compared with the normal group; this experiment has revealed that in nociception stress, many FR are generated, which reduced the enzymatic defense mechanisms and activated lipid peroxidation mechanisms [35]. Visceral pain increases oxidative stress in animals, which could be prevented by prior administration of antioxidants; however, antioxidants did not attenuate signs of visceral pain caused by CRD [36]

Free Radicals Produced After Pain Are a Risk for Brain Damage

Pain can have ominous effects on the brain, and a possible reason is the FR production due to hypoxia. In fact (Fig. 16.1), acute pain produces hypoxia, tachycardia, and vasoconstriction due to several mechanisms among which is the release of noradrenaline [37]. Blood hypoxia (Fig. 16.2) increases adenosine trisphosphate (ATP) utilization in the face of decreased ATP synthesis and enhance the degradation of ATP to adenosine diphosphate (ADP) and adenosine monophosphate (AMP). AMP degradation leads to dramatic increase in adenosine hypoxanthine, xanthine, and uric acid generating FR when reperfusion occurs [38]. The enzyme xanthine oxidase is activated catalyzing the conversion of hypoxanthine to xanthine and xanthine to

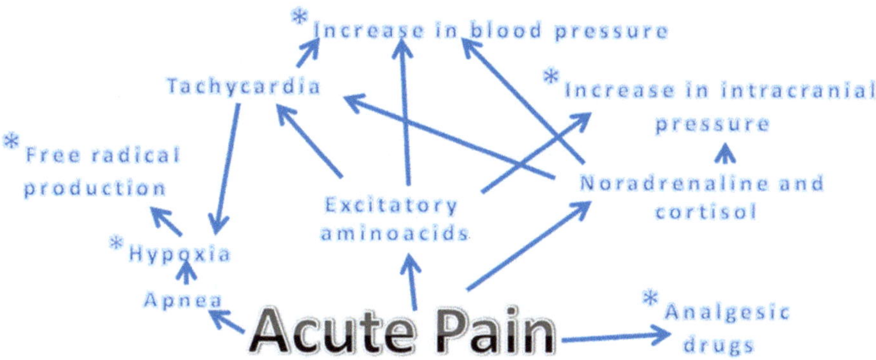

Fig. 16.1 Possible mechanisms leading to pain damage after pain. Features highlighted with * are an outstanding risk for brain damage

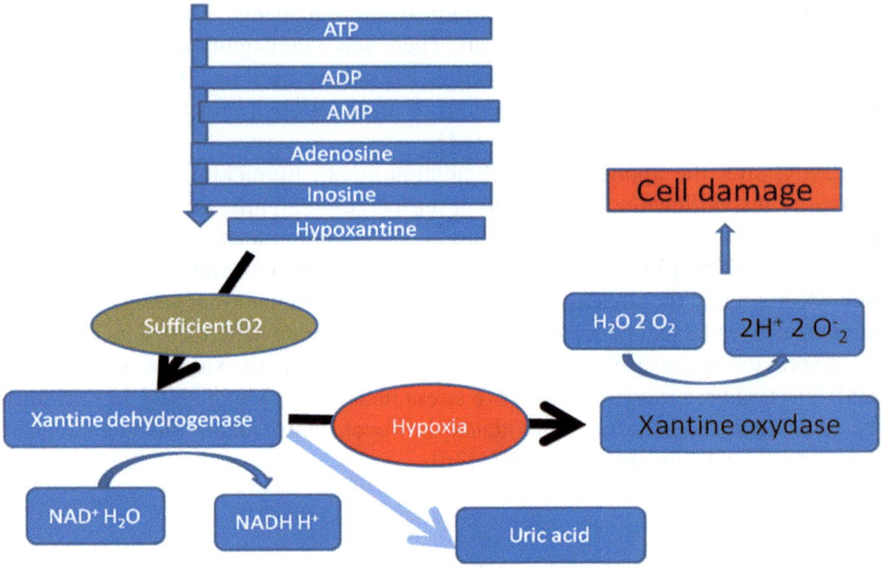

Fig. 16.2 Molecular mechanisms of cell damage after relevant hypoxia. Xanthine dehydrogenase, which needs nicotinamide adenine dinucleotide (NAD-) as an electron acceptor, is converted to xanthine oxygenase when ischemia occurs in the endothelia. This conversion changes the normal degradation of hypoxanthine to uric acid into a source of oxygen radicals (From Ref [49] modif.)

uric acid, while generating reactive oxygen species [39, 40]. Several studies have shown that non-treated acute pain can induce brain damage [41]; this can be due to several mechanisms including FR production causing microglia damage and neuron death [42]. Common procedural painful stimuli are not so long lasting to induce a significant hypoxemia to release FR enough to be a risk for the baby, but several procedures when repeated and prolonged induce threat to the baby. Two primary mechanisms lead to enhanced neuronal cell death in the immature brain: (a) NMDA-mediated excitotoxicity and (b) enhanced naturally occurring neuronal apoptosis due to multiple metabolic stresses or lack of social stimulation [43]. The pattern and magnitude of abnormalities will depend on genetic variability as well as the timing, intensity, and duration of adverse environmental experiences. It is not easy to correlate procedural pain with brain damage in newborns; nevertheless, since babies undergo multiple painful events per day during their stay in neonatal intensive wards [44], an abnormal production of FR can occur inducing the FR-related disease of the newborn [45–47].

To our knowledge, little data exist on possible FR generation after a short painful procedure in the newborn. Slater et al. found significant positive correlations in preterm neonates between pain scores and MDA, suggesting a significant relationship between procedural pain and oxidative stress [48].

Pain provoked by a routine heel prick can generate an increase in potentially harmful FR in neonatal blood [45]. We measured two markers of FR production –

advanced oxidation protein products (AOPP) and total hydroperoxides (TH) – at the beginning and at the end of a heel prick procedure, to assess a possible increase after this painful stimulus. Sixty-four newborns underwent heel prick blood sampling for routine tests. The test was divided into three phases: (a) heel prick with immediate collection of 100 μl of blood in a microprovette (sample A), (b) collection of 100 μl of blood for the diagnostic necessity by glass micro-capillary, and (c) collection of 100 μl of blood in a microtube (sample B). Samples A and B were stored and analyzed to determine blood levels of AOPP and TH. During the blood collection, pain was scored using the ABC pain scale. Considering the population as whole, no increase in AOPP or in TH was observed between samples A and B, as well as when we considered only babies with little or no pain (ABC<4). But when babies' pain was high (ABC score ≥4), mean AOPP and TH blood levels increased significantly. Linear regression analysis performed between the increase in AOPP blood levels from A to B sample and pain score showed a significant correlation between the entity of OS and pain level. This study shows that without analgesia, even a simple heel prick can induce OS, and since newborns undergo many heel pricks and many painful maneuvers, all these FR-producing events can be a further risk for the integrity of the baby.

Conclusion

Pain can induce several alteration in the physiologic behavior of a baby, some of which can threat the integrity of the brain. We do not yet know whether a single painful experience can be a pure risk for the brain, but since every day most premature babies in any NICU undergo many painful procedures – some long lasting – we should set a precaution principle and protect babies from this possible risk using drugs but also non-pharmacologic strategies [50, 51]. OS is one of the ways this risk can act, and since FR are produced during pain, this is to be considered. Thus, pain is ominous for the baby not only from a moral or psychological point of view and cannot be underestimated anymore, with the thought that rapid procedural pain has no impact on babies' health.

Acknowledgment Research support was provided by an unrestricted grant from the Italian Ministry of Health, Research Project RF-2009-1499651 – and Fondazione EURAIBI Onlus, Department of Molecular and Developmental Medicine, University of Siena, Siena, Italy.

References

1. Anand KJ, Coskun V, Thrivikraman KV, Nemeroff CB, Plotsky PM. Long-term behavioral effects of repetitive pain in neonatal rat pups. Physiol Behav. 1999;66:627–37.
2. Anand KJ, Scalzo FM. Can adverse neonatal experiences alter brain development and subsequent behavior? Biol Neonate. 2000;77:69–82.

3. Carbajal R, Rousset A, Danan C, Coquery S, Nolent P, Ducrocq S, Saizou C, Lapillonne A, Granier M, Durand P, Lenclen R, Coursol A, Hubert P, de Saint Blanquat L, Boëlle PY, Annequin D, Cimerman P, Anand KJ, Bréart G. Epidemiology and treatment of painful procedures in neonates in intensive care units. JAMA. 2008;300(1):60–70.
4. Blanquat L, Boëlle PY, Annequin D, Cimerman P, Anand KJ, Bréart G. Epidemiology and treatment of painful procedures in neonates in intensive care units. JAMA. 2008;300:60–70.
5. Cignacco E, Hamers JP, Stoffel L, van Lingen RA, Schütz N, Müller R, Zimmermann LJ, Nelle M. Routine procedures in NICUs: factors influencing pain assessment and ranking by pain intensity. Swiss Med Wkly. 2008;138:484–91.
6. Juul SE, Beyer RP, Bammler TK, Farin FM, Gleason CA. Effects of neonatal stress and morphine on murine hippocampal gene expression. Pediatr Res. 2011;69(4):285–92.
7. Grunau RE. Neonatal pain in very preterm infants: long-term effects on brain, neurodevelopment and pain reactivity. Rambam Maimonides Med J. 2013;29;4(4):e0025.
8. Currie JM. Stress and pain relief in the care of the surgical neonate. Semin Pediatr Surg. 2008;17:285–9.
9. Bellieni CV, Buonocore G. Neonatal pain treatment: ethical to be effective. J Perinatol. 2008;28:87–8.
10. Walker SM. Pain in children: recent advances and ongoing challenges. Br J Anaesth. 2008;101:101–10.
11. Anand KJ, Garg S, Rovnaghi CR, Narsinghani U, Bhutta AT, Hall RW. Ketamine reduces the cell death following inflammatory pain in newborn rat brain. Pediatr Res. 2007;62:283–90.
12. Anand KJ, Hickey PR. Pain and its effects in the human neonate and fetus. N Engl J Med. 1987;317:1321–9.
13. Bajic D, Commons KG, Soriano SG. Morphine-enhanced apoptosis in selective brain regions of neonatal rats. Int J Dev Neurosci. 2013;31(4):258–66.
14. Asmerom Y, Slater L, Boskovic DS, Bahjri K, Holden MS, Phillips R, Deming D, Ashwal S, Fayard E, Angeles DM. J Pediatr. 2013;163(1):29–35.e1.
15. Bhutta AT, Anand KJ. Vulnerability of the developing brain. Neuro Mech Clin Perinatol. 2002;29:357–72.
16. Buonocore G, Perrone S, Longini M, Paffetti P, Vezzosi P, Gatti MG, Bracci R. Non protein bound iron as early predictive marker of neonatal brain damage. Brain. 2003;126:1224–30.
17. Mishra OP, Delivoria-Papadopoulos M. Cellular mechanisms of hypoxic injury in the developing brain. Brain Res Bull. 1999;48:233–8.
18. Goplerud JM, Mishra OP, Delivoria-Papadopoulos M. Brain cell membrane dysfunction following acute asphyxia in newborn piglets. Biol Neonate. 1992;61:33–41.
19. Palmer C, Menzies SL, Roberts RL, Pavlick G, Connor JR. Changes in iron histochemistry after hypoxic–ischemic brain injury in the neonatal rat. J Neurosci Res. 1999;56:60–71.
20. Saugstad OD. Oxidative stress in the newborn – a 30-year perspective. Biol Neonate. 2005;88:228–36.
21. Sorg O, Horn TFW, Yu NC, Gruol DL, Bloom FE. Inhibition of astrocyte glutamate uptake by reactive oxygen species: role of antioxidant enzymes. Mol Med. 1997;3:431–40.
22. Buonocore G, Perrone S, Longini M, Terzuoli L, Bracci R. Total hydroperoxide and advanced oxidation protein products in preterm hypoxic babies. Pediatr Res. 2000;47:221–4.
23. Buonocore G, Zani S, Perrone S, Caciotti B, Bracci B. Intraerythrocyte nonprotein-bound iron and plasma malondialdehyde in the hypoxic newborn. Free Radic Biol Med. 1998;25:766–70.
24. Dorrepaal CA, Berger HM, Benders MJN, van Zoeren-Grobben D, Van de Bor M, Van Bel F. Non protein-bound iron in postasphyxial reperfusion injury of the newborn. Pediatrics. 1996;98:883–9.
25. Kaindl AM, Sifringer M, Koppelstaetter A, Genz K, Loeber R, Boerner C, Stuwe J, Klose J, Felderhoff-Mueser U. Erythropoietin protects the developing brain from hyperoxia-induced cell death and proteome changes. Ann Neurol. 2008;64:523–34.

26. Kakita H, Hussein MH, Daoud GA, Kato T, Murai H, Sugiura T, Mizuno K, Yamada Y, Ito T, Fukuda S, Kato I, Suzuki S, Togari H. Total hydroperoxide and biological antioxidant potentials in a neonatal sepsis model. Pediatr Res. 2006;60:675–9.
27. Ziskoven C, Jäger M, Zilkens C, Bloch W, Brixius K, Krauspe R. Oxidative stress in secondary osteoarthritis: from cartilage destruction to clinical presentation? Orthop Rev (Pavia). 2010;2(2):e23.
28. Rokyta R, Holecek V, Pekárkova I, Krejcová J, Racek J, Trefil L, Yamamotová A. Free radicals after painful stimulation are influenced by antioxidants and analgesics. Neuro Endocrinol Lett. 2003;24(5):304–9.
29. Kim HK, Zhang YP, Gwak YS, Abdi S. Phenyl N-tert-butylnitrone, a free radical scavenger, reduces mechanical allodynia in chemotherapy-induced neuropathic pain in rats. Anesthesiology. 2010;112(2):432–9.
30. Ohmichi Y, Sato J, Ohmichi M, Sakurai H, Yoshimoto T, Morimoto A, Hashimoto T, Eguchi K, Nishihara M, Arai YC, Ohishi H, Asamoto K, Ushida T, Nakano T, Kumazawa T. Two-week cast immobilization induced chronic widespread hyperalgesia in rats. Eur J Pain. 2012; 16(3):338–48.
31. Choi JI, Kim WM, Lee HG, Kim YO, Yoon MH. Role of neuronal nitric oxide synthase in the antiallodynic effects of intrathecal EGCG in a neuropathic pain rat model. Neurosci Lett. 2012;510(1):53–7.
32. Mathew NT. Pathophysiology of chronic migraine and mode of action of preventive medications. Headache. 2011;51 Suppl 2:84–92.
33. Fricova J, Vejražka M, Stopka P, Krizova J, Běláček J, Rokyta R. The influence of pre-emptive analgesia on postoperative analgesia and its objective evaluation. Arch Med Sci. 2010;6(5):764–71.
34. Rokyta R, Stopka P, Holeček Křikava K, Pekárková I. Direct measurement of free radicals in the brain cortex and the blood serum after nociceptive stimulation. Neuro Endocrinol Lett. 2004;25:252–6.
35. Păduraru I, Păduraru O, Manolidis G, Bild W, Haulică I. Antioxidant activity in rat models of nociceptive stress. Rev Med Chir Soc Med Nat Iasi. 2010;114(1):175–9.
36. Vaculin S, Franek M, Vejrazka M. Role of oxidative stress in animal model of visceral pain. Neurosci Lett. 2010;477(2):82–5.
37. Smith RP, Gitau R, Glover V, Fisk NM. Pain and stress in the human fetus. Eur J Obstet Gynecol Reprod Biol. 2000;92(1):161–5.
38. Collard CD, Gelman S. Pathophysiology, clinical manifestations, and prevention of ischemia-reperfusion injury. Anesthesiology. 2001;94(6):1133–8.
39. Johnson MA, Tekkanat K, Schmaltz SP, Fox IH. Adenosine triphosphate turnover in humans. Decreased degradation during relative hyperphosphatemia. J Clin Invest. 1989;84:990–5.
40. Slater L, Asmerom Y, Boskovic DS, Bahjri K, Plank MS, Angeles KR, Phillips R, Deming D, Ashwal S, Hougland K, Fayard E, Angeles DM. Procedural pain and oxidative stress in premature neonates. J Pain. 2012;13(6):590–7.
41. Bhutta AT, Anand KJ. Vulnerability of the developing brain. Neuro Mech Clin Perinatol. 2002;29(3):357–72.
42. Wu LJ. Microglial voltage-gated proton channel Hv1 in ischemic stroke. Transl Stroke Res. 2014;5:99–108.
43. Bhutta AT, Anand KJ. Vulnerability of the developing brain. Neuro Mech Clin Perinatol. 2002;29:357–72.
44. Carbajal R, Rousset A, Danan C, Coquery S, Nolent P, Ducrocq S, Saizou C, Lapillonne A, Granier M, Durand P, Lenclen R, Coursol A, Hubert P, de Saint BL, Boëlle PY, Annequin D, Cimerman P, Anand KJ, Bréart G. Epidemiology and treatment of painful procedures in neonates in intensive care units. JAMA. 2008;300(1):60–70.
45. Perrone S, Tataranno LM, Stazzoni G, Ramenghi L, Buonocore G. Brain susceptibility to oxidative stress in the perinatal period. J Matern Fetal Neonatal Med. 2013 Aug 23. [Epub ahead of print].

46. Perrone S, Tataranno ML, Stazzoni G, Buonocore G. J Matern Fetal Neonatal Med. 2012;25(12):2575–8. Early identification of the risk for free radical-related diseases in preterm newborns.
47. Perrone S, Tataranno ML, Negro S, Longini M, Marzocchi B, Proietti F, Iacoponi F, Capitani S, Buonocore G. Early Hum Dev. 2010;86(4):241–4.
48. Bellieni CV, Iantorno L, Perrone S, Rodriguez A, Longini M, Capitani S, Buonocore G. Even routine painful procedures can be harmful for the newborn. Pain. 2009;147(1–3):128–31.
49. den Hengst WA, Gielis JF, Lin JY, Van Schil PE, De Windt LJ, Moens AL. Lung ischemia-reperfusion injury: a molecular and clinical view on a complex pathophysiological process. Am J Physiol Heart Circ Physiol. 2010;299(5):H1283–99.
50. Bellieni CV, Alagna MG, Buonocore G. Analgesia for infants' circumcision. Ital J Pediatr. 2013;39:38. doi:10.1186/1824-7288-39-38.
51. Bellieni CV, Tei M, Coccina F, Buonocore G. Sensorial saturation for infants' pain. J Matern Fetal Neonatal Med. 2012;25 Suppl 1:79–81.

Chapter 17
Ontogeny of Antioxidant Systems

Richard L. Auten Jr.

Ontogeny of Antioxidant Systems

Planetary Ontogeny of Oxidative/Nitrosative Evolutionary Pressures

Mammalian ontogeny of antioxidant defenses has evolved under the selective pressure of the chemistry of highly reactive molecular species comprised of oxygen and nitrogen atoms in what became the biosphere: reactive oxygen species (ROS) and reactive nitrogen species (RNS) [1]. There is scarcely an aspect of biology that is not dependent on the interactions of ROS/RNS and the defense systems which regulate them [2–4].

Fixation of atmospheric nitrogen through the likely intermediate of nitric oxide formation in the CO_2-rich atmosphere led to the eventual evolution of life forms able to generate their own NO without relying on atmospheric NO which became decreasingly available with the fall in atmospheric CO_2. Formation of excess NO levels may have been ameliorated by the development of globins (globin, flavohemoglobin, hemoglobin) in bacteria, yeast, protozoa, and fungi which bind NO more avidly than they bind O_2 [4–6]. The emergence of prokaryotic hemoglobins before the evolution of the ability to transport O_2 supports this view. The ability of hemoglobins to scavenge O_2 would be advantageous to organisms contending with rising O_2 concentrations produced by photosynthetic bacteria, although this role for globins across taxa is not a settled question.

R.L. Auten Jr., M.D. (✉)
Neonatal-Perinatal Research Institute, Duke University Medical Center, 365 Sands Laboratory Building, DUMC Box 3373, Research Drive, Durham, NC 27710, USA
e-mail: auten@duke.edu

P.A. Dennery et al. (eds.), *Perinatal and Prenatal Disorders*,
Oxidative Stress in Applied Basic Research and Clinical Practice,
DOI 10.1007/978-1-4939-1405-0_17

In the Precambrian Era, cyanobacteria developed the capacity to fix CO_2 and release O_2 into the atmosphere, attaining O_2 concentrations of ~1 % of the present conditions [1]. Over the succeeding 1.5 billion years, this process continued, coincident with the development of aquatic metazoans and ultimately land plants and animals which had developed the ability to exchange respiratory gases available by diffusion.

Evolutionary Ontogeny of Antioxidant Enzyme (AOE) Systems (and Antinitrosant) Systems

The increasingly recognized important roles of RNS and ROS in the natural history of the evolution of the biosphere is partially recapitulated during mammalian development, which begins at zygote implantation in a low oxygen environment [7]. Cellular processes, from zygote implantation to fetal development, encounter and respond to temporospatially restricted oxidative and nitrosative milieux and stressors. In general, for several organ systems, expression of antioxidant enzymes that include superoxide dismutases (SOD1-3), catalase (CAT), and glutathione peroxidases (GPX) tends to rise during gestation [8], along with vasculogenesis, angiogenesis, and the burden of oxidative stress that accompanies rising metabolic demands.

The ontogeny of the antioxidant systems must be considered in the context of NO signaling as well. The rapid, diffusion limited inactivation of NO by superoxide to form peroxynitrite terminates both NO signaling and potentially ROS-mediated signaling. This chemistry suggests a requirement for tightly restricted temporospatial conditions under which these species coexist in biological systems and may serve as reciprocal signaling pathways [9, 10]. The restricted expression of NOS and SOD isoforms in various organ systems, as suggested below, conforms to this idea.

In addition to activation of soluble guanylyl cyclase, its canonical target, NO affects many biological regulatory pathways through the oxygen-dependent *S*-nitrosylation of critical thiol residues to form S-nitrosothiol (SNO). This post-translational modification can affect protein structure and function, affecting a wide variety of pathways ranging from apoptosis to cell differentiation (see Ref. [11] for review). Physiologic modulation of endogenous SNO signaling would therefore logically include a mechanism to regulate this modification. SNO denitrosylation [12, 13] can be achieved by the action of thioredoxin-1 (Trx-1) [14], shown to be inducible in the newborn rat brain by oxidative stress [15]. Trx-1 functional effects on *S*-nitrosothiol modifications are strictly dependent on primary sequence and tertiary structure of proteins that determine the access of Trx-1 to the putative catalytic thiol. Structural selectivity for Trx-1 "access" may have arisen early in evolution, since bacterial Trx uses both single-electron and S_N2 nucleophilic mechanisms [16]. The same "rules" for Trx-1 expression (and thioredoxin reductase [17]) as for AOE expression should therefore apply: tight control requires tightly regulated temporospatial expression. However, at present, we lack cell-specific and developmental stage-specific data on Trx-1/Trx reductase expression.

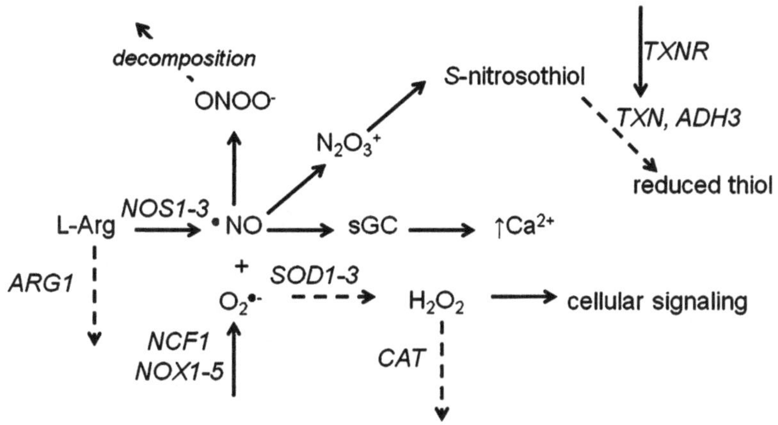

Fig. 17.1 Interrelationships among classical AOE systems, endogenous ROS production, and endogenous NO production. *ARG1* arginase-1, *NCF1* neutrophil cytosolic factor-1 (p47^{phox}), *sGC* soluble guanylyl cyclase, *NOX1-5* NAD(P)H oxidases 1–5, *SOD1-3* superoxide dismutases 1–3, *CAT* catalase, *TXN* thioredoxin, *TXNR* thioredoxin reductase, *ADH3* alcohol dehydrogenase 3 (*S*-nitrosoglutathione reductase), *NOS 1–3* nitric oxide synthases 1–3

Similar caveats apply to *S*-nitrosoglutathione reductase (also known as alcohol dehydrogenase-3, ADH3), which also metabolizes SNO: it is likely to be critical to the overall functional contribution of endogenously produced or pharmacologically induced SNO accumulation and pharmacologic effects, but very little is known about its temporospatial pattern of expression in mammalian systems. The interrelationships of these systems are depicted in Fig. 17.1. Not depicted is the wider network of the so-called phase 2 or indirect systems that include detoxifying enzymes through the biosynthesis or recycling of thiols (e.g., glutathione *S*-transferase, NAD(P)H:quinone oxidoreductase (NQO1), γ-glutamyl cysteine ligase, peroxiredoxins) or the mechanisms that augment clearance of potentially ROS-enhancing metabolites (e.g., quinones, aldehydes, peroxides). These systems are addressed at the end of this chapter, in consideration of antioxidant and antinitrosant development in the pulmonary system.

This chapter will therefore focus on what is known about the temporospatial ontogeny of the classical antioxidant (and antinitrosant) systems in three organ systems for which we have some information: placenta, brain, lung, as well as on some of the disorders that challenge these systems.

Antioxidant Enzyme (AOE) System Ontogeny in Placenta

The contribution of the systems governing ROS/RNS bioavailability and signaling during human gestation is necessarily inferential. The relevance to human biology of information gained with widely used animal model systems may be even more

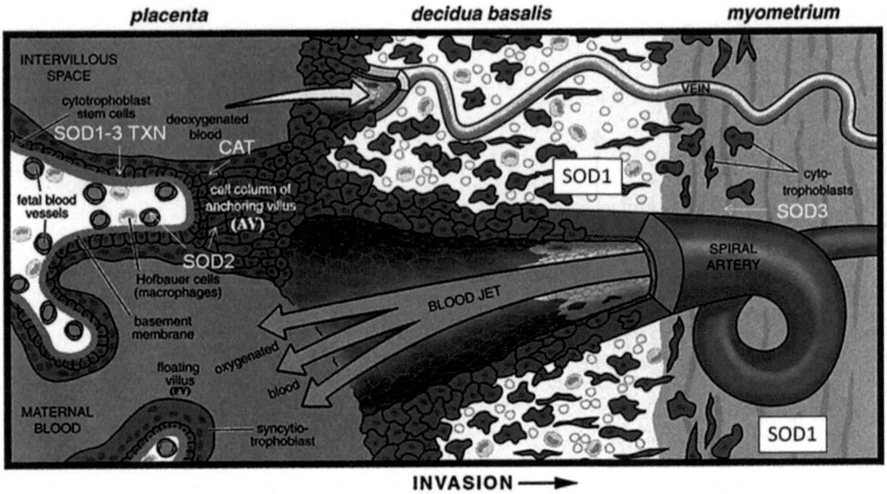

Fig. 17.2 Temporospatial expression of antioxidants in the placenta. Superoxide dismutases (SOD 1, 2, 3) are expressed in the cytotrophoblast, along with thioredoxin. At term, SOD1 is abundant in myometrium, as well as syncytiotrophoblast and decidua. SOD2 expression is similar except higher labeling is observed in fetal villous endothelium. Expression is not affected by labor. Intracellular SOD3 expression is abundant in villous trophoblast in the first trimester but appears to be relocated to extracellular matrix within villi after 17 weeks. Since SOD expression supports NO survival and angiogenic signaling, it may be that intracellular SOD3 expression is important for early placental vascular development. Thioredoxin (TXN) expression in trophoblast is increased in women with preeclampsia. Catalase (CAT), which metabolizes superoxide to hydrogen peroxide, is found in trophoblast with increasing immunolabel intensity as pregnancy progresses (Reprinted from Davis and Auten [20] with permission from Elsevier)

limited in the case of the placenta since mammalian reproduction has significant diversity of placental systems [18]. Despite morphologic distinctions, the functional similarities among mammalian placental systems afford some rational basis for systems biological approaches to better understand conserved biochemical and genetic networks [19].

In humans as in most animal models, the cellular and tissue sites of rising oxidative stress appear to be balanced by accompanying increases in antioxidant enzyme expression, as recently reviewed [20]. With advancing gestational age, superoxide dismutase, catalase, glutathione peroxidase, and glutathione reductase activities rise in placental homogenates obtained from abortuses [21]. Cellular and subcellular expression of superoxide dismutase 3 (SOD3) increases in villous trophoblast in the second and third trimester, which is accompanied by an extracellular shift of SOD3 immunoreactivity to extracellular matrix and fetal vessels [22]. The spatial expression of AOE in placenta is depicted in Fig. 17.2.

In experimental model systems, murine *Sod3* mRNA expression in extraembryonic tissues also gradually rose from E8.5 to E18.5, with roughly parallel changes in protein expression determined by immunoblotting. In situ RNA hybridization showed highest expression in spongiotrophoblast at E16 and labyrinthine trophoblast by E18 [23]. In contrast, the similar trend toward rising SOD1 activity throughout gestation in ovine placenta is accompanied by a decline in SOD2 and catalase activity [24]. Species differences in the structure and function of the placenta [18], and the metabolic demands of altricial versus precocial species [25] may account for some of these disparities in AOE temporospatial expression/activity patterns.

The effects of ROS on placental function depend not only on AOE expression but necessarily on the contribution of endogenous ROS production. Myatt and colleagues have found that NAD(P)H oxidase (NOX) isoform expression is region- and cell-specific in the placenta (see Ref. [26] for review), but the contribution of these isoforms to normal placental development and disease is largely inferential. Isolated explant studies implicate NOX-derived ROS contributions to human decidual cell differentiation [27], but studies linking NOX-derived ROS to placental development are restricted to animal studies, particularly mice [28].

Despite the long-standing observation that a number of common conditions in pregnancy are accompanied by placental oxidative stress—pregnancy-induced hypertension, caloric restriction [29], and ethanol use—among others, experimental studies aimed at supplementing antioxidant defenses have yielded decidedly mixed results, which will be discussed in more detail elsewhere in this volume.

Placental Defenses Against Maladaptive Reactive Nitrogen Species

Placentation is dependent on adaptive NO signaling, which governs trophoblast invasion [30], placental vascular development [31], and vascular tone. Surprisingly little is known about the precise role of abnormal amounts of NO species in this process or the contribution of NOS isoform functions in the placenta under physiologic or pathologic conditions [32].

Indirect evidence has implicated peroxynitrite formation in connection with conditions typically afflicted by placental insufficiency, such as preeclampsia and maternal diabetes [33] as well as other abnormal conditions [34], with vascular explants from placentae of affected pregnancies demonstrating impaired vasoreactivity. Pathway analysis of proteomic studies of placentae from preeclamptic pregnancies has shown *S*-nitrosothiol modifications in systems governing cell metabolism, replication, and signaling [35].

Because of the profound effect that posttranslational RNS-derived modifications can have on enzyme function, it would be expected that the placenta would have denitrosation mechanisms—e.g., thioredoxin (TXN)/thioredoxin reducatse (TXNR) and *S*-nitrosoglutathione reductase (ADH3)—to regulate *S*-nitrosation. Although TXN is known to be expressed in placenta, there is little known about temporospatial relationships, so inference of potential functional roles in disease is necessarily limited. TXN is upregulated in placentae of preterm deliveries, along with markers of inflammation and oxidative stress [36], and its overexpression in mice promotes fetal growth [37]. Consistent with this observation, mouse models of assisted reproduction (intracytoplasmic sperm injection) yielded placentae with diminished activities of SOD and TXNR (which would tend to abrogate NO-based signaling/*S*-nitrosation) [38].

Antioxidant/Nitrosant System Ontogeny in the Perinatal Central Nervous System

Because of the ubiquitous involvement of ROS and RNS with normal cellular processes and development, it follows that aberrations in defenses against ROS/RNS could contribute to abnormal CNS development. For example, cerebral palsy is most commonly unaccompanied by identifiable risk factors during pregnancy, yet it has been linked to elevations of proinflammatory cytokines and coagulation factors measured at birth [39]. Since many effects of neuroinflammation are mediated through endogenous ROS/RNS production, developmentally distinct patterns of AOE expression and function in the CNS will likely dictate the regions and cell types that are vulnerable to oxidative stress.

Normal organogenesis takes place in a "Goldilocks zone" of redox-dependent processes such as cellular proliferation, differentiation, and apoptosis [40]. Hypoxia, somewhat nebulously used in this context, refers to bioavailable oxygen that is too low, to support adaptive metabolic processes, and may develop as a result of abnormalities of placentation. Furthermore, hypoxia promotes formation of ROS in mitochondria and NO by activation of guanylyl cyclase. The fetal brain has specific features that may increase vulnerability to oxidative stress. High metabolic demands present during organogenesis, along with particular vulnerability of oligodendrocytes to oxidative stress [41, 42]. The developmental susceptibility of the developing fetal brain to this particular mechanism declines with maturation toward term because of the cellular locations of oligodendrocytes in the more peripheral, mature brain, rather than in the germinal matrix.

Development of the CNS Antioxidant Repertoire

The anatomic vulnerabilities during fetal brain development are accompanied by potential biochemical vulnerabilities. Analysis of AOE expression in homogenates of human pre-myelinated, telencephalic white matter obtained at 20–35 weeks' gestation from autopsy materials (non-neurologic diagnoses) suggests a steady rise in SOD2 and catalase in direct proportion to gestation/postnatal development between 20 weeks' gestation and ~2–3 months postnatal, but a fall in SOD1 until ~32 weeks' gestation followed by a sharp rise continuing for several postnatal months. Expression of AOEs in developing brain was found to be spatially restricted, with SOD2 expression prominent in vessels at 20–27 weeks, and then including glial cells by 35 weeks. In contrast, CAT and GPX expressions were observed in both compartments beginning at 20 weeks. This apparent mismatch between peroxide generating and peroxide degrading systems has been suggested as a mechanism underlying mid-gestation vulnerability of fetal oligodendrocytes in particular [43], an idea supported by the observed protection of neonatal rat oligodendrocytes by overexpression of SOD2 [44]. In contrast, the expression of SOD1,2, CAT, and GPX in murine brain was observed to rise from E18 to P21, although the activity levels fell, illustrating the difficulty of interpreting likely biological effects in whole organ analyses [45], since differing regions of the brain experience different redox milieux and have disparate, cell-specific vulnerabilities to ROS/RNS [46].

Development of CNS Nitrosative Stress Defenses

Since ROS and RNS interact to terminate endogenous NO signaling, it follows that temporospatial expression of tissue-specific NOS and SOD isoforms in the developing brain could be coordinate [47]. In human autopsy material ranging from 13 weeks' gestation to 2 years of age, immunohistochemical studies of the expression of SOD1, SOD2, NOS1, and NOS2 suggest that SOD isoform expression precedes NOS isoform expression throughout gestation in neuroblasts and mature neurons in cerebrum, with NOS1 staining in neurons only evident in brainstem after 15 weeks of gestation and in cerebrum after 28 weeks of gestation. Since NO is believed to affect neuronal differentiation, widespread expression of functional SOD1 would serve to restrict the degradation of NO signaling (and formation of peroxynitrite) by superoxide production.

On the other hand, NO biological signaling, including regulation of brainstem ventilatory hypoxia responses [48], suppression of NF-κB activation [49], augmentation of cerebral blood flow [50], and calcium channel function [51], as well as apoptosis [52] and many others [53], are substantially mediated by formation of *S*-nitrosothiols (SNO), which are formed by the oxygen-dependent interaction of NO with susceptible regulatory protein thiol residues (see Ref. [54] for review). During experimental stroke, loss of NO signaling that accompanies increased peroxynitrite formation and oxidative stroke can be ameliorated by supplementation with *S*-nitrosoglutathione without increasing nitrosative stress [55], probably through effects on endothelial function.

S-nitrosoglutathione reductase (ADH3) also serves to terminate SNO-based signaling [11] and may participate in the downregulation of brainstem ventilatory responses to hypoxia-induced increases in circulating SNO [56], inferred by effects observed in $Adh3^{-/-}$ mice. ADH3 has been shown by in situ RNA hybridization to be expressed in embryonic rat brain beginning at embryonic day ~12.5 and in mouse at embryonic day 19.5 [57]. At present, there is no published information about its expression in developing human brain.

Prenatal Challenges to AOE Defenses in the CNS

Experimental umbilical cord occlusion in fetal sheep, which admittedly does not directly mimic the abnormalities in gas exchange that would be expected in clinical conditions affected by abnormal placentation, nevertheless induces increases in ROS in white and gray matter [58, 59]. Likewise, postnatal acute hypoxia in newborn piglets activates cortical apoptosis in a manner dependent on induction of neuronal-NOS [60].

Oxidative stress during gestation may suppress otherwise potentially adaptive induction of antioxidant systems. A mouse model of fetal alcohol syndrome showed that daily ethanol administration to pregnant dams from E7-E11 suppressed the mRNA expression of fetal brain superoxide dismutase (*Sod1*), catalase, and glutathione peroxidase measured at E18 [61]. In a similar rat model of ethanol in drinking water, some of the adverse effects of fetal exposure on brain glutathione in offspring were ameliorated by postnatal treatment with ω-3 fatty acids, but it must be emphasized that this model system has selective effects on glutathione metabolism, with no effect on SOD or catalase expression [62]. These findings raise the question of whether impaired AOE induction could be remedied by supplementation, which is addressed in detail elsewhere in this volume.

Antioxidant, Antinitrosant System Development in the Lung

Several animal model systems demonstrate that, in general, pulmonary AOE expression is developmentally regulated, with rises in expression and activity [63–68]. SOD3 expression rises after birth in preterm and term rabbit [69], and its expression appears to be induced by oxidative stress, although there may be a lack of concordance with activity [70]. In general, animal studies of oxidative stress in neonatal lung injury models show that oxidative stress impairs lung development and that antioxidant treatments ameliorate this effect (see Ref. [71] for review).

AOE expression and function in the fetal and neonatal lung are under hormonal control of both glucocorticoids and thyroid hormones [72]. Interestingly, combined treatment with thyrotropin releasing hormone (TRH) and glucocorticoids during late gestation suppresses fetal rat lung AOE expression and activity, but pups born to rat dams after such dual hormone treatment paradoxically exhibit a greater postnatal induction of AOE activity and expression (at the whole lung level) and slightly enhanced survival when exposed to hyperoxia [72]. These findings suggest additional mechanisms besides glucocorticoids and TRH that control the AOE system.

Nuclear factor (erythroid-derived 2)-like 2, also known as Nrf2, encoded by the *NFE2L2* gene, is one candidate for this additional control mechanism, although both AP-1 and NF-κB are redox sensitive. Nrf2 is a transcription factor that controls expression of several antioxidant genes by binding the antioxidant response element and is redox sensitive owing to the thiol of its repressor, Keap1 [73] (see Ref. [74] for review). Deletion of *Nfe2l2* in adult [75] and neonatal [76] mice increases susceptibility to oxidative stress induced by hyperoxia with commensurate depression of AOE expression. Unpublished findings suggest that lung expression of Nrf2 peaks during late fetal and early neonatal life in rats [77]. Surprisingly, there is nothing yet published on the cell-specific regulation of Nrf2 in fetal and neonatal development.

Premature delivery subjects the immature lung to elevated redox stress at a time when endogenous antioxidant systems are being poorly developed [64]. Indirect measurements of redox stress suggested that premature newborns have impaired antioxidant capacity [78, 79]. SOD1 and SOD2 expression in fetal lung has been

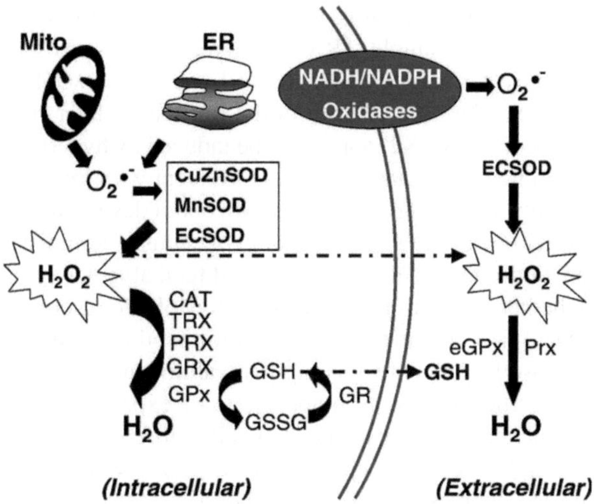

Fig. 17.3 Intracellular and extracellular distribution of various antioxidant enzymes and their interrelationship. *Trx* thioredoxin, *Prx* peroxiredoxins, *Grx* glutathione reductases, *GPx* glutathione peroxidase, *GSH/GSSG* oxidized, reduced glutathione (Reprinted from Rahman et al. [87] with permission from Elsevier)

shown to rise in peripheral developing airways in mid-gestation [80], as does CAT [81]. Although autopsy studies show comparable levels of inducible AOE expression in lungs from babies that died with BPD compared with those without lung disease [82], we lack information about the level of expression at the initiation of the injury process.

A number of environmental exposures may also contribute to altered regulation of AOE defenses through epigenetic effects. SOD2 regulation by microRNA has been reported in the context of intestinal neoplasia [83], and SOD2 histone methylation in the retina has been implicated in the development of diabetic retinopathy [84]. NRF2 regulates AOE transcription and has itself been shown to be regulated by microRNAs and DNA methylation, chiefly in the context of neoplasia but also in association with xenobiotic exposure (see Ref. 85 for review). Although these mechanisms have not been studied directly in the context of the ontogeny of antioxidant systems, circumstantial evidence demonstrates vulnerability of circulating glutathione and cysteine levels in autism in association with redox-relevant epigenetic alterations [86].

Antinitrosant System: Lung

Somewhat more is known about the control of NO-driven biological responses in developing lung than in placenta or brain. As depicted in Fig. 17.3, the AOE function is spatially restricted depending on the cellular, subcellular, and tissue compartment: this effectively controls, to some extent, the antinitrosant regulation of NO-driven processes. For example, TRX, which regulates *S*-nitrosothiol abundance, is in turn regulated by peroxiredoxins (PRDX) 1 and 2. The whole lung expression of PRDX1 peaks near the time of birth in the rat, with a later rise in whole lung mRNA. PRDX1 (but not PRDX2) appears to be induced by hyperoxia in neonatal rat lung [88] as it is in newborn baboons [89]. Consistent with the state of knowledge in placenta and brain, surprisingly little is known about developmental TRX expression in lung. Indirect evidence suggests that exogenously administered TRX could serve to augment antioxidant defenses in isolated fetal alveolar epithelium [90], but it is unknown whether it plays such a role as a physiologic response.

Although ADH3 is a potent reductase of *S*-nitrosothiol in the lung, capable of terminating *S*-nitrosothiol-based signaling [91], almost nothing is known about its pulmonary cell-specific expression. It appears to be expressed throughout the gastrointestinal tract in adult rat, but with cell-specific differences in the magnitude of expression, with higher levels in epithelium [92]. One would expect similar cell specificity in ADH3 expression in the lung, but at present, there are no published studies addressing this question.

Summary

The development of the classical antioxidant enzyme response systems in perinatal development takes place in tightly orchestrated, temporospatially specified patterns, similar to those patterns typical of embryonic morphogens, and parallels in part the expected spatially restricted increases in redox stress observed at the interfaces with the circulatory and respiratory systems. Emerging evidence suggests similarly tight regulation of systems that produce ROS and RNS as well as those that degrade them, in keeping with their nearly ubiquitous role in the control of gene expression as well as posttranslational effects on biological responses.

References

1. Van Der Giezen M, Lenton TM. The rise of oxygen and complex life. J Eukaryot Microbiol. 2012;59(2):111–3.
2. Dennery PA. Role of redox in fetal development and neonatal diseases. Antioxid Redox Signal. 2004;6(1):147–53.
3. Pacher P, Beckman JS, Liaudet L. Nitric oxide and peroxynitrite in health and disease. Physiol Rev. 2007;87(1):315–424.
4. Derakhshan B, Hao G, Gross SS. Balancing reactivity against selectivity: the evolution of protein S-nitrosylation as an effector of cell signaling by nitric oxide. Cardiovasc Res. 2007;75(2):210–9.
5. Angelo M, et al. Interactions of NO with hemoglobin: from microbes to man. Methods Enzymol. 2008;436:131–68.
6. Forrester MT, Foster MW. Protection from nitrosative stress: a central role for microbial flavohemoglobin. Free Radic Biol Med. 2012;52(9):1620–33.

7. Burton GJ, Hempstock J, Jauniaux E. Oxygen, early embryonic metabolism and free radical-mediated embryopathies. Reprod Biomed Online. 2003;6(1):84–96.
8. Frank L, Groseclose EE. Preparation for birth into an O2-rich environment: the antioxidant enzymes in the developing rabbit lung. Pediatr Res. 1984;18(3):240–4.
9. Beckman JS, Koppenol WH. Nitric oxide, superoxide, and peroxynitrite: the good, the bad, and ugly. Am J Physiol. 1996;271(5 Pt 1):C1424–37.
10. Bonaventura J, Gow A. NO and superoxide: opposite ends of the seesaw in cardiac contractility. Proc Natl Acad Sci U S A. 2004;101(47):16403–4.
11. Foster MW, Hess DT, Stamler JS. Protein S-nitrosylation in health and disease: a current perspective. Trends Mol Med. 2009;15(9):391–404.
12. Benhar M, Forrester MT, Stamler JS. Protein denitrosylation: enzymatic mechanisms and cellular functions. Nat Rev Mol Cell Biol. 2009;10(10):721–32.
13. Anand P, Stamler JS. Enzymatic mechanisms regulating protein S-nitrosylation: implications in health and disease. J Mol Med (Berl). 2012;90(3):233–44.
14. Wu C, et al. Redox regulatory mechanism of transnitrosylation by thioredoxin. Mol Cell Proteomics. 2010;9(10):2262–75.
15. Bendix I, et al. Hyperoxia changes the balance of the thioredoxin/peroxiredoxin system in the neonatal rat brain. Brain Res. 2012;1484:68–75.
16. Perez-Jimenez R, et al. Diversity of chemical mechanisms in thioredoxin catalysis revealed by single-molecule force spectroscopy. Nat Struct Mol Biol. 2009;16(8):890–6.
17. Sengupta R, Holmgren A. Thioredoxin and thioredoxin reductase in relation to reversible S-Nitrosylation. Antioxid Redox Signal. 2013;18(3):259–69.
18. Carter AM. Animal models of human placentation – a review. Placenta. 2007;28(Suppl A): S41–7.
19. Cox B, et al. Comparative systems biology of human and mouse as a tool to guide the modeling of human placental pathology. Mol Syst Biol. 2009;5:279.
20. Davis JM, Auten RL. Maturation of the antioxidant system and the effects on preterm birth. Semin Fetal Neonatal Med. 2010;15(4):191–5.
21. Qanungo S, Mukherjea M. Ontogenic profile of some antioxidants and lipid peroxidation in human placental and fetal tissues. Mol Cell Biochem. 2000;215(1–2):11–9.
22. Boggess KA, et al. Differential localization of placental extracellular superoxide dismutase as pregnancy progresses. Am J Obstet Gynecol. 2000;183(1):199–205.
23. Kim MR, et al. Expression profiles of extracellular superoxide dismutase during mouse organogenesis. Gene Expr Patterns. 2011;11(3–4):207–15.
24. Garrel C, Fowler PA, Al-Gubory KH. Developmental changes in antioxidant enzymatic defences against oxidative stress in sheep placentomes. J Endocrinol. 2010;205(1):107–16.
25. Carter AM. Evolution of factors affecting placental oxygen transfer. Placenta. 2009; 30(Suppl A):S19–25.
26. Myatt L. Review: Reactive oxygen and nitrogen species and functional adaptation of the placenta. Placenta. 2010;31(Suppl):S66–9.
27. Al-Sabbagh M, et al. NADPH oxidase-derived reactive oxygen species mediate decidualization of human endometrial stromal cells in response to cyclic AMP signaling. Endocrinology. 2011;152(2):730–40.
28. Bevilacqua E, et al. NADPH oxidase as an important source of reactive oxygen species at the mouse maternal-fetal interface: putative biological roles. Reprod Biomed Online. 2012;25(1):31–43.
29. Vieira-Filho LD, et al. Placental oxidative stress in malnourished rats and changes in kidney proximal tubule sodium ATPases in offspring. Clin Exp Pharmacol Physiol. 2009;36(12): 1157–63.
30. Ayling LJ, et al. Dimethylarginine dimethylaminohydrolase (DDAH) regulates trophoblast invasion and motility through effects on nitric oxide. Hum Reprod. 2006;21(10):2530–7.
31. Teichert AM, et al. Endothelial nitric oxide synthase gene expression during murine embryogenesis: commencement of expression in the embryo occurs with the establishment of a unidirectional circulatory system. Circ Res. 2008;103(1):24–33.

32. Krause BJ, Hanson MA, Casanello P. Role of nitric oxide in placental vascular development and function. Placenta. 2011;32(11):797–805.
33. Kossenjans W, et al. Role of peroxynitrite in altered fetal-placental vascular reactivity in diabetes or preeclampsia. Am J Physiol Heart Circ Physiol. 2000;278(4):H1311–9.
34. Webster RP, Roberts VH, Myatt L. Protein nitration in placenta – functional significance. Placenta. 2008;29(12):985–94.
35. Zhang HH, Wang YP, Chen DB. Analysis of nitroso-proteomes in normotensive and severe preeclamptic human placentas. Biol Reprod. 2011;84(5):966–75.
36. Song J, et al. The expression of thioredoxin-1 in preterm delivery placenta. Redox Rep. 2012;17(5):187–93.
37. Umekawa T, et al. Overexpression of thioredoxin-1 reduces oxidative stress in the placenta of transgenic mice and promotes fetal growth via glucose metabolism. Endocrinology. 2008;149(8):3980–8.
38. Raunig JM, et al. Placental inflammation and oxidative stress in the mouse model of assisted reproduction. Placenta. 2011;32(11):852–8.
39. Nelson KB, et al. Neonatal cytokines and coagulation factors in children with cerebral palsy. Ann Neurol. 1998;44(4):665–75.
40. Buonocore G, Perrone S, Tataranno ML. Oxygen toxicity: chemistry and biology of reactive oxygen species. Semin Fetal Neonatal Med. 2010;15(4):186–90.
41. Mishra OP, Delivoria-Papadopoulos M. Cellular mechanisms of hypoxic injury in the developing brain. Brain Res Bull. 1999;48(3):233–8.
42. Volpe JJ, et al. The developing oligodendrocyte: key cellular target in brain injury in the premature infant. Int J Dev Neurosci. 2011;29(4):423–40.
43. Folkerth RD, et al. Developmental lag in superoxide dismutases relative to other antioxidant enzymes in premyelinated human telencephalic white matter. J Neuropathol Exp Neurol. 2004;63(9):990–9.
44. Baud O, et al. Developmental up-regulation of MnSOD in rat oligodendrocytes confers protection against oxidative injury. Eur J Neurosci. 2004;20(1):29–40.
45. Khan JY, Black SM. Developmental changes in murine brain antioxidant enzymes. Pediatr Res. 2003;54(1):77–82.
46. Xu L, Sapolsky RM, Giffard RG. Differential sensitivity of murine astrocytes and neurons from different brain regions to injury. Exp Neurol. 2001;169(2):416–24.
47. Takikawa M, et al. Temporospatial relationship between the expressions of superoxide dismutase and nitric oxide synthase in the developing human brain: immunohistochemical and immunoblotting analyses. Acta Neuropathol. 2001;102(6):572–80.
48. Lipton AJ, et al. S-nitrosothiols signal the ventilatory response to hypoxia. Nature. 2001; 413(6852):171–4.
49. Marshall HE, Stamler JS. Inhibition of NF-kappa B by S-nitrosylation. Biochemistry. 2001;40(6):1688–93.
50. Sheng H, et al. Pharmacologically augmented S-nitrosylated hemoglobin improves recovery from murine subarachnoid hemorrhage. Stroke. 2011;42(2):471–6.
51. Joksovic PM, et al. Functional regulation of T-type calcium channels by s-nitrosothiols in the rat thalamus. J Neurophysiol. 2007;97(4):2712–21.
52. Mannick JB, et al. Fas-induced caspase denitrosylation. Science. 1999;284(5414):651–4.
53. Forrester MT, et al. Thioredoxin-interacting protein (Txnip) is a feedback regulator of S-nitrosylation. J Biol Chem. 2009;284(52):36160–6.
54. Foster MW, McMahon TJ, Stamler JS. S-nitrosylation in health and disease. Trends Mol Med. 2003;9(4):160–8.
55. Khan M, et al. The inhibitory effect of S-nitrosoglutathione on blood-brain barrier disruption and peroxynitrite formation in a rat model of experimental stroke. J Neurochem. 2012;123 Suppl 2:86–97.
56. Palmer LA, et al. Ventilatory responses during and following exposure to a hypoxic challenge in conscious mice deficient or null in S-nitrosoglutathione reductase. Respir Physiol Neurobiol. 2013;185(3):571–81.

57. Westerlund M, et al. Tissue- and species-specific expression patterns of class I, III, and IV Adh and Aldh 1 mRNAs in rodent embryos. Cell Tissue Res. 2005;322(2):227–36.
58. Castillo-Melendez M, Chow JA, Walker DW. Lipid peroxidation, caspase-3 immunoreactivity, and pyknosis in late-gestation fetal sheep brain after umbilical cord occlusion. Pediatr Res. 2004;55(5):864–71.
59. Miller SL, et al. Melatonin provides neuroprotection in the late-gestation fetal sheep brain in response to umbilical cord occlusion. Dev Neurosci. 2005;27(2–4):200–10.
60. Mishra OP, Delivoria-Papadopoulos M. Mechanism of tyrosine phosphorylation of procaspase-9 and Apaf-1 in cytosolic fractions of the cerebral cortex of newborn piglets during hypoxia. Neurosci Lett. 2010;480(1):35–9.
61. Drever N, et al. The expression of antioxidant enzymes in a mouse model of fetal alcohol syndrome. Am J Obstet Gynecol. 2012;206(4):358.e19-22.
62. Patten AR, Brocardo PS, Christie BR. Omega-3 supplementation can restore glutathione levels and prevent oxidative damage caused by prenatal ethanol exposure. J Nutr Biochem. 2013;24(5):760–9.
63. Frank L, Sosenko IR. Prenatal development of lung antioxidant enzymes in four species. J Pediatr. 1987;110(1):106–10.
64. Frank L, Sosenko IR. Development of lung antioxidant enzyme system in late gestation: possible implications for the prematurely born infant. J Pediatr. 1987;110(1):9–14.
65. Sosenko IR, Frank L. Guinea pig lung development: antioxidant enzymes and premature survival in high O2. Am J Physiol. 1987;252(4 Pt 2):R693–8.
66. Yuan HT, Bingle CD, Kelly FJ. Differential patterns of antioxidant enzyme mRNA expression in guinea pig lung and liver during development. Biochim Biophys Acta. 1996;1305(3):163–71.
67. Asayama K, et al. Immunohistochemical study on perinatal development of rat superoxide dismutases in lungs and kidneys. Pediatr Res. 1991;29(5):487–91.
68. Sharma S, et al. Lung antioxidant enzymes are regulated by development and increased pulmonary blood flow. Am J Physiol Lung Cell Mol Physiol. 2007;293(4):L960–71.
69. Nozik-Grayck E, et al. Secretion of extracellular superoxide dismutase in neonatal lungs. Am J Physiol Lung Cell Mol Physiol. 2000;279(5):L977–84.
70. Mamo LB, et al. Discordant extracellular superoxide dismutase expression and activity in neonatal hyperoxic lung. Am J Respir Crit Care Med. 2004;170(3):313–8.
71. Auten RL, White CW. Oxidative and nitrosative stress and bronchopulmonary dysplasia. In: Abman SH, editor. Bronchopulmonary dysplasia. Philadelphia: Informa Health Care; 2009. p. 499.
72. Chen Y, Whitney PL, Frank L. Negative regulation of antioxidant enzyme gene expression in the developing fetal rat lung by prenatal hormonal treatments. Pediatr Res. 1993;33(2):171–6.
73. Buckley BJ, Li S, Whorton AR. Keap1 modification and nuclear accumulation in response to S-nitrosocysteine. Free Radic Biol Med. 2008;44(4):692–8.
74. Cho HY, Reddy SP, Kleeberger SR. Nrf2 defends the lung from oxidative stress. Antioxid Redox Signal. 2006;8(1–2):76–87.
75. Cho HY, et al. Role of NRF2 in protection against hyperoxic lung injury in mice. Am J Respir Cell Mol Biol. 2002;26(2):175–82.
76. McGrath-Morrow S, et al. Nrf2 increases survival and attenuates alveolar growth inhibition in neonatal mice exposed to hyperoxia. Am J Physiol Lung Cell Mol Physiol. 2009;296(4):L565–73.
77. Kadkodayan, R., et al. Developmental expression and interaction of Nrf2 with ECSOD in developing lung. In: American Thoracic Society International Conference. San Diego: American Thoracic Society; 2009.
78. Saugstad OD, Rognum TO. High postmortem levels of hypoxanthine in the vitreous humor of premature babies with respiratory distress syndrome. Pediatrics. 1988;81(3):395–8.
79. Jain A, et al. Glutathione metabolism in newborns: evidence for glutathione deficiency in plasma, bronchoalveolar lavage fluid, and lymphocytes in prematures. Pediatr Pulmonol. 1995;20(3):160–6.

80. Dobashi K, et al. Immunohistochemical study of copper-zinc and manganese superoxide dismutases in the lungs of human fetuses and newborn infants: developmental profile and alterations in hyaline membrane disease and bronchopulmonary dysplasia. Virchows Arch A Pathol Anat Histopathol. 1993;423(3):177–84.
81. Asikainen TM, et al. Expression and developmental profile of antioxidant enzymes in human lung and liver. Am J Respir Cell Mol Biol. 1998;19(6):942–9.
82. Kaarteenaho-Wiik R, Kinnula VL. Distribution of antioxidant enzymes in developing human lung, respiratory distress syndrome, and bronchopulmonary dysplasia. J Histochem Cytochem. 2004;52(9):1231–40.
83. Meng X, et al. Genetic and epigenetic down-regulation of microRNA-212 promotes colorectal tumor metastasis via dysregulation of MnSOD. Gastroenterology. 2013;145(2):426–36.e1-6.
84. Zhong Q, Kowluru RA. Epigenetic modification of Sod2 in the development of diabetic retinopathy and in the metabolic memory: role of histone methylation. Invest Ophthalmol Vis Sci. 2013;54(1):244–50.
85. Bryan HK, et al. The Nrf2 cell defence pathway: Keap1-dependent and -independent mechanisms of regulation. Biochem Pharmacol. 2013;85(6):705–17.
86. Waly MI, et al. Prenatal and postnatal epigenetic programming: implications for GI, immune, and neuronal function in autism. Autism Res Treat. 2012;2012:190930.
87. Rahman I, Biswas SK, Kirkham PA. Regulation of inflammation and redox signaling by dietary polyphenols. Biochem Pharmacol. 2006;72(11):1439–52.
88. Kim HS, et al. Rat lung peroxiredoxins I and II are differentially regulated during development and by hyperoxia. Am J Physiol Lung Cell Mol Physiol. 2001;280(6):L1212–7.
89. Das KC, et al. Induction of peroxiredoxin gene expression by oxygen in lungs of newborn primates. Am J Respir Cell Mol Biol. 2001;25(2):226–32.
90. Chen Y, et al. Thioredoxin protects fetal type II epithelial cells from hyperoxia-induced injury. Pediatr Pulmonol. 2010;45(12):1192–200.
91. Que LG, et al. Protection from experimental asthma by an endogenous bronchodilator. Science. 2005;308(5728):1618–21.
92. Westerlund M, et al. High and complementary expression patterns of alcohol and aldehyde dehydrogenases in the gastrointestinal tract: implications for Parkinson's disease. FEBS J. 2007;274(5):1212–23.

Chapter 18
Antioxidant Therapies for Preterm Infants

Jennifer W. Lee and Jonathan M. Davis

Introduction

Under normal conditions, a delicate balance exists between the production of reactive oxygen species (ROS) and the antioxidant defenses that protect cells in vivo. The balance may be disturbed by increased ROS production or an inability to quench production because of inadequate antioxidant defenses. Increased generation of ROS can occur as a result of many conditions affecting newborn infants, including hyperoxia, reperfusion, and/or inflammation (Fig. 18.1). There is increasing evidence that links early exposure to oxidative stress with potentially lifelong consequences [1, 2]. The premature infant is especially susceptible to ROS-induced damage for two major reasons. First, adequate concentrations of antioxidants may be absent at birth (Fig. 18.2). Increases in antioxidant capacity are known to occur in the latter part of gestation in preparation for the transition to the higher oxygen environment that is encountered in extrauterine life [3]. Second, the ability to increase synthesis of antioxidants in response to hyperoxia or other oxidant challenges is relatively impaired [4, 5]. This can lead to an increased risk for the development of ROS-induced diseases of the newborn, such as bronchopulmonary dysplasia (BPD), retinopathy of prematurity (ROP), necrotizing enterocolitis (NEC), and periventricular leukomalacia (PVL) [6]. The focus of this chapter will be on current therapeutic interventions using antioxidants to reduce ROS-mediated damage known to

J.W. Lee, M.D., MS
Boston Children's Hospital, Harvard Medical School, Boston, MA, USA

J.M. Davis, M.D. (✉)
Division of Newborn Medicine, Department of Pediatrics, The Floating Hospital for Children at Tufts Medical Center, Tufts University School of Medicine,
TMC #44, 750 Washington Street, Boston, MA 02111, USA
e-mail: jdavis@tuftsmedicalcenter.org

P.A. Dennery et al. (eds.), *Perinatal and Prenatal Disorders*,
Oxidative Stress in Applied Basic Research and Clinical Practice,
DOI 10.1007/978-1-4939-1405-0_18

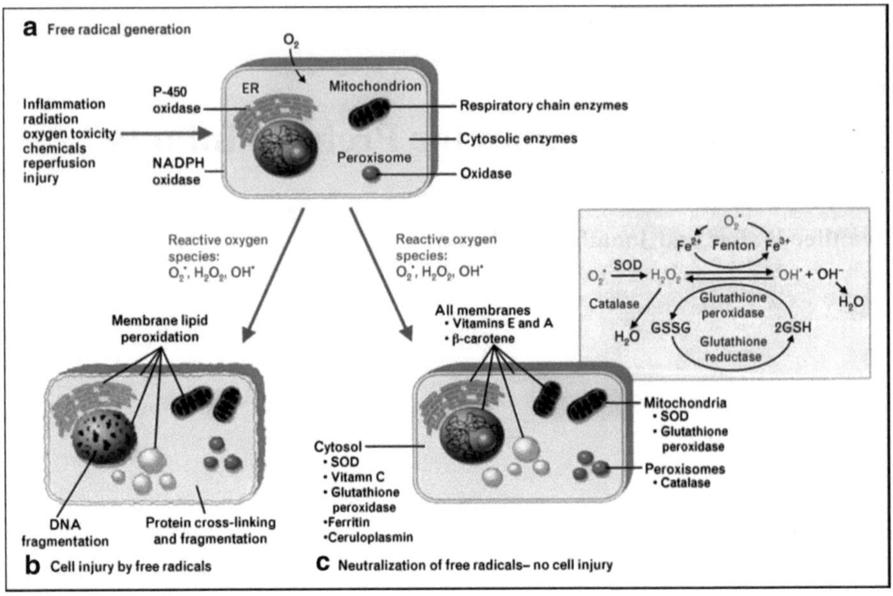

Fig. 18.1 Generation of ROS in various cellular compartments and the antioxidant defenses that protect cells in vivo. (**a**) Inflammation, radiation, oxygen toxicity, chemicals, and reperfusion injury incite the generation of reactive oxygen species. (**b**) The resulting free radicals can cause cellular injury by various mechanisms. (**c**) With SOD, glutathione peroxidase, peroxisomes, vitamins, and trace elements, free radicals are neutralized and no cell injury occurs

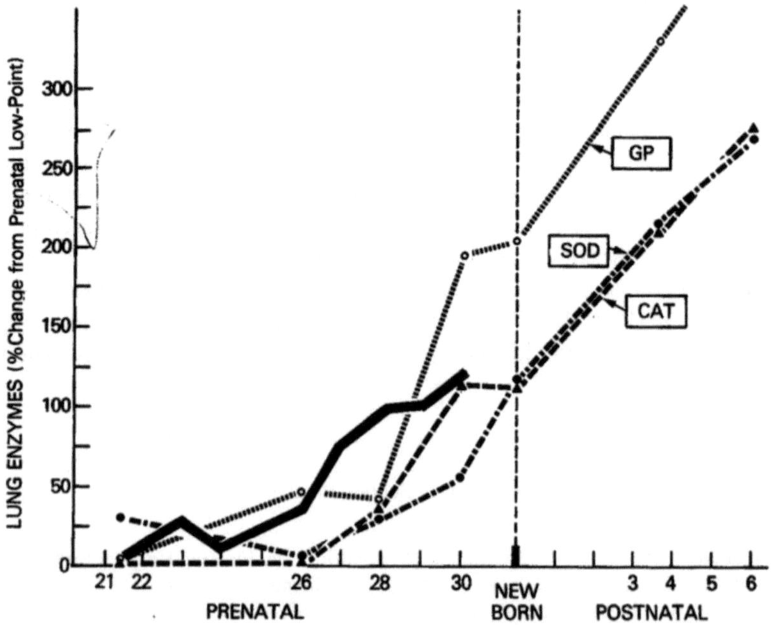

Fig. 18.2 Developmental changes in antioxidant levels and activity during gestation in rabbits. The increases in superoxide dismutase (*SOD*), catalase (*CAT*), and glutathione peroxidase (*GP*) late in gestation are similar to those seen for pulmonary surfactant (*dark, thick line*) [3]

contribute to common neonatal disorders. Since many ROS act as important cell signaling molecules, attempts to mitigate ROS production and injury must be balanced against any interference with normal cell homeostatic mechanisms.

Antioxidants

Enzymes

Superoxide Dismutase (SOD)

Superoxide dismutases (SODs) are enzymes that have significant antioxidant properties. Three forms of SOD have been identified in mammals: CuZnSOD (SOD1) is located primarily in the cytoplasm, MnSOD (SOD2) in the mitochondria, and extracellular (EC)-SOD (SOD3) which is a CuZn-containing protein in extracellular spaces. The only known function of SOD is to convert extremely toxic and reactive superoxide (O_2^-) radicals to hydrogen peroxide (H_2O_2) and water. Infants who are born prematurely have been found to be relatively deficient in these enzymes, making them much more susceptible to oxidant injury [3]. The preterm lung is particularly at risk for acute and chronic injury due to direct exposure to supraphysiologic oxygen levels after birth and the excess production of ROS [7]. In 1984, Rosenfeld and colleagues randomized infants to receive bovine CuZnSOD or saline subcutaneously to determine if SOD could influence the development of BPD. Of the survivors, both radiographic signs and clinical symptoms of BPD were improved in those who received SOD, suggesting that this enzyme might play a role in decreasing the severity of BPD [8].

The effects of recombinant human CuZn superoxide dismutase (rhSOD) were first studied in neonatal piglets developing acute lung injury from exposure to hyperoxia and mechanical ventilation for 48 h. Those piglets who received a single intratracheal (IT) dose of rhSOD (5 mg/kg suspended in 2 ml/kg of saline) at baseline had significantly less inflammatory changes in tracheal aspirates and bronchoalveolar lavage as well as improved histology when the lung was examined by light and electron microscopy. Labeling studies indicated that the rhSOD was taken up by virtually every cell type in the lung within 30 min of administration, was present in the serum within 4–6 h, and excreted in the urine by 12 h [9]. This was then followed by pilot studies evaluating intratracheal (IT) rhSOD administration in preterm infants. The investigators found that both single and multiple doses of rhSOD were well tolerated and resulted in significant increases in antioxidant activity in serum, tracheal aspirates, and urine as well as a decrease in inflammatory markers in the lung [10, 11]. The rhSOD excreted in the urine was largely intact (by western blot) and still active (by activity gel) for a significant amount of time after IT administration.

Davis and colleagues then administered IT rhSOD (5 mg/kg) or placebo every 48 h (as long as intubation was required) for up to 1 month of age to 302 premature infants [12]. Although there were no differences in the incidence of death or BPD,

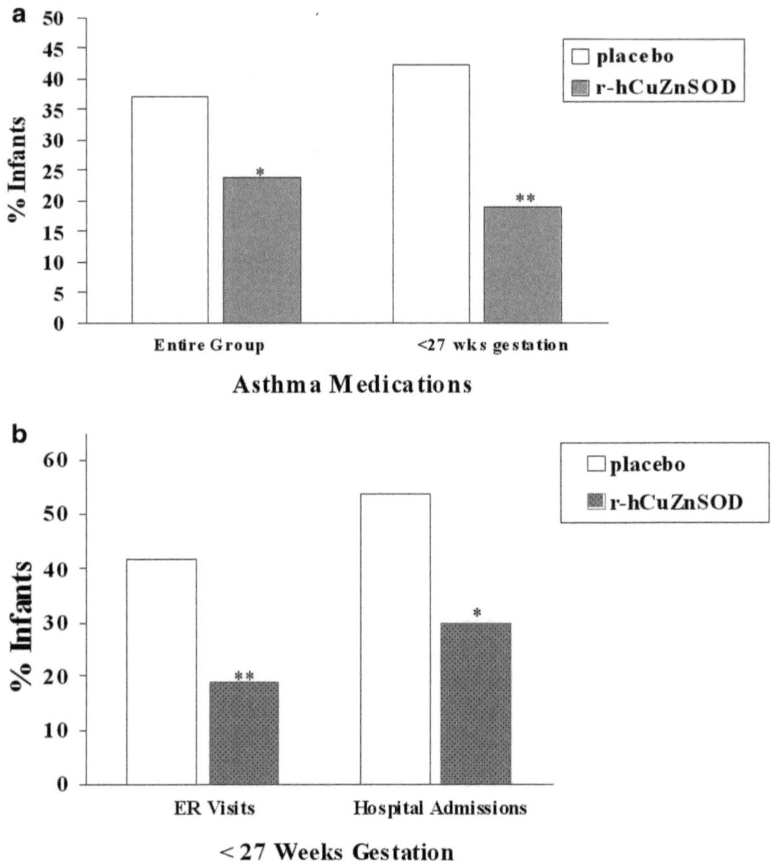

Fig. 18.3 (**a**) The use of asthma medications to treat significant respiratory illness in infants followed out to 1 year corrected gestational age in infants receiving recombinant human CuZn superoxide dismutase (r-hCuZnSOD). The entire group is presented as well as a subset of higher-risk infants born at <27 weeks' gestation (*$p=0.05$, **$p<0.05$). (**b**) The number of emergency room visits and hospital readmissions (all causes) in infants <27 weeks' gestation at birth receiving r-hCuZnSOD or placebo (*$p=0.05$, **$p=0.01$) [12]

there was approximately a 50 % reduction in the incidence of severe intraventricular hemorrhage (IVH) and periventricular leukomalacia (PVL) which did not quite reach statistical significance since the study was not powered to evaluate this. When followed to 1 year corrected gestational age (CGA), 37 % of placebo-treated infants had repeated episodes of wheezing or other respiratory illness severe enough to require treatment with asthma medications compared with 24 % of rhSOD-treated infants ($p<0.05$). In infants <27 weeks' gestation, 42 % receiving placebo had received asthma medications compared with 19 % of rhSOD-treated infants ($p<0.05$). This subgroup of rhSOD-treated infants also had a 55 % decrease in emergency room visits ($p<0.01$) and a 44 % decrease in subsequent hospitalizations ($p<0.05$), (Fig. 18.3) [12]. Economic analyses examining the cost-effectiveness of rhSOD treatment also found favorable results [13]. These data indicate that treatment at birth with rhSOD may reduce ROS-induced lung, brain, and retinal injury, resulting in improved outcome and significantly reduced costs when measured at 1 year CGA.

Glutathione Peroxidase (GPX)

Glutathione peroxidase (GPX) is an antioxidant enzyme contained in erythrocytes that detoxifies peroxides generated from metabolism of the superoxide anion. GPX activity has been found to be significantly lower in preterm infants developing BPD compared to preterm infants without BPD [14]. Since reduced GPX activity has also been associated with the development of ROP [15], GPX lends itself as a potential treatment option. In laboratory studies measuring oxidative stress in the neonatal rat intestine, those rats that were supplemented with oral glutamine or the combination of glutamine and arginine had significantly higher levels of GPX than those that were not provided glutamine. This suggests that glutamine and arginine supplementation may help decrease oxidative stress by increasing levels of GPX [16].

Nonenzymatic Proteins

Transferrin and Ceruloplasmin

Transferrin and ceruloplasmin participate in the metabolism of iron. Since iron is a potent oxidizing agent, diminished function or bioavailability would be expected to increase susceptibility to oxidative stress [17]. Lindeman and associates studied postnatal changes in plasma transferrin, ferroxidase, and ceruloplasmin iron-binding antioxidant activity in ten healthy preterm infants [18]. Ceruloplasmin levels and ferroxidase activity were low at birth and did not increase until 3–6 weeks of life. Transferrin levels were also initially low and did not increase at all when measured at 6 weeks of age. Reductions in the concentrations of these nonenzymatic antioxidants may further predispose the preterm infant to increased production of ROS.

Recombinant Erythropoietin (rEPO)

Erythropoietin (EPO) is a glycoprotein hormone that stimulates red blood cell production. It has been postulated that by stimulating erythropoiesis, EPO mobilizes iron, decreasing its availability to promote oxidant injury and thus serves as an antioxidant. Administration of rEPO to rabbits exposed to hyperoxia resulted in inhibition of lipid peroxidation and decreased nonsedimentable protein in bronchoalveolar lavage. rEPO treatment also decreased alveolar thickening and proteinaceous exudate in animals in the prolonged hyperoxia group [19]. Systemic treatment of neonatal rodents with rEPO has also been shown to provide neuroprotection against hyperoxia-induced apoptosis [20]. There have been no reports of a reduction in lung injury in preterm infants receiving rEPO for prevention/treatment of anemia of prematurity.

Trace Elements

Zinc (Zn), copper (Cu), iron (Fe), and selenium (Se) are trace elements which serve as important cofactors in many antioxidant enzymatic reactions [21]. Supplementation with these nutrients could potentially optimize total antioxidant capacity and improve outcome in preterm infants [22]. In a recent study, concentrations of these trace elements or their associated antioxidant enzymes did not influence ROS-induced disease processes of the preterm infant [21]. A Cochrane review analyzed three trials administering Se to preterm infants. While infants receiving supplemental Se had lower rates of sepsis, there were no differences in survival or the incidence of BPD or ROP, suggesting that supplementation with trace elements alone did not significantly improve outcome [23].

Oxidizable Molecules

Glutathione

Glutathione depletion or inadequate synthesis is believed to contribute to the developmental susceptibility of the preterm newborn to oxidative stress. Low levels in tracheal aspirate samples have been associated with the later development of BPD in premature infants [24]. Pharmacologic supplementation was tested in preterm infants who received supplemental cysteine in an attempt to stimulate glutathione synthesis, since glutathione acts as an antioxidant and is an important co-factor for GPX activity [25]. Despite significant increases in cysteine, glutathione concentrations and synthesis rates did not increase, and ROS-induced injury was not affected.

N-Acetylcysteine (NAC)

NAC is a precursor of the antioxidant glutathione and is readily taken up into multiple cell types within numerous organs. A large multicenter trial showed no reduction in survival, incidence of BPD, or improved pulmonary function at term following NAC administration [26]. NAC has been shown to decrease lipid hydroperoxide formation in a rat model but was not found to significantly reduce avascularity or neovascularization which are important markers of ROP [27]. NAC pretreatment attenuates lipopolysaccharide (LPS)-induced cerebral white matter injury by replenishment of reduced glutathione, scavenging of ROS, and maintenance of peroxisomal proliferation/function via a peroxisome proliferators activated receptor-α (PPAR-α)-dependent mechanisms [28]. LPS-activated microglia induce cell death and greatly impair oligodendrocyte development through ROS-dependent mechanisms, which may cause selective white matter damage and hypomyelination in PVL [29].

Vitamins

Vitamin E and BPD

Vitamin E has long been known to be a potent antioxidant, and several studies have shown that preterm infants are vitamin E deficient at birth [30]. Supplementation with vitamin E has been studied in multiple trials as a treatment for disorders associated with oxidative stress in newborns (e.g., BPD). Ehrenkranz and associates treated 20 preterm infants with intramuscular vitamin E and 20 infants served as controls. No infant required prolonged supplemental oxygen, died, or developed radiographic changes consistent with BPD in the treatment group compared to four deaths and six with abnormal radiographic changes in the control group [31]. This study supported the notion that vitamin E may decrease oxygen-induced lung damage and paved the way for subsequent clinical trials in preterm newborns. A randomized, double-blind trial then treated infants within the first 24 h of life with vitamin E (20 mg/kg) or placebo with repeat dosing while the infant was receiving supplemental oxygen. There were no differences in the duration of oxygen or ventilatory support, radiographic findings, or the development of BPD with vitamin E treatment [32]. Subsequent studies of vitamin E supplementation have also failed to show a reduction in the incidence of BPD [33–35]. It is important to note the significant side effects in infants treated with high doses of vitamin E when serum tocopherol levels were >3.5 mg/dl, such as an increased incidence of sepsis, NEC, and unexpected death [36].

High-dose antenatal vitamins E and C have also been studied in maternal preeclampsia, with the hope that higher levels of antioxidants at birth may reduce associated respiratory morbidity in the newborns. Although respiratory outcomes at 2 years of age were not significantly different, treated infants had more healthcare visits (all causes) compared to those who did not receive high-dose vitamin E and C prenatally [37].

Vitamin E and ROP

Given that the development of ROP is also influenced by excessive oxygen exposure and toxicity, vitamin E has also been studied in preventing the development and progression of ROP. Multiple studies have failed to demonstrate significant reductions in the incidence of ROP between vitamin E-treated infants and placebo controls [38]. In a clinical trial examining prophylactic vitamin E in preventing ROP, an increased incidence of sepsis and NEC was also found, indicating the need for extreme caution when using pharmacologic doses of these antioxidants in preterm infants [39].

Vitamin E and IVH

Since the development of IVH is also thought to be associated with ROS-induced injury, studies have been performed to examine prevention by antioxidant supplementation. In one study, the incidence of IVH was compared between neonates receiving additional vitamin E via intramuscular injection. The incidence and severity of IVH were significantly reduced in the vitamin E group compared to placebo controls [40]. A Cochrane review also examined the effects of vitamin E supplementation in preterm infants and concluded that although vitamin E reduced the risk of intracranial hemorrhage, there was a significantly increased risk of sepsis which precluded the routine use of this treatment [34].

Vitamin A and BPD

The role of vitamin A likely is mediated through its action on retinol-binding protein and the retinoic acid receptor, rather than direct antioxidant effects. Normal lung growth is dependent upon vitamin A and the protection it offers against repeated oxidative damage. However, many preterm infants are vitamin A deficient at birth which may contribute to a reduction in antioxidant defenses and promote the development of BPD in preterm infants [41, 42]. Given this, multiple trials have been performed to determine if supplementation with vitamin A can prevent or reduce the severity of BPD. A small clinical trial found similar concentrations of vitamin A in the placebo and vitamin A treatment groups at 4 weeks of age, also noting no change in the incidence of BPD with the treatment [43].

A Cochrane review in 2011 aimed "to evaluate vitamin A supplementation on the incidence of death and/or neonatal chronic lung disease and long-term neurodevelopmental disability in very low birth weight (VLBW) infants; and to consider the effect of the supplementation, route, dose, and timing [44]." Nine trials met inclusion criteria, and compared to controls, vitamin A appeared to be beneficial in reducing death or BPD at 36 weeks' postmenstrual age (RR 0.87, 95 %CI 0.77–0.98). At 1 year corrected age, there were no benefits of vitamin A on clinical respiratory status. Neurodevelopmental assessment of 88 % of surviving infants in the largest trial showed no differences between the groups at 18–22 months corrected gestational age, with different dosages of vitamin A showing similar results. The authors concluded that whether clinicians decide to utilize repeat intramuscular doses of vitamin A to prevent BPD may depend upon the local incidence of this outcome and the value attached to achieving a modest reduction in this outcome, balanced against the lack of other proven benefits and the acceptability of treatment. Information on long-term neurodevelopmental status and pulmonary outcome suggests no evidence of either benefit or harm from the intervention, indicating that short-term outcomes may not necessarily correlate effectively with longer-term measures of clinical neurologic or pulmonary status.

Vitamin C

Vitamin C, also known as ascorbic acid, is an antioxidant supplied to neonates primarily through multivitamins and parenteral nutrition. Despite its antioxidant properties, it has also been shown to have pro-oxidant activity as well depending on the dose. High concentrations of ascorbic acid in premature infants at birth have been associated with adverse outcomes, likely by inhibition of ferroxidase activity and decreased antioxidant activity in these premature infants. The inhibition is dependent on the ratio of ascorbic acid to ceruloplasmin (see section "Transferrin and Ceruloplasmin") [45]. A large randomized controlled trial was performed to determine if maintaining a lower plasma ascorbic acid level (supplementation with 10 mg/kg/day) during the first week of life and then higher levels (supplementation with 30 mg/kg/day) in weeks 3–4 would be accompanied by decreased morbidity in premature infants. Although there were no significant differences in the incidence of BPD or ROP between the groups, a trend was noted toward less BPD in infants with higher ascorbic acid levels [46].

Others

Melatonin

Melatonin is a pineal hormone that exhibits antioxidant effects by supporting SOD and GPX activities while mitigating lipid peroxidation and ROS-induced cell injury. It is an endogenously produced indoleamine and is a highly effective antioxidant and free radical scavenger. Melatonin has been studied as an antioxidant in hypoxic brain injury, postsurgical procedures, and sepsis. Administration of melatonin in neonates undergoing surgery was found to reduce cytokine production and nitrite/nitrate concentrations as markers of oxidative stress [47]. Reduced levels of pro-inflammatory cytokines were also seen in infants treated with melatonin who had respiratory distress syndrome and required treatment with mechanical ventilation [48]. Melatonin has also been studied as a neuroprotective factor in mouse models of PVL. In a recent report, agomelatine and melatonin did not prevent the initial appearance of white matter lesions but did promote secondary lesion repair. The effects of melatonin were only observed when given within the first 2 h following the insult [49]. In a neonatal rat model, melatonin reduced ROS production, increased antioxidant levels, and decreased lung damage in animals exposed to prolonged hyperoxia, indicating a potential protective effect in BPD [50].

Antenatal Glucocorticoids

Antenatal corticosteroids are administered routinely to expectant mothers at risk for delivering a premature infant <34 weeks' gestation. These medications help to reduce the severity of respiratory distress syndrome. In addition to stimulating surfactant production, antenatal steroids also have been found to reduce oxidative stress by enhancing the expression of endogenous antioxidant enzymes. Vento and colleagues conducted a prospective study of extremely low gestational age newborns whose mothers either received or did not receive a complete course of antenatal steroids [51]. Infants who were exposed to antenatal steroids had less severe lung disease and less oxidative stress (reduced-to-oxidized glutathione ratio) at birth and on day 1 of life. Expression of SOD and catalase (CAT) was significantly increased in those infants whose mother's received steroids, with greater activity noted in females as compared to males. Antenatal steroids have also been shown to significantly reduce the incidence of PVL in extremely preterm infants presumably through a similar inhibition of inflammatory changes and production of ROS [52].

Allopurinol

Allopurinol is an inhibitor of xanthine oxidase (XO) which is an enzyme that produces superoxide radicals following hypoxic-ischemic insults, contributing to cell death, especially within the brain. Therefore, research has been directed to reducing the extent of this injury with allopurinol. A study of 400 preterm infants randomized to receive allopurinol or placebo failed to show protection against the primary outcome of development of PVL [53]. A recent Cochrane meta-analysis analyzed available research involving the impact of allopurinol treatment in reducing adverse cellular effects of hypoxic neuronal injury [54]. Three trials which included 114 infants were studied, but the results remained insufficient to conclude whether treatment with allopurinol in the setting of hypoxic-ischemic encephalopathy has clinically significant benefits for these infants.

Breast Milk

Lutein and the isomer zeaxanthin are carotenoids located in the neuronal retina that protect against oxidative and light damage. Human milk is the only dietary source in neonates. Studies have shown that lutein is well absorbed in premature neonates after oral administration [55]. However, studies of oral lutein and zeaxanthin administration failed to show a significant benefit in the incidence or severity of ROP, NEC, or BPD [56]. Although further studies with larger sample sizes are needed, formula companies are routinely adding this compound to preterm formula in an attempt to more closely simulate breast milk and improve ROS-induced disease processes.

Since upregulation of antioxidants does not seem to occur in preterm infants, some groups have studied whether increased ROS could be scavenged by feeding human milk [57]. Stable preterm infants were fed exclusively human milk or preterm formula and compared with healthy term controls. As expected, preterm infants eliminated greater amounts of urinary oxidative metabolites which correlated with birth weight and gestational age compared to term infants. However, formula-fed preterm infants excreted significantly higher concentrations of 8-oxodG and o-tyrosine (indirect markers of ROS production) than did infants fed human milk. These data confirm the notion that preterm infants are exposed to greater oxidant stress than term controls. Enteral feeds of human milk are partially protective, which reinforces the beneficial effects of this mode of feeding.

Summary and Conclusions

A delicate balance exists between the production of ROS and the antioxidant defenses that protect cells in vivo. The balance may be disturbed by increased ROS production or an inability to quench production because of inadequate antioxidant defenses. Increased generation of ROS can occur as a result of many conditions affecting newborn infants, including hyperoxia, reperfusion, and/or inflammation. The premature infant is especially susceptible to ROS-induced damage due to inadequate concentrations of antioxidants at birth and an impaired ability to increase synthesis of antioxidants in response to hyperoxia or other oxidant challenges. This can lead to an increased risk for the development of ROS-induced diseases of the newborn, such as BPD, ROP, NEC, and PVL. Therapies aimed at preventing this damage have been developed and are actively being studied in this population. Antioxidant enzymes, proteins, vitamins, trace elements, and even human breast milk are among the most commonly utilized to minimize the effects of these ROS. Many of these agents have been studied independently, and it is unlikely that any one agent will be completely effective in improving outcomes. Administration of adequate amounts of various antioxidants to premature infants to achieve physiologic sufficiency for many of these substances may be difficult. There is also the possibility of achieving supraphysiological concentrations, which may promote pro-oxidant activity and actually cause tissue injury. It may be necessary for several approaches to be used to have a significant impact on survival and outcome in this vulnerable and high-risk population. A better understanding of these relationships will be necessary for the development of rational treatments aimed at improving pregnancy outcomes and reducing the burden of oxidative stress in premature infants.

References

1. Sola A, Rogido MR, Deulofeut R. Oxygen as a neonatal health hazard: call for detente in clinical practice. Acta Paediatr. 2007;96:801–12.
2. Giussani DA, Camm EJ, Niu Y, Richter HG, Blanco CE, Gottschalk R, et al. Developmental programming of cardiovascular dysfunction by prenatal hypoxia and oxidative stress. PLoS One. 2012;7:e31017.
3. Frank L, Groseclose EE. Preparation for birth into an O_2-rich environment: the antioxidant enzymes in the developing rabbit lung. Pediatr Res. 1984;18:240–4.

4. Georgeson GD, Szony BJ, Streitman K, Varga IS, Kovacs A, Kovacs L, et al. Antioxidant enzyme activities are decreased in preterm infants and in neonates born via caesarean section. Eur J Obstet Gynecol Reprod Biol. 2002;103:136–9.
5. Perrone S, Negro S, Tataranno ML, Buonocore G. Oxidative stress and antioxidant strategies in newborns. J Matern Fetal Neonatal Med. 2010;23 Suppl 3:63–5.
6. Davis JM, Auten RL. Maturation of the antioxidant system and the effects on preterm birth. Semin Fetal Neonatal Med. 2010;15:191–5.
7. Frank L. Developmental aspects of experimental pulmonary oxygen toxicity. Free Radic Biol Med. 1991;11:463–94.
8. Rosenfeld W, Evans H, Concepcion L, Jhaveri R, Schaeffer H, Friedman A. Prevention of bronchopulmonary dysplasia by administration of bovine superoxide dismutase in preterm infants with respiratory distress syndrome. J Pediatr. 1984;105:781–5.
9. Davis JM, Rosenfeld WN, Sanders RJ, Gonenne A. Prophylactic effects of recombinant human superoxide dismutase in neonatal lung injury. J Appl Physiol. 1993;74:2234–41.
10. Davis JM, Rosenfeld WN, Richter SE, Parad MR, Gewolb IH, Spitzer AR, et al. Safety and pharmacokinetics of multiple doses of recombinant human CuZn superoxide dismutase administered intratracheally to premature neonates with respiratory distress syndrome. Pediatrics. 1997;100:24–30.
11. Rosenfeld WN, Davis JM, Parton L, Richter SE, Price A, Flaster E, et al. Safety and pharmacokinetics of recombinant human superoxide dismutase administered intratracheally to premature neonates with respiratory distress syndrome. Pediatrics. 1996;97:811–7.
12. Davis JM, Parad RB, Michele T, Allred E, Price A, Rosenfeld W. Pulmonary outcome at 1 year corrected age in premature infants treated at birth with recombinant human CuZn superoxide dismutase. Pediatrics. 2003;111:469–76.
13. McBride JA, Parad RB, Davis JM, Zheng Z, Zupancic JA. Economic evaluation of recombinant human copper zinc superoxide dismutase administered at birth to premature infants. J Perinatol. 2009;29:364–71.
14. Fu RH, Chiu TH, Chiang MC, Lien R, Chou YH, Chiang CC, et al. Lower erythrocyte glutathione peroxidase activity in bronchopulmonary dysplasia in the first week of neonatal life. Neonatology. 2008;93:269–75.
15. Fu RH, Chiu TH, Chiang MC, Cho YH, Lien R, Chiang CC, et al. Erythrocyte anti-oxyenzyme activity in preterm infants with retinopathy of prematurity. Neonatology. 2007;92:59–63.
16. Kul M, Vurucu S, Demirkaya E, Tunc T, Aydinoz S, Meral C, et al. Enteral glutamine and/or arginine supplementation have favorable effects on oxidative stress parameters in neonatal rat intestine. J Pediatr Gastroenterol Nutr. 2009;49:85–9.
17. Collard KJ. Iron homeostasis in the neonate. Pediatrics. 2009;123(4):1208–16.
18. Lindeman JH, Lentjes EG, van Zoeren-Grobben D, Berger HM. Postnatal changes in plasma ceruloplasmin and transferrin antioxidant activities in preterm babies. Biol Neonate. 2000;78:73–6.
19. Bany-Mohammed FM, Slivka S, Hallman M. Recombinant human erythropoietin: possible role as an antioxidant in premature rabbits. Pediatr Res. 1996;40:381–7.
20. Kaindl AM, Sifringer M, Koppelstaetter A, Genz K, Loeber R, Boerner C, et al. Erythropoietin protects the developing brain from hyperoxia-induced cell death and proteome changes. Ann Neurol. 2008;64:523–34.
21. Loui A, Raab A, Maier RF, Bratter P, Obladen M. Trace elements and antioxidant enzymes in extremely low birth weight infants. J Trace Elem Med Biol. 2010;24(2):111–8.
22. Biniwale MA, Ehrenkranz RA. The role of nutrition in the prevention and management of bronchopulmonary dysplasia. Semin Perinatol. 2006;30:200–8.
23. Darlow BA, Austin NC. Selenium supplementation to prevent short-term morbidity in preterm neonates. Cochrane Database Syst Rev. 2003;(4):CD003312.
24. Grigg J, Barber A, Silverman M. Bronchoalveolar lavage fluid glutathione in intubated premature infants. Arch Dis Child. 1993;69:49–51.
25. te Braake FW, Schierbeek H, Vermes A, Huijmans JG, van Goudoever JB. High-dose cysteine administration does not increase synthesis of the antioxidant glutathione preterm infants. Pediatrics. 2009;124:e978–84.

26. Sandberg K, Fellman V, Stigson L, Thiringer K, Hjalmarson O. N-acetylcysteine administration during the first week of life does not improve lung function in extremely low birth weight infants. Biol Neonate. 2004;86:275–9.
27. Saito Y, Geisen P, Uppal A, Hartnett ME. Inhibition of NAD(P)H oxidase reduces apoptosis and avascular retina in an animal model of retinopathy of prematurity. Mol Vis. 2007;13: 840–53.
28. Paintlia MK, Paintlia AS, Contreras MA, Singh I, Singh AK. Lipopolysaccharide-induced peroxisomal dysfunction exacerbates cerebral white matter injury: attenuation by N-acetyl cysteine. Exp Neurol. 2008;210:560–76.
29. Pang Y, Campbell L, Zheng B, Fan L, Cai Z, Rhodes P. Lipopolysaccharide-activated microglia induce death of oligodendrocyte progenitor cells and impede their development. Neuroscience. 2010;166:464–75.
30. Kositamongkol S, Suthutvoravut U, Chongviriyaphan N, Feungpean B, Nuntnarumit P. Vitamin A and E status in very low birth weight infants. J Perinatol. 2011;31:471–6.
31. Ehrenkranz RA, Bonta BW, Ablow RC, Warshaw JB. Amelioration of bronchopulmonary dysplasia after vitamin E administration. A preliminary report. N Engl J Med. 1978;299: 564–9.
32. Ehrenkranz RA, Ablow RC, Warshaw JB. Effect of vitamin E on the development of oxygen-induced lung injury in neonates. Ann N Y Acad Sci. 1982;393:452–66.
33. Zoberlein HG, Freudenberg V, Wehinger H. Influence of vitamin E on the development of bronchopulmonary dysplasia in neonates with artificial ventilation. A prospective randomized study. Monatsschr Kinderheilkd. 1982;130:706–9.
34. Brion LP, Bell EF, Raghuveer TS. Vitamin E supplementation for prevention of morbidity and mortality in preterm infants. Cochrane Database Syst Rev. 2003;(3):CD003665.
35. Watts JL, Milner R, Zipursky A, Paes B, Ling E, Gill G, et al. Failure of supplementation with vitamin E to prevent bronchopulmonary dysplasia in infants less than 1,500 g birth weight. Eur Respir J. 1991;4:188–90.
36. Mino M. Use and safety of elevated dosages of vitamin E in infants and children. Int J Vitam Nutr Res Suppl. 1989;30:69–80.
37. Greenough A, Shaheen SO, Shennan A, Seed PT, Poston L. Respiratory outcomes in early childhood following antenatal vitamin C and E supplementation. Thorax. 2010;65:998–1003.
38. Puklin JE, Simon RM, Ehrenkranz RA. Influence on retrolental fibroplasia of intramuscular vitamin E administration during respiratory distress syndrome. Ophthalmology. 1982;89:96–103.
39. Johnson L, Bowen Jr FW, Abbasi S, Herrmann N, Weston M, Sacks L, et al. Relationship of prolonged pharmacologic serum levels of vitamin E to incidence of sepsis and necrotizing enterocolitis in infants with birth weight 1,500 grams or less. Pediatrics. 1985;75:619–38.
40. Speer ME, Blifeld C, Rudolph AJ, Chadda P, Holbein ME, Hittner HM. Intraventricular hemorrhage and vitamin E in the very low-birth-weight infant: evidence for efficacy of early intramuscular vitamin E administration. Pediatrics. 1984;74:1107–12.
41. Vogelsang A, van Lingen RA, Slootstra J, Dikkeschei BD, Kollen BJ, Schaafsma A, et al. Antioxidant role of plasma carotenoids in bronchopulmonary dysplasia in preterm infants. Int J Vitam Nutr Res. 2009;79:288–96.
42. Shenai JP, Chytil F, Stahlman MT. Vitamin A status of neonates with bronchopulmonary dysplasia. Pediatr Res. 1985;19:185–8.
43. Pearson E, Bose C, Snidow T, Ransom L, Young T, Bose G, et al. Trial of vitamin A supplementation in very low birth weight infants at risk for bronchopulmonary dysplasia. J Pediatr. 1992;121:420–7.
44. Darlow BA, Graham PJ. Vitamin A supplementation to prevent mortality and short- and long-term morbidity in very low birthweight infants. Cochrane Database Syst Rev. 2011;(10):CD000501.
45. Powers HJ, Loban A, Silvers K, Gibson AT. Vitamin C at concentrations observed in premature babies inhibits the ferroxidase activity of ceruloplasmin. Free Radic Res. 1995;22:57–65.
46. Darlow BA, Buss H, McGill F, Fletcher L, Graham P, Winterbourn CC. Vitamin C supplementation in very preterm infants: a randomised controlled trial. Arch Dis Child Fetal Neonatal Ed. 2005;90:F117–22.

47. Gitto E, Romeo C, Reiter RJ, Impellizzeri P, Pesce S, Basile M, et al. Melatonin reduces oxidative stress in surgical neonates. J Pediatr Surg. 2004;39(2):184–9.
48. Gitto E, Reiter RJ, Sabatino G, Buonocore G, Romeo C, Gitto P, et al. Correlation among cytokines, bronchopulmonary dysplasia and modality of ventilation in preterm newborns: improvement with melatonin treatment. J Pineal Res. 2005;39:287–93.
49. Bonnefont-Rousselot D, Collin F. Melatonin: action as antioxidant and potential applications in human disease and aging. Toxicology. 2010;278:55–67.
50. Pan L, Fu JH, Xue XD, Xu W, Zhou P, Wei B. Melatonin protects against oxidative damage in a neonatal rat model of bronchopulmonary dysplasia. World J Pediatr. 2009;5:216–21.
51. Vento M, Aguar M, Escobar J, Arduini A, Escrig R, Brugada M, et al. Antenatal steroids and antioxidant enzyme activity in preterm infants: influence of gender and timing. Antioxid Redox Signal. 2009;11:2945–55.
52. Baud O, Foix-L'Helias L, Kaminski M, Audibert F, Jarreau PH, Papiernik E, et al. Antenatal glucocorticoid treatment and cystic periventricular leukomalacia in very premature infants. N Engl J Med. 1999;341:1190–6.
53. Russell GA, Cooke RW. Randomised controlled trial of allopurinol prophylaxis in very preterm infants. Arch Dis Child Fetal Neonatal Ed. 1995;73:F27–31.
54. Chaudhari T, McGuire W. Allopurinol for preventing mortality and morbidity in newborn infants with hypoxic-ischaemic encephalopathy. Cochrane Database Syst Rev. 2012;7, CD006817.
55. Romagnoli C, Tirone C, Persichilli S, Gervasoni J, Zuppi C, Barone G, et al. Lutein absorption in premature infants. Eur J Clin Nutr. 2010;64:760–1.
56. Manzoni P, Guardione R, Bonetti P, Priolo C, Maestri A, Mansoldo C, et al. Lutein and zeaxanthin supplementation in preterm very low-birth-weight neonates in neonatal intensive care units: a multicenter randomized controlled trial. Am J Perinatol. 2013;30:25–32.
57. Ledo A, Arduini A, Asensi MA, Sastre J, Escrig R, Brugada M, et al. Human milk enhances antioxidant defenses against hydroxyl radical aggression in preterm infants. Am J Clin Nutr. 2009;89:210–5.

Chapter 19
Postnatal Oxidative Stress and the Role of Enteral and Parenteral Nutrition

James Friel

The Newborn Infant

Premature infants are vulnerable to the effects of early childbirth, which includes immature defenses against the early novel adaptation to oxygen. Developmental progress initiated early in the life cycle pushes the normal programming sequence in ways we only partially understand. In addition to antioxidant capabilities that are pushed possibly too soon, the exposure to excess oxygen is almost routine for the smaller younger infant. What role nutrition plays in the mosaic of prematurity is not certain, but we do know that what the infant is fed is clinically important perhaps more so than at any other time in the life cycle. Every attempt is made to introduce human milk as early as possible, the only food made by humans for humans. It is clear human milk can be tailored for the premature and perhaps for each and every infant. When the premature is too young for suckling, parenteral nutrition is needed which unexpectedly may increase the oxidative load of the premature infant. Technology while extending the life of many premature infants brings its own problems in need of resolution. The unfortunate creation of free radicals in solution while unintended and in fact only recently recognized creates iatrogenic issues. We describe nutritional advances in feeding the premature infant and the important role of nutrition as medicine has on outcome.

Childbirth is accompanied by an increase in oxidative stress, as birth is a hyperoxic challenge. The fetus is born from an intrauterine hypoxic environment (pO_2 of 20–25 mmHg) to an extrauterine normoxic yet relatively hyperoxic environment

J. Friel (✉)
Richardson Center for Functional Foods, University of Manitoba,
196 Innovation Drive, Room 203, Winnipeg, MB R3T 6C5, Canada
e-mail: james.friel@umanitoba.ca

P.A. Dennery et al. (eds.), *Perinatal and Prenatal Disorders*,
Oxidative Stress in Applied Basic Research and Clinical Practice,
DOI 10.1007/978-1-4939-1405-0_19

with a pO_2 of 100 mmHg [86]. Increased exposure to oxygen at high concentrations compared with the womb can be accommodated by neonatal animals of many species because of the newborn lungs' ability to increase its normal battery of protective antioxidant enzymes during O_2 exposure [31]. The evolutionary adaptation to extrauterine aerobic existence necessitated the development of efficient cellular electron transport systems to produce energy. Along with this challenge of energy-producing oxidative metabolism, biochemical defenses including antioxidant enzymes evolved to protect against oxidation of cellular constituents by oxygen radicals [32].

Antioxidant enzymes mature during late gestation [95] accompanied by an increased transfer of antioxidants across the placenta, including vitamins E and C, betacarotenes, and ubiquinone during the latter stage of gestation [38, 39, 94]. While disease in newborns have been extensively studied, particularly in the premature infant [2] little is known about neonatal adaptation to physiologic stress of delivery and early postnatal life in normal full-term healthy infants.

Not all free radicals are "bad." ROS play a major and important role in signal transduction and are required for development. How the newborn infant can cope with possible excess exposure to ROS is not yet clear. Developing antioxidant defense mechanisms may be overcome by the generation of excessive ROS during the neonatal period. We and others have shown that human milk provides antioxidant protection in early life with the ability to scavenge free radicals, a function not seen in artificial infant feeds [13, 38, 39]. Indeed, van Zoeren-Grobben reported that infants fed human milk had higher resistance to oxidative stress, than did control infants who were formula fed [114]. This may be due to the presence of antioxidant enzymes glutathione peroxidase (GPx), catalase (Cat), and superoxide dismutase (SOD) in human milk, but not in formula [63]. Further, in addition to their antioxidant effect in the gut, these enzymes may pass intact through the porous neonatal intestine early in infancy [38, 39].

We hypothesize that early infancy would be a time of oxidative stress due to the major shift in oxygen exposure and the difficulty of adapting to ambient oxygen. Therefore, we assessed lipid peroxidation, the activity of antioxidant enzymes, and the ability to resist oxidative stress in full-term healthy breast-fed infants during the 1st year of life. As a measure of lipid peroxidation we measured F_2-isoprostanes, which are prostaglandin F_2-like compounds produced by free radical-catalyzed peroxidation of arachidonic acid [85].

We found markedly elevated levels of F_2-isoprostanes in plasma in early infancy and a decreasing ability to resist oxidative stress [36]. The rapid decline in F_2-isoprostanes between 1 and 3 and 3 and 6 months to normal adult levels [85] suggests that these infants adjusted to oxidative stress due to birth itself over time. In support of these findings, both SOD and CAT in red blood cells increased between 1 and 3 months and then declined, suggesting a response to oxidative stress that appears to normalize with age. Another interpretation might be that metabolites of arachidonic acid or F2-isoprostanes themselves are required for cell signaling or other processes in the rapidly growing child. Growth rate is the greatest at that time of life.

The most likely candidates for early stress are the transition from a hypoxic environment in the womb to a relatively hyperoxic extrauterine environment and a high metabolic rate requiring a high level of mitochondrial respiration and subsequent increased mitochondrial superoxide formation [81]. High levels of inspired oxygen are required to maintain arterial oxygen tension necessary for postnatal life and are substantially higher than those normally present during fetal existence. Therefore, newborns are exposed to more reactive oxygen species (ROS) than they would be had they remained in utero. The transition from fetus to newborn can be stressful as seen from supportive evidence concerning the mortality rate in the first 28 days of life compared with the remainder of infancy [79]. In North America, 67 % of all deaths during the 1st year occur in the 1st month of life. This data suggests that the transition from the womb to the extrauterine environment may be an oxidative challenge that may overwhelm the antioxidant capacity of the organism and be one possible reason that not all infants can survive this event. That this challenge involves an oxidative stress in coping with ambient oxygen pressure has been shown in several studies [12, 15, 96, 117].

There are numerous reports in the literature of oxidative stress associated with birth. Higher lipid peroxidation reflected by increased malondialdehyde levels (MDA) in cord blood compared to the neonatal period suggests oxidative stress during the birth process [96, 117]. Other studies report increased lipid peroxidation in newborn infants; however, these results are difficult to interpret as the methods are not universally accepted.

Collectively these data suggest that newborn infants are experiencing oxidative stress that resolves only with age. Possible mechanisms may include the degradation of fetal erythrocytes that are present in early infancy. In vitro, fetal erythrocytes produce more superoxide and hydrogen peroxide than do adult red blood cells [53]. It is known that during the fetal–neonatal transition period, dramatic changes are occurring in the pO_2 in the lung and blood cells with a more gradual change in liver and brain. These changes may result in increased oxidative stress to cells.

Role of Nutrition

Gonzalez et al. [44] suggested that changes in antioxidant defenses could be affected by beginning food intake after birth, which entails higher hepatic metabolism rate as well as increased oxygen consumption. We reported no differences in any of our measurements of full-term infants according to type of feed, whether by bottle or by breast-feeding [36]. Others [47] reported that bottle-fed infants from 2–4 months of age had lower MDA than did breast-fed infants and attributed these findings to higher levels of long-chain fatty acids, present in human milk (HM) but not formulas (F). Infant formulas have been fortified with DHA (*all-cis*-docosa-4,7,10,13,16,19-hexaenoic acid) and EPA (all-cis-5,8,11,14,17-icosapentanenoic acid) long-chain fatty acids (Mead Johnson, Evansville, IN). The effect on oxidative stress seen in early infancy by formula-fed infants with these new fatty acid supplements remains to be determined. An alternative explanation could be the presumed higher energy cost of sucking on the breast compared to the bottle.

HM is the ideal first food during infancy. In addition to being the best source of nutrients, it also supplies a number of beyond nutrition defense factors for the growing infant. Protection by HM resides in a complex system of host defense factors that are distinct from other mammalian milks [43]. Buescher and McIlheran [13] reported that human colostrum manifests antioxidant properties, being capable of spontaneous reduction of cytochrome c, depletion of polymorphonuclear leukocyte-produced H_2O_2, and protection of epithelial cells from polymorphonuclear leukocyte-mediated detachment. We do not know the complete array of active antioxidant components in HM. It is known, however, that scavengers of free radicals, which include (alpha)-tocopherol, cysteine, and ascorbate, are considerably higher in HM than in cow's milk [62]. In addition, GPx (EC 1.161.9), SOD (EC 1.15.1.1) [90], and CAT (EC 1.11.1.6) are also present in HM [62, 63] to assist in the destruction of H_2O_2.

Cow's milk is not routinely fed to human infants but is modified into humanized formulas that are more comparable to HM. These formulas routinely have excess chain-breaking antioxidants compared with HM [42]. Goldman et al. [42] state that many factors, including antioxidants, are either absent or poorly represented in cow's milk or other artificial feedings. Van Zoeren-Grobben et al. [114] reported that premature infants who were fed HM had higher plasma peroxyl radical-trapping ability in vitro than did control infants who were formula fed. Milk from mothers of PT infants is also known to vary in composition from milk from mothers of FT infants [78].

The role of nonenzymatic, bioactive antioxidants in HM is important because endogenous infant enzyme maturation appears not before 28 weeks of gestation, and the transfer of vitamin E from mother to fetus occurs primarily during the last trimester [39]. An undeveloped antioxidant defense system in both premature and full-term newborns leaves them susceptible to oxidative stress [92]. We postulate that HM with bioactive agents is crucial to newborns and will assist in reducing oxidative stress, while physiologic defenses are underdeveloped. The identification of these naturally occurring bioactives in HM is underway in our and other laboratories.

As an example, we reported that tryptophan (Trp) appears to be a powerful free radical scavenger naturally present in HM [110]. HM contains a higher content of Trp than do both soy- and milk-based infant formulas [100]. Trp content of HM protein is about 2.2–2.4 % compared with 1.3 % in cow milk protein and approximately 1.7 % in whey protein-supplemented infant formulas. Trp is a precursor in the synthesis of the neurotransmitter serotonin, which is involved in the regulation of appetite, circadian rhythm, and affective reaction control. Steinberg et al. [108] fed healthy term infant formulas supplemented with 294, 588, and 882 μM/L of free Trp and reported shorter sleep latency in the Trp-fortified formula groups. Trp may be a potent antioxidant for consideration of addition to infant formulas to relive oxidative stress.

Introduction to Solid Food

With the recent recommendation from Health Canada [46] to extend exclusive breast-feeding to 6 months of age, there has arisen concern about what is the best solid food to introduce at that time. This is because traditionally solids were introduced in Canada at 4–6 months [37], and usually iron-fortified rice cereal was the first food of choice. By extending exclusive breast-feeding to 6 months, this has pushed ahead the time of introduction of solid food from 4 to 6 months. In addition to meeting iron needs, we are concerned about the possible generation of reactive oxygen species (ROS) in the gut of the infant fed traditional iron-fortified cereals. Infant cereals are fortified at 25–30 mg iron per 100 g dry-weight [37]. Absorption of the nonheme electrolytic iron ranges from 5 to 10 % [30], so that most of the residual iron enters the colon. Normally excess iron is sequestered by a variety of mechanisms in the body, but there is no such system for the sequestering of iron in the gut lumen. Iron supplements in adults where the majority of the iron is unabsorbed and passes through the digestive tract can lead to the generation of ROS in the colon [90]. These effects are seen in adults receiving 1 mg/kg/day supplemental iron. By 5–6 months of age, infants consuming iron-fortified cereals will receive the same dose and are likely producing ROS in their digestive tract [90]. This may cause inflammation and make infants more susceptible to disease. We are conducting a study to examine this hypothesis where we provide meats and infant cereals with phenolic antioxidants available from fruits that may reduce the generation of ROS in vivo.

The Premature Infant

Introduction

Premature infants (<37 weeks' gestation) are a very vulnerable group in the population. While they account for 5–7 % of all live births, they are responsible for 40 % of all infant deaths. Those that survive are at a high risk for handicap and medical complications that are extremely costly to the health care system as well as compromising the quality of their lives. VLBW infants suffer from a multitude of diseases—including: chronic lung disease (CLD) or bronchopulmonary dysplasia (BPD) and respiratory distress syndrome (RDS); necrotizing enterocolitis (NEC), an inflammation of the small intestine; intraventricular–periventricular hemorrhage (IVH–PVH) a brain injury often leading to developmental abnormalities; and retinopathy of prematurity (ROP), damage to blood vessels in the retina—and the risk of abnormal development due to high inspired oxygen concentrations, among other factors.

Due to immature lung development, immediately after birth, premature infants do not get enough oxygen (hypoxia) and therefore require supplemental oxygen, often as high as 100 %. Because of the high levels of inspired oxygen required to maintain arterial oxygen tension necessary for postnatal life, which are substantially higher than those that would normally be present during fetal existence, these infants are far more exposed to reactive oxygen species (ROS) than they would be had they remained in utero. Saugstad [101] suggests that all the factors, conditions, and problems affecting premature infants are the outcome of one unifying disease, "oxygen radical disease." He suggests that there is higher production and lower protection against free radicals, and both molecular and genetic rationales have been reported [102]. Clearly there is a need to reduce oxidative stress and/or boost antioxidant defenses in these vulnerable infants. Data from our group [38, 39] and others [13, 114] suggests that HM has antioxidant properties that will assist the premature in coping with increased oxidative stress.

Feeding Practices

Very low birthweight premature infants (VLBW) are defined as those infants <1,500 g birthweight. Almost all infants <1,000 g birthweight require mechanical ventilation and supplemental oxygen. Because oxygen must be delivered in excess of room air, often by mechanical means, this may play a role in their sequellae of diseases.

Feeding practices for the VLBW infant during the first week of life vary from unit to unit [82, 89]. VLBW infants usually receive electrolytes and dextrose soon after birth. They routinely start parenteral nutrition (PN) consisting of amino acids, lipids, and micronutrients including vitamins and trace elements, within 2–3 days of birth. Fluid intake starts at 65–80 mL/kg bodyweight/day increasing to 150–180 mL/kg/day by 5–7 days of life. The aim is full oral feeds with partial feeds started slowly to prime the gut [89].

Ideally premature infants should be fed their own mother's milk [89]. VLBW infants often are not able to suckle due to neurodevelopmental immaturity or due to their medical condition requiring orogastric feeding of expressed breastmilk. Difficulty in expressing breastmilk (manually or by lactation pumps) also contributes to the full or partial formula feeding in this population compared to full-term infants.

Reactive Oxygen Species and the Premature Infant

ROS are extremely reactive species that are essential in many biological processes. However, they also have the potential to cause damage when generated in excessive amounts or when acting on inadequately protected tissues. ROS play a role in injury caused by hypoxia and ischemia (reperfusion injury) and are involved in the inflammatory process. Oxidative injury has been proposed to be an essential condition explaining some of the major sequellae of premature birth, such as BPD, ROP, NEC, and IVH [49].

The superoxide anion is formed when molecular oxygen acquires an additional electron: $O_2 + e^- \rightarrow O_2^{-\cdot}$

Normally, the superoxide anion is short lived and is converted to hydrogen peroxide by the enzyme superoxide dismutase. Superoxide can then go on to form hydrogen peroxide, H_2O_2:

$$2O_2^{-\cdot} + 2H^+ \rightarrow H_2O_2 + O_2$$

Hydrogen peroxide can react with Fe^{2+}, resulting in the formation of hydroxyl radical (HO^-):

$$O_2^{-\cdot} + Fe^{3+} \rightarrow O2 + Fe^{2+}$$

$$Fe^{2+} + H_2O_2 \rightarrow Fe^{3+} + HO^- + \underline{HO^{\cdot}}\,(\text{Fenton})$$

Global reaction (Haber-Weiss):

$$O_2^{-\cdot} + H_2O_2 \rightarrow O_2 + HO^- + \underline{HO^{\cdot}}$$

The hydroxyl free radical is a very energetic, short-lived toxic oxygen species [1]. An excess of oxidant generation may overwhelm body defenses resulting in tissue injury so that some investigators recommend direct antioxidant therapy [14].

Increased Postnatal Oxidative Stress in Premature Infants

Premature infants are routinely exposed to supplemental oxygen (>21 % room air) due to the immaturity of their lungs. Intercurrent problems of lung disease, sepsis and poor respiratory control, generate situations where infants are transiently exposed to high inspired oxygen concentrations (FiO_2) during resuscitation [10]. This results in alternating periods of hypoxia and hyperoxia. These may be associated with depressed cardiac output, vasoconstriction of critical vascular beds, and low blood pressure, resulting in ischemia as well as hypoxia of tissues. Ischemia–reperfusion injury is well characterized in heart disease leading to cellular damage and dysfunction. Free radical generation has been shown to occur in ischemia–reperfusion injury and is likely responsible for ensuing tissue damage [103]. Within and adjacent to cells throughout the body, oxygen can react with pro-oxidants to produce ROS that can produce inflammation as well as cell death. The most important sequellae of prematurity appear to be due to inadequate protection against oxidant stress [101, 102]. Post-hypoxic reoxygenating injury caused by ROS may be a key factor. Ischemia–reperfusion injury is now recognized as a probable contributing factor to much of the morbidity of premature infants as described below.

VLBW infants have an increased susceptibility to brain damage as 5–15 % develop cerebral palsy and an additional 25–50 % have less severe neurological deficits [116]. Due to higher pO_2 and the developmental immaturity of their free radical defenses, the brain appears to be susceptible to oxidative stress. Premature infants who showed more oxidative stress at birth were more likely to develop brain damage as periventricular leukomalacia [55]. Lung injury and eye damage due to hyperoxia and free radical-mediated damage are strongly implicated [4, 49].

Respiratory distress due to RDS occurs in the majority of cases because of a surfactant deficiency at the gas–liquid interface. About 14 % of all infants <2,500 g at birth and 60 % of infants born at 29 weeks or less will suffer from RDS. Because of this condition, oxygen requirements may increase from 21 % (room air) and may reach 100 % after 48 h. BPD or CLD is often related to mechanical ventilation that is often required with supplemental oxygen. BPD is defined as either an oxygen requirement at 28 days of life plus confirmation by radiography or supplemental oxygen at 36 weeks [101]. Inhaled nitric oxide, a major player in the physiology of ROS, given to infants at risk for BPD does not alter biomarkers of oxidative stress and appears to be safe [10].

Necrotizing enterocolitis (NEC) is an inflammation of the small intestine and bowel surface, with infiltration of epithelial cells by bacteria. While its etiology is unclear, it may be precipitated by ischemia–reperfusion injury [101]. Nutrition may be both a causative and preventative factor.

Intraventricular–periventricular hemorrhage (IVH–PVH) is a brain injury, usually occurring within 24–48 h after birth (95 % occur by 3 days) or may occur later. It is believed to be an ischemic injury followed by reperfusion injury. IVH–PVH in its severest form carries a high risk of poor neurodevelopmental outcome (≥50 %).

ROP is oxygen-induced damage to blood vessels in the retina that are undergoing neovascularization. Oxygen therapy in the early 1940s led to an epidemic of ROP [49], ending with a more judicious use of oxygen. Increased oxygen can be administered at controlled levels without inducing ROP, but high levels of oxygen before 32 weeks are still a serious concern for the genesis of ROP.

Saugstad [101] has proposed that the above diseases are expressions of an inability to cope with an overexposure to ROS. Hypoxia at birth leads to a breakdown of ATP to hypoxanthine and a conversion of xanthine dehydrogenase to xanthine oxidase (XOD), which circulates systematically after release from the liver. XOD is also highly concentrated in the lung and in the gut of the newborn. Once reperfusion occurs, oxygen is available to form the superoxide radical. From here, tissue damage is inevitable.

Evidence for the Role of Damage Done by ROS in Prematurity

There is increasing experimental and clinical data that ROS are formed too rapidly to be detoxified by the immature defenses of premature infants. Tissue damage is increased in premature infants exposed to supplemental oxygen. Plasma F_2 isoprostanes were more than threefold higher in the cord blood of premature infants than in adults [24] and higher than those of term infants. Berger et al. [12] found increased oxidative stress in premature infants due to unbound iron in the blood suggesting that premature infants are at risk of ROS damage. Vitamin C and E levels are low in preterm infants [114]. Berger et al. [12] proposed that low vitamin C might increase the risk of oxidant injury, manifested as BPD and IVH-PH. Increased urinary o-tyrosine at 1 week of life was associated with increased inspired oxygen [23]. Buonocore et al. found increased oxidation in the cord blood of hypoxic newborn infants and in premature infants at birth and 7 days of life [16]. However, an increase in biomarkers of oxidative stress does not confirm or imply causality. Indeed several clinical trials with antioxidants have not produced positive results [51].

Defense Against ROS Attack

Normally we defend against ROS with antioxidants and several essential nutrients act as antioxidants. Enzymatic defenses include superoxide dismutase (SOD), glutathione peroxidase (GHSPx), and catalase (CAT). PUFA can protect against oxygen toxicity [109], and/or the premature can cope with potential oxidative stress form lipid nutrition [52]. The maturity of the antioxidant enzymes may peak in late gestation in different species. It may be that preterm babies have prompter involvement of antioxidant defenses than term babies. As well, this group reported that GSH concentration in VLBW infants increases significantly after birth. A concomitant increased synthesis rate was not found, suggesting that GSH consumption decreases upon amino acid administration [98].

The Role of Nutrition in Prematurity

Given the growing role of oxidative stress in newborn preterm morbidity, one of the goals of modern neonatology is to minimize free radical production and promote the development of adequate antioxidant systems through an adequate nutritional strategy. Appropriate administration of total parenteral solutions and lipid emulsions with light protection can minimize the risk of peroxidation. Providing the baby with amino acid substrates for cellular glutathione synthesis immediately after birth promotes antioxidant defenses at the early stages of life. Minimal enteral nutrition has been shown to improve the adaptation of the immature intestine to oral feeds [99]. Breast milk has been found to have many advantages over formula, including the potential to provide antioxidant protection to infants. It is conceivable that these antioxidants in breast milk help to eliminate free radicals in infants. The role of vitamin administration in preterm nutrition has not yet been established. Clinical trials carried out to test the efficacy of antioxidant drugs or vitamins were inconclusive. At present, there are no evidence-based recommendations with the use of nutritional strategies or antioxidant drugs to minimize oxidative stress in the management of preterm infants [92, 101]. Of further concern, Laborie et al. [64] reported that liquid vitamin preparations for enteral administration are contaminated with high level of peroxides. The administration of these oral vitamin solutions to premature infants (33 ± 1 weeks) who are in transition between parenteral and enteral nutrition (at 9 ± 1 days of life) was associated with a higher concentration of peroxides in their urine.

Antioxidant Properties of Human Milk to Defend Against ROS

Human milk is the ideal first food during infancy. In fact human milk may provide some of the defenses needed for the premature infant that are not available due to the immature development of antioxidant systems [13]. In addition to GPx in mother's milk [63], there is another peroxidase in human milk that may protect the mammary gland [6]. Breast milk also seems to contain substances that reduce hydroxyl radical formation [3, 77].

Early feeding of human milk compared to cow's milk-based infant formulas reduces the incidence of NEC in the premature infant [105] and appears to protect against sepsis [40]. Premature infants who were human milk fed had higher plasma peroxyl radical-trapping ability in vitro than did control formula-fed infants [114]. Premature infants may absorb intact apolactoferrin from human milk into their plasma sufficient to boost plasma antioxidant capacity by sequestering iron, as digestion of proteins in human milk is not fully developed in the newborn. It needs to be considered that positive results from in vitro testing may not be reflected in vivo.

We hypothesize that CAT, SOD, and GPx which exist in human milk may exert their influence directly in the gut or even be absorbed whole in the "leaky" gut of the infant to provide systemic protection [38, 39]. Several groups [7, 60, 112] have corroborated these findings and confirm that mother's milk can assist in preventing oxidative stress.

Developing enterocytes are subjected to oxidative stress as part of their normal metabolism. It may be that human milk of the premature infant compensates for underdevelopment as it does for other nutrients [5]. There is clinical evidence that mother's milk may be protective against mucosal injury in developing intestine [25]. As described below, we have found that HM is better able to scavenge free radicals than are artificial humanized formulas [6]. The exact mechanism how this is done has not been elucidated. This data has important consequences for the developing premature.

Human Milk Fortification

Normal standard of care for the premature infant is to add human milk fortifier (HMF) to expressed breast milk [25, 35]. However, little work with HMF beyond tolerance and growth of premature infants has been carried out. In order to assess the effect of HMF on oxidant/antioxidant status, we studied 65 PT infants (birth weight, 1,146 ± 261 g; GA, 29 ± 2.5 weeks; mean ± SD) biweekly, once oral feeds were introduced [35]. Feeding groups—F (>75 % total feeds) and HM (>75 % total feeds)—were further subdivided according to human milk fortifier (HMF) content of 0–19, 20–49, and ≥50 %. Oxidative stress was quantified by F2-isoprostanes (F2-IsoPs) in urine, protein carbonyls, and oxygen radical absorbance capacity (ORAC) in plasma. Highest isoprostane values occurred in infants with >50 % of their mother's milk fortified. This was unexpected as isoprostanes were higher in formula-fed infants than those fed HM only. Isoprostanes are the breakdown product of lipid peroxidation and are seen as a sign of oxidative stress [85]. Chen et al. [17] have identified a novel physiological role for isoprostanes during postnatal vascular transition and provide evidence that oxidative stress may act on membrane lipids to produce vasoactive mediators that stimulate physiological DA closure at birth or induce pathological patency of the preterm DA. The possible mechanisms for the generation of lipid peroxidation in the HMF-fed infant are unknown.

Iron

Iron supplementation may be associated with oxidative stress particularly in premature infants. In a previous study, our purpose was to examine (1) early supplemental iron during treatment with erythropoietin (EPO) and oxidative stress and (2) enhanced iron absorption during EPO in those infants receiving human milk. Therefore, we determined the effect of erythropoietin plus supplemental iron intakes (4 mg/kg/day) on antioxidant status and iron incorporation. Data suggest that during erythropoietin therapy, antioxidant defense in VLBW infants are capable of dealing with early supplemental iron during treatment with EPO [34].

Parenteral Nutrition

Introduction

Because the smallest premature neonates (<31 weeks of gestation) require intravenous nutrition to compensate for their gastrointestinal immaturity, parenteral nutrition may be a safe way to enhance their weak antioxidant capacity. While parenteral nutrition provides antioxidant vitamins such as vitamins C and E, it is contaminated by high concentrations of oxidant molecules such as peroxides [11, 65, 66, 68, 75]. As a consequence, parenteral nutrition is associated with oxidative stress and with several pathological complications, as demonstrated in newborn animal studies and in premature infants themselves.

In the following sections, we will present evidence documenting the impact of parenteral nutrition, and its specific components on the oxidant status of neonates, how

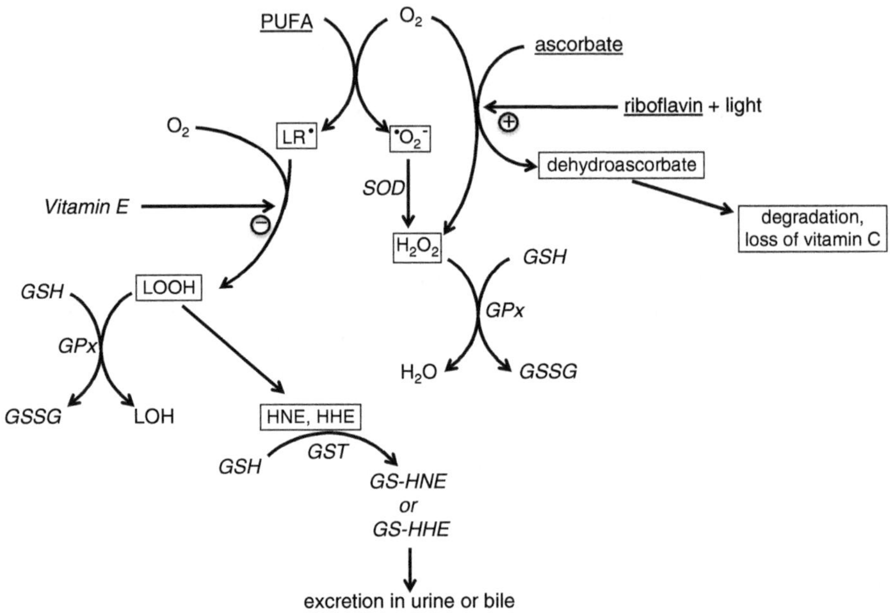

Fig. 19.1 Interactions between nutrients (PUFA, ascorbate, riboflavin) and dissolved oxygen in TPN. In squares: molecules generated by these interactions. In italics: endogenous antioxidant defenses. Underlined: nutrients from TPN. *PUFA* polyunsaturated fatty acids, *LR* lipid free radical species, *LOOH* lipid hydroperoxides, *HNE* 4-hydroxynonenal, *HHE* 4-hydroxyhexanal. $^{\cdot}O_2^-$ anion superoxide, *SOD* superoxide dismutases, *GSH* reduced form of glutathione, *GSSG* disulfide form of glutathione, *GPx* glutathione peroxidase, *GST* glutathione S-transferase

these nutrients generate oxidant molecules, and how they might enhance antioxidant defenses of the newborn. Figure 19.1 summarizes some of the nutrient–oxidant interactions presented in the present section of this chapter. The impacts of these reactions on biological or clinical phenotypes will be addressed in newborn animal models as well as in premature newborn infants. Finally, potential solutions will be discussed.

TPN as Source of Oxidant Molecules

Multivitamin Preparation

Generation of H_2O_2

Total parenteral nutrition (TPN) contains the following elements that are required to sustain basal metabolism and growth: glucose, amino acids, lipids, vitamins, and trace elements. Several of these nutrients have the potential to influence the generation of oxidant molecules in solution. The presence in the same solution of strong electron donors such as polyunsaturated fatty acids (PUFA) from the lipid emulsion and vitamin C from multivitamin preparation (MVP), combined with a strong

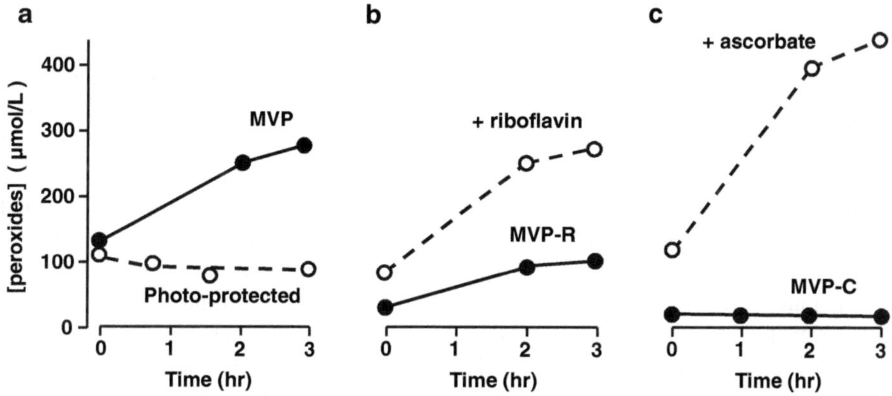

Fig. 19.2 Peroxide generation in function of time in different multivitamin preparations. Panel (**a**) *Dark symbols* (*full line*): MVP (1 % (v,v) Multi-12 pediatrics in water; without light protection). *Open symbols* (*dashed line*): light-protected MVP. Panel (**b**) *Dark symbols* (*full line*): MVP-R* (MVP devoid of riboflavin). *Open symbols* (*dashed line*): MVP-R in which riboflavin was added at same concentration as labeled on the product information on the multivitamin bottle. Panel (**c**) *Dark symbols* (*full line*): MVP-C* (MVP devoid of vitamin C). *Open symbols* (*dashed line*): MVP-C in which ascorbate was added at same concentration than labeled on the multivitamin bottle. Mean ± SEM ($n = 3$), symbols are greater than SEM. * Graciously prepared and provided by Sabex (Sandoz), Boucherville, Qc, Canada

electron acceptor such as dissolved oxygen, will lead to oxidation of PUFA and vitamin C and reduction of oxygen. These interactions result in (1) a production of lipid peroxides [88, 104], (2) oxidation of ascorbate in dehydroascorbate that is rapidly degraded leading to the lost of this antioxidant vitamin [9, 27, 67], and (3) production of hydrogen peroxide (H_2O_2) from the reduction of oxygen [65, 66]. In fact, close to 80 % of peroxides generated in TPN are in the form of H_2O_2 [66]. Because ascorbate is the strongest reducer present in TPN, it is the primary contributor to the generation of peroxides [65]. Furthermore, riboflavin, a component of MVP, is a light-sensitive molecule that uses energy from ambient light to accelerate the reaction between vitamin C and O_2 leading to the generation of H_2O_2 [65]. Adequate photo-protection of TPN reduces by half the concentration of peroxides [11, 66, 75]. Figure 19.2 portrays the components taking part in the interactions between riboflavin, ascorbate, and light on the generation of peroxides. Thus, parenteral multivitamin preparations and ambient light are considered the main contributors to the generation of peroxides in TPN [11, 66].

Impact of H_2O_2 Infused with TPN

In isolated cells, the levels of peroxides as measured in TPN or MVP solution exposed to ambient light are toxic. These concentrations of peroxides are sufficient to induce cell death within 3–4 h incubation in endothelial cells from human umbilical vein in primary culture. The toxic effect is even significantly more pronounced if cells are derived from male infants [67]. Incubation of cultured human skin fibroblasts with a solution containing MVP induces damages to the DNA (single- and double-strand breaks) leading to cell death if the MVP solution was previously exposed to ambient light [118]. These observations were reproduced with H_2O_2 [118]. However, this toxicity may well be very different if the oxidant solutions are administrated into the blood stream where red blood cells have a strong catalase activity. Hence, the importance of documenting if infants can detoxify the oxidants generated and infused in TPN. Two days after the introduction of MVP in intravenous nutrition of premature infants, their urinary concentration of peroxides rises [11], independently of the presence of lipid emulsion or amino acids [18]. Full photo-protection of TPN against ambient light prevents the increase of peroxides in urine of these neonates [11, 18, 70]. This data suggests that neonates are unable to quench infused peroxides, at least if TPN is not adequately photo-protected.

Infusion of solutions containing MVP or hydrogen peroxide to newborn guinea pigs is associated in the lungs with a reduced level of glutathione [69] and a lower catabolism of PGE_2 and PGF_2 [69], which is known to be sensitive to oxidation [8]. In the liver of these animals, H_2O_2 induces (1) a lower level of glutathione; (2) a lower ratio of prostacyclin/thromboxane, an index of radical injury [18, 19, 48]; and (3) a higher level of $F_2\alpha$ isoprostanes. In contrast to the lung, these effects of H_2O_2 are not reproduced in the liver by infusion of a solution containing MVP [18, 19]. Antioxidant vitamins from MVP seem to quench free radicals derived from H_2O_2 in the liver, but not in the lungs. Could this be related to the fact that unlike vitamin C, vitamin E may accumulate in the liver [71]?

Effect of Light

To reproduce in premature infants studies similar to those performed in animals is challenging because of the absence of a control group, namely, premature infants who are not exposed to intravenous peroxides. Indeed, parenteral nutrition free of peroxide does not yet exist. As well, finding those extremely premature infants who are not receiving some form of TPN in the early days of life is very rare. Thus, in neonates it is only the impact of differential level of peroxides that may be investigated by applying photo-protection to TPN. Infants receiving the complete formulation of parenteral nutrition are infused with some 350–400 μM peroxides, whereas those infused with photo-protected TPN receive around 175–200 μM peroxides [75]. Even if this appears as a limited reduction of peroxide content, photo-protection of TPN does influence the metabolism of these infants. Indeed, the usual increase in plasma triacylglycerol observed during the first week of life of premature infants on TPN does not occur if the parenteral solution is shielded from light [56]. Similar results have been observed in newborn guinea pigs infused with a solution containing dextrose with MVP or with TPN, as the absence of photo-protection led to a higher level of plasma triacylglycerol and hepatic steatosis [22]. These effects were not reproduced by infusion of H_2O_2 [22]. The apparent free radical protection against MVP in the liver [18, 19], as mentioned in section "Impact of H_2O_2 infused with TPN"), and the absence of an effect of H_2O_2 suggest that another compound generated during light exposure of MVP may be responsible for the observed disorder in lipid metabolism. A different line of investigation has suggested that an ascorbylperoxide [57, 70, 72], deriving from the interaction between H_2O_2 and dehydroascorbate generated in TPN [57], might well be the active agent leading to a higher triacylglycerol synthesis [80].

Further evidence that shielding TPN against ambient light has a beneficial impact in premature infants is provided by the observed reduction in the incidence of chronic lung disease [11] or bronchopulmonary dysplasia [21]. Incidences are lower in preterm newborn infants receiving photo-protected TPN. A similar protective effect was observed in lungs of newborn guinea pigs. The number of alveoli was lower in animals infused with TPN or MVP lacking adequate photo-protection [70, 72, 75]. Reduced alveolar development is a characteristic feature of bronchopulmonary dysplasia [50]. The lungs having a lower number of alveoli presented a higher number of apoptotic events [70, 72]. Thus, the simple fact of shielding TPN from ambient light seems to be sufficient to prevent or reduce the impact of oxidative stress in premature infants. In contrast, a recent study failed to demonstrate a beneficial impact of light protection of TPN on oxidative stress when the multivitamin preparation was added to the lipid emulsion [23]. In this study, TPN was administered as a binary solution (see section "Lipid emulsion"), in which the lipid emulsion is administered separately from the other components (dextrose, amino acids, trace elements), and only the lipid moiety of TPN was subjected to light protection. Adding multivitamins to the amino acids part (unprotected against light) or to the lipid moiety of TPN led to no difference at 1 week of age in three markers of oxidative stress: urinary F2α-isoprostanes, urinary di-tyrosine, and blood redox potential of glutathione [23]. The fact that the concentration of these three markers was higher than those reported in adults suggests that beyond the photo-protection, oxidative stress remains a distinguishing feature in this population of very immature infants.

The positive relationship found in this same study [23] between the redox potential of glutathione and the severity of BPD (the more oxidized the redox, the more severe was BPD) points to the importance of glutathione as a central antioxidant, an anti-peroxide (via glutathione peroxidase), or a key player in cellular redox signaling. The low glutathione levels reported in premature neonates [67] may explain this oxidized redox potential [23]. Furthermore, as presented in section "Impact of H_2O_2 infused with TPN," the infusion of MVP or H_2O_2 does induce also a lower level of glutathione in the lungs [69].

Lipid Emulsion

TPN is administered as a binary solution or as all in one. The binary mode consists in separating the lipid emulsion from the other components of TPN (dextrose, amino acids, trace elements); the lipid emulsion is mixed with the other components close to the site of infusion. The multivitamin preparation is generally added to the dextrose–amino acids moiety of TPN or separately. Water-soluble vitamins are mixed with the dextrose–amino acid preparation, while the liposoluble vitamins are mixed with the lipid emulsion. The all-in-one modality consists in a mix of all components of TPN, together, in a same solution. Thus, the modality of TPN administration as well as the photo-protection of the different parts of the TPN solution may lead to variation in oxidant molecules generated in these solutions.

Lipids in TPN are subject to auto-oxidation [88, 104], which is more pronounced if the emulsion is exposed to light [83, 88, 104]. The main products of oxidation are lipid peroxides and aldehydes. The latter are derived from fragmentation of lipid peroxides. These molecules are biologically active and have potential deleterious effects [61, 113]. Adding the multivitamin preparation, at least the liposoluble antioxidant vitamins, with the lipid emulsion rather than with the amino acids moiety of TPN, is an alternative protocol that may limit oxidation of lipids [88, 104]. It has been shown [104] that the addition of liposoluble vitamins to the lipid emulsion prevented the generation of lipid peroxides, whereas [83] reported that the addition of the complete multivitamin preparation in the lipid emulsion rather than to the dextrose–amino acids part of TPN led to the generation of a greater amount of aldehydes and total peroxides. This last study made no discrimination between organic (lipid peroxides) and inorganic peroxides (H_2O_2) after reconstitution of TPN (lipid emulsion + dextrose–amino acids).

Impact of Co-administration of Multivitamin Preparation with Lipid Emulsion

In vivo, co-administration of the multivitamin solution with lipid emulsion rather than with the dextrose–amino acids moiety of TPN has induced a more oxidized redox potential in the blood of newborn guinea pigs [75]. In the lungs, however, in spite of the detrimental impact of light exposure of TPN, the alveolarization index was better in animals infused with TPN in which multivitamins were co-administered with lipid emulsion in the absence of photo-protection [73, 75]. In premature infants receiving TPN in which multivitamins were mixed with the dextrose–amino acid fraction, those requiring oxygen supplementation ($FiO_2 \geq 0.25$) exhibited greater oxidation of the redox potential [23] and an elevation of IL-6 and IL-8 in the blood compared to those not requiring oxygen supplementation [76]. The impact of elevated oxygen was not observed in infants receiving TPN in which the multivitamin preparation was co-administered with the lipid emulsion.

Lipid Emulsion Rich in Omega-3, Omega-6, or Omega-9 Fatty Acids

Different lipid emulsions are available for parenteral nutrition. The major differences between them are linked to the saturation of their polyunsaturated fatty acids (PUFAs) (Table 19.1). The susceptibility of these fatty acids to oxidation is dependent on the position of the first unsaturated bond. PUFAs from the omega-3 series are more prone to oxidation than the omega 6 series [83]. Thus, the level of peroxides and aldehydes in TPN may differ according to the type of lipid emulsion used.

Table 19.1 Proportion (in % of total fatty acids) of principal fatty acids in parenteral lipid emulsions

	Intralipid (Fresenius Kabi)	ClinOleic (Baxter)	Omegaven (Fresenius Kabi)	SMOFlipid (Fresenius Kabi)
	Soybean oil	Soybean oil: olive oil (1:4)	Fish oil	Soybean oil: medium chain: triglycerides: olive oil: fish oil (2:2:1.67:1)
% Total fatty acids				
Oleic acid (18:1n9)	22	58	14	31
Linoleic acid (18:2n6)	50	18	3.5	20
Linolenic acid (18:3n3)	7	2.1	1.3	2.2
Eicosapentaenoic acid EPA (20:5n3)		0	19	2.9
Docosahexaenoic acid DHA (22:6n3)		0.1	25	20
Tocopherol[a] (mg/L)				
dl alpha			15–30	200
Alpha	~15	~30		
Gamma	~120	~20		
Delta	~ 60	~15		

[a]Intralipid and ClinOleic are not supplemented with tocopherol; the reported concentration is derived from the one found in natural oils. Omegaven and SMOFlipid are supplemented with dl-alpha tocopherol [41, 58, 87, 107]

On the other hand, omega-3 series are reported to have antioxidant properties [106]. This group reported that after 14 days on TPN prepared with SMOFlipid (high omega-3) rather than with Intralipid (high omega-6), preterm infants had a higher "total antioxidant potential" (TAP, reported as uric acid equivalent). In newborn guinea pigs, the redox potential of glutathione in the liver was more reduced (less oxidized) in animals infused 4 days with TPN made with Omegaven rather than with Intralipid [83, 84]. Lipid emulsion rich in omega-9 fatty acids (ClinOleic) does not seem to bear an advantage in terms of antioxidant/oxidant status over Intralipid. Indeed, indices of antioxidant status such as total radical-trapping antioxidant potential (TRAP) [97] or total antioxidant capacity (TAC) [58] as well as urinary level of $F_2\alpha$ isoprostanes [97] or MDA [41] did not differ between newborn infants who have received ClinOleic or Intralipid for 1 week.

As shown in Table 19.1, differences in vitamin E content cannot explain these results. Moreover, considering the quantity of vitamin E from the multivitamin preparation (1 % Multi-12 (Sandoz) provides 14 mg of tocopherol per liter of TPN), the fraction of vitamin E in TPN deriving from the lipid emulsions (Intralipid, ClinOleic, or Omegaven) is low and varies between 10 and 30 % of total tocopherol content. In contrast, SMOFlipid provides 60–75 % of all tocopherol in TPN. Thus, the beneficial effect of lipid emulsions containing high levels of omega-3 (Omegaven vs. Intralipid) on TPN-related cholestasis [59] excludes the participation of tocopherol in the development of this liver disease [45]. However, the proportion of vitamin E from the lipid emulsion that is infused in an active form is unknown. Tocopherol present in the multivitamin preparation is in an acetate form. This form of vitamin E will become active only after hydrolysis of the acetate in vivo; however, it is unclear if this actually does take place in premature infants.

Amino Acids and Trace Elements

Some amino acids such as tryptophan, tyrosine, or cysteine have free radical scavenging capacity. At the concentration usually used in neonatal TPN (~2 %, w/v), the amino acid preparations have a marginal impact [66]. On the other hand, cysteine and methionine, which are essential for glutathione synthesis, may play an important role in the building of an antioxidant defense based on glutathione as discussed in section "Glutathione."

Despite containing free iron and copper, the trace element preparation does not interfere with peroxide concentration in TPN [11].

TPN and Antioxidant Defenses

Glutathione

Oxygen supplementation and TPN are important sources of oxidants in premature infants. Oxygen is a double free radical (·O–O·) that reacts with electron donors such as vitamin C, PUFA, or NADPH to generate superoxide anion (·O–O·· or (·O–O$^-$) (Fig. 19.1) in the presence of enzymatic catalysts (e.g., NADPH oxidase) or photosensitive molecules (e.g., riboflavin). This reactive oxygen species will be dismutated into H_2O_2 and O_2 spontaneously or catalyzed by superoxide dismutase (SOD). H_2O_2 is reduced in water by Peroxiredoxins, catalase, or glutathione peroxidase (GPx). Peroxiredoxins detoxify H_2O_2 by oxidation of their own thiol residues that will be reduced back by the action of thioredoxin–thioredoxin reductase using NADPH as electron donor [93]. Catalase, except in erythrocytes, is found in peroxisomes and mitochondria [91]. It is specialized in reducing large concentrations of H_2O_2 present in peroxisomes [54, 91]. GPx is a cytosolic enzyme with high affinity (low km) for H_2O_2 as well as for organic peroxides. The activity of GPx can be overwhelmed in the presence of large concentration of H_2O_2. In this case, catalase helps in the detoxification of the peroxide [91]. The roles of SOD and GPx fundamental in the regulation of the intracellular concentration of H_2O_2 [32, 33] have shown in several animal models (mice, rats, hamsters, and guinea pigs) that pulmonary activity of catalase, superoxide dismutase, and glutathione peroxidase increases significantly in the last fifth of gestation. By extrapolation we can assume that the same phenomenon occurs in humans, i.e., before 32 weeks' gestation. The activity of these antioxidant enzymes is very low, between 10 and 15 % of the activity measured in the lung of term newborn infants [33]. Anion superoxide is slowly transformed to H_2O_2, and peroxides from TPN are not well detoxified. In addition, the level of glutathione (γ-Glu–Cys–Gly), which is the essential cofactor for GPx activity, is dependent on gestational age and the sex of the infants [67]. At <32 weeks' gestation, the glutathione level in isolated leukocytes from endotracheal secretions is approximately 15 % of the level measured in leukocytes from the cord blood of term infants and is even lower for the subgroup of males. We reported that in such cells, the level of glutathione remained low until at least 3 weeks of life [67]. These low glutathione levels appeared to be explained by a lack of substrate for glutathione synthesis rather than an impaired capacity to produce it. Indeed, the activity of synthesis is well developed in leukocytes from preterm infants [68], whereas the capacity of these cells for cysteine uptake is immature, limiting the increase in cellular glutathione concentration [74].

The importance of glutathione in neonates has been emphasized by, among other things, the increased survival of newborn guinea pig exposed to >95 % O_2. Animals infused with TPN had a higher level of glutathione in the lung and liver as well as a better rate of survival than those infused with a solution devoid of amino acids or those fed enterally [20]. The fact that the hepatic rate of glutathione synthesis was greater in animals infused with TPN compared to those fed enterally [20] suggested that the level of glutathione in newborn guinea pig was dependant on intravenous supplementation of amino acid substrates. Because the amino acid blend infused in this study [20] did not contain cysteine, an essential amino acid for glutathione synthesis, it would highlight the essential role of hepatic transformation of methionine to cysteine.

However, in premature infants, the transmethylation pathway (Fig. 19.3) leading to the formation of cysteine from methionine is weak because cystathionase activity is immature [115, 119]. In addition, a recent publication reports that, in newborn guinea pigs, infusion of H_2O_2 or TPN inhibited the activity of methionine adenosyltransferase, the first enzyme of the cascade leading to the conversion of methionine

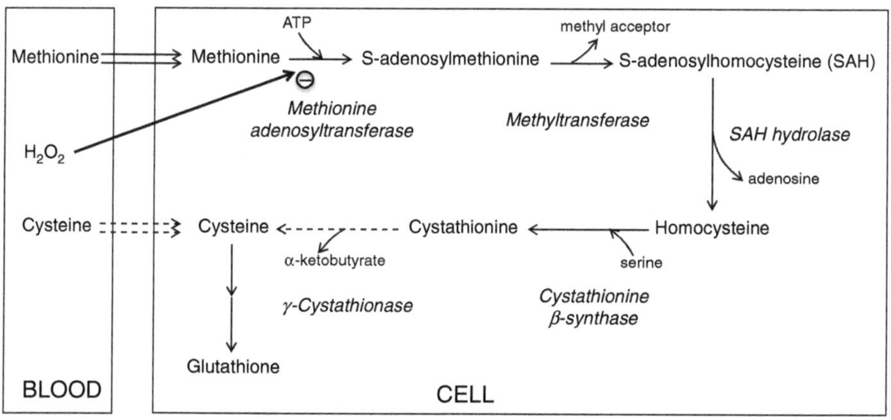

Fig. 19.3 Impact of nutrients (methionine, cysteine) and peroxide infused with TPN on glutathione. In *italics*: enzymes. *Dashed line*: immaturity in premature newborns. H_2O_2 from TPN inhibits the activity of methionine adenosyltransferase leading to a lower level of glutathione

into cysteine [29]. This inhibition led to a lower glutathione level in the blood and liver and to a more oxidized redox potential of glutathione in whole blood [29]. Thus, in comparison to oxygen that has a positive effect on the activity of glutathione synthesis as shown in newborn animals [20] as well as in preterm neonates [65], peroxides infused with TPN have a different and negative impact on glutathione levels by lowering the substrate availability.

Vitamin C and Vitamin E

The main antioxidant vitamins in TPN are ascorbate and tocopherol. As discussed in section "Generation of H_2O_2," vitamin C is unstable in TPN. During its reaction with dissolved oxygen in generating H_2O_2, ascorbate is transformed to dehydroascorbate, which is rapidly hydrolyzed to diketogulonate which is the first step in degradation of vitamin C. In TPN, the concentration of ascorbate decreases rapidly without equivalent increase in dehydroascorbate [9, 27]. The half-life of this vitamin in TPN is reported to be 3 h at 21 °C [27]. Its stability can be improved for up to several days if TPN is stocked in a multi-plastic layer bag with low oxygen permeability [9, 27, 28].

Vitamin E in TPN is derived from the lipid emulsion and mainly from the multivitamin preparation (section "Lipid emulsion rich in omega-3, omega-6, or omega-9 fatty acids"). Within the multivitamin preparation vitamin E is stable because it is in acetate form. Studies conducted by Drott et al. [26] demonstrated that the level of vitamin E in TPN infused over 5–7 days does not influence its plasma concentration in neonates born term or preterm. The level was increased, in both studies, after introduction of enteral alimentation. These two studies suggest that the acetate form of tocopherol for intravenous administration is not adequate for newborn infants.

Prevention of the Generation of Oxidant Molecules in TPN

In premature neonates, who have immature antioxidant defenses, several severe pathological complications may be associated with the presence of an oxidative stress. Although antioxidant capacity is dependent on nutrition, as discussed above, parenteral nutrition contributes more to the oxidant load than to antioxidant defenses. It is important to find ways to administer safer TPN solutions, free of undesirable oxidant molecules. In this section, four alternatives are discussed: (a) photo-protection of TPN, (b) the use of bags made with a plastic material that limits oxygen diffusion, (c) lipid emulsions rich in omega-3 PUFAs, and (d) the addition of antioxidant hexapeptides derived from human milk.

While waiting for the manufacture of a new intravenous nutritional formulation free of light-induced oxidants, photo-protection remains an efficient approach. It is very important to adequately protect the solution by minimizing exposure to ambient

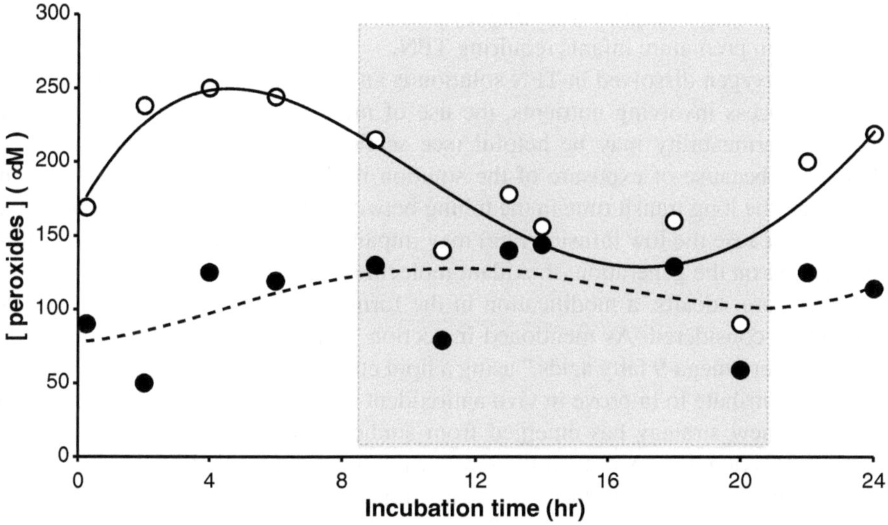

Fig. 19.4 Concentration of peroxides in function of time. Solution of 1 % (v,v) multivitamin preparation (Multi-12, Sandoz, Boucherville, Qc, Canada) in water, protected (*close symbol, dashed line*) or not (*open symbol, continuous line*) against ambient light incubated at room temperature for 24 h. *White area*: light exposure (about 70 foot candles). *Gray area*: without light. The concentration of peroxides in solution devoid of light protection varies in function of time and light exposure ($r^2=0.72$, $p<0.01$), whereas there is no significant ($r^2=0.26$) variation in solution protected from light. Mean ± SEM ($n=3$); symbols are greater than SEM

light all the way from the compounding process through its delivery and administration at the bedside. The light-induced generation of peroxide takes place quickly as shown in Fig. 19.4. In less than 10 min exposure to a light solution of 1 % multivitamin preparation in water generates >160 μM peroxides. The fact that the peroxide level decreases in the solution following exposure to darkness (Fig. 19.4) suggests that these peroxides are transformed by auto-degradation or when reacting with other components of the multivitamin preparation. The peroxide concentration measured in the solution benefiting from full protection against the light (about 100 μM) may be explained, at least in part, by the high level of hydroperoxides measured in newly reconstituted vial of multivitamin preparation: 3,280 ± 110 μM (mean ± SEM $n=6$) in MVI Pediatric from Rhône Poulenc Rorer [66]. However, in the clinical setting, achieving full photo-protection of TPN, from the time of compounding to its delivery to the newborn in the neonatal unit, is impractical. Even if this goal was achievable, some 150 μM peroxide would remain in solution. These peroxides stem from the multivitamin preparation and the lipid emulsion in which we reported 134 ± 16 μM (mean ± SEM $n=6$) in newly open bottles of Intralipid [66]. In absence of a pertinent clinical study in which a control group of neonates would receive TPN free of peroxides, we speculate that the infusion with TPN of peroxide concentrations as low as 150 μM are not desirable for premature newborns.

Even complete photo-protection might be insufficient to avoid some degree of oxidant load in premature infants requiring TPN.

Since oxygen dissolved in TPN solution is an important component for the oxidative process involving nutrients, the use of multi-plastic layer bag having low oxygen permeability may be helpful (see section "Vitamin C and vitamin E"). However, because of exposure of the solution in the tubing to oxygen present in room air, the long transit time in the tubing between the bag and the site of infusion (3–4 h, because the low infusion rate) may impair the beneficial impact of the multilayer bag on the generation of oxidant molecules.

Therefore, ideally, a modification in the formulation and composition of TPN should be considered. As mentioned in section "Lipid emulsion rich in omega-3, omega-6, or omega-9 fatty acids," using a lipid emulsion rich in omega-3 fatty acids should contribute to improve in vivo antioxidant capacity [83, 84, 106]. Recently, a potential new strategy has emerged from studies investigating why human milk exerts antioxidant properties in the newborn. Tsopmo et al. [110] have discovered that two *hexapeptides*, produced following digestion of human milk, exhibit strong free radical scavenging properties. These peptides, which reduced in vitro levels of peroxides, were added to TPN and infused into animals. They were subsequently associated with increased hepatic glutathione levels and a reduction in two markers of inflammation in liver, TNFα and IL-1 mRNA [83, 84]. Much remains to be done before these peptides become a clinically applicable solution for children. However, hope remains as the scientific community and clinicians are well aware of the oxidant risks associated with TPN and are investigating new ways to administer safer products to this frail population.

References

1. Acworth IN. The handbook of oxidative metabolism. Chelmsford: ESA, Incorporated; 1995.
2. Allen RG. Oxygen-reactive species and antioxidant responses during development: the metabolic paradox of cellular differentiation. Proc Soc Exp Biol Med. 1991;196(2):117–29. doi:10.3181/00379727-196-43171A.
3. Almaas R, Rootwelt T, Øyasæter S, et al. Ascorbic acid enhances hydroxyl radical formation in iron-fortified infant cereals and infant formulas. Eur J Pediatr. 1997;156(6):488–92. doi:10.1007/s004310050645.
4. Ates O, Alp HH, Caner I, et al. Oxidative DNA damage in retinopathy of prematurity. Eur J Ophthalmol. 2009;19(1):80–5.
5. Atkinson SA. E. – Effects of gestational stage at delivery on human milk components. In: Handbook of milk composition. San Diego: Academic; 1995. p. 222–37. Retrieved from http://www.sciencedirect.com/science/article/pii/B9780123844309500135.
6. Atkinson SA, Lönnerdal B. B. – Nonprotein nitrogen fractions of human milk. In: Handbook of milk composition. San Diego: Academic; 1995. p. 369–87. Retrieved from http://www.sciencedirect.com/science/article/pii/B9780123844309500172.
7. Aycicek A, Erel O, Kocyigit A, et al. Breast milk provides better antioxidant power than does formula. Nutrition. 2006;22(6):616–9. doi:10.1016/j.nut.2005.12.011.
8. Bakhle YS. Synthesis and catabolism of cyclo-oxygenase products. Br Med Bull. 1983; 39(3):214–8.

9. Balet A, Cardona D, Jané S, Molins-Pujol AM, Sánchez Quesada JL, et al. Effects of multilayered bags vs ethylvinyl-acetate bags on oxidation of parenteral nutrition. JPEN J Parenter Enteral Nutr. 2004;28(2):85–91.
10. Ballard PL, Truog WE, Merrill JD, Gow A, Posencheg M, et al. Plasma biomarkers of oxidative stress: relationship to lung disease and inhaled nitric oxide therapy in premature infants. Pediatrics. 2008;121(3):555–61. doi:10.1542/peds.2007-2479.
11. Bassiouny MR, Almarsafawy H, Abdel-Hady H, Nasef N, et al. A randomized controlled trial on parenteral nutrition, oxidative stress, and chronic lung diseases in preterm infants. J Pediatr Gastroenterol Nutr. 2009;48(3):363–9.
12. Berger TM, Polidori MC, Dabbagh A, Evans PJ, et al. Antioxidant activity of vitamin C in iron-overloaded human plasma. J Biol Chem. 1997;272(25):15656–60.
13. Buescher ES, McIlheran SM. Antioxidant properties of human colostrum. Pediatr Res. 1988;24(1):14–9.
14. Bulkley GB. Free radicals and other reactive oxygen metabolites: clinical relevance and the therapeutic efficacy of antioxidant therapy. Surgery. 1993;113(5):479–83.
15. Buonocore G, Zani S, Perrone S, Caciotti B, Bracci R. Intraerythrocyte nonprotein-bound iron and plasma malondialdehyde in the hypoxic newborn. Free Radic Biol Med. 1998;25(7):766–70.
16. Buonocore G, Perrone S, Longini M, Vezzosi P, Marzocchi B, et al. Oxidative stress in preterm neonates at birth and on the seventh day of life. Pediatr Res. 2002;52(1):46–9.
17. Chen JX, O'Mara PW, Poole SD, Brown N, Ehinger NJ, Slaughter JC, Paria BC, Aschner JL, Reese J. Isoprostanes as physiological mediators of transition to newborn life: novel mechanisms regulating patency of the term and preterm ductus arteriosus. Pediatr Res. 2012 Aug;72(2):122–8. PubMed PMID: 22565502.
18. Chessex P, Laborie S, Lavoie JC, Rouleau T. Photoprotection of solutions of parenteral nutrition decreases the infused load as well as the urinary excretion of peroxides in premature infants. Semin Perinatol. 2001;25(2):55–9.
19. Chessex P, Lavoie JC, Laborie S, Rouleau T. Parenteral multivitamin supplementation induces both oxidant and antioxidant responses in the liver of newborn guinea pigs. J Pediatr Gastroenterol Nutr. 2001;32(3):316–21.
20. Chessex P, Lavoie JC, Laborie S, Vallée J. Survival of guinea pig pups in hyperoxia is improved by enhanced nutritional substrate availability for glutathione production. Pediatr Res. 1999;46(3):305–10. doi:10.1203/00006450-199909000-00009.
21. Chessex P, Harrison A, Khashu M, Lavoie J-C. In preterm neonates, is the risk of developing bronchopulmonary dysplasia influenced by the failure to protect total parenteral nutrition from exposure to ambient light? J Pediatr. 2007;151(2):213–4. doi:10.1016/j.jpeds.2007.04.029.
22. Chessex P, Lavoie J-C, Rouleau T, Brochu P, St-Louis P, Lévy E, Alvarez F. Photooxidation of parenteral multivitamins induces hepatic steatosis in a neonatal guinea pig model of intravenous nutrition. Pediatr Res. 2002;52(6):958–63. doi:10.1203/00006450-200212000-00023.
23. Chessex P, Watson C, Kaczala GW, Rouleau T, Lavoie M-E, Friel J, Lavoie J-C. Determinants of oxidant stress in extremely low birth weight premature infants. Free Radic Biol Med. 2010;49(9):1380–6. doi:10.1016/j.freeradbiomed.2010.07.018.
24. Comporti M, Signorini C, Leoncini S, Buonocore G, Rossi V, Ciccoli L. Plasma F2-isoprostanes are elevated in newborns and inversely correlated to gestational age. Free Radic Biol Med. 2004;37(5):724–32. doi:10.1016/j.freeradbiomed.2004.06.007.
25. Diehl-Jones WL, Askin DF. Nutritional modulation of neonatal outcomes. AACN Clin Issues. 2004;15(1):83–96.
26. Drott P, Ewald U, Meurling S. Plasma levels of fat-soluble vitamins A and E in neonates, after administration of two different vitamin solutions. Clin Nutr. 1993;12(2):96–102.
27. Dupertuis YM, Morch A, Fathi M, Sierro C, Genton L, Kyle UG, Pichard C. Physical characteristics of total parenteral nutrition bags significantly affect the stability of vitamins C and B1: a controlled prospective study. JPEN J Parenter Enteral Nutr. 2002;26(5):310–6.
28. Dupertuis YM, Ramseyer S, Fathi M, Pichard C. Assessment of ascorbic acid stability in different multilayered parenteral nutrition bags: critical influence of the bag wall material. JPEN J Parenter Enteral Nutr. 2005;29(2):125–30.

29. Elremaly W, Rouleau T, Lavoie J-C. Inhibition of hepatic methionine adenosyltransferase by peroxides contaminating parenteral nutrition leads to a lower level of glutathione in newborn Guinea pigs. Free Radic Biol Med. 2012;53(12):2250–5. doi:10.1016/j.freeradbiomed.2012.10.541.
30. Fomon SJ, Nelson SE, Ziegler EE. Retention of iron by infants. Annu Rev Nutr. 2000;20(1):273–90. doi:10.1146/annurev.nutr.20.1.273.
31. Frank L. Effects of oxygen on the newborn. Fed Proc. 1985;44(7):2328–34.
32. Frank L, Sosenko IR. Development of lung antioxidant enzyme system in late gestation: possible implications for the prematurely born infant. J Pediatr. 1987;110(1):9–14.
33. Frank L, Ilene Sosenko RS. Prenatal development of lung antioxidant enzymes in four species. J Pediatr. 1987;110(1):106–10. doi:10.1016/S0022-3476(87)80300-1.
34. Friel JK, Aziz K, Andrews WL, Serfass RE. Iron absorption and oxidant stress during erythropoietin therapy in very low birth weight premature infants: a cohort study. BMC Pediatr. 2005;5:29. doi:10.1186/1471-2431-5-29.
35. Friel JK, Diehl-Jones B, Cockell KA, Chiu A, et al. Evidence of oxidative stress in relation to feeding type during early life in premature infants. Pediatr Res. 2011;69(2):160–4. doi:10.1203/PDR.0b013e3182042a07.
36. Friel JK, Friesen RW, Harding SV, Roberts LJ. Evidence of oxidative stress in full-term healthy infants. Pediatr Res. 2004;56(6):878–82. doi:10.1203/01.PDR.0000146032.98120.43.
37. Friel JK, Hanning RM, Isaak CA, Prowse D, Miller AC. Canadian infants' nutrient intakes from complementary foods during the first year of life. BMC Pediatr. 2010;10(1):43. doi:10.1186/1471-2431-10-43.
38. Friel JK, Martin SM, Langdon M, Herzberg GR, Buettner GR. Milk from mothers of both premature and full-term infants provides better antioxidant protection than does infant formula. Pediatr Res. 2002;51(5):612–8.
39. Friel JK, Widness JA, Jiang T, Belkhode SL, Rebouche CJ, Ziegler EE. Antioxidant status and oxidant stress may be associated with vitamin E intakes in very low birth weight infants during the first month of life. Nutr Res. 2002;22(1):55–64.
40. Furman L, Taylor G, Minich N, Hack M. The effect of maternal milk on neonatal morbidity of very low-birth-weight infants. Arch Pediatr Adolesc Med. 2003;157(1):66–71.
41. Göbel Y, Koletzko B, Böhles H-J, et al. Parenteral fat emulsions based on olive and soybean oils: a randomized clinical trial in preterm infants. J Pediatr Gastroenterol Nutr. 2003;37(2):161–7.
42. Goldman AS, Goldblum RM, Hanson LA. Anti-inflammatory systems in human milk. Adv Exp Med Biol. 1990;262:69–76.
43. Goldman AS, Thorpe LW, Goldblum RM, Hanson LA. Anti-inflammatory properties of human milk. Acta Paediatr Scand. 1986;75(5):689–95.
44. Gonzalez MM, Madrid R, Arahuetes RM. Physiological changes in antioxidant defences in fetal and neonatal rat liver. Reprod Fertil Dev. 1995;7(5):1375–80.
45. Goulet O, Joly F, Corriol O, Colomb-Jung V. Some new insights in intestinal failure-associated liver disease. Curr Opin Organ Transplant. 2009;14(3):256–61. doi:10.1097/MOT.0b013e32832ac06f.
46. Government of Canada HC. Nutrition for healthy term infants: Statement of the Joint Working Group: Canadian Paediatric Society, Dietitians of Canada and Health Canada [Health Canada, 2005]. publication. 2006. Retrieved from http://www.hc-sc.gc.ca/fn-an/pubs/infant-nourrisson/nut_infant_nourrisson_term-eng.php. Accessed 17 July 2012.
47. Granot E, Golan D, Rivkin L, Kohen R. Oxidative stress in healthy breast fed versus formula fed infants. Nutr Res. 1999;19(6):869–79. doi:10.1016/S0271-5317(99)00047-0.
48. Ham EA, Egan RW, Soderman DD, Gale PH, Kuehl Jr FA. Peroxidase-dependent deactivation of prostacyclin synthetase. J Biol Chem. 1979;254(7):2191–4.
49. Hay Jr WW, Bell EF. Oxygen therapy, oxygen toxicity, and the STOP-ROP trial. Pediatrics. 2000;105(2):424–5.
50. Husain AN, Siddiqui NH, Stocker JT. Pathology of arrested acinar development in postsurfactant bronchopulmonary dysplasia. Hum Pathol. 1998;29(7):710–7.

51. Iannitti T, Palmer B. antioxidant therapy effectiveness: an up to date. Eur Rev Med Pharmacol Sci. 2009;13:245–78.
52. Jacobs NJ, van Zoeren-Grobben D, Drejer GF, Bindels JG, Berger HM. Influence of long chain unsaturated fatty acids in formula feeds on lipid peroxidation and antioxidants in preterm infants. Pediatr Res. 1996;40(5):680–6.
53. Jain SK. The neonatal erythrocyte and its oxidative susceptibility. Semin Hematol. 1989;26(4):286–300.
54. Jones P, Suggett A. The catalase-hydrogen peroxide system. Kinetics of catalatic action at high substrate concentration. Biochem J. 1968;110:617–20.
55. Kakita H, Hussein MH, Yamada Y, Henmi H, Kato S, et al. High postnatal oxidative stress in neonatal cystic periventricular leukomalacia. Brain Devel. 2009;31(9):641–8. doi:10.1016/j.braindev.2008.10.008.
56. Khashu M, Harrison A, Lalari V, Lavoie J-C, Chessex P. Impact of shielding parenteral nutrition from light on routine monitoring of blood glucose and triglyceride levels in preterm neonates. Arch Dis Child Fetal Neonatal Ed. 2009;94(2):F111–5. doi:10.1136/adc.2007.135327.
57. Knafo L, Chessex P, Rouleau T, Lavoie J-C. Association between hydrogen peroxide-dependent byproducts of ascorbic acid and increased hepatic acetyl-CoA carboxylase activity. Clin Chem. 2005;51(8):1462–71. doi:10.1373/clinchem.2005.050427.
58. Köksal N, Kavurt AV, Cetinkaya M, Ozarda Y, Ozkan H. Comparison of lipid emulsions on antioxidant capacity in preterm infants receiving parenteral nutrition. Pediatr Int. 2011;53(4):562–6. doi:10.1111/j.1442-200X.2011.03335.x.
59. Koletzko B, Goulet O. Fish oil containing intravenous lipid emulsions in parenteral nutrition-associated cholestatic liver disease. Curr Opin Clin Nutr Metab Care. 2010;13(3):321–6. doi:10.1097/MCO.0b013e3283385407.
60. Korchazhkina O, Jones E, Czauderna M, Spencer SA. Effects of exclusive formula or breast milk feeding on oxidative stress in healthy preterm infants. Arch Dis Child. 2006;91(4):327–9. doi:10.1136/adc.2005.084798.
61. Kowalczyk P, Cieśla JM, Komisarski M, Kuśmierek JT, Tudek B. Long-chain adducts of trans-4-hydroxy-2-nonenal to DNA bases cause recombination, base substitutions and frameshift mutations in M13 phage. Mutat Res. 2004;550(1–2):33–48. doi:10.1016/j.mrfmmm.2004.01.007.
62. L'Abbe MR, Friel JK. Superoxide dismutase and glutathione peroxidase content of human milk from mothers of premature and full-term infants during the first 3 months of lactation. J Pediatr Gastroenterol Nutr. 2000;31(3):270–4.
63. L'Abbé MR, Friel JK. Copper status of very low birth weight infants during the first 12 months of infancy. Pediatr Res. 1992;32(2):183–8.
64. Laborie S, Lavoie JC, Rouleau T, Chessex P. Multivitamin solutions for enteral supplementation, a source of peroxides. Nutrition 2002;18:470–3.
65. Laborie S, Lavoie JC, Chessex P. Paradoxical role of ascorbic acid and riboflavin in solutions of total parenteral nutrition: implication in photoinduced peroxide generation. Pediatr Res. 1998;43(5):601–6. doi:10.1203/00006450-199805000-00007.
66. Lavoie JC, Bélanger S, Spalinger M, Chessex P. Admixture of a multivitamin preparation to parenteral nutrition: the major contributor to in vitro generation of peroxides. Pediatrics. 1997;99(3):E6.
67. Lavoie JC, Chessex P. Gender and maturation affect glutathione status in human neonatal tissues. Free Radic Biol Med. 1997;23(4):648–57.
68. Lavoie JC, Chessex P. Development of glutathione synthesis and gamma-glutamyltranspeptidase activities in tissues from newborn infants. Free Radic Biol Med. 1998;24(6):994–1001.
69. Lavoie JC, Laborie S, Rouleau T, Spalinger M, Chessex P. Peroxide-like oxidant response in lungs of newborn guinea pigs following the parenteral infusion of a multivitamin preparation. Biochem Pharmacol. 2000;60(9):1297–303.
70. Lavoie J-C, Chessex P, Rouleau T, Migneault D, Comte B. Light-induced byproducts of vitamin C in multivitamin solutions. Clin Chem. 2004;50(1):135–40. doi:10.1373/clinchem.2003.025338.
71. Lavoie J-C, Chessex P, Rouleau T, Tsopmo A, Friel J. Shielding parenteral multivitamins from light increases vitamin A and E concentration in lung of newborn guinea pigs. Clin Nutr. 2007;26(3):341–7. doi:10.1016/j.clnu.2006.12.006.

72. Lavoie J-C, Rouleau T, Chessex P. Interaction between ascorbate and light-exposed riboflavin induces lung remodeling. J Pharmacol Exp Ther. 2004;311(2):634–9. doi:10.1124/jpet.104.070755.
73. Lavoie J-C, Rouleau T, Chessex P. Effect of coadministration of parenteral multivitamins with lipid emulsion on lung remodeling in an animal model of total parenteral nutrition. Pediatr Pulmonol. 2005;40(1):53–6. doi:10.1002/ppul.20216.
74. Lavoie J-C, Rouleau T, Truttmann AC, Chessex P. Postnatal gender-dependent maturation of cellular cysteine uptake. Free Radic Res. 2002;36(8):811–7.
75. Lavoie J-C, Rouleau T, Tsopmo A, Friel J, Chessex P. Influence of lung oxidant and antioxidant status on alveolarization: role of light-exposed total parenteral nutrition. Free Radic Biol Med. 2008;45(5):572–7. doi:10.1016/j.freeradbiomed.2008.04.018.
76. Lavoie PM, Lavoie J-C, Watson C, Rouleau T. Inflammatory response in preterm infants is induced early in life by oxygen and modulated by total parenteral nutrition. Pediatr Res. 2010;68(3):248–51. doi:10.1203/PDR.0b013e3181eb2f18.
77. Ledo A, Arduini A, Asensi MA, Sastre J, Escrig R. Human milk enhances antioxidant defenses against hydroxyl radical aggression in preterm infants. Am J Clin Nutr. 2009;89(1):210–5. doi:10.3945/ajcn.2008.26845.
78. Lönnerdal B, Atkinson S. A. – Human milk proteins. In: Handbook of milk composition. San Diego: Academic; 1995. p. 351–68. Retrieved from http://www.sciencedirect.com/science/article/pii/B9780123844309500160.
79. MacDorman MF, Minino AM, Strobino DM, Guyer B. Annual summary of vital statistics—2001. Pediatrics. 2002;110(6):1037–52.
80. Maghdessian R, Côté F, Rouleau T, Ben Djoudi Ouadda A, Levy E, Lavoie J-C. Ascorbylperoxide contaminating parenteral nutrition perturbs the lipid metabolism in newborn guinea pig. J Pharmacol Exp Ther. 2010;334(1):278–84. doi:10.1124/jpet.110.166223.
81. McCord JM. The evolution of free radicals and oxidative stress. Am J Med. 2000; 108(8):652–9.
82. Meyerink RO, Marquis GS. Breastfeeding initiation and duration among low-income women in Alabama: the importance of personal and familial experiences in making infant-feeding choices. J Hum Lact. 2002;18(1):38–45.
83. Miloudi K, Comte B, Rouleau T, Montoudis A, Levy E, Lavoie J-C. The mode of administration of total parenteral nutrition and nature of lipid content influence the generation of peroxides and aldehydes. Clin Nutr. 2012;31(4):526–34. doi:10.1016/j.clnu.2011.12.012.
84. Miloudi K, Tsopmo A, Friel JK, Rouleau T, Comte B, Lavoie J-C. Hexapeptides from human milk prevent the induction of oxidative stress from parenteral nutrition in the newborn guinea pig. Pediatr Res. 2012;71(6):675–81. doi:10.1038/pr.2012.29.
85. Morrow JD, Roberts LJ. Mass spectrometric quantification of F2-isoprostanes in biological fluids and tissues as measure of oxidant stress. Methods Enzymol. 1999;300: 3–12.
86. Muller DP. Free radical problems of the newborn. Proc Nutr Soc. 1987;46(1):69–75.
87. Nanhuck RM, Doublet A, Yaqoob P. Effects of lipid emulsions on lipid body formation and eicosanoid production by human peripheral blood mononuclear and polymorphonuclear cells. Clin Nutr. 2009;28(5):556–64.
88. Neuzil J, Darlow BA, Inder TE, Sluis KB, Winterbourn CC, Stocker R. Oxidation of parenteral lipid emulsion by ambient and phototherapy lights: potential toxicity of routine parenteral feeding. J Pediatr. 1995;126(5 Pt 1):785–90.
89. Nutrient needs and feeding of premature infants. Nutrition Committee, Canadian Paediatric Society. CMAJ. 1995;152(11):1765–85.
90. Orozco MN, Solomons NW, Schümann K, Friel JK, de Montenegro ALM. Antioxidant-rich oral supplements attenuate the effects of oral iron on in situ oxidation susceptibility of human feces. J Nutr. 2010;140(6):1105–10. doi:10.3945/jn.109.111104.
91. Oshino N, Chance B. Properties of glutathione release observed during reduction of organic hydroperoxide, demethylation of aminopyrine and oxidation of some substances in perfused rat liver, and their implications for the physiological function of catalase. Biochem J. 1977; 162:509–25.

92. Perrone S, Salvi G, Bellieni CV, Buonocore G. Oxidative stress and nutrition in the preterm newborn. J Pediatr Gastroenterol Nutr. 2007;45 Suppl 3:S178–82. doi:10.1097/01.mpg.0000302968.83244.d2.
93. Rhee SG, Woo HA, Kil IS, Bae SH. Peroxiredoxin functions as a peroxidase and a regulator and sensor of local peroxides. J Biol Chem. 2012;287:4403–10.
94. Ripalda MJ, Rudolph N, Wong SL. Developmental patterns of antioxidant defense mechanisms in human erythrocytes. Pediatr Res. 1989;26(4):366–9.
95. Robles R, Palomino N, Robles A. Oxidative stress in the neonate. Early Hum Dev. 2001;65 Suppl 2(0):S75–81. doi:10.1016/S0378-3782(01)00209-2.
96. Rogers MS, Mongelli JM, Tsang KH, Wang CC, Law KP. Lipid peroxidation in cord blood at birth: the effect of labour. Br J Obstet Gynaecol. 1998;105(7):739–44.
97. Roggero P, Mosca F, Giannì ML, Orsi A, Amato O, et al. F2-isoprostanes and total radical-trapping antioxidant potential in preterm infants receiving parenteral lipid emulsions. Nutrition. 2010;26(5):551–5. doi:10.1016/j.nut.2009.06.018.
98. Rook D, Te Braake FWJ, Schierbeek H, Longini M, Buonocore G, Van Goudoever JB. Glutathione synthesis rates in early postnatal life. Pediatr Res. 2010;67(4):407–11. doi:10.1203/PDR.0b013e3181d22cf6.
99. Sangild PT. Gut responses to enteral nutrition in preterm infants and animals. Exp Biol Med (Maywood). 2006;231(11):1695–711.
100. Sarwar G, Botting HG, Peace RW. Amino acid rating method for evaluating protein adequacy of infant formulas. J Assoc Off Anal Chem. 1989;72(4):622–6.
101. Saugstad OD. Oxygen toxicity in the neonatal period. Acta Paediatr Scand. 1990;79(10):881–92.
102. Saugstad OD, Sejersted Y, Solberg R, Wollen EJ, Bjørås M. Oxygenation of the newborn: a molecular approach. Neonatology. 2012;101(4):315–25. doi:10.1159/000337345.
103. Sharma MK, Buettner GR, Spencer KT, Kerber RE. Ascorbyl free radical as a real-time marker of free radical generation in briefly ischemic and reperfused hearts. An electron paramagnetic resonance study. Circ Res. 1994;74(4):650–8.
104. Silvers KM, Sluis KB, Darlow BA, McGill F, Stocker R, Winterbourn CC. Limiting light-induced lipid peroxidation and vitamin loss in infant parenteral nutrition by adding multivitamin preparations to Intralipid. Acta Paediatr. 2001;90(3):242–9.
105. Sisk PM, Lovelady CA, Dillard RG, Gruber KJ, O'Shea TM. Early human milk feeding is associated with a lower risk of necrotizing enterocolitis in very low birth weight infants. J Perinatol. 2007;27(7):428–33. doi:10.1038/sj.jp.7211758.
106. Skouroliakou M, Konstantinou D, Koutri K, Kakavelaki C, et al. A double-blind, randomized clinical trial of the effect of omega-3 fatty acids on the oxidative stress of preterm neonates fed through parenteral nutrition. Eur J Clin Nutr. 2010;64(9):940–7. doi:10.1038/ejcn.2010.98.
107. Steger PJ, Mühlebach SF. Lipid peroxidation of i.v. lipid emulsions in TPN bags: the influence of tocopherols. Nutrition. 1998;14(2):179–85.
108. Steinberg LA, O'Connell NC, Hatch TF, Picciano MF, Birch LL. Tryptophan intake influences infants' sleep latency. J Nutr. 1992 Sep;122(9):1781–91. PubMed PMID: 1512627.
109. Stier C, Schweer H, Jelinek J, Watzer B, Seyberth HW, Leonhardt A. Effect of preterm formula with and without long-chain polyunsaturated fatty acids on the urinary excretion of F2-isoprostanes and 8-epi-prostaglandin F2alpha. J Pediatr Gastroenterol Nutr. 2001;32(2):137–41.
110. Tsopmo A, Diehl-Jones BW, Aluko RE, Kitts DD, Elisia I, Friel JK. Tryptophan released from mother's milk has antioxidant properties. Pediatr Res. 2009;66(6):614–8. doi:10.1203/PDR.0b013e3181be9e7e.
111. Tsopmo A, Romanowski A, Banda L, Lavoie JC, Jenssen H, Friel JK. Novel anti-oxidative peptides from enzymatic digestion of human milk. Food Chem. 2011;126(3):1138–43. doi:10.1016/j.foodchem.2010.11.146.
112. Turoli D, Testolin G, Zanini R, Bellù R. Determination of oxidative status in breast and formula milk. Acta Paediatr. 2004;93(12):1569–74.

113. Uchida K, Szweda LI, Chae HZ, Stadtman ER. Immunochemical detection of 4-hydroxynonenal protein adducts in oxidized hepatocytes. Proc Natl Acad Sci U S A. 1993;90(18):8742–6.
114. Van Zoeren-Grobben D, Lindeman JH, Houdkamp E, Brand R, Schrijver J, Berger HM. Postnatal changes in plasma chain-breaking antioxidants in healthy preterm infants fed formula and/or human milk. Am J Clin Nutr. 1994;60(6):900–6.
115. Viña J, Vento M, García-Sala F, Puertes IR, Gascó E, et al. L-cysteine and glutathione metabolism are impaired in premature infants due to cystathionase deficiency. Am J Clin Nutr. 1995;61(5):1067–9.
116. Volpe JJ. Brain injury in the premature infant: is it preventable? Pediatr Res. 1990;27 (6 Suppl):S28–33.
117. Wispe JR, Bell EF, Roberts RJ. Assessment of lipid peroxidation in newborn infants and rabbits by measur without long-chain polyunsaturated fatty acids on the urinary excretion ements of expired ethane and pentane: influence of parenteral lipid infusion. Pediatr Res. 1985;19(4):374–9.
118. Zaniolo K, St-Laurent J-F, Gagnon SN, Lavoie J-C, Desnoyers S. Photoactivated multivitamin preparation induces poly(ADP-ribosyl)ation, a DNA damage response in mammalian cells. Free Radic Biol Med. 2010;48(8):1002–12. doi:10.1016/j.freeradbiomed.2010.01.001.
119. Zlotkin SH, Anderson GH. The development of cystathionase activity during the first year of life. Pediatr Res. 1982;16(1):65–8.

Chapter 20
Oxidative Stress in Epilepsy

Salvatore Grosso and Ursula Geronzi

List of Abbreviation

8-dG	8-2′-deoxyguanosine
8-oxo-dG (8-OH-dG)	8-hydroxy-2′-deoxyguanosine
AEDs	antiepileptic drugs
ATP	adenosine triphosphate
BBB	blood-brain barrier
CBZ	carbamazepine
CoQ10	ubiquinone or coenzymeQ10
CSF	cerebral spinal fluid
Cu/ZnSOD	copper-zinc SOD
DNA	deoxyribonucleic acid
F2-Iso-Ps	F2-isoprostanes
FBM	felbamate
GABA	γ-aminobutyric acid
GPx	glutathione peroxidase
GPx-GR	glutathione peroxidase-glutathione reductase
GSH	glutathione tripeptide
H_2O_2	hydrogen peroxide
LEV	levetiracetam
LMT	lamotrigine
MLT	melatonin

S. Grosso, M.D., Ph.D. (✉) • U. Geronzi, M.D.
Pediatric Neurology-Immunology and Endocrinology Unit, Department of Pediatrics, Gynecology, Obstetrics, and Reproductive Medicine, University of Siena, Via M. Bracci, Le Scotte, 53100 Siena, Italy
e-mail: grosso@unisi.it

P.A. Dennery et al. (eds.), *Perinatal and Prenatal Disorders*, Oxidative Stress in Applied Basic Research and Clinical Practice, DOI 10.1007/978-1-4939-1405-0_20

MnSOD	manganese SOD
mtDNA	mitochondrial DNA
NAC	N-acetylcysteine
NADH	nicotinamide adenine dinucleotide (reduced status)
NMDA receptor	N-methyl-D-aspartate receptor
NO	nitric oxide
PHT	phenytoin
POLG1	mitochondrial DNA polymerase γ
ROS	reactive oxygen species
SOD	superoxide dismutase
TPM	topiramate
tRNAPhe	phenylalanine transfer RNA
TRPM2	transient receptor potential M2 channel
TSPO	translocation protein
VPA	valproic acid

Introduction

An epileptic seizure is a paroxysmal disorder characterized by an abnormal, excessive hypersynchronous discharge of neurons which results in an alteration of function of the patient. Epilepsy is a condition characterized by repeated, unprovoked seizures. It was probably first described in ancient Egyptian writings around 2000 BC and was a popular topic of the Greek and Roman scholars. Epilepsy was closely identified with supernatural forces and was considered a manifestation of the gods and spirits. The understanding of the underlying pathophysiology of epilepsy has greatly expanded since the introduction of electroencephalography.

Epidemiological investigations suggest that epilepsy is one of the most common neurological disorders in the world, affecting at least 50 million people worldwide. That figure doubles when people with one epileptic seizure during their lifetime are considered. The median prevalence of lifetime epilepsy for developed countries is 5.8 per 1,000 and 10.3 per 1,000 for developing countries [1]. Epilepsy may provoke serious physical, psychological, socioeconomic consequences and can be classified as idiopathic, provoked, or symptomatic. Idiopathic epilepsy has presumably genetic causes, and there are no significant neuroanatomical or neuropathological anomalies [1, 2]. Progress in the field of genetics allowed the identification of several genes and genetic conditions including epilepsy in their phenotypes [3]. Provoked seizures are predominantly caused by specific environmental or systemic factors, and there are no significant neuroanatomical or neuropathological anomalies. Symptomatic epilepsies may recognize several etiologies including trauma, infections, and brain malformations. In this context neuroimaging evaluation plays a central role in the identification of structural or anatomic anomalies associated with epilepsy, such as tumors, hydrocephalus, congenital lesions, vascular accidents, and hippocampal sclerosis. Neuropharmacological studies provided insight into the role of neurotransmitters (mainly GABA and glutamate), alterations in membrane functions, receptors, ionic changes, and alteration of neural networks in the epileptogenic process [4].

From a pathophysiological point of view, the initiation and propagation of a seizure depends on an imbalance between depolarizing and hyperpolarizing processes in a large interconnected network of neurons. Epileptogenicity depends on changes in intrinsic neuronal properties with enhanced connectivity, enhanced excitatory transmission, and failure of inhibitory functions [5]. The prolonged excitation of neurons occurring during seizures determines, at the cellular level, a cascade of events such as activations of glutamate receptors and cytokines, changes in the composition of glutamate and gamma-aminobutyric acid receptors, and modifications in neuronal plasticity [6–8]. A number of studies also provide strong evidence that the prolonged seizure-related neuronal excitation results in increased reactive oxygen species (ROS) production [9] which contributes to seizure-induced brain damage. ROS formation occurs when unpaired electrons escape the electron transport chain and react with molecular oxygen, thus generating superoxide. Superoxide can react with DNA, proteins, and lipids resulting in functional cellular disruption, cellular damage, and subsequent cell death [8]. In particular, lipid peroxidation causes membrane structure alterations that affect membrane fluidity and permeability and membrane protein activity [10].

The brain is particularly susceptible to oxidative stress being the most aerobically active organ in the body due to its high metabolic demands. The brain consumes 20 % of total oxygen even though it accounts for the only 2 % of the total body mass. Brain also contains high concentrations of polyunsaturated fatty acids that are prone to lipid peroxidation and is rich in iron which can catalyze hydroxyl radical formation. The maintenance of low ROS levels is therefore critical to normal neuronal cell function. However, the brain has only 10 % of the hepatic catalase activity [11–13], and the abnormally prolonged increases in ROS occurring during seizures carry an inherent risk of increasing neurodegeneration [8, 13].

Improved understanding has resulted in the development of a large number of effective and well-tolerated drugs for various seizure types. Effective AEDs are needed to prevent seizure recurrence-related brain damage. In this perspective, the use of AEDs as possible neuroprotective compounds has been receiving increased attention. In fact, most AEDs currently in use have been tested for a possible neuroprotective effect both in human and animal models of epilepsy [8]. However, despite the increasing number and variety of antiepileptic drugs, more than 30 % of patients with epilepsy still remain drug resistant [14].

The present review outlines studies which address the relationship between oxidative stress and epilepsy. Evidence of damage to lipids, DNA, and proteins as the consequence of seizure-related oxidative stress, occurring in various animal models of epilepsy, is reviewed. In addition, the influence of AEDs in the regulation of oxidative stress is also discussed.

Oxidative Stress and Epilepsy in Experimental Models

Oxidative stress markers have been found to be increased in different experimental models of epilepsy [15]. Although experimental seizure models have been designed to define the role of various endogenous antioxidants and markers of oxidative stress, the site(s) of the generation of seizures has not been uniform in various models [15].

Increased lipid peroxidation has been found in diverse brain areas of rats after picrotoxin- and pentylenetetrazole-induced seizures as well as in the kainic acid-induced seizure model [16, 17]. In the electrically induced convulsion model, lipid peroxidation content was found to be increased in the whole brain [18] but reduced in the hippocampus after both acute and chronic electroshock [19]. In particular, in kainic acid and pilocarpine models, lipid peroxidation was reported to increase during the first 12 h of treatment followed by a "return" to the pretreatment period during chronic time measurements, which may suggest hypometabolism, neuronal loss, and/or compensatory mechanisms related to ROS catabolism [20–22].

Regarding DNA oxidation, experimental studies detected significant seizure-related increase in the levels of 8-hydroxy-2′-deoxyguanosine (8-oxo-dG) in both acute and chronic models of epilepsy [23, 24]. High levels of H_2O_2 and O_2 may trigger the Haber-Weiss reaction with production of highly reactive species hydroxyl radical. The latter molecule has a very great affinity toward guanine in DNA and in the nucleotide pool, increasing the formation of 8-oxo-dG [25]. These findings strongly suggest that mitochondrial oxidative stress and genomic instability may play a key role in the epileptogenesis [26]. In fact, in the kainate animal model of temporal lobe epilepsy, acute seizure activity caused a time-dependent increase in mitochondrial 8-OHdG levels and a greater frequency of mtDNA lesions associated with increased mitochondrial H_2O_2 production and transient decrease in mtDNA repair capacity. A gradual recovery in the mtDNA repair capacity was noted in the following period. By contrast, progressive accumulation of mtDNA damage, increased mitochondrial H_2O_2 levels, and impairment in the mtDNA repair capacity were detected during recurrent seizures in the context of the chronic phase of epilepsy [26].

Studies on transgenic animals further corroborate the importance of free radical in epileptogenesis. Transgenic mice overexpressing mitochondrial SOD (the enzyme that dismutates $O_2^{-\cdot}$) are resistant to seizure-induced neurodegeneration, whereas mice with partial SOD deficiency (SOD–/+) demonstrated intensification of these findings resulting in an increased seizure susceptibility. On that basis, it can be stated that O_2- may play a key role in seizure-induced brain changes [23, 24, 27]. However, experimental studies produced contradictory results about how SOD levels change both in acute and chronic experimental seizure models. An increase in SOD levels was reported in the majority of studies [20, 28–30]. However, when compared to normal controls, other studies reported no significant changes [31] or even an SOD level reduction [32, 33]. Similar contradictory results were observed in chronic experimental epilepsy models with either increased [34] or decreased [37]

SOD levels. Of interest, although the widely held opinion is that oxidative stress during seizures is age dependent and does not occur in immature brain [28, 36, 37], recent findings indicated that seizures induced in immature rats by dl-homocysteic acid are associated with the presence of several indirect markers of oxidative stress, such as the decreased activities of aconitase and respiratory chain complex I [38]. In the same epilepsy model, it was recently observed that the activity of total SOD, SOD1, and particularly SOD2 was significantly increased at 20 h and 6 days of survival. Importantly, upregulation of SOD2 was also confirmed in mitochondria at the protein level by immunoblotting. The latter finding suggests enhanced ROS levels in the mitochondrial matrix [39].

Although catalase levels are physiologically lower in brain tissue than in other organs, it is able to metabolize great amounts of H_2O_2 very rapidly. A significant increase in catalase levels have been found in acute [15, 30, 31, 40–42] and chronic [34] models of epilepsy. According to Barros et al. [42], the increased catalase activity may be the consequence of an enzymatic antioxidant response to increased basal-free radical production, which may potentially lead to a damage of neural tissue. It can by hypothesized that free radical scavenging may take part in controlling seizure-induced damage [42]. However, it has been recently observed that in the immature rat's brain with seizures induced by dl-homocysteic, catalase activity was decreased when compared to the control groups, suggesting limited antioxidant defense [39].

The system GPx-GR is highly efficient in metabolizing small amounts of H_2O_2. Experimental studies about GPx showed conflicting results. In fact GPx levels have been found to be increased [20, 29, 30, 35, 43, 44], decreased [43, 45, 46], or unchanged [28–30] when compared to normal controls. By contrast, several studies agreed in showing decreased GR levels in laboratory animals [22, 44].

Oxidative Stress and Epilepsy in Clinical Studies

When globally considered clinical studies suggest that recurrent seizures may produce brain degeneration and contribute to progressive functional and cognitive declines despite adequate AED treatment [47, 48]. The emerging perspective is that prolonged epileptic seizures or status epilepticus in both humans [49, 50] and animal models [51, 52] may determine cerebral damage represented by characteristic patterns of neuronal cell loss preferentially in the hippocampus. This, in turn, increases the risk of subsequent seizures. This perspective provides a rationale for development of neuroprotective treatments to prevent adverse long-term effects of recurrent seizures [53]. Oxidative stress and mitochondrial dysfunction may play a key role in the pathophysiological mechanisms of seizure-induced neuronal cell death and subsequent epileptogenesis [54]. In particular, these processes may be interconnected being linked, either directly or indirectly, to respiratory chain impairments, depolarization of the inner mitochondrial membrane, ATP depletion, and the opening of the mitochondrial permeability transition pore.

However, as in the experimental studies, controversies still exist in clinical investigations on the relationship between oxidative stress and epilepsy. In fact, redox status in children with newly diagnosed epilepsy has been reported to be either altered [55] or normal [56, 57]. A possible limiting factor, in the majority of these investigations, is the nonhomogeneous series of patients included in the study in terms of epilepsy duration, seizure frequency, seizure types, epilepsy syndromes, and therapeutic strategy. It has been reported that both febrile and afebrile seizures were able to induce oxidative stress with respect to healthy subjects [58]. Oxidant and antioxidant states of blood and cerebrospinal fluid were evaluated by determining erythrocyte arginase and erythrocyte catalase, plasma and cerebrospinal fluid malondialdehyde, and plasma and cerebrospinal fluid nitric oxide levels. The greater levels of oxidative stress markers observed in patients with afebrile seizures might affect prognosis adversely [58]. Patients with symptomatic West syndrome showed higher cerebrospinal fluid nitrate and nitrite levels than those with cryptogenic West syndrome. CSF nitrate and nitrite levels rise during the first 40 days of symptoms. Although they cannot be used as a marker to estimate symptoms duration or to predict the prognosis of mental development, they may help in differentiating symptomatic from cryptogenic etiologies of West syndrome [59]. Patients affected by epileptic encephalopathy, but not those affected by idiopathic epilepsy syndromes, showed higher plasma levels of F2-isoprostanes (F2-IsoPs) and advanced oxidation protein products. By contrast, there were no differences in the levels of nonprotein binding iron, suggesting that free-iron is not involved in generating oxidative stress in these patients [60].

There is strong evidence that mitochondrial dysfunctions may affect neuronal excitability and synaptic transmission. Decreased intracellular ATP levels and neuronal calcium homeostasis changes represent valid contributing factors to epilepsy-susceptibility associated with mitochondrial impairment [24]. The experimentally proven involvement of mitochondrial dysfunctions in epileptogenesis is further corroborated by several clinical observations including epileptic encephalopathic disorders related to mitochondrial DNA (mtDNA) mutations. Myoclonic epilepsy with ragged red fibers, which is a syndrome characterized by myoclonus, myopathy, cerebellar ataxia, and dementia, may be considered a prototype of epilepsy phenotype related to mtDNA mutations [49]. In particular, an A to G transition mutation of nucleotide pair 8344 in mtDNA affects the biosynthesis of mitochondrial oxidative phosphorylation proteins [50] with inefficient ATP production, increased ROS levels, and unbalanced genetic expression of antioxidant enzymes [50]. Epileptic phenotypes have also been associated with other mtDNA mutations affecting the respiratory chain function or mitochondrial ATP synthesis [61] as well as with genes mutations compromising the activity of mitochondrial DNA polymerase γ (POLG1) [51] and mitochondrial tRNAPhe [52]. In children affected by mitochondrial encephalopathies, the determination of NADH and cytochrome components of the respiratory chain in mitochondria from a vastus lateralis muscle biopsy showed that those affected by epilepsy had higher incidence of complex I defects than children without epilepsy, further indicating the relationship between mitochondrial oxidative stress dysfunctions and epileptogenic mechanisms [62].

Other epilepsy-related genes involved in oxidative stress processes include EFHC1 gene, whose mutation causes juvenile myoclonic epilepsy. EFHC1 protein product, which is co-expressed with transient receptor potential M2 channel (TRPM2) in hippocampal neurons and ventricle cells, enhances TRPM2-conferred susceptibility to H(2)O(2)-induced cell death. This suggests that TRPM2 plays a role in the phenotypic expression of juvenile myoclonic epilepsy by mediating disruptive effects of EFHC1 mutations through oxidative stress-mediated cell dysfunction up to cell death [63].

Epilepsy is also a major clinical finding of Menkes disease, an X-linked recessive neurodegenerative disorder, related to mutations in the ATP7A gene encoding for the copper-transporting ATPase. Mutations of the latter gene lead to deficiencies of key copper-containing enzymes as cytochrome c oxidase resulting in widespread changes in oxidative metabolism, as proved by the presence of significant lactic acidosis in brain and cerebrospinal fluid of Menkes patients [64]. Changes in the redox status have also been demonstrated in autopsy cases of dentatorubral-pallidoluysian atrophy, a disorder belonging to the group of progressive myoclonus epilepsies. Oxidative products to nucleosides such as 8-oxo-dG and 8-dG were accumulated in the lenticulate nucleus. Cytoplasmic immunoreactivity for Cu/ZnSOD was reduced in the external segment of globus pallidus, dentate nucleus, and cerebellar cortex while mitochondrial immunoreactivity for MnSOD was reduced in the lenticulate nucleus and cerebellum [65].

Antioxidant Strategies in Patients with Epilepsy

Several compounds have been proved to have antioxidant properties in animal models. In Table 20.1 are listed all molecules with potential or proved antioxidant effects. The combination of AEDs with *neuroprotective* compounds may provide beneficial effects [39, 53], leading to new therapeutic strategies in patients affected by epilepsy. Compounds used in clinical trials are briefly discussed.

It has been found that chronic supplementation with N-acetylcysteine (NAC) may lead to a gradual increase in serum glutathione levels [66, 67]. NAC therapy resulted effective in improving and stabilizing the chronic neurological deterioration in patients affected by Unverricht-Lundborg disease. In these patients myoclonus was markedly improved and the frequency of generalized seizure significantly reduced [68]. The effectiveness of NAC remains to be proved in other types of epilepsy [68]. However, the long-term use of NAC in the clinical practice is not free of possible side effects, as sensorineural deafness and epigastric pain may develop [67]. Massive overdose of intravenous NAC are reported to cause seizures, cerebral dysfunction, and life-threatening effects [69]. Controlled clinical trials are therefore required to define NAC doses in the clinical practice and to confirm the effectiveness and tolerability of NAC in humans.

Preliminary clinical trials provide evidences that ubiquinone (or CoQ10) may have a therapeutic role in many brain disorders such as Parkinson's and Alzheimer's

Table 20.1 Antioxidant molecules in experimental studies

Compound	Antioxidant activity	Antiepileptic activity	(Ref)
Melatonin	It is a potent free radical scavenger such as ROS and RNS; it protects nuclear DNA and membrane lipids from damage by lipid peroxidation, upregulates antioxidant enzyme activities, and downregulates pro-oxidant enzymes. It induces the activity of gamma-glutamylcysteine synthetase, stimulating the production of glutathione	Not proved	[127]
Lipoic acid	It scavenges free radicals, chelates transition metals, raises intracellular levels of glutathione, and protects biomembranes from lipid peroxidations. LA may also inhibit TNF-alpha-induced ROS generation, GSH reduction, and cell apoptosis	NA	[128]
Ascorbate	Neuroprotective actions related to decreased lipid peroxidation and increased catalase activity in seizures induced by pilocarpine. It is a potent scavenger of ROS/RNS	Not proved	[129]
a-Tocopherol	(1) It acts as chain-breaking antioxidant and protects cell membrane against oxidative damage; (2) it reduces markers of oxidative stress; and (3) it is an effective free radical scavenger in the brain	Not proved	[130]
NAC	It is an antioxidant and a free radical-scavenging agent that increases intracellular GSH acting as a precursor for glutathione synthesis and as a stimulator of the cytosolic enzymes involved in glutathione regeneration. NAC protects against oxidative stress-induced neuronal death in cultured granule neurons	Anticonvulsant properties as adjunctive therapy in both human and animal epilepsy model	[131]
Curcumin	Curcumin attenuates oxidative stress neuronal death. It maintains glutathione levels and inhibits lipid peroxidation. Curcumin manganese complex has more potent anticonvulsive and neuroprotective effects than curcumin	It attenuates KA-induced seizure	[132]
Ginsenosides	They (1) block mitochondrial dysfunction and improve mitochondrial antioxidant capacity and (2) attenuate KA-induced ultrastructural mitochondrial damage and mitochondrial oxidative stress	They reduce seizure activity induced by KA, PTZ, and pilocarpine	[13]
Honeybee propolis	It attenuated KA-induced neuronal damage through maintenance of glutathione homeostasis and adenosine A1 receptor activation	It reduces KA-induced seizure activity	[133]
Ginkgo biloba (EGb71)	It exerts both neuroprotective and antioxidant effects by reducing excessive ROS formation	It suppresses PTZ-induced seizure activity	[134]

EUK-134	This is a salen SOD mimetic able to block KA-induced neuronal death, lipid peroxidation, nitrite formation, and nucleic acid oxidation in vivo and in vitro	No influence on latency and duration of KA-induced seizures	[13]
Tempol	This is an SOD mimetic which inhibits apoptotic changes and formation of superoxide and nitrite after intrahippocampal microinjection of KA	No influence on seizure-like hippocampal EEG activity in animal models	[135]
Aspalatone	It is a GPx mimetic that inhibits KA-induced oxidative stress and hippocampal neuronal death. It scavenges hydroxyl radicals in vitro models and enhances anti-peroxidative enzyme activity, such as GPx and CAT in animal models	It inhibits KA-induced seizure activity	[136]
Buspirone	It produces and reduces lipid peroxidation level and nitrite content; it increases SOD and catalase activities in rat hippocampus. It inhibits cell apoptosis in several neuronal models	Both antiepileptic and pro-convulsant activities have been reported	[137, 138]
Huperzine A	In animal models it reduces memory impairment and neuronal degeneration in the CA1 region compared to controls. It inhibits cell apoptosis through its effects on intrinsic caspase-3 pathway. However, further study on the neuroprotective effects of huperzine A is needed	Proved on experimental models	[139]
YKP3089	NPA has been documented in the hypoxia-induced lethality mice model	Proved	[140]
Ubiquinone	It is a powerful antioxidant that buffers adverse consequences of free radicals produced in the inner mitochondrial membrane. It inhibits the mitochondrial permeability transition and oxidative stress-induced cell apoptosis. It also exerts neuroprotective effects by regulating ATP production and reducing free radical damage		
Riluzole	NPA in TBI and SCI models. Protects against excitotoxic/nonexcitotoxic oxidative neuronal damage	NA	[129]
Ketogenic diet	NPA by enhancing ATP synthesis and suppressing ROS production	Proved	[141]

Legend. *NPA* neuroprotective activity, *NA* not assessed, *KA* kainic acid, *PTZ* pentylenetetrazol, *NAC* n-acetylcysteine diseases, with tolerated doses ranging across 300–2.400 mg/day. However, clinical trials assessing the efficacy of ubiquinone as a clinical neuroprotectant in patients with epilepsy are still lacking.

Regarding melatonin (MLT), it has been found that children with epilepsy preserve MLT circadian rhythms when evaluated by diurnal patterns of salivary MLT and urinary metabolite 6-sulphatoxymelatonin. There are no associations between MLT secretion/excretion parameters and seizure findings such as time, type of seizures, and antiepileptic medications [70]. Immediate-release MLT in children with neurodevelopmental delay may be effective in improving total duration of night-time sleep and in reducing sleep-onset latency in children with neurodevelopmental delay [71]. The effect of MLT on seizures, sleep quality, and behavior was also evaluated in a pilot study in a series of patients with intractable epilepsy. Compared to placebo group, patients under MLT showed a significant decrease of diurnal seizures while the maximal number of seizures and seizure duration remained unchanged. No major adverse events were registered, suggesting that MLT may be effective and safe for decreasing daytime seizure frequency in patients with intractable epilepsy [72]. Children with epilepsy often experience sleep problems, including insomnia, and MLT has been considered in the treatment of that disorder. In this context, the effect of MLT therapy on the pattern of sleep and characteristics of seizure disorder has been evaluated in a group of children with intractable epilepsy. A 3-month trial of MLT was associated with improvement of both many sleep-related phenomena and severity of seizures [73]. MLT treatment was shown to exert definite effects on human brain activity including a slowing down of the EEG rhythm, an increase of REM, and a rise of the convulsive threshold [74].

Ardura et al. [75] showed that children with generalized idiopathic epilepsy and simple fever convulsions showed lower peak MLT values than healthy controls at 04:00 h. They also demonstrated the presence of diurnal rhythm in blood melatonin concentrations in this population [75].

Finally, a Cochrane review has been recently published on the efficacy and tolerability of melatonin as add-on treatment for epilepsy. Only four publications, including a total of 102 patients, were eligible to be analyzed. According to the authors [76], it was not possible to perform any meta-analysis. Of the four randomized controlled trials, it was only in one study that the exact number of seizures during the trial compared to the baseline was evaluated. None of the patients with seizures during the trial showed modifications in seizure frequency when compared to the baseline. No adverse events were reported, neither was significant improvement in quality of life in patients assuming the melatonin. The authors concluded that it is not possible to draw any conclusion about the role of melatonin in reducing seizure frequency or improving quality of life in patients with epilepsy [76]. By contrast, high doses of melatonin have proven to be useful as adjunctive therapy in seizure control in a girl with severe infantile myoclonic epilepsy [77].

The lipid-soluble vitamin E, alpha-tocopherol, is able to penetrate the blood-brain barrier (BBB) and accumulate at high concentration in the brain [78]. The concentration of vitamin E in erythrocyte or plasma was lower in children with epilepsy than in healthy controls. Among epileptic children, those under polytherapy had lower levels than children under monotherapy [79, 80]. Sudha et al. [81] confirmed that patients with epilepsy have significantly lower serum levels of alpha-tocopherol than controls, suggesting possible pharmacological effects of a-tocopherol supplementation in controlling seizures and in preventing neuronal damage. Clinical trials on the therapeutic use of a-tocopherol, as an adjunctive anti-epileptic agent, have yielded conflicting results. However, trials had diverse study designs and were nonhomogeneous in terms of age at the therapy onset and duration of treatment. According to Ogunmekan and Hwang [82], children with drug-resistant epilepsy placed under 400 IU/day d-alpha-tocopheryl acetate showed a significant reduction in seizure frequency compared to those under placebo. Treatment with a-tocopherol had no effect on plasma levels of anticonvulsant medications. Alpha-tocopherol (250 IU(day) was also administered to 18 patients affected by epilepsy for a period of 3 months, while a remaining group of seven received placebo. Anticonvulsants were continued as previously. Globally, seizure frequency was reduced in ten out of 18 patients under alpha-tocopherol. No changes were observed in patients under placebo [83]. Forty-three adolescents and adults affected by drug-resistant epilepsy were randomly placed under vitamin E or placebo for a 3-month period; then, after a 1-week washout period, each patient received the alternate treatment for an additional 3 months. AEDs remained unchanged during the study period. No significant change in seizure frequency was observed with vitamin E as compared with placebo [84]. All the previous data seem to suggest further trials are necessary to define the effectiveness of a-tocopherol as add-on therapy in patients with epilepsy.

When globally considered, clinical trials on the therapeutic use of antioxidant compounds in patients with epilepsy remain disappointing. That may be because animal models do not adequately epitomize time-dependent epileptogenic process occurring in humans. In animal models supplementation is usually given in the early stages of the epileptic disorder, whereas in patients the supplementation often occurs later in the course of the disease. Moreover, pharmacokinetics and pharmacodynamics of the antioxidants used in animals may differ from those in humans. Finally, doses of antioxidant compounds should also be adequately defined both in animals and in humans. In fact, it should be taken into account that low concentrations of ROS are essential for a number of neurophysiological processes and that excessive antioxidant levels may paradoxically produce pathological effects [85, 86].

Antiepileptic Drugs and Oxidative Stress in Experimental and Clinical Studies

Despite the plethora of antiepileptic drugs (AEDs) currently available, 30 % of patients continue having seizures. This group of patients requires a more aggressive treatment, since monotherapy, the first choice scheme, fails to control seizures. Nevertheless, polytherapy often results in a number of unwanted effects including neurologic disturbances, psychiatric and behavioral symptoms, and metabolic alteration. The need for better tolerated AEDs is even more urgent in this group of patients [76].

Several AEDs may interfere with the mitochondrial oxidative process. In peripheral white blood cells from children under therapy, it has been noted that several AEDs were able to impair the oxidative phosphorylation process when evaluated by enzymatic activity of respiratory chain complexes (II-IV) and mitochondrial ATP production [87].

Barbiturates

Barbiturates may have neuroprotective effects exerted by mechanisms which remain to be elucidated. It is thought that the compounds' ability to inhibit lipid peroxidation may play a central role in the antioxidant effect. Experimental studies demonstrated that phenobarbital may decrease cerebral metabolic rate to a more profound extent than the associated reductions in cerebral blood flow [88]. The effect of high-dose phenobarbital on lipid peroxidation and antioxidant enzymes has also recently been evaluated in full-term inborn neonates with severe birth asphyxia. In a group of them, receiving 40 mg/kg of phenobarbital intravenously within the first 2 h of life, cerebrospinal fluid examination showed decreased CSF levels of lipid peroxides and antioxidant enzymes compared to the control group [89].

Benzodiazepines

Diazepam was reported to prevent seizure-induced increase in nitric oxide (NO) and lipid peroxidation [43]. In models of selective neuronal death in CA1 sector of hippocampus, both in vivo and in vitro, the drug showed significant neuroprotective property by reducing over 50 % the number of cell death, even in the presence of GABA(A) receptor antagonist bicuculline. In addition, diazepam may also reduce the efflux of cytochrome C out of mitochondria. The latter finding indicates that the neuroprotective action of diazepam is exerted not only through its known hypothermic effect but also through the GABA(A) receptor complex and, possibly, through its peripheral receptor, the translocator protein TSPO located in the outer mitochondrial membrane [90]. Zolpidem also acts at the GABA(A) receptor site accounting for its sedative and anticonvulsant activities. Experimental studies confirm that as other benzodiazepines, zolpidem does not trigger production of ROS and oxidative stress, which is in line with its minor side effects [91]. Tanguy et al. [92] recently compared the effects of midazolam to propofol on cerebral biomarkers at the acute phase of severe traumatic brain injury patients. A cerebral microdialysis catheter was used to measure the lactate/pyruvate ratio, glutamate, glycerol, and glucose for 72 h. No difference between groups was observed, confirming the neuroprotective properties of both drugs [92]. Alprazolam has been found to reduce the level of intracellular ROS in lymphocytes, granulocytes, and monocytes in the peripheral blood of mice under provoked stress, suggesting that therapeutic efficacy of alprazolam may be mediated, at least partially, by the reversal of oxidative damage [93]. Acute model of immobilization (restraint) stress in rats produces enhanced striatal levels of lipid peroxidation, decreased SOD activity, reduced levels of mitochondrial functions, and increased corticosterone serum levels. Pretreatment of stressed rats with diazepam decreases the striatal lipid peroxidation levels and improves mitochondrial function, providing evidence that drugs exhibiting anxiolytic and antioxidant properties might be useful for the design of therapies against early acute phases of physical stress [94].

Carbamazepine

Carbamazepine (CBZ), an iminostilbene derivative commonly used for treatment of epilepsy, neuralgic pain, and bipolar affective disorders, showed some ability to control free radical-related seizures and the level of trance elements when compared to valproic acid (VPA) and phenytoin (PHT) [95]. In CBZ-treated red blood cells, lipid peroxidation, SOD, and free radical-scavenging activities were unmodified. In contrast, that drug demonstrated beneficial effects by increasing release of ATP and NO-derived metabolites from erythrocytes to lumen, leading to an increased NO pool in the vasculature [96].

Felbamate

Felbamate (FBM) is an antiepileptic drug that had been associated with minimal toxicity in preclinical species but showed an unacceptable incidence of serious idiosyncratic reactions in man. However, the drug showed neuroprotective properties in rat hippocampal slice preparations [97]. FBM acts on NMDA receptor reducing excitatory neurotransmission and decreasing glutamate release, accounting for the ability of the drug to reduce sustained repetitive firing in mouse spinal cord neurons [98] and the neuroprotective action [99]. Leone et al. [100] compared FBM to several other antiepileptic drugs, including VPA, CBZ, phenobarbital, and PHT, to evaluate oxidative stress/reactive metabolite potential. FBM produced robust effects on an established oxidative stress/reactive metabolite gene expression. Pronounced effects were also observed after CBZ, phenobarbital, and PHT administration. By contrast, VPA had only minor effects on the oxidative stress/reactive metabolite indicator genes [100].

Lamotrigine

Lamotrigine (LMT) is approved as monotherapy or adjunctive therapy in patients older than 12 years affected by partial onset seizures, generalized tonic-clonic seizures, or Lennox-Gastaut syndrome. It is also approved as monotherapy in children older than 2 years affected by absences. LMT has the ability to decrease sodium and calcium channel activity. The drug may have neuroprotective property as corroborated by Kumar et al. [101], who showed that LMT significantly restores oxidative defense levels and mitochondrial complex enzyme dysfunctions induced by 3-Nitropropionic acid, a neurotoxic compound. These effects are mediated by GABAergic mechanisms and significantly potentiated by the coadministration of gabapentin, another antiepileptic drug [101]. The effects of LMT and CBZ on cognitive function and oxidative stress were evaluated in an animal epilepsy model. In contrast to CBZ, LMT did not produce cognitive dysfunctions and decreased oxidative stress in the pentylenetetrazole-kindled group as compared to the pentylenetetrazole-kindled CBZ-treated group [102]. Agarwal et al. [103] confirmed that LMT, as well as oxcarbazepine, do not impair cognitive functions and do not determine significant alteration in oxidative stress parameters when compared to topiramate (TPM) [103].

Levetiracetam

Levetiracetam (LEV) is commonly used in the treatment of partial seizures in adults, children, and infants older than one month [104, 105]. Mechanisms of actions of LEV differ from that of conventional antiepileptic drugs including the ability to act as a histone deacetylase inhibitor, which explains the LEV anti-inflammatory and antioxidative properties [105].

LEV neuroprotective effects have been substantiated in the rat middle cerebral artery model of stroke [88]. Hippocampus of adult mice brain pretreated with LEV 60 min before pilocarpine-induced seizures showed no changes in the levels of lipid peroxidation, nitrite-nitrate, and catalase activity. Moreover, LEV administration prevented the pilocarpine-induced loss of GSH in this cerebral area [106]. Gibbs et al. [107] evaluated the effects of the drug on the known pattern of mitochondrial dysfunctions occurring after status epilepticus, showing that LEV determined significant reduction of glutathione, alpha-ketoglutarate dehydrogenase, aconitase, citrate synthase, and complex I activities. These features may well explain both LEV antiepileptic activity and neuroprotective effects [107]. However, urinary 15F-2-isoprostane levels, evaluated before and after 3 months of LEV therapy, resulted to be increased suggesting that LEV may actually induce oxidative stress in epileptic patients [108].

Phenytoin

Phenytoin (PHT) is an antiepileptic drug able to stabilize neuronal membranes and known to be potentially teratogenic. It may have antioxidant properties by slowing the release of K+ from ischemic neurons and reducing free fatty acid accumulation during complete global ischemia [88]. PHT may exert effects on neuroprotection-related genes, such as survival-promoting and antioxidant genes, as demonstrated by using DNA microarrays on rat brain genes after prolonged PHT administration [109]. In a murine embryo culture model used to investigate PHT-initiated embryonic DNA oxidation and dysmorphogenesis, PHT-treated embryos had significantly higher 8-OH-2′-dG formation than that of controls. Interestingly, the PHT-initiated dysmorphological anomalies may be prevented by SOD and catalase which virtually eliminate PHT-initiated 8-OH-2′-dG formation. This finding provides evidence that free radical-mediated oxidative stress plays a key role in PHT teratogenesis [110]. Bioactivation of PHT and CBZ metabolites (4-OH-PHN and 3-OH-CBZ, respectively) generates ROS by peroxidases and may play a role in the ability of these drugs to cause idiosyncratic drug reactions [111]. Clinical studies reported that patients affected by epilepsy and placed under PHT therapy had significantly higher lipid hydroperoxide concentrations compared to controls. Moreover, the total antioxidant capacity of sera of epileptic patients was lower than the antioxidant capacity of control sera [112]. Female with epilepsy and under PHT monotherapy had serum malondialdehyde concentration and copper/zinc-SOD activity significantly increased, but GSH level significantly decreased. The enhanced oxidative stress is therefore involved in the PHT-mediated toxicity. Supplement of GSH and modification of copper/zinc-SOD enzyme activity might prevent the incidence of fetal hydantoin syndrome during pregnancy [113].

Topiramate

Topiramate is approved for the treatment of partial onset seizures as adjunctive therapy in patients older than 4 years and as monotherapy in those older than 16 years [114]. TPM has diverse mechanisms of actions including the ability to increase brain GABA levels and block kainate/AMPA subtype of the glutamate receptor [88]. Neuroprotective properties of TPM are related to inhibition of fast Na+ and HVA Ca2+ conductances as demonstrated in studies on in vitro model of cerebral ischemia [115]. However, few studies have been reported on the effects of TPM on the oxidative status. Kubera et al. [116] showed that 80 mg/kg of TPM was able to significantly reduce levels of lipid peroxidation in piriform cortex and frontal cortex of rats, following kainate-induced status epilepticus. TPM is able to reduce levels of lipid peroxidation in animal models [117, 118] and to increase the erythrocyte glutathione peroxidase activity and GSH levels [118]. In particular, the association of TPM and vitamin E determined an increase of kidney SOD and catalase enzyme activities [119]. Pro-oxidant effects, by contrast, have been observed in astrocytes [120] and in kindled and non-kindled animals [103]. When compared to reference molecules, TPM demonstrates significant scavenging capacity accounting for the neuroprotective effects attributed to this compound and suggesting its use as a possible chemopreventive agent [121].

Valproic Acid

Valproic acid (VPA), a branched-chain fatty acid, is the most widely used AED for the treatment of epilepsy, bipolar psychiatric disorders, and migraine [60]. A large number of studies evaluated the influence of VPA on oxidative balance with contrasting results. In patients affected by idiopathic epilepsy, VPA treatment may determine changes in the erythrocyte lipid peroxidation and antioxidant enzyme levels when compared to those of the pretreatment period suggesting a clear VPA pro-oxidant effect [55]. These findings have been further corroborated by the observation of increased neutrophils' oxidative metabolism and oxidant status in children with idiopathic epilepsy treated with VPA [122]. Oxidative stress indexes, evaluated in the cerebral and cerebellar cortices of young rats placed under VPA load, have been found significantly abnormal as the consequence of a VPA-related impairment of the antioxidant status of the neuronal tissue [123]. On the contrary, VPA showed a neuroprotective effect against ischemia/reperfusion injury in the rat retina by enhancing antioxidative effects and inhibiting apoptosis of retinal cells [124]. Controversies of the VPA effects still exist also in clinical studies. It was recently observed that patients affected by epilepsy present significantly increased levels of oxidative markers when compared to control group. However, several AEDs, including valproate, do not influence oxidative marker changes which, therefore, are mainly seizure related [125]. These findings are in line with our data detected in children affected by epileptic encephalopathy in whom increased oxidative marker levels were related to seizures rather than to AEDs [60]. Moreover, it has been reported that after 1-year VPA treatment, in patients with epilepsy, increased oxidative markers could be noted only in those who developed obesity but not in the remaining group of nonobese patients. It was clear therefore that increased levels of oxidative markers were related to weight gain rather than VPA therapy [56]. In addition, the well-known VPA teratogenicity may be exerted by VPA-induced ROS formation and apoptosis [126]. The drug has also been implicated in favoring oxidative stress and sperm DNA damage in the testes of mice [127] and in increasing oxidation of low-density lipoprotein [128].

Conclusion

This review provides an overview of the pathophysiological mechanisms associated with oxidative stress in epileptic seizures in both experimental and clinical studies. When globally considered, they seem to show that lipid peroxidation, DNA oxidation, mitochondrial dysfunctions, redox-active metals, and an impaired antioxidant system may be crucial factors for epileptogenesis.

Therapeutic intervention with free radical scavengers may represent a key strategy to counteract the epilepsy-related neurodegenerative process. However, in spite of the incredible development of new drugs for epilepsy treatment, definite evidence about the neuroprotective ability of the existing compounds is still lacking.

Therefore, further studies are required to understand molecular mechanisms underlying oxidative stress. New molecules potentially active in the treatment of epilepsy wait to be discovered and delineated.

References

1. Ngugi AK, Bottomley C, Kleinschmidt I, Sander JW, Newton CR. Estimation of the burden of active and life-time epilepsy: a meta-analytic approach. Epilepsia. 2010;51(5):883–90.
2. Shorvon SD. The etiologic classification of epilepsy. Epilepsia. 2011;52(6):1052–57.
3. Grosso S, Orrico A, Galli L, Di Bartolo R, Sorrentino V, Balestri P. SCN1A mutation associated with atypical Panayiotopoulos syndrome. Neurology. 2007;69(6):609–11.
4. Shorvon SD. The causes of epilepsy: changing concepts of etiology of epilepsy over the past 150 years. Epilepsia. 2011;52(6):1033–44.
5. McCormick DA, Contreras D. On the cellular and network bases of epileptic seizures. Annu Rev Physiol. 2001;63:815–46.
6. Chuang YC. Mitochondrial dysfunction and oxidative stress in seizure-induced neuronal cell death. Acta Neurol Taiwan. 2010;19(1):3–15.
7. Fujikawa DG. Prolonged seizures and cellular injury: understanding the connection. Epilepsy Behav. 2005;7(Suppl3):S3–S11.
8. Azam F, Prasad M, Thangavel N. Targeting oxidative stress component in the therapeutics of epilepsy. Curr Top Med Chem. 2012;12(9):994–1007.
9. Ashrafi MR, Shams S, Nouri M, Mohseni M, Shabanian R, Yekaninejad MS, Chegini N, Khodadad A, Safaralizadeh R. A probable causative factor for an old problem: selenium and glutathione peroxidase appear to play important roles in epilepsy pathogenesis. Epilepsia. 2007;48(9):1750–5.
10. Wong-Ekkabut J, Xu Z, Triampo W, Tang IM, Tieleman DP, Monticelli L. Effect of lipid peroxidation on the properties of lipid bilayers: a molecular dynamics study. Biophys J. 2007;93(12):4225–36.
11. Mariani E, Polidori MC, Cherubini A, Mecocci P. Oxidative stress in brain aging, neurodegenerative and vascular diseases: an overview. J Chromatogr B Analyt Technol Biomed Life Sci. 2005;827(1):65–75.
12. Jellinger KA. General aspects of neurodegeneration. J Neural Transm. 2003;65:101–44.
13. Shin EJ, Jeong JH, Chung YH, Kim WK, Ko KH, Bach JH, Hong JS, Yoneda Y, Him HC. Role of oxidative stress in epileptic seizures. Neurochem Int. 2011;59(2):122–37.

14. Kwan P, Brodie MJ. Early identification of refractory epilepsy. N Eng J Med. 2000; 342(5):314–9.
15. Devi PU, Manocha A, Vohora D. Seizures, antiepileptics, antioxidants and oxidative stress: an insight for researchers. Expert Opin Pharmacother. 2008;9:3169–77.
16. Candelario-Jalil E, Ajamieh HH, Sam S, Martínez G, León Fernández OS. Nimesulide limits kainate-induced oxidative damage in the rat hippocampus. Eur J Pharmacol. 2000;390:295–8.
17. Patel M, Liang LP, Roberts II LJ. Enhanced hippocampal F2-isoprostane formation following kainite-induced seizures. J Neurochem. 2001;79:1065–9.
18. Rola R, Swiader M, Czuczwar SJ. Electroconvulsions elevate the levels of lipid peroxidation products in mice. Pol J Pharmacol. 2002;54:521–4.
19. Barichello T, Bonatto F, Agostinho FR, Reinke A, Moreira JC, Dal-Pizzol F, Izquierdo I, Quevedo J. Structure-related oxidative damage in rat brain after acute and chronic electroshock. Neurochem Res. 2004;24(9):1749–53.
20. Gilberti EA, Trombetta LD. The relationship between stress protein induction and the oxidative defense system in the rat hippocampus following kainic acid administration. Toxicol Lett. 2000;116(1–2):17–26.
21. Dal-Pizzol F, Klamt F, Vianna MM, Schröder N, Quevedo J, Benfato MS, Moreira JC, Walz R. Lipid peroxidation in hippocampus early and late after status epilepticus induced by pilocarpine or kainic acid in Wistar rats. Neurosci Lett. 2000;291(3):179–82.
22. Waldbaum S, Patel M. Mitochondria, oxidative stress, and temporal lobe epilepsy. Epilepsy Res. 2010;88(1):23–45.
23. Liang LP, Ho YS, Patel M. Mitochondrial superoxide production in kainate-induced hippocampal damage. Neuroscience. 2000;101(3):563–70.
24. Patel M. Mitochondrial dysfunction and oxidative stress: cause and consequence of epileptic seizures. Free Radic Biol Med. 2004;37(12):1951–62.
25. Gutteridge JM, Halliwell B. Free radicals and antioxidants in the year 2000. A historical look to the future. Ann N Y Acad Sci. 2000;899:136–47.
26. Jarrett SG, Liang LP, Hellier JL, Staley KJ, Patel M. Mitochondrial DNA damage and impaired base excision repair during epileptogenesis. Neurobiol Dis. 2008;30(1):130–8.
27. Liang LP, Patel M. Mitochondrial oxidative stress and increased seizure susceptibility in Sod2 (–/+) mice. Free Radic Biol Med. 2004;36(5):542–54.
28. Bruce AJ, Baudry M. Oxygen free radicals in rat limbic structures after kainite-induced seizures. Free Radic Biol Med. 1995;18(6):993–1002.
29. Cheng H, Fu YS, Guo JW. Ability of GDNF to diminish free radical production leads to protection against kainate-induced excitotoxicity in hippocampus. Hippocampus. 2004; 14(1):77–86.
30. Tejada S, Sureda A, Roca C, Gamundi A, Esteban S. Antioxidant response and oxidative damage in brain cortex after high dose of pilocarpine. Brain Res Bull. 2007;71(4):372–75.
31. Freitas RM, Vasconcelos SM, Souza FC, Viana GS, Fonteles MM. Oxidative stress in the hippocampus after pilocarpine-induced status epilepticus in Wistar rats. FEBS J. 2005; 272(6):1307–12.
32. Candelario-Jalil E, Al-Dalain SM, Castillo R, Martínez OS. Selective vulnerability to kainate-induced oxidative damage in different rat brain regions. J Appl Toxicol. 2001;21(5):403–7.
33. Rauca C, Wiswedel I, Zerbe R, Keilhoff G, Krug M. The role of superoxide dismutase and alpha-tocopherol in the development of seizures and kindling induced by pentylenetetrazol-influence of the radical scavenger alpha-phenyl-N-tert-butyl nitrone. Brain Res. 2004;1009(1–2):203–12.
34. Freitas RM. Investigation of oxidative stress involvement in hippocampus in epilepsy model induced by pilocarpine. Neurosci Lett. 2009;462(3):225–9.
35. Bellissimo MI, Amado D, Abdalla DS, Ferreira EC, Cavalheiro EA, Naffah-Mazzacoratti MG. Superoxide dismutase, glutathione peroxidase activities and the hydroperoxide concentration are modified in the hippocampus of epileptic rats. Epilepsy Res. 2001;461(2):121–8.
36. Patel M, Li QY. Age dependence of seizure-induced oxidative stress. Neuroscience. 2003;118(2):431–7.

37. Sullivan PG, Dubé C, Dorenbos K, Steward O, Baram TZ. Mitochondrial uncoupling protein-2 protects the immature brain from excitotoxic neuronal death. Ann Neurol. 2003;53:711–7.
38. Folbergrová J, Jesina P, Drahota Z, Lisý V, Haugvicová R, Vojtísková A, Houstěk J. Mitochondrial complex I inhibition in cerebral cortex of immature rats following homocysteic acid-induced seizures. Exp Neurol. 2007;204(2):597–609.
39. Folbergrová J, Otáhal J, Druga R. Brain superoxide anion formation in immature rats during seizures: protection by selected compounds. Exp Neurol. 2012;233(1):421–9.
40. Freitas RM, Nascimento VS, Vasconcelos SM, Sousa FC, Viana GS, Fonteles MM. Catalase activity in cerebellum, hippocampus, frontal cortex and striatum after status epilepticus induced by pilocarpine in Wistar rats. Neurosci Lett. 2004;365(2):102–5.
41. Xavier SM, Barbosa CO, Barros DO, Silva RF, Oliveira AA, Freitas RM. Vitamin C antioxidant effects in hippocampus of adult Wistar rats after seizures and status epilepticus induced by pilocarpine. Neurosci Lett. 2007;420(1):76–9.
42. Barros DO, Xavier SM, Barbosa CO, Silva RF, Maia FD, Oliveira AA, Freitas RM, Takahashi RN. Effects of the vitamin E in catalase activities in hippocampus after status epilepticus induced by pilocarpine in Wistar rats. Neurosci Lett. 2007;416(3):227–30.
43. Rajasekaran K. Seizure-induced oxidative stress in rat brain regions: blockade by nNOS inhibition. Pharmacol Biochem Behav. 2005;80(2):263–72.
44. Liang LP, Patel M. Seizure-induced changes in mitochondrial redox status. Free Rad Biol Med. 2006;40(2):316–22.
45. Eraković V, Zupan G, Varljen J, Radosević S, Simonić A. Electroconvulsive shock in rats: changes in superoxide dismutase and glutathione peroxidase activity. Brain Res Mol Brain Res. 2000;76(2):266–74.
46. Eraković V, Zupan G, Varljen J, Simonić A. Pentylenetetrazol-induced seizures and kindling: changes in free fatty acids, superoxide dismutase, and glutathione peroxidase activity. Neurochem Int. 2003;42(2):173–8.
47. Cendes F. Progressive hippocampal and extrahippocampal atrophy in drug resistant epilepsy. Curr Opin Neurol. 2005;18(2):173–7.
48. Sutula TP, Hagen J, Pitkänen A. Do epileptic seizures damage the brain? Curr Opin Neurol. 2003;16(2):189–95.
49. Wallace DC, Zheng X, Lott MT, et al. Familial mitochondrial encephalomyophathy (MERRF): genetic, pathophysiological, and biochemical characterization of a mitochondrial DNA disease. Cell. 1988;55(4):601–10.
50. Wu SB, Ma YS, Wu YT, Chen YC, Wei YH. Mitochondrial DNA mutation-elicited oxidative stress, oxidative damage, and altered gene expression in cultured cells of patients with MERRF syndrome. Mol Neurobiol. 2010;41(2–3):256–66.
51. Zsurka G, Baron M, Stewart JD, et al. Clonally expanded mitochondrial DNA mutations in epileptic individuals with mutated DNA polymerase γ. J Neuropath Exp Neurol. 2008;67(9):857–66.
52. Zsurka G, Hampel KG, Nelson I, et al. Severe epilepsy as the major symptom of new mutations in the mitochondrial tRNAPhe gene. Neurology. 2010;74(6):507–12.
53. Ezz HS, Khadrawy YA, Noor NA. The neuroprotective effect of curcumin and Nigella sativa oil against oxidative stress in the pilocarpine model of epilepsy: a comparison with valproate. Neurochem Res. 2011;36(11):2195–204.
54. Júnior HV, de França Fonteles MM, Mendes de Freitas R. Acute seizure activity promotes lipid peroxidation, increased nitrite levels and adaptive pathways against oxidative stress in the frontal cortex and striatum. Oxid Med Cell Longev. 2009;2(3):130–7.
55. Yi□ U, Seçkin E, Kurul SH, Kuralay F, Dirik E. Effects of epilepsy and valproic acid on oxidant status in children with idiopathic epilepsy. Epilepsy Res. 2009;84:232–7.
56. Verrotti A, Scardapane A, Franzoni E, Manco R, Chiarelli F. Increased oxidative stress in epileptic children treated with valproic acid. Epilepsy Res. 2008;78:171–7.
57. Michoulas A, Tong V, Teng XW, Chang TK, Abbott FS, Farrell K. Oxidative stress in children receiving valproic acid. J Pediatr. 2006;149:692–6.

58. Akarsu S, Yilmaz S, Ozan S, Kurt A, Benzer F, Gurgoze MK. Effects of febrile and afebrile seizures on oxidant state in children. Pediatr Neurol. 2007;36:307–11.
59. Vanhatalo S, Riikonen R. Markedly elevated nitrate/nitrite levels in the cerebrospinal fluid of children with progressive encephalopathy with edema, hypsarrhythmia, and optic atrophy (PEHO syndrome). Epilepsia. 2000;41:705–8.
60. Grosso S, Longini M, Rodriguez A, Proietti F, Piccini B, Balestri P, Buonocore G. Oxidative stress in children affected by epileptic encephalopathies. J Neuro Sci. 2011;300(1–2):103–6.
61. Zsurka G, Kunz WS. Mitochondrial dysfunction in neurological disorders with epileptic phenotypes. J Bioenerg Biomembr. 2010;42(6):443–8.
62. Khurana DS, Salganicoff L, Melvin JJ, Hobdell EF, Valencia I, Hardison HH, Marks HG, Grover WD, Legido A. Epilepsy and respiratory chain defects in children with mitochondrial encephalopathies. Neuropediatrics. 2008;39(1):8–13.
63. Katano M, Numata T, Aguan K, Hara Y, Kiyonaka S, Yamamoto S, Miki T, Sawamura S, Suzuki T, Yamakawa K, Mori Y. The juvenile myoclonic epilepsy-related protein EFHC1 interacts with the redox-sensitive TRPM2 channel linked to cell death. Cell Calcium. 2012;51(2):179–85.
64. Prasad AN, Levin S, Rupar CA, Prasad C. Menkes disease and infantile epilepsy. Brain Dev. 2011;30(10):866–76.
65. Miyata R, Hayashi M, Tanuma N, Shioda K, Fukatsu R, Mizutani S. Oxidative stress in neurodegeneration in dentatorubral-pallidoluysian atrophy. J Neurol Sci. 2008;264(1–2):133–9.
66. Shin EJ, Suh SK, Lim YK, Jhoo WK, Hjelle OP, Ottersen OP, Shin CY, Ko KH, Kim WK, Kim DS, Chun W, Ali S, Kim HC. Ascorbate attenuates trimethyltin-induced oxidative burden and neuronal degeneration in the rat hippocampus by maintaining glutathione homeostasis. Neuroscience. 2005;33:715–27.
67. Edwards MJ, Hargreaves IP, Heales SJ, Jones SJ, Ramachandran V, Bhatia KP, Sisodiya S. N-acetylcysteine and Unverricht-Lundborg disease: variable response and possible side effects. Neurology. 2002;59(9):1447–9.
68. Ben-Menachem E, Kyllerman M, Marklund S. Superoxide dismutase and glutathione peroxidase function in progressive myoclonus epilepsies. Epilepsy Res. 2000;40:33–9.
69. Heard K, Schaeffer TH. Massive acetylcysteine overdose associated with cerebral edema and seizures. Clin Toxicol (Phila). 2011;49(5):423–5.
70. Praninskiene R, Dumalakiene I, Kemezys R, Mauricas M, Jucaite A. Melatonin secretion in children with epilepsy. Epilepsy Behav. 2012;25(3):315–22.
71. Appleton RE, Jones AP, Gamble C, Williamson PR, Wiggs L, Montgomery P, Sutcliffe A, Barker C, Gringras P. The use of MElatonin in children with neurodevelopmental disorders and impaired sleep: a randomised, double-blind, placebo-controlled, parallel study (MENDS). Health Technol Assess. 2012;16(40):i–239.
72. Goldberg-Stern H, Oren H, Peled N, Garty BZ. Effect of melatonin on seizure frequency in intractable epilepsy: a pilot study. J Child Neurol. 2012;27(12):1524–8.
73. Paprocka J, Dec R, Jamroz E, Marszał E. Melatonin and childhood refractory epilepsy–a pilot study. Med Sci Monit. 2010;16(9):CR389–96.
74. Antón-Tay F, Díaz JL, Fernández-Guardiola A. On the effect of melatonin upon human brain. Its possible therapeutic implications. Life Sci I. 1971;10(15):841–50.
75. Ardura J, Andres J, Garmendia JR, Ardura F. Melatonin in epilepsy and febrile seizures. J Child Neurol. 2010;25(7):888–91.
76. Brigo F, Del Felice A. Melatonin as add-on treatment for epilepsy. Cochrane Database Syst Rev. 2012;6:CD006967.
77. Molina-Carballo A, Muñoz-Hoyos A, Reiter RJ, Sánchez-Forte M, Moreno-Madrid F, Rufo-Campos M, Molina-Font JA, Acuña-Castroviejo D. Utility of high doses of melatonin as adjunctive anticonvulsant therapy in a child with severe myoclonic epilepsy: two years' experience. J Pineal Res. 1997;23(2):97–105.
78. Veinbergs I, Mallory M, Sagara Y, Masliah E. Vitamin E supplementation prevents spatial learning deficits and dendritic alterations in aged apolipoprotein E-deficient mice. Eur J Neurosci. 2000;12(12):4541–6.

79. Tamai H, Wakamiya E, Mino M, Iwakoshi M. Alpha-tocopherol and fatty acid levels in red blood cells in patients treated with antiepileptic drugs. J Nutr Sci Vitaminol (Tokyo). 1988;34(6):627–31.
80. Ogunmekan AO. Vitamin E deficiency and seizures in animals and man. Can J Neurol Sci. 1979;6(1):43–5.
81. Sudha K, Rao AV, Rao A. Oxidative stress and antioxidants in epilepsy. Clin Chim Acta. 2001;303(1–2):19–24.
82. Ogunmekan AO, Hwang PA. A randomized, double-blind, placebo-controlled, clinical trial of D-alpha-tocopheryl acetate (vitamin E), as add-on therapy, for epilepsy in children. Epilepsia. 1989;30(1):84–9.
83. Sullivan C, Capaldi N, Mack G, Buchanan N. Seizures and natural vitamin E. Med J Aust. 1990;152(11):613–4.
84. Raju GB, Behari M, Prasad K, Ahuja GK. Randomized, double-blind, placebo-controlled, clinical trial of D-alpha-tocopherol (vitamin E) as add-on therapy in uncontrolled epilepsy. Epilepsia. 1994;35(2):368–72.
85. Kishida KT, Klann E. Sources and targets of reactive oxygen species in synaptic plasticity and memory. Antioxid Redox Signal. 2007;9(2):233–44.
86. Fransen M. Cyclophilin D: a therapeutic target to counteract reactive oxygen species-mediated damage in neurodegenerative disease? Brain. 2012;135(Pt 12):3525–6.
87. Berger I, Segal I, Shmueli D, Saada A. The effect of antiepileptic drugs on mitochondrial activity: a pilot study. J Child Neurol. 2010;25(5):541–5.
88. Jain KK. The handbook of neuroprotection. New York: Humana Press; 2011.
89. Gathwala G, Marwah A, Gahlaut V, Marwah P. Effect of high-dose phenobarbital on oxidative stress in perinatal asphyxia: an open label randomized controlled trial. Indian Pediatr. 2011;48(8):613–7.
90. Sarnowska A, Beresewicz M, Zabłocka B, Domańska-Janik K. Diazepam neuroprotection in excitotoxic and oxidative stress involves a mitochondrial mechanism additional to the GABAAR and hypothermic effects. Neurochem Int. 2009;55(1–3):164–73.
91. Kovacic P, Somanathan R. Zolpidem, a clinical hypnotic that affects electronic transfer, alters synaptic activity through potential GABA receptors in the nervous system without significant free radical generation. Oxid Med Cell Longev. 2009;2(1):52–7.
92. Tanguy M, Seguin P, Laviolle B, Bleichner JP, Morandi X, Malledant Y. Cerebral microdialysis effects of propofol versus midazolam in severe traumatic brain injury. J Neurotrauma. 2012;29(6):1105–10.
93. Núñez MJ, Novío S, Amigo G, Freire-Garabal M. The antioxidant potential of alprazolam on the redox status of peripheral blood leukocytes in restraint-stressed mice. Life Sci. 2011;89(17–18):650–4.
94. Méndez-Cuesta LA, Márquez-Valadez B, Pérez-De La Cruz V, Escobar-Briones C, Galván-Arzate S, Alvarez-Ruiz Y, Maldonado PD, Santana RA, Santamaría A, Carrillo-Mora P. Diazepam blocks striatal lipid peroxidation and improves stereotyped activity in a rat model of acute stress. Basic Clin Pharmacol Toxicol. 2011;109(5):350–6.
95. Hamed SA, Abdellah MM. Trace elements and electrolytes homeostasis and their relation to antioxidant enzyme activity in brain hyperexcitability of epileptic patients. J Pharma Sci. 2004;96(4):349–59.
96. Ficarra S, Misiti F, Russo A, Carelli-Alinovi C, Bellocco E, Barreca D, Laganà G, Leuzzi U, Toscano G, Giardina B, Galtieri A, Tellone E. Antiepileptic carbamazepine drug treatment induces alteration of membrane in red blood cells: possible positive effects on metabolism and oxidative stress. Biochimie. 2012; pii: S0300-9084(12)00481-6.
97. Wallis RA, Panizzon KL. Felbamate neuroprotection against CA1 traumatic neuronal injury. Eur J Pharmacol. 1995;294(2–3):475–82.
98. White HS, Wolf HH, Swinyard EA, Skeen GA, Sofia RD. A neuropharmacological evaluation of felbamate as a novel anticonvulsant. Epilepsia. 1992;33(3):564–72.
99. Pellock JM, Perhach JL, Sofia RD. In: Levy RH, Mattson RH, Meldrum BS, Perucca E, editors. Antiepileptic drugs. 5th ed. Philadelphia: Lippincott Williams & Wilkins; 2002. p. 301–18.

100. Leone AM, Kao LM, McMillian MK, Nie AY, Parker JB, Kelley MF, Usuki E, Parkinson A, Lord PG, Johnson MD. Evaluation of felbamate and other antiepileptic drug toxicity potential based on hepatic protein covalent binding and gene expression. Chem Res Toxicol. 2007;20(4):600–8.
101. Kumar P, Kalonia H, Kumar A. Possible GABAergic mechanism in the neuroprotective effect of gabapentin and lamotrigine against 3-nitropropionic acid induced neurotoxicity. Eur J Pharmacol. 2012;674(2–3):265–74.
102. Arora T, Mehta AK, Sharma KK, Mediratta PK, Banerjee BD, Garg GR, Sharma AK. Effect of carbamazepine and lamotrigine on cognitive function and oxidative stress in brain during chemical epileptogenesis in rats. Basic Clin Pharmacol Toxicol. 2010;106(5):372–7.
103. Agarwal NB, Agarwal NK, Mediratta PK, Sharma KK. Effect of lamotrigine, oxcarbazepine and topiramate on cognitive functions and oxidative stress in PTZ-kindled mice. Seizure. 2011;20(3):257–62.
104. Grosso S, Franzoni E, Coppola G, Iannetti P, Verrotti A, Cordelli DM, Marchiani V, Pascotto A, Spalice A, Acampora B, Morgese G, Balestri P. Efficacy and safety of levetiracetam: an add-on trial in children with refractory epilepsy. Seizure. 2005;14(4):248–53.
105. Grosso S, Cordelli DM, Franzoni E, Coppola G, Capovilla G, Zamponi N, Verrotti A, Morgese G, Balestri P. Efficacy and safety of levetiracetam in infants and young children with refractory epilepsy. Seizure. 2007;16(4):345–50.
106. Oliveira AA, Almeida JP, Freitas RM, Nascimento VS, Aguiar LM, Júnior HV, Fonseca FN, Viana GS, Sousa FC, Fonteles MM. Effects of levetiracetam in lipid peroxidation level, nitrite-nitrate formation and antioxidant enzymatic activity in mice brain after pilocarpine-induced seizures. Cell Mol Neurobiol. 2007;27(3):395–406.
107. Gibbs JE, Walker MC, Cock HR. Levetiracetam: antiepileptic properties and protective effects on mitochondrial dysfunction in experimental status epilepticus. Epilepsia. 2006;47(3):469–78.ef.
108. Ozden H, Kabay SC, Toker A, Ustüner MC, Ozbayer C, Ustüner D, Günes HV. The effects of levetiracetam on urinary 15f-2t-isoprostane levels in epileptic patients. Seizure. 2010; 19(8):514–6.
109. Mariotti V, Melissari E, Amar S, Conte A, Belmaker RH, Agam G, Pellegrini S. Effect of prolonged phenytoin administration on rat brain gene expression assessed by DNA microarrays. Exp Biol Med (Maywood). 2010;235(3):300–10.
110. Winn LM, Wells PG. Phenytoin-initiated DNA oxidation in murine embryo culture, and embryo protection by the antioxidative enzymes superoxide dismutase and catalase: evidence for reactive oxygen species-mediated DNA oxidation in the molecular mechanism of phenytoin teratogenicity. Mol Pharmacol. 1995;48(1):112–20.
111. Lu W, Uetrecht JP. Peroxidase-mediated bioactivation of hydroxylated metabolites of carbamazepine and phenytoin. Drug Metab Dispos. 2008;36(8):1624–36.
112. Mahle C, Dasgupta A. Decreased total antioxidant capacity and elevated lipid hydroperoxide concentrations in sera of epileptic patients receiving phenytoin. Life Sci. 1997;61(4):437–43.
113. Liu CS, Wu HM, Kao SH, Wei YH. Phenytoin-mediated oxidative stress in serum of female epileptics: a possible pathogenesis in the fetal hydantoin syndrome. Hum Exp Toxicol. 1997;16(3):177–81.
114. Grosso S, Galimberti D, Farnetani MA, Cioni M, Mostardini R, Vivarelli R, Di Bartolo RM, Bernardoni E, Berardi R, Morgese G, Balestri P. Efficacy and safety of topiramate in infants according to epilepsy syndromes. Seizure. 2005;14(3):183–9.
115. Costa C, Martella G, Picconi B, Prosperetti C, Pisani A, Di Filippo M, Pisani F, Bernardi G, Calabresi P. Multiple mechanisms underlying the neuroprotective effects of antiepileptic drugs against in vitro ischemia. Stroke. 2006;37(5):1319–26.
116. Kubera M, Budziszewska B, Jaworska-Feil L, Basta-Kaim A, Leśkiewicz M, Tetich M, Maes M, Kenis G, Marciniak A, Czuczwar SJ, Jagła G, Nowak W, Lasoń W. Effect of topiramate on the kainate-induced status epilepticus, lipid peroxidation and immunoreactivity of rats. Pol J Pharmacol. 2004;56(5):553–61.

117. Naziroğlu M, Uğuz AC, Gokçimen A, Bülbül M, Karatopuk DU, Türker Y, Cerçi C. Tenoxicam modulates antioxidant redox system and lipid peroxidation in rat brain. Neurochem Res. 2008;33(9):1832–7.
118. Kutluhan S, Naziroğlu M, Celik O, Yilmaz M. Effects of selenium and topiramate on lipid peroxidation and antioxidant vitamin levels in blood of pentylentetrazol-induced epileptic rats. Biol Trace Elem Res. 2009;129(1–3):181–9.
119. Armagan A, Kutluhan S, Yilmaz M, Yilmaz N, Bülbül M, Vural H, Soyupek S, Naziroglu M. Topiramate and vitamin e modulate antioxidant enzyme activities, nitric oxide and lipid peroxidation levels in pentylenetetrazol-induced nephrotoxicity in rats. Basic Clin Pharmacol Toxicol. 2008;103(2):166–70.
120. Pavone A, Cardile V. An in vitro study of new antiepileptic drugs and astrocytes. Epilepsia. 2003;44 Suppl 10:34–9.
121. Cárdenas-Rodríguez N, Coballase-Urrutia E, Huerta-Gertrudis B, García-Cruz ME, Pedraza-Chaverri J, Coria-Jiménez R, Bandala C, Ruíz-García M. Antioxidant activity of topiramate: an antiepileptic agent. Neurol Sci. 2012;4.
122. Zhang YJ, Zhang M, Wang XC, Yu YH, Jin PJ, Wang Y. Effects of sodium valproate on neutrophils' oxidative metabolism and oxidant status in children with idiopathic epilepsy. Zhonghua Er Ke Za Zhi. 2011;49(10):776–81.
123. Chaudhary S, Parvez S. An in vitro approach to assess the neurotoxicity of valproic acid-induced oxidative stress in cerebellum and cerebral cortex of young rats. Neuroscience. 2012;225:258–68.
124. Zhang Z, Qin X, Zhao X, Tong N, Gong Y, Zhang W, Wu X. Valproic acid regulates antioxidant enzymes and prevents ischemia/reperfusion injury in the rat retina. Curr Eye Res. 2012;37(5):429–37.
125. Menon B, Ramalingam K, Kumar RV. Oxidative stress in patients with epilepsy is independent of antiepileptic drugs. Seizure. 2012;21(10):780–4.
126. Tung EW, Winn LM. Valproic acid increases formation of reactive oxygen species and induces apoptosis in postimplantation embryos: a role for oxidative stress in valproic acid-induced neural tube defects. Mol Pharmacol. 2011;80(6):979–87.
127. Khan S, Ahmad T, Parekh CV, Trivedi PP, Kushwaha S, Jena G. Investigation on sodium valproate induced germ cell damage, oxidative stress and genotoxicity in male Swiss mice. Reprod Toxicol. 2011;32(4):385–94.
128. Yildiz M, Simsek G, Uzun H, Uysal S, Sahin S, Balci H. Assessment of low-density lipoprotein oxidation, paraoxonase activity, and arterial distensibility in epileptic children who were treated with antiepileptic drugs. Cardiol Young. 2010;20(5):547–54.
129. Reiter RJ, Korkmaz A, Paredes SD, Manchester LC, Tan DX. Melatonin reduces oxidative/nitrosative stress due to drugs, toxins, metals, and herbicides. Neuro Endocrinol Lett. 2008;29(5):609–13.
130. Freitas RM. The evaluation of effects of lipoic acid on the lipid peroxidation, nitrite formation and antioxidant enzymes in the hippocampus of rats after pilocarpine-induced seizures. Neurosci Lett. 2009;455(2):140–4.
131. Santos LF, Freitas RL, Xavier SM, Saldanha GB, Freitas RM. Neuroprotective actions of vitamin C related to decreased lipid peroxidation and increased catalase activity in adult rats after pilocarpine-induced seizures. Pharmacol Biochem Behav. 2008;89(1):1–5.
132. Traber MG. Vitamin E, regulatory mechanisms. Annu Rev Nutr. 2007;27:347–62.
133. Arakawa M, Ito Y. N-acetylcysteine and neurodegenerative diseases: basic and clinical pharmacology. Cerebellum. 2007;6(4):308–14.
134. Shin HJ, Lee JY, Son E, Lee DH, Kim HJ, Kang SS, Cho GJ, Choi WS, Roh GS. Curcumin attenuates the kainic acid-induced hippocampal cell death in the mice. Neurosci Lett. 2007;416:49–54.
135. Kwon YS, Park DH, Shin EJ, Kwon MS, Ko KH, Kim WK, Jhoo JH, Jhoo WK, Wie MB, Jung BD, Kim HC. Antioxidant propolis attenuates kainate-induced neurotoxicity via adenosine A1 receptor modulation in the rat. Neurosci Lett. 2004;355(3):231–5.

136. Ilhan A, Iraz M, Kamisli S, Yigitoglu R. Pentylenetetrazol-induced kindling seizure attenuated by Ginkgo biloba extract (EGb 761) in mice. Prog Neuropsychopharmacol Biol Psychiatry. 2006;30(8):1504–10.
137. Chuang YC, Chen SD, Liou CW, Lin TK, Chang WN, Chan SH, Chang AY. Contribution of nitric oxide, superoxide anion, and peroxynitrite to activation of mitochondrial apoptotic signaling in hippocampal CA3 subfield following experimental temporal lobe status epilepticus. Epilepsia. 2009;50:731–46.
138. Kim HC, Choi DY, Jhoo WK, Lee DW, Koo CH, Kim C. Aspalatone, a new antiplatelet agent, attenuates the neurotoxicity induced by kainic acid in the rat. Life Sci. 1997;61(24):PL 373–81.
139. McConnell H, Duncan D. In: McConnell H, Snyder PJ, editors. Psychiatric comorbidity in epilepsy: basic mechanisms, diagnosis, and treatment. Washington, DC: American Psychiatric Press; 1998. p. 245–362.
140. de Freitas RL, Santos IM, de Souza GF, Tomé Ada R, Saldanha GB, de Freitas RM. Oxidative stress in rat hippocampus caused by pilocarpine-induced seizures is reversed by buspirone. Brain Res Bull. 2010;81(4–5):505–9.
141. Zhou J, Tang XC. Huperzine A attenuates apoptosis and mitochondria-dependent caspase-3 in rat cortical neurons. FEBS Lett. 2002;526(1–3):21–5.
142. Bialer M, Johannessen SI, Levy RH, Perucca E, Tomson T, White HS. Progress report on new antiepileptic drugs: a summary of the Tenth Eilat Conference (EILAT X). Epilepsy Res. 2010;92(2–3):89–124.
143. Acharya MM, Hattiangady B, Shetty AK. Progress in neuroprotective strategies for preventing epilepsy. Prog Neurobiol. 2008;84(4):363–404.

Chapter 21
The Oxidative Stress in the Fetus and in the Newborn

Serafina Perrone, Maria Luisa Tataranno, Antonino Santacroce, and Giuseppe Buonocore

Introduction

OS occurs when the production of FRs exceeds the capacity of antioxidant defenses [16]. It is due to an imbalance between the production of reactive species and the biological system's ability to readily detoxify the reactive intermediates or to repair the resulting damage. Each cell is characterized by a fixed number of electrons stored in many different organelles, and the cellular redox state with its delicate balance allows cellular functioning [20]. Disturbances in the normal redox state of tissues can cause toxic effects through the production of peroxides and FRs that damage all components of the cell, including proteins, lipids, and DNA. Some reactive oxidative species can even act as second messengers: at low levels, they are signaling molecules, and at high levels, they can damage organelles, particularly the mitochondria. Oxidative damage and the associated mitochondrial dysfunction may result in energy depletion, accumulation of cytotoxic mediators, and cell death. Understanding the interface between stress adaptation and cell death is important to clarify redox biology and disease pathogenesis [57].

OS is an unavoidable consequence of life in an O_2-rich atmosphere. It occurs at birth in all newborns as a consequence of the hyperoxic challenge due to the transition from the hypoxic intrauterine environment to extrauterine life. FR sources such as hyperoxia, hypoxia, ischemia, hypoxia–reperfusion, inflammation, and high levels of NPBI increase OS during the perinatal period. In addition, preterm babies have reduced antioxidant defenses; thus, they are not able to counteract the harmful effects of FRs that lead to cellular, tissue, and organ damages (kidney, retina, lung, and bowel injury) [82].

S. Perrone, M.D., Ph.D. • M.L. Tataranno, M.D. • A. Santacroce, M.D. • G. Buonocore, M.D. (✉)
Department of Molecular and Developmental Medicine, University of Siena,
V.le Bracci 36, 53100 Siena, Italy
e-mail: saraspv@yahoo.it; giuseppe.buonocore@unisi.it

P.A. Dennery et al. (eds.), *Perinatal and Prenatal Disorders*,
Oxidative Stress in Applied Basic Research and Clinical Practice,
DOI 10.1007/978-1-4939-1405-0_21

Oxidative Stress

The increased generation of FRs in cells directly relates to the fact that each oxygen atom has one unpaired electron in its outer valence shell; thus, molecular oxygen (O_2) has two unpaired electrons. If the oxygen atom does not find a twin atom, it can accept hydrogen to form water (H_2O), but there may be one electron less or more, resulting in no stability. This complex represents an FR. FRs are capable of giving rise to chain reactions, i.e., reactions that involve a number of steps, each of which forms an FR that triggers the next step. Three different phases (initiation, propagation, and termination) and different FR species (oxygen-centered radicals (ROS), nitrogen-centered radicals (RNS), carbon-centered radicals, and sulfur-centered radicals) can be recognized [82]. Superoxide anion ($O_2^{-\cdot}$) is considered the primary ROS and it can generate secondary ROS. It is generated primarily within the mitochondrial respiratory chain that is fundamental for the ATP production in mammalian cells. During the respiratory process, O_2 is utilized as an electron acceptor and completely reduced to water through the acquisition of four electrons:

$$O_2 + 4e^- + 4H^+ \rightarrow 2H_2O$$

When this process is realized through subsequent steps, radical formation is then possible:

$$\begin{aligned}
&O_2 + 1e^- + H^+ \leftrightarrow HO_2\cdot \leftrightarrow H^+ + O_2^{-\cdot} \\
&HO_2\cdot + 1e^- + H^+ \leftrightarrow H_2O_2 \\
&H_2O_2 + 1e^- + H^+ \leftrightarrow [H_3O_2] \leftrightarrow H_2O + \cdot OH \\
&\cdot OH + 1e^- + H^+ \leftrightarrow H_2O
\end{aligned}$$

Three intermediate products are generated from O_2 reduction: $O_2^{-\cdot}$, hydrogen peroxide (H_2O_2), and hydroxyl radical (·OH). The second one is the most stable and it can be accumulated in large quantities.

Besides mitochondrial respiratory chain, there are many other sources of FR generation, both endogenous and exogenous.

In cells, one-electron abstraction from molecules yields sulfur-, oxygen-, carbon-, and nitrogen-centered FRs. For example, the removal of one electron (and a proton, H^+) from a –SH group of a protein by a radical species (R·) yields a sulfur-centered FR. If R· had just one unpaired electron, the reaction would have converted it to a non-radical. Another example is the one-electron removal from bis-allylic C–H bonds of polyunsaturated fatty acids (PUFAs) that yields a carbon-centered FR.

Nitric oxide (NO) can also be an FR source because it contains an unpaired electron in the outer orbital. NO interacts with the heme prosthetic group of the soluble guanylate cyclase, prompting cyclic guanosine-monophosphate (cGMP) formation and activating cGMP-dependent ion channels and kinases. Nitric oxide synthase

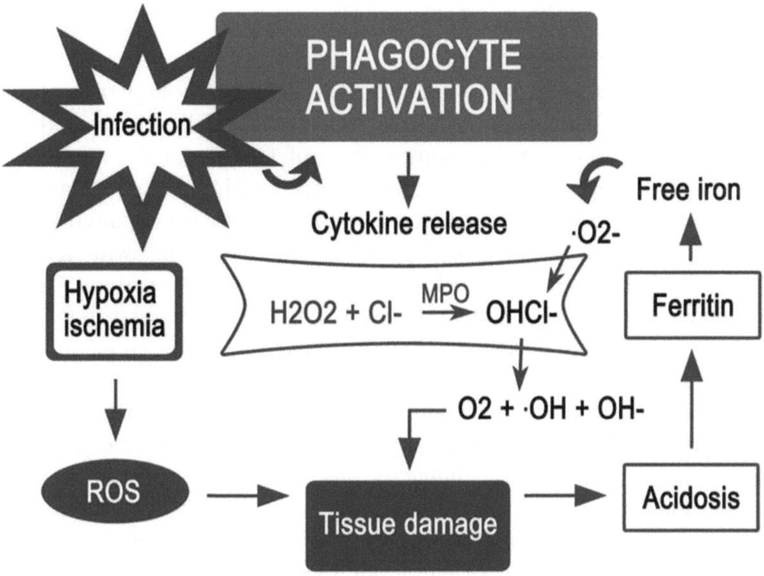

Fig. 21.1 Free radical generation and the tissue damage from infection

(NOS) catalyzes the formation of NO. It reacts relatively slowly with O_2, producing the orange–brown gas nitrogen dioxide ($\cdot NO_2$), a highly reactive FR [3].

Other potential endogenous sources include inflammatory cell activation (through nicotinamide adenine dinucleotide phosphate (NADPH)-reduced oxidase of phagocytes and some endothelial cells), monooxygenase system, nitric oxide synthase, and several other enzymes involved in the inflammatory process [5]. The burden of ROS can be further amplified by the presence of "free" metals, such as iron, copper, and manganese, that are released by metalloprotein complexes [63] (Fig. 21.1). Iron can damage tissues by catalyzing the conversion of superoxide and hydrogen peroxide to FR species through the Haber–Weiss and Fenton reactions when it is unbound to plasma proteins [43].

Additional endogenous sources of cellular FRs are activated neutrophils, eosinophils, and macrophages [29]. FRs can also be produced through many exogenous processes such as environmental agents and xenobiotics (metal ions, radiations, barbiturates) [8]. Stress factors, such as tumor necrosis factor alpha (TNF-α), can induce an increase in FR production with a redox cascade leading to the activation of both pro-survival and pro-cell-death factors [82]. Whatever the source of FRs, they are really dangerous because of their toxic effects able to damage all cells' components, including proteins, lipids, and DNA.

OS also causes a very wide spectrum of genetic, metabolic, and cellular responses. Many oxidative conditions are able to modulate gene expression, stimulate cell growth, or cause a protective temporary growth arrest [31]. Necrosis is the most extreme outcome and involves direct cell destruction.

Oxidative Stress in Intrauterine Life

Many studies confirmed the important role of ROS in the etiology of fetal diseases. The relative immaturity of the antioxidant system facilitates the exposure of embryos and fetuses to the damaging effects of OS. Prenatal hypoxic/ischemic injury and inflammatory/infective insults are specific triggers for an acute increase in FRs, generating an adverse intrauterine environment with impaired fetal development [18, 19]. OS has been demonstrated in pregnancies with fetal growth restriction [62]. Fetal growth restriction is often complicated by intrauterine hypoxia and impaired blood flow to the fetus. Intrauterine hypoxia may induce FR generation and fetal OS. It has been demonstrated that increased isoprostane concentrations, reliable markers of lipid peroxidation, in amniotic fluid indicate fetal growth restriction and also induce damage to amniotic epithelium and chorioamniotic collagen. This aspect is clarified by recent data demonstrating that F2-isoprostane concentrations are significantly higher in pregnancies with premature rupture of membranes than in normal ones [62]. FRs may disrupt amino acid binding in proteins and polyunsaturated fatty acids of the membrane lipid bilayers, causing cell dysfunction, modification of chorioamniotic biology, and predisposition to premature rupture of membranes (Fig. 21.2).

By favoring intracellular release of NPBI into plasma, asphyxia and acidosis supply redox-cycling iron, predisposing to OS [17, 21, 26, 28]. NPBI leads to the catalysis of $O_2^{-\cdot}$, H_2O_2, and the generation of the damaging $\cdot OH$. In the presence of free iron, huge increases in FR generation are possible, with the potential to cause

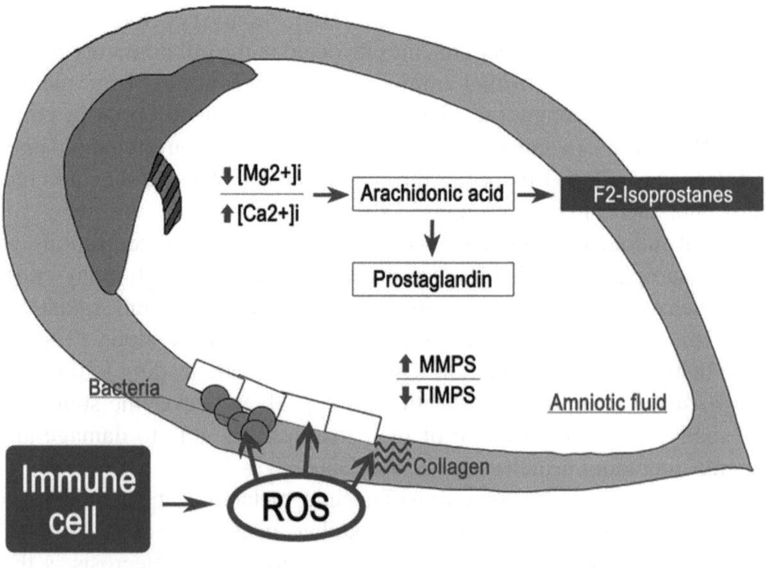

Fig. 21.2 Oxidative stress and premature rupture of membranes (*MMPS* matrix metalloproteinases, *TIMPS* tissue inhibitor of metalloproteinases)

tissue damage. Plasma NPBI may leak into the brain through a damaged barrier and is particularly damaging, being taken up directly by cells. When NPBI gains access to the extracellular space, its uptake by cells is enhanced by intracellular calcium and paradoxically also by increased levels of intracellular iron. Differentiating oligodendrocytes are particularly vulnerable to FR damage, being rich in iron which is required for differentiation [74].

A recent in vivo and ex vivo rat model of intrauterine growth restriction (IUGR) evidences the delays in oligodendrocyte differentiation and myelination probably due to bone morphogenetic protein 4 (BMP4) upregulation induced by OS. When BMP4 expression in oligodendrocyte increases, impaired differentiation occurs. A normal myelination has been observed abrogating BMP signaling [89].

Down syndrome comes from one chromosome 21 in excess in cellular karyotype. Superoxide dismutase (SOD) gene is localized on chromosome 21. This enzyme has the capacity to detoxify cells from superoxide anion in vivo with the participation of catalase and glutathione peroxidase. Consequently, an increased SOD production leads to high H_2O_2 generation, which can itself be toxic and also interfere with SOD activity [15]. An increased level of 8-iso-PGF2a isoprostane was found in the amniotic fluid of pregnancies with a Down syndrome fetus [81]. The immature oligodendroglial cells are glutathione peroxidase and catalase deficient so overexpression of SOD can be dangerous, instead of being protective. The early occurrence of OS in pregnancies with trisomy 21 and their subsequent oxidative damage as a major contributing factor in brain aging and cognitive function decline are likely due to the overexpression of SOD, coming from the supernumerary chromosome. SOD is also overexpressed in the immature brain, especially under stressful conditions (such as hypoxia) [72].

Furthermore, there are many data indicating that maternal diabetes during pregnancy may also induce OS in the newborn probably through the biochemical disturbance of the fetus [32, 68, 91]. In gestational diabetes mellitus (GDM) pregnancies, an FR overproduction was demonstrated; in addition, placental oxidation reactions are accelerated, and the radical scavenger function mechanisms are impaired due to the relative immaturity of the antioxidant system that facilitates the exposure of embryos and fetuses to the damaging effects of OS [56]. Interestingly, alterations in embryonic and fetoplacental development in experimental models of diabetes have been associated with FR increase in intrauterine tissues [56]. Evidence of increased FRs has been found in embryos, fetuses, and placentas in diabetes experimental models [34, 49, 73]. There is also evidence that several teratogens affect the developing embryo by increasing its OS and FR levels. There is a large number of pathways contributing to OS injury in diabetic pregnancies; hyperglycemia induces OS and causes cell and tissue damage through the polyol pathway, the formation of advanced glycation end products (AGE), the activation of protein kinase C (PKC), and the hexosamine pathway that leads to the formation of glucosamine 6-phosphate which competes with glucose-6 phosphate dehydrogenase limiting the synthesis of the reduced form of nicotinamide adenine dinucleotide phosphate (NADPH), necessary for reduced glutathione rebuilding. The enhanced FR production is also due to an incorrectly coupled electron transport in the mitochondrial respiratory chain by oxidative phosphorylation [56]. Finally, gestational diabetes pregnancies experience chronic hypoxia as reflected by increased erythropoietin concentrations in amniotic fluid near term. Moreover, erythropoietin levels significantly correlated with the concentration of oxidative and nitrosative stress biomarkers in amniotic fluid [35]. This pro-oxidant status may predispose newborns to poor postnatal adaptation and early neonatal complications. All these data clearly demonstrate the relevance of FRs as teratogenic agents in diabetic pregnancies, leading to oxidative and nitrosative stress in many fetal organs and to an impairment of antioxidant enzymes [56].

Finally, many studies demonstrated a delay of development or developmental arrest in mammalian preimplantation embryos cultured in vitro when compared to embryos that had developed in vivo [98]. The difference in O_2 concentration can be the cause of this phenomenon. In vivo, O_2 concentration in oviductal and uterine environment is lower (2–8 %) than those used for the culture of embryos in vitro, that is, atmospheric O_2 (approx. 20 %) [98].

FRs produced through different mechanisms highly affect embryo and fetus development, causing several different diseases with a common pathophysiology based on the antioxidant impairment and FR overproduction.

Oxidative Stress and Free Radical-Related Diseases in the Newborn

Newborns are particularly susceptible to OS; this is in part due to the rapid environment change from a poor (womb) to a relatively rich O_2 habitat. There are also other predisposing conditions such as the relative antioxidant systems' deficiency, the increased release of NPBI, and the exposure to hyperoxia, to hypoxic–ischemic conditions, and to inflammation events. The term "free radical disease" (FRD) was used to pull together all neonatal pathologies with a probable oxidative etiology.

Intraventricular Hemorrhage and Ischemic–Hypoxic Brain Injury

Intraventricular hemorrhage (IVH) in very preterm infants is a common disease associated with long-term consequences [107]. In Italy 20–25 % of all very low birth weight (VLBW) infants suffer from IVH. Importantly, 10–15 % of neonates of <1,500 g birth weight will experience the more severe grades of hemorrhage, and over three-quarters of these will develop mental retardation and/or cerebral palsy [65].

IVH has been attributed to changes in cerebral blood flow due to the immature germinal matrix microvasculature and secondary periventricular venous infarction [65]. Recently, more detailed analyses have demonstrated the role of OS in this context [82].

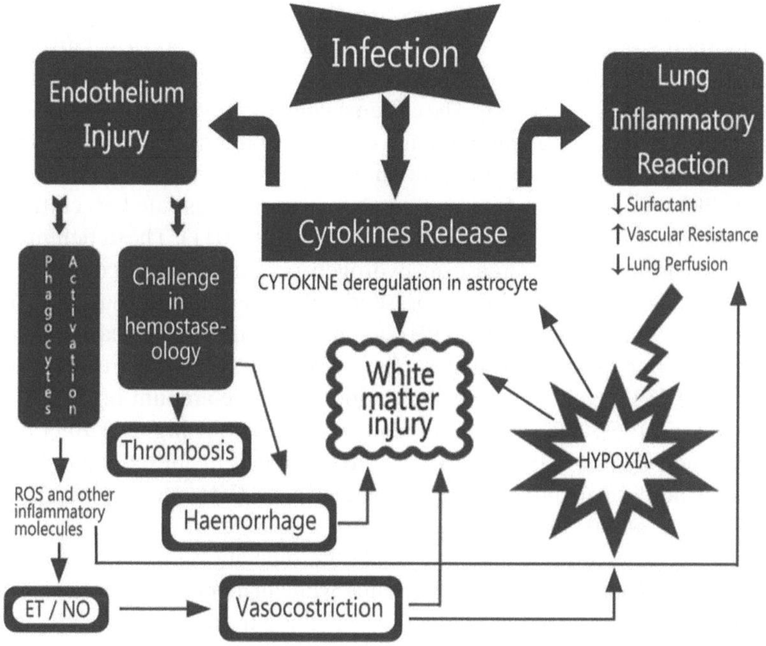

Fig. 21.3 Complex mechanisms involved in perinatal brain injury (*ET* endothelin, *NO* nitric oxide)

Important modulators of cerebral blood flow in the developing brain include the cyclooxygenase 2 (COX-2) system and prostaglandins (PGs). COX-2 expression is induced by hypoxia; hypotension; growth factors such as epidermal growth factor receptor and transforming growth factor β (TGFβ); and inflammatory modulators including IL-6, IL-1β, TNF-α, and NFkappaB [2]. The resultant prostanoids promote the production and release of vascular endothelial growth factor (VEGF), a potent angiogenic factor [55].

Those same triggers, which initiate hemorrhage into the germinal matrix, promote a cascade of processes leading to the disruption of tight junctions, increased blood–brain barrier permeability, and microglial activation within the developing periventricular white matter (Fig. 21.3). These events are mediated by cytokines (IL-1β and TNF-α), VEGF, and NO. Furthermore, hypoxia alone has been shown to alter the blood–brain barrier proteins. Finally, reactive microglia release FRs, which not only contribute to endothelial damage but also alter hemostasis and increase anaerobic metabolism [102].

The preterm brain is more susceptible to FRs than adult brain because of the immaturity of detoxifying enzyme systems but also for its high lipid content with an increasing amount of polyunsaturated fatty acids, high O_2 consumption, and consequently high capacity to generate FRs [7, 71]. In addition, FRs are released by activated microglia and are also generated following the activation of the COX-2 system. Because of their multifaceted effects on the developing vasculature, FRs are believed to play a significant role in periventricular parenchymal infarction [38].

The nature of the cells involved in OS-associated brain injury is currently unclear. In postmortem examination of premature infants, the cerebral oxidative damage particularly targeted the oligodendrocyte lineage, suggesting that these cells are the most vulnerable to OS injury [10]. The extreme vulnerability of pre-oligodendrocytes to FR attack is based on human brain and experimental models. A delay in production of SOD and catalase was found in human brains by Folkerth et al. [38], and a deficit in glutathione peroxidase and catalase was demonstrated in cultured pre-oligodendrocytes by Rosenberg et al. ([14], Volpe et al. 2011). The deficit in antioxidant systems leads to an increase in hydrogen peroxide that is converted to the dangerous hydroxyl radical in the presence of ferrous ion through the Fenton reaction in the white matter [74]. NPBI was also found to be increased in cerebral white matter after hypoxia–ischemia ([96], Volpe et al. 2011). Furthermore, the occurrence of IVH leads to the accumulation of a large amount of NPBI in cerebral spinal fluid [95]. All these data indicates that there is higher vulnerability of pre-oligodendrocytes to OS primarily due to the deficit in antioxidant systems but also to the increase in NPBI as a consequence of IVH and hypoxic–ischemic injury (Volpe et al. 2011).

Other vulnerable cell populations include the subplate neurons. Oxidative tissue damage was documented in the periventricular white matter of premature neonates, as well as the cerebral cortex of term infants succumbing from severe hypoxic–ischemic insults [10]. The cerebrospinal fluid 15-F2t-IsoP levels were significantly increased in one-third of infants with white matter injury [47].

In our previous study, we found increased plasma levels of total hydroperoxides (THs), advanced oxidation protein products (AOPPs), and particularly NPBI in newborns who developed IVH, suggesting that OS markers are direct indexes of increased production of FRs in the central nervous system as a response to oxidative neuronal damage [82]. Perinatal hypoxia alters mechanisms that regulate cerebral blood flow and triggers a cascade of biochemical events that begins with a shift from oxidative to anaerobic metabolism leading to oxidative brain damage [80]. Polyunsaturated fatty acids are constituents of lipidic membrane in white matter, and they are highly susceptible to FR attack. The lipid peroxidation that occurs following acute hypoxia in the fetal brain [23, 46] may be due to the generation of peroxynitrite (OONO), following the nonenzymatic combination of NO and superoxide [99]. Lipid peroxidation affects immature oligodendrocytes in particular in the preterm fetal brain [9] and may be a major factor in the white matter damage that can arise from hypoxic insult to the developing brain [7, 9, 14].

The propensity to brain oxidative injury in immature babies is related to deficient antioxidant defenses as already stated before but also to several pro-oxidant characteristics. Basically, the developing tissues have a high metabolic rate supported almost exclusively by oxidative metabolism, an excellent source for FRs, and a relatively high concentration of NPBI [66]. The fetus accumulates iron from maternal stores during pregnancy [40], suggesting a role of iron as a trophic factor at an early developmental stage, when growth is especially active. Despite the brain's highly regulated system for iron utilization and metabolism, iron overload is associated with saturation in protein involved in iron transport and storage, causing an increase in free ferrous iron [39]. In this case, iron is capable of causing degeneration of endothelial cells [108]. Endothelial cell injury and dysfunction may additionally contribute to the inflammatory response and alteration in coagulation, through loss of normal endothelial nitric oxide production [109]. Other potential implications of iron overload are acute impairment of endothelium-dependent flow-mediated vasodilation [104], loss of tight junction proteins and degeneration of endothelial cells, and opening of the blood–brain barrier [108]. The separation of endothelial tight junctions, the loss of endothelial attachment to the basement membrane, the endothelial blebbing, and endothelial necrosis have been described in the cerebral vasculature following ischemic injury [86]. The progression of endothelial dysregulation can contribute to the ongoing pathogenesis of IVH.

Retinopathy of Prematurity

Retinopathy of prematurity (ROP) is the major cause of visual impairment and blindness in premature neonates [90]. The overall incidence rate of the condition is about 68 % among infants born with birth weight less than 1,200 g and 98 % among children born with birth weight less than 750 g. The pathogenesis is multifactorial [97].

Normally the vascularization of the human retina is largely complete by the 4th month of gestation, but peripheral retinal vascularization is still immature until the fetus is near term. ROP occurs in two phases. The first phase starts when premature infants are exposed to a high O_2 concentration inside the incubator after birth. This condition of relative hyperoxia leads to the downregulation of VEGF expression and to a subsequent regression of developed retinal vessels. After the cessation of O_2 therapy, infants are returned to normal O_2 tension. In addition, the metabolic demand of the relatively vascular depleted retina is now higher. The increased requirements of O_2 pose the newborn's retina in a state of hypoxic injury that hesitates in the abnormal proliferation of vessels and leads to neovascularization; this is the second phase of ROP progression [90].

During the two phases of ROP, the retina is subjected to the overproduction of FRs, which activates NADPH oxidase and contributes to intravitreal neovascularization by the activation of signaling pathways such as JAK/STAT [45, 90]. NADPH oxidase is a major enzyme responsible for the release of superoxide radicals from macrophages to fight invading microorganisms [37]. This enzyme is activated by a number of stimuli relevant to ROP, including hypoxia [103], hyperoxia [75], and FRs [70, 77, 79, 103], and its activation can trigger disparate signaling pathways from endothelial apoptosis [50] to endothelial proliferation and angiogenesis [103]. Therefore, OS plays a crucial role in the pathogenesis of ROP [59]. Nitro-oxidation is also involved in the pathogenesis of ROP. A higher level of NOS, which contributes to the increase of NO production, is observed in neonatal retina exposed to hypoxia [77].

Nowadays, hyperoxia is considered a fundamental factor, with a direct relationship between high oxygen saturation and ROP [94]. Penn et al. have shown how the exposure to alternating hypoxia and hyperoxia causes severe proliferative retinopathy in the newborn rat [78]. Many other clinical studies describe how fluctuations in O_2 and increased inspired O_2 of infants are associated with a higher risk for severe ROP [25, 30, 64, 92, 100].

Our previous studies showed how O_2 administration in the delivery room was significantly associated with the development of ROP [84]. We also demonstrated elevated levels of TH, AOPP, and NPBI in cord blood of babies at risk for FRD and consequently of ROP [82].

Furthermore, it has been shown that the retinal antioxidant and antioxyenzyme contents are low in ROP cases [24]. Some protective effects have been shown by giving the potent antioxidant D-penicillamine and vitamin E [88]. In a later clinical study, vitamin E was proven to be effective in the prophylaxis of ROP development, and vitamin E supplementation is related to lower incidence of ROP [61]. Recently, interventions such as increasing retinal EPO and vitamin E supplements have been suggested [87].

Although we and others have found that the VEGF signaling pathway is not the only mechanism involved in the development of intravitreous neovascularization [33, 51, 54], it can be considered the dominant pathway in conditions in which hypoxia occurs. VEGF is produced by astrocytes in response to hypoxia [85], giving rise to retinal neovascularization at the junction between vascularized and nonvascularized retina [94]. VEGF levels are suppressed by hyperoxia with a stop of the normal retinal vascularization and loss of some developing vessels. In addition, fetal insulin-like growth factor-1 (IGF-1) precipitously falls after premature birth [60]. The action of VEGF depends on IGF-1. Subsequently, with the retinal development, the O_2 need increases, and the result is a hypoxic environment with a consequent VEGF production. This fact, assisted by IGF-1, leads to the neovascularization process.

Bronchopulmonary Dysplasia

Bronchopulmonary dysplasia (BPD) of the newborn is one of the most important diseases influencing perinatal mortality and morbidity. BPD occurs almost exclusively in preterm infants. O_2 toxicity, FRs, prematurity, surfactant deficit, and inflammation are major factors contributing to its pathogenesis [52]. Inadequate nutrition and how the baby is ventilated contribute to the increase of OS which may trigger oxidative changes leading to permanent lung damage [93].

Phagocytic cells in the lung mediate their antimicrobial functions through the release of lysozymes, peroxidases, and proteases, in addition to ROS and NOS.

Activated neutrophils and pulmonary type II cells are also important inducers of the Fenton reaction and of the increase of adhesiveness to the endothelium. The release of inflammatory mediators can stimulate the endothelium to promote transendothelial migration, facilitating the passage of cytokines [36]. The increase in phagocyte number and interleukin concentrations in bronchoalveolar lavage fluid of infants with BPD indicates that O_2 toxicity and inflammation are involved in the development of lung injury [42].

BPD is characterized by a tissue remodeling process, which has been divided into different phases, ending up in the chronic phase with an increased number of fibroblasts and fibrotic areas. Matrix metalloproteins (MMPs) are important in regulating fibrotic processes; they degrade extracellular matrix proteins and fibrillar collagen. In human cells, the balance between MMPs and their inhibitors is of great importance in the development and regulation of fibrosis. The expression of MMPs is regulated by cytokines, growth factors, and extracellular matrix components. OS increases both MMPs and their inhibitors, so it may lead to lung damage by increasing collagenase activity, causing disruption of the extracellular matrix [93]. OS is also able to inactivate surfactant, as demonstrated in rats treated with the combination of hyperoxia and hypoxanthine infusion; as a result of inflammation and edema, transudated plasma proteins and inflammatory cells impair extracellular surfactant. Hyperoxia results in an increase of the surfactant proteins (SP) SP-A, SP-B, and SP-C85 while oxidized SP-A loses its surfactant and immune defense functions. Hyperoxia also reduces surfactant phospholipid production [44]. It seems that glycerol-3 phosphate acyltransferase, which catalyzes the first reaction in phosphoglyceride synthesis, is a rate-limiting enzyme, extremely sensitive to oxidative damage [93].

An increased concentration of products of lipid peroxidation, pentane and ethane, in exhaled gas was demonstrated a few days after birth in infants who subsequently developed BPD [105]. A positive correlation was found between the duration of O_2 ventilatory therapy and lipid peroxidation by Inder et al. [48].

Protein oxidation was also demonstrated by Gladstone and Levine with higher carbonyl content in lung lavage samples from preterm infants with an O_2 requirement of over 40 % or who were ventilated for longer than 72 h compared with those requiring less O_2 or less time on mechanical ventilation [41]. In tracheal aspirates from premature infants requiring O_2 therapy, Varsila et al. showed that those who subsequently developed BPD had a significantly higher carbonyl concentration during the first 6 days of life than those who did not [106].

Necrotizing Enterocolitis

Necrotizing enterocolitis (NEC) is a gastrointestinal surgical emergency in preterm neonates [12]. The overall incidence is 1–5 per 1,000 live births, dramatically increasing to 100 per 1,000 in very low birth weight and to 200 per 1,000 in extremely low birth weight infants [76]. NEC has a multifactorial etiology: ischemia, infections, cytokines, enteral feeding, and FRs may contribute to the disruption of the protective gut barrier.

In particular, FRs are generated as a result of various injury to the gut, in premature infants. Ischemia, hypoxia–reperfusion, infection, and inflammation are mechanisms capable of producing high levels of FRs, perturbing the normal redox balance and shifting cells into a state of OS [6].

Some authors recently proposed that the underlying initial condition is the reduced ability of the neonatal gut epithelial cells to reduce OS and that when epithelial gut cells are exposed to enteral feeding, the increased metabolic OS tips the population toward apoptosis, inflammation, bacterial activation, and eventual necrosis [53]. Recent findings suggested that the mitochondria are the major sources of intestinal apoptotic signaling during OS and that by modulating mitochondrial apoptosis, it is possible to prevent NEC damaging effects [13].

Recently, we found a strong association between NEC and cord blood concentration of AOPP, NPBI, and TH, showing a clear correlation between intrauterine OS events and the risk of developing NEC [83].

The immaturity of the gastrointestinal tract in preterm infants may also contribute to NEC development [58]. The decreased intestinal peristalsis exposes the intestinal epithelium to noxious substances, for example, Toll-like receptor-4 expression is downregulated in the mature intestinal epithelium under stimulation by gram-negative lipopolysaccharides but is increased in the immature intestinal epithelium, leading to an exaggerated proinflammatory response through the upregulation of the NF-kB pathway [27]. Moreover, prolonged antibiotic exposure is associated with an increased risk of NEC [58], and this association persisted in multivariate analyses excluding confounding factors (gestational age, birth weight, and sepsis) [4]. In addition, preterm infants frequently suffer from ischemic events, such as hypotension, hypothermia, anemia, and patent ductus arteriosus, during the intensive care stay. Severe anemia results in insufficient O_2 supply for the increased requirements of the mesenteric vessels after enteral feeding. Ischemia leads to endothelial cell dysfunction and alters the endothelin-1/nitric oxide balance in favor of vasoconstriction, causing expansion of ischemic intestinal lesions [69]. Other risk factors are sensitization to cow milk proteins [1] and polymorphisms, such as mutations in carbamoyl phosphate synthetase [67], vascular endothelial growth factor [11], and interleukins 10 and 12 [101].

Further studies providing a rational basis for the formulation of interventions to interrupt those dangerous mechanisms and elucidating the role of perinatal oxidative insults in injured intestinal epithelial cells are needed.

Conclusion

The existence of a redox homeostasis is essential for normal health and survival of the cell. OS occurs when there is an imbalance between pro-oxidant and antioxidant factors; this process leads to cellular and tissue damage. The newborn, especially if preterm, is highly prone to OS and to the toxic effect of FRs. The challenge for the future is to develop new effective antioxidant therapies and to demonstrate their benefits for patients' treatment. Many efforts have been already done in elucidating the differential responses to initial insults in the fetus and newborns. Nevertheless, a lot of work still has to be done to identify the exact target population for the new antioxidant strategies (moderate or severe HIE, premature infants, mother during pregnancy, etc.) as well as for the proper timing and dosing. However, longitudinal studies evaluating the panel of OS biomarkers and elucidating the molecular mechanisms that engender OS in the perinatal period are needed before antioxidant therapies can be accepted in clinical practice.

References

1. Abdelhamid AE, Chuang SL, Hayes P, Fell JM. In vitro cow's milk protein-specific inflammatory and regulatory cytokine responses in preterm infants with necrotizing enterocolitis and sepsis. Pediatr Res. 2011;69:165–9.
2. Ackerman 4th WE, Rovin BH, Kniss DA. Epidermal growth factor and interleukin-1beta utilize divergent signaling pathways to synergistically upregulate cyclooxygenase-2 gene expression in human amnion-derived WISH cells. Biol Reprod. 2004;71:2079–86.
3. Alderton WK, Cooper CE, Knowles RG. Nitric oxide synthases: structure, function and inhibition. Biochem J. 2001;357:593–615.
4. Alexander VN, Northrup V, Bizzarro MJ. Antibiotic exposure in the newborn intensive care unit and the risk of necrotizing enterocolitis. J Pediatr. 2011;159:392–7.
5. Auten RL, Davis JM. Oxygen toxicity and reactive oxygen species: the devil is in the details. Pediatr Res. 2009;66:121–7.
6. Aydemir C, Dilli D, Uras N, et al. Total oxidant status and oxidative stress are increased in infants with necrotizing enterocolitis. J Pediatr Surg. 2011;46:2096–100.
7. Baburamani AA, Ek CJ, Walker DW, Castillo-Melendez M. Vulnerability of the developing brain to hypoxic-ischemic damage: contribution of the cerebral vasculature to injury and repair? Front Physiol. 2012;3:424.
8. Bachowski S, Kolaja KL, Xu Y, et al. Role of oxidative stress in the mechanism of dieldrin's hepatotoxicity. Ann Clin Lab Sci. 1997;27:196–209.
9. Back SA, Gan X, Li Y, Rosenberg PA, Volpe JJ. Maturation-dependent vulnerability of oligodendrocytes to oxidative stress-induced death caused by glutathione depletion. J Neurosci. 1998;18(16):6241–53.
10. Back SA, Luo NL, Mallinson RA, et al. Selective vulnerability of preterm white matter to oxidative damage defined by F2-isoprostanes. Ann Neurol. 2005;58:108–20.
11. Bányász I, Bokodi G, Vásárhelyi B, et al. Genetic polymorphisms for vascular endothelial growth factor in perinatal complications. Eur Cytokine Netw. 2006;17:266–70.
12. Baregamian N, Song J, Bailey CE, Papaconstantinou J, Evers BM, Chung DH. Tumor necrosis factor-alpha and apoptosis signal-regulating kinase 1 control reactive oxygen species release, mitochondrial autophagy, and c-Jun N-terminal kinase/p38 phosphorylation during necrotizing enterocolitis. Oxid Med Cell Longev. 2009;2:297–306.
13. Baregamian N, Song J, Papaconstantinou J, Hawkins HK, Evers BM, Chung DH. Intestinal mitochondrial apoptotic signaling is activated during oxidative stress. Pediatr Surg Int. 2011;27:871–7.
14. Baud O, Greene AE, Li J, Wang H, Volpe JJ, Rosenberg PA. Glutathione peroxidase-catalase cooperativity is required for resistance to hydrogen peroxide by mature rat oligodendrocytes. J Neurosci. 2004;24(7):1531–40.
15. Bray RC, Cockle SA, Fielden EM, Roberts PB, Rotilio G, Calabrese L. Reduction and inactivation of superoxide dismutase by hydrogen peroxide. Biochem J. 1974;139(1):43–8.
16. Buonocore G, Groenendaal F. Anti-oxidant strategies. Semin Fetal Neonatal Med. 2007; 12:287–95.
17. Buonocore G, Perrone S. Biomarkers of hypoxic brain injury in the neonate. Clin Perinatol. 2004;31(1):107–16.
18. Buonocore G, Perrone S, Longini M, et al. Oxidative stress in preterm neonates at birth and on the seventh day of life. Pediatr Res. 2002;52:46–9.

19. Buonocore G, Perrone S, Longini M, Terzuoli L, Bracci R. Total hydroperoxide and advanced oxidation protein products in preterm hypoxic babies. Pediatr Res. 2000;47:221–4.
20. Buonocore G, Perrone S, Tataranno ML. Oxygen toxicity: chemistry and biology of reactive oxygen species. Semin Fetal Neonatal Med. 2010;15:186–90.
21. Buonocore G, Zani S, Perrone S, Caciotti B, Bracci R. Intraerythrocyte nonprotein-bound iron and plasma malondialdehyde in the hypoxic newborn. Free Radic Biol Med. 1998;25:766–70.
22. Byfield G, Budd S, Hartnett ME. The role of supplemental oxygen and JAK/STAT signaling in intravitreous neovascularization in a ROP rat model. Invest Ophthalmol Vis Sci. 2009; 50:3360–5.
23. Castillo-Meléndez M, Chow JA, Walker DW. Lipid peroxidation, caspase-3 immunoreactivity, and pyknosis in late-gestation fetal sheep brain after umbilical cord occlusion. Pediatr Res. 2004;55(5):864–71.
24. Chen W, Hunt DM, Lu H, Hunt RC. Expression of antioxidant protective proteins in the rat retina during prenatal and postnatal development. Invest Ophthalmol Vis Sci. 1999; 40:744–51.
25. Chow LC, Wright KW, Sola A. Can changes in clinical practice decrease the incidence of severe retinopathy of prematurity in very low birth weight infants? Pediatrics. 2003; 111:339–45.
26. Ciccoli L, Rossi V, Leoncini S, et al. Iron release in erythrocytes and plasma non protein-bound iron in hypoxic and non hypoxic newborns. Free Radic Res. 2003;37:51–8.
27. Claud EC, Zhang X, Petrof EO, Sun J. Developmentally regulated tumor necrosis factor-alpha induced nuclear factor-kappaB activation in intestinal epithelium. Am J Physiol Gastrointest Liver Physiol. 2007;292:G1411–9.
28. Comporti M, Signorini C, Buonocore G, Ciccoli L. Iron release, oxidative stress and erythrocyte ageing. Free Radic Biol Med. 2002;32(7):568–76.
29. Conner EM, Grisham MB. Inflammation, free radicals, and antioxidants. Nutrition. 1996; 12:274–7.
30. Cunningham S, Fleck BW, Elton RA, McIntosh N. Transcutaneous oxygen levels in retinopathy of prematurity. Lancet. 1995;346:1464–5.
31. Davies KJ. Oxidative stress, antioxidant defenses, and damage removal, repair, and replacement systems. IUBMB Life. 2000;50:279–89.
32. Dennery PA. Effects of oxidative stress on embryonic development. Birth Defects Res C Embryo Today. 2007;81:155–62.
33. Dorrell MI, Aguilar E, Scheppke L, Barnett FH, Friedlander M. Combination angiostatic therapy completely inhibits ocular and tumor angiogenesis. Proc Natl Acad Sci U S A. 2007;104:967–72.
34. Eriksson UJ, Cederberg J, Wentzel P. Congenital malformations in offspring of diabetic mothers—animal and human studies. Rev Endocr Metab Disord. 2003;4:79–93.
35. Escobar J, Teramo K, Stefanovic V, Andersson S, Asensi MA, Arduini A, et al. Amniotic fluid oxidative and nitrosative stress biomarkers correlate with fetal chronic hypoxia in diabetic pregnancies. Neonatology. 2013;103(3):193–8.
36. Fellman V. Respiratory distress syndrome of neonates today. Duodecim. 1997;113:1024–31.
37. Fialkow L, Wang Y, Downey GP. Reactive oxygen and nitrogen species as signaling molecules regulating neutrophil function. Free Radic Biol Med. 2007;42:153–64.
38. Folkerth RD, Keefe RJ, Haynes RL, Trachtenberg FL, Volpe JJ, Kinney HC. Interferon-gamma expression in periventricular leukomalacia in the human brain. Brain Pathol. 2004; 14:265–74.
39. Gaasch JA, Lockman PR, Geldenhuys WJ, Allen DD, Van der Schyf CJ. Brain iron toxicity: differential responses of astrocytes, neurons, and endothelial cells. Neurochem Res. 2007;32:1196–208.
40. Gazzolo D, Perrone S, Paffetti P, et al. Non protein bound iron concentrations in amniotic fluid. Clin Biochem. 2005;38:674–7.
41. Gladstone Jr IM, Levine RL. Oxidation of proteins in neonatal lungs. Pediatrics. 1994; 93:764–8.

42. Groneck P, Speer CP. Interleukin-8 in pulmonary effluent fluid of preterm infants. J Pediatr. 1993;123:839–40.
43. Gutteridge JM. The role of superoxide and hydroxyl radicals in phospholipid peroxidation catalysed by iron salts. FEBS Lett. 1982;150:454–8.
44. Haagsman HP. Interactions of surfactant protein A with pathogens. Biochim Biophys Acta. 1998;1408:264–77.
45. Hartnett ME. The effects of oxygen stresses on the development of features of severe retinopathy of prematurity: knowledge from the 50/10 OIR model. Doc Ophthalmol. 2010; 120:25–39.
46. Ikeda K, Toda M, Tanaka K, Tokumaru S, Kojo S. Increase of lipid hydroperoxides in liver mitochondria and inhibition of cytochrome oxidase by carbon tetrachloride intoxication in rats. Free Radic Res. 1998;28(4):403–10.
47. Inder T, Mocatta T, Darlow B, Spencer C, Volpe JJ, Winterbourn C. Elevated free radical products in the cerebrospinal fluid of VLBW infants with cerebral white matter injury. Pediatr Res. 2002;52:213–8.
48. Inder TE, Graham P, Sanderson K, Taylor BJ. Lipid peroxidation as a measure of oxygen free radical damage in the very low birthweight infant. Arch Dis Child Fetal Neonatal Ed. 1994;70:F107–11.
49. Jawerbaum A, Gonzalez E. Diabetic pregnancies: the challenge of developing in a pro-inflammatory environment. Curr Med Chem. 2006;13:2127–38.
50. Jay Forman H, Torres M. Redox signaling in macrophages. Mol Aspects Med. 2001;22: 189–216.
51. Jo N et al. Inhibition of platelet-derived growth factor B signaling enhances the efficacy of anti-vascular endothelial growth factor therapy in multiple models of ocular neovascularization. Am J Pathol. 2006;168:2036–53.
52. Joung KE, Kim HS, Lee J, et al. Correlation of urinary inflammatory and oxidative stress markers in very low birth weight infants with subsequent development of bronchopulmonary dysplasia. Free Radic Res. 2011;45:1024–32.
53. Kim M, Christley S, Alverdy JC, Liu D, An G. Immature oxidative stress management as a unifying principle in the pathogenesis of necrotizing enterocolitis: insights from an agent-based model. Surg Infect. 2012;13(1):18–32.
54. Kramerov AA et al. Expression of protein kinase CK2 in astroglial cells of normal and neovascularized retina. Am J Pathol. 2006;168:1722–36.
55. Kuwano T, Nakao S, Yamamoto H, et al. Cyclooxygenase 2 is a key enzyme for inflammatory cytokine-induced angiogenesis. FASEB J. 2004;18:300–10.
56. Lappas M, Hiden U, Desoye G, Froehlich J, Hauguel-de Mouzon S, Jawerbaum A. The role of oxidative stress in the pathophysiology of gestational diabetes mellitus. Antioxid Redox Signal. 2011;15(12):3061–100.
57. Lee J, Giordano S, Zhang J. Autophagy, mitochondria and oxidative stress: cross-talk and redox signalling. Biochem J. 2012;441:523–40.
58. Lee JH. An update on necrotizing enterocolitis: pathogenesis and preventive strategies. Korean J Pediatr. 2011;54:368–72.
59. Li SY, Fu ZJ, Lo AC. Hypoxia-induced oxidative stress in ischemic retinopathy. Oxid Med Cell Longev. 2012;2012:426769.
60. Lineham JD, Smith RM, Dahlenburg GW, et al. Circulating insulin-like growth factor I levels in newborn premature and full-term infants followed longitudinally. Early Hum Dev. 1986; 13:37–46.
61. Liu PM, Fang PC, Huang CB, et al. Risk factors of retinopathy of prematurity in premature infants weighing less than 1600 g. Am J Perinatol. 2005;22:115–20.
62. Longini M, Perrone S, Kenanidis A, Vezzosi P, Marzocchi B, Petraglia F, et al. Isoprostanes in amniotic fluid: a predictive marker for fetal growth restriction in pregnancy. Free Radic Biol Med. 2005;38(11):1537–41.
63. McCarthy SM, Bove PF, Matthews DE, Akaike T, van der Vliet A. Nitric oxide regulation of MMP-9 activation and its relationship to modifications of the cysteine switch. Biochemistry. 2008;47:5832–40.

64. McColm JR, Cunningham S, Wade J, et al. Hypoxic oxygen fluctuations produce less severe retinopathy than hyperoxic fluctuations in a rat model of retinopathy of prematurity. Pediatr Res. 2004;55:107–13.
65. McCrea HJ, Ment LR. The diagnosis, management, and postnatal prevention of intraventricular hemorrhage in the preterm neonate. Clin Perinatol. 2008;35:777–92.
66. McQuillen PS, Ferriero DM. Selective vulnerability in the developing central nervous system. Pediatr Neurol. 2004;30:227–35.
67. Moonen RM, Paulussen AD, Souren NY, Kessels AG, Rubio-Gozalbo ME, Villamor E. Carbamoyl phosphate synthetase polymorphisms as a risk factor for necrotizing enterocolitis. Pediatr Res. 2007;62:188–90.
68. Myatt L, Cui X. Oxidative stress in the placenta. Histochem Cell Biol. 2004;122:369–82.
69. Nankervis CA, Giannone PJ, Reber KM. The neonatal intestinal vasculature: contributing factors to necrotizing enterocolitis. Semin Perinatol. 2008;32:83–91.
70. Niesman MR, Johnson KA, Penn JS. Therapeutic effect of liposomal superoxide dismutase in an animal model of retinopathy of prematurity. Neurochem Res. 1997;22:597–605.
71. O'Brien JS, Sampson EL. Lipid composition of the normal human brain: gray matter, white matter, and myelin. J Lipid Res. 1965;6(4):537–44.
72. Okado-Matsumoto A, Fridovich I. Subcellular distribution of superoxide dismutases (SOD) in rat liver: Cu, Zn-SOD in mitochondria. J Biol Chem. 2001;276:38388–93.
73. Ornoy A. Embryonic oxidative stress as a mechanism of teratogenesis with special emphasis on diabetic embryopathy. Reprod Toxicol. 2007;24:31–41.
74. Ozawa H, Nishida A, Mito T, Takashima S. Development of ferritin-containing cells in the pons and cerebellum of the human brain. Brain Dev. 1994;16(2):92–5.
75. Parinandi NL, Kleinberg MA, Usatyuk PV, et al. Hyperoxia-induced NAD(P)H oxidase activation and regulation by MAP kinases in human lung endothelial cells. Am J Physiol Lung Cell Mol Physiol. 2003;284:L26–38.
76. Parravicini E, Fromm F. Necrotizing enterocolitis. In: Buonocore G, editor. Neonatology a practical approach to neonatal management. Milano: Springer; 2011. p. 724–30.
77. Penn JS. Oxygen-induced retinopathy in the rat: vitamins C and E as potential therapies. Invest Ophthalmol Vis Sci. 1992;33:1836–45.
78. Penn JS, Henry MM, Tolman BL. Exposure to alternating hypoxia and hyperoxia causes severe proliferative retinopathy in the newborn rat. Pediatr Res. 1994;36:724–31.
79. Penn JS, Tolman BL, Bullard LE. Effect of a water-soluble vitamin E analog, Trolox C, on retinal vascular development in an animal model of retinopathy of prematurity. Free Radic Biol Med. 1997;22:977–84.
80. Perrone S, Bracci R, Buonocore G. New biomarkers of fetal-neonatal hypoxic stress. Acta Paediatr Suppl. 2002;91:135–8.
81. Perrone S, Longini M, Bellieni CV, Centini G, Kenanidis A, De Marco L, et al. Early oxidative stress in amniotic fluid of pregnancies with Down syndrome. Clin Biochem. 2007;40:177–80.
82. Perrone S, Tataranno ML, Negro S, et al. Early identification of the risk for free radical-related diseases in preterm newborns. Early Hum Dev. 2010;86:241–4.
83. Perrone S, Tataranno ML, Negro S, et al. May oxidative stress biomarkers in cord blood predict the occurrence of necrotizing enterocolitis in preterm infants? J Matern Fetal Neonatal Med. 2012;1:128–31.
84. Perrone S, Vezzosi P, Longini M, et al. Biomarkers of oxidative stress in babies at high risk for retinopathy of prematurity. Front Biosci. 2009;1:547–52.
85. Pierce EA, Foley ED, Smith LE. Regulation of vascular endothelial growth factor by oxygen in a model of retinopathy of prematurity. Arch Ophthalmol. 1996;114:1219–28.
86. Pomfy M, Húska J. The state of the microcirculatory bed after total ischaemia of the brain. An experimental ultrastructural study. Funct Dev Morphol. 1992;2:253–8.
87. Raghuveer TS, Bloom BT. A paradigm shift in the prevention of retinopathy of prematurity. Neonatology. 2011;100:116–29.
88. Raju TN, Langenberg P, Bhutani V, Quinn GE. Vitamin E prophylaxis to reduce retinopathy of prematurity: a reappraisal of published trials. J Pediatr. 1997;131:844–50.

89. Reid MV, Murray KA, Marsh ED, Golden JA, Simmons RA, Grinspan JB. Delayed myelination in an intrauterine growth retardation model is mediated by oxidative stress upregulating bone morphogenetic protein 4. J Neuropathol Exp Neurol. 2012;71(7):640–53.
90. Rivera JC, Sapieha P, Joyal JS, et al. Understanding retinopathy of prematurity: update on pathogenesis. Neonatology. 2011;100:343–53.
91. Ryu S, Kohen R, Samuni A, Ornoy A. Nitroxide radicals protect cultured rat embryos and yolk sacs from diabetic-induced damage. Birth Defects Res A Clin Mol Teratol. 2007; 79:604–11.
92. Saito Y, Omoto T, Cho Y, et al. The progression of retinopathy of prematurity and fluctuation in blood gas tension. Graefes Arch Clin Exp Ophthalmol. 1993;231:151–6.
93. Saugstad OD. Bronchopulmonary dysplasia-oxidative stress and antioxidants. Semin Neonatol. 2003;8:39–49.
94. Saugstad OD. Oxygen and retinopathy of prematurity. J Perinatol. 2006;26:S46–50.
95. Savman K, Nilsson UA, Blennow M, Kjellmer I, Whitelaw A. Non-protein-bound iron is elevated in cerebrospinal fluid from preterm infants with posthemorrhagic ventricular dilatation. Pediatr Res. 2001;49(2):208–12.
96. Sävman K, Nilsson UA, Thoresen M, Kjellmer I. Non-protein-bound iron in brain interstitium of newborn pigs after hypoxia. Dev Neurosci. 2005;27(2–4):176–84.
97. Shastry BS. Genetic susceptibility to advanced retinopathy of prematurity (ROP). J Biomed Sci. 2010;17:69.
98. Takahashi M. Oxidative stress and redox regulation on in vitro development of mammalian embryos. J Reprod Dev. 2012;58(1):1–9.
99. Tan S, Zhou F, Nielsen VG, Wang Z, Gladson CL, Parks DA. Sustained hypoxia-ischemia results in reactive nitrogen and oxygen species production and injury in the premature fetal rabbit brain. J Neuropathol Exp Neurol. 1998;57(6):544–53.
100. Tin W, Milligan DWA, Pennefather PM, Hey E. Pulse oximetry, severe retinopathy, and outcome at one year in babies of less than 28 weeks gestation. Arch Dis Child Fetal Neonatal Ed. 2001;84:106–10.
101. Treszl A, Kaposi A, Hajdú J, Szabó M, Tulassay T, Vásárhelyi B. The extent to which genotype information may add to the prediction of disturbed perinatal adaptation: none, minor, or major? Pediatr Res. 2007;62:610–4.
102. Ulfig N, Bohl J, Neudörfer F, Rezaie P. Brain macrophages and microglia in human fetal hydrocephalus. Brain Dev. 2004;26:307–15.
103. Ushio-Fukai M. Redox signaling in angiogenesis: role of NADPH oxidase. Cardiovasc Res. 2006;71:226–35.
104. Van Buren P, Velez RL, Vaziri ND, Zhou XJ. Iron overdose: a contributor to adverse outcomes in randomized trials of anemia correction in CKD. Int Urol Nephrol. 2011;44(2): 499–507.
105. Varsila E, Hallman M, Andersson S. Free-radical-induced lipid peroxidation during the early neonatal period. Acta Paediatr. 1994;83:692–5.
106. Varsila E, Pesonen E, Andersson S. Early protein oxidation in the neonatal lung is related to development of chronic lung disease. Acta Paediatr. 1995;84:1296–9.
107. Volpe JJ. Perinatal brain injury: from pathogenesis to neuroprotection. Ment Retard Dev Disabil Res Rev. 2001;7:56–64.
108. Won SM, Lee JH, Park UJ, Gwag J, Gwag BJ, Lee YB. Iron mediates endothelial cell damage and blood-brain barrier opening in the hippocampus after transient forebrain ischemia in rats. Exp Mol Med. 2011;43:121–8.
109. Yang Y, Loscalzo J. Regulation of tissue factor expression in human microvascular endothelial cells by nitric oxide. Circulation. 2000;101:2144–8.

Chapter 22
New Antioxidant Drugs

Giuseppe Buonocore, Serafina Perrone, and Maria Luisa Tataranno

Introduction

Oxidative stress (OS) in vivo is a degenerative process due to overproduction of free radicals (FR) and propagation of their reactions. OS exists and tissue damage is possible when there are low levels of antioxidants or increased FR activity [1].

Newborns and particularly preterm infants are at high risk for OS and damage due to their organs' structural and functional immaturity, the overloading of aerobic metabolism with rapidly growing energy demand, and the presence of conditions leading to increased free iron levels with excessive FR production (i.e., hypoxia, inflammation, need of blood transfusions) (Fig. 22.1). Neonatal plasma has profoundly disturbed antioxidant profiles with low levels of glutathione peroxidase activity, superoxide dismutase, β-carotene, riboflavin, α-proteinase, vit. E, selenium, copper, zinc, ceruloplasmin, transferrin, and other antioxidant plasma factors [2].

In the developing fetus, hypoxia results in an increased anaerobic metabolism, leading to a rapid rise in levels of lactic acid and reduced forms of electron transport chain components in the mitochondria, with subsequent production of oxygen FR [3]. Other mechanisms contributing to reactive oxygen FR formation and membrane lipid peroxidation include phagocyte activation, metabolism of arachidonic acid through the cyclooxygenase and lipoxygenase pathways, and reactions catalyzed by increased free intracellular Fe^{++} and increased xanthine oxidase activity as a result of increased degradation of ATP [4]. Excess free iron and deficient iron-binding metabolizing capacity are additional features favoring OS in the premature infants [5].

G. Buonocore, M.D. (✉) • S. Perrone, M.D., Ph.D. • M.L. Tataranno, M.D.
Department of Molecular and Developmental Medicine,
University of Siena, V.le Bracci 36, 53100 Siena, Italy
e-mail: giuseppe.buonocore@unisi.it; saraspv@yahoo.it; marialuisatataranno@yahoo.it

P.A. Dennery et al. (eds.), *Perinatal and Prenatal Disorders*,
Oxidative Stress in Applied Basic Research and Clinical Practice,
DOI 10.1007/978-1-4939-1405-0_22

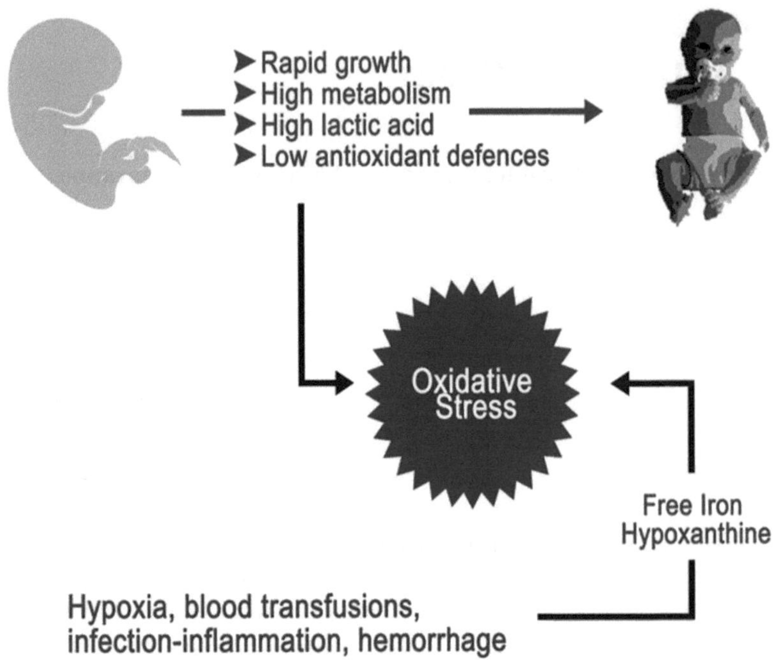

Fig. 22.1 Neonatal susceptibility to oxidative stress and damage

The properties and complex role of reactive oxygen species in the development of diseases make antioxidant therapy very difficult to be realized. Virtually, every cellular and extracellular molecular component is potentially sensitive to damage caused by OS. A critical aspect of protection is the prevention of the hypoxic-ischemic insult, and because most infants appear to experience the primary insult in utero, prevention of intrauterine asphyxia is of great importance.

Insights into the biochemical and cellular mechanisms of cellular injury with perinatal hypoxic-ischemic-reperfusion insults provide a rational basis for the formulation of interventions to interrupt those mechanisms and thereby prevent or ameliorate the injury. The principal mechanisms of cell death following a hypoxic-ischemic event involve the accumulation of cytosolic calcium and activation of a variety of calcium-mediated deleterious events, especially including the generation of FR.

Antioxidant substances may act by decompartmentalizing metal complexes, limiting FR production, modifying antiradical defenses, enhancing intracellular or extracellular antioxidant levels, incorporating lipophilic antioxidants into membranes, or scavenging superoxide [6] (Fig. 22.2). Some drugs inhibit phagocyte activation or xanthine oxidase and arachidonic acid metabolism or directly scavenge FR or repair FR-induced membrane injury, like calcium antagonists and beta blockers. Elimination of transition metals and especially non-protein bound iron (NPBI)

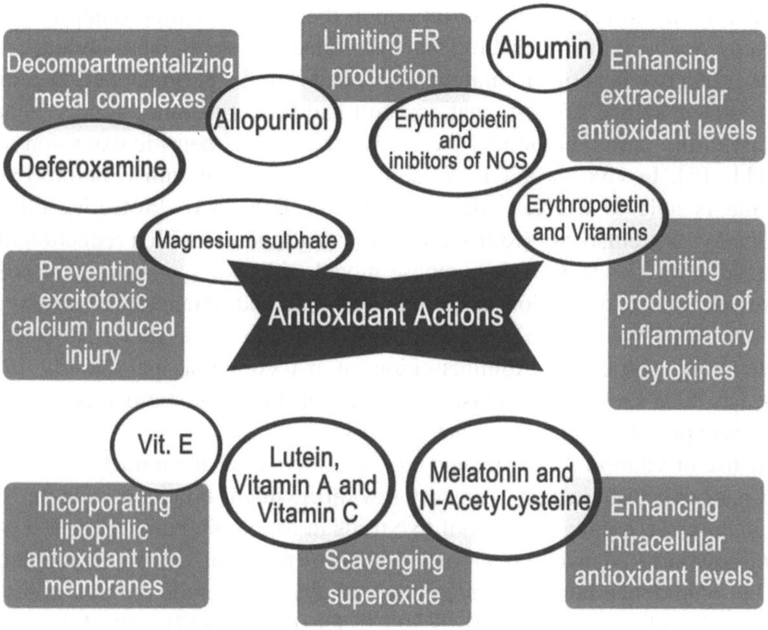

Fig. 22.2 Mechanism of action of some antioxidant substances

is crucial for interrupting the formation of FR. It can be achieved by iron chelators, e.g., deferoxamine, which have been successfully used in neonatal animals to limit oxidative damage [7, 8].

Although many antioxidant agents have been shown to be protective in animal models of hypoxia-ischemia, only few substances have been used in pilot studies for newborns. At the moment, therapeutic options are only a few more than what seems obvious for oxidative protection: maintain adequate oxygenation, ventilation, and perfusion pressure. Emerging therapeutic approaches for perinatal ischemia provide an encouraging outlook on new antioxidant protective opportunities.

Vitamins A, C, and E

Vitamins A, C, and E are essential nutrients and are considered the most important antioxidants obtained through the diet. Vitamin A acts as an antioxidant by neutralizing the reactive molecules without modifying and interrupting oxidant chain reactions [9].

The antioxidant actions of vitamin E (the α, β, γ tocopherols) lie in its ability to be incorporated into biological membranes to stabilize and protect against lipid peroxidation [10], while the antioxidant properties of vitamin C (ascorbic acid) arise from the ability to act as an electron donor, thereby preventing other agents from becoming oxidized and quenching an overproduction of FR. The literature regarding the protective benefits of vitamins A, C, and E in the perinatal period is limited, but in vitro evidence suggests that, in adult and fetal rat brain cultures, vitamin E decreases lipid peroxidation and increases survival and neuritic extension of neurons [11, 12]. In vivo, prophylactic administration of vitamin E before hypoxia-ischemia is able to decrease the incidence of IVH [13]. Protective effects on retinopathy of prematurity (ROP) have also been reported with a reduction of ROP III+from 5.3 to 2.8 % [14]. In a mouse model of Down syndrome, α -tocopherol suppresses lipid peroxidation in the hippocampus and ameliorates behavioral and cognitive impairments [15]. α-Tocopherol has also been shown to have anti-inflammatory properties. Administration of α-tocopherol, particularly in large doses, has been shown to decrease the release of pro-inflammatory cytokines from cell lines exposed to lipopolysaccharide [16].

The use of vitamin A supplementation is effective for the reduction of bronchopulmonary dysplasia incidence in very low birth weight infants in modern NICU's [17]. In contrast, a meta-analysis of five trials in developing countries, in a population of newborns, showed no survival benefit of neonatal vitamin A supplementation, but confidence intervals were wide and did not exclude the possibility of an important benefit (relative risk 0·92, 95 % CI 0·75—1·12) [18]. A large cluster randomized controlled trial, in rural Bangladesh, reports that the use of weekly vitamin A or beta carotene in pregnant women, compared with placebo, did not reduce all-cause of maternal, fetal, or infant mortality [19].

Ascorbate deficiency in the postnatal mouse brain (in the presence of normal GSH levels) leads to diminished motor functions, yet an exaggerated response to a dopaminergic agonist [20]. Ascorbate antioxidant effects are enhanced in conjunction with vit E. When vitamin E is oxidized, it forms α -tocopherol radical which is harmful, but vitamin C is able to mediate the return of α-tocopherol radical to α-tocopherol, thus regenerating α-tocopherol concentrations in plasma [21]. In support to these findings, a study of transient intrauterine ischemia in pregnant rats showed that either vitamin E or vitamin C treatment alone, started before the ischemic insult, was able to decrease oxidative mitochondrial impairment in the fetal brain, but the improvement was greater when vitamins were administered together [22].

On a cautionary note vitamin C has both pro-oxidant or antioxidant effects in vitro; it is able to inhibit protein synthesis and induce late neuronal death [23–25]. Similarly, vitamin E may induce apoptosis in the absence of OS [26], potentially limiting its protective effects only to situations when OS is established. It is hypothesized that protecting the fetus through the pretreatment of the mother could in itself be beneficial and without any additional risk burden on either the mother or her baby, although the possibility of toxicity of these agents in the absence of OS was also postulated. Prenatal antioxidant treatment with vitamin C protects against placental OS in rats' hypoxic pregnancies [27]. It may provide a useful intervention to improve placental function and protect fetal growth in pregnancy complicated by fetal hypoxia [27]. Markers of OS and placental dysfunction are significantly reduced in the maternal circulation following supplementation of vitamins C and E [28].

A large trial also showed that administration of vitamins C and E in the second half of pregnancy, to women at risk for preeclampsia, significantly decreases biochemical markers of preeclampsia and decreases the proportion of women who develop the disease when compared to placebo [29]. However, a recent systematic review and meta-analysis demonstrates that there is no difference in the rates of preeclampsia in women receiving antioxidant vitamin supplements, nor were there any differences in the maternal, fetal, or neonatal complications [30]. No trial has yet reported adverse effects on long-term neurodevelopmental outcome of these babies.

Erythropoietin

Erythropoietin (Epo) is directly involved in the prevention of OS with generation of antioxidant enzymes, inhibition of nitric oxide production, and decrease of lipid peroxidation [31]. Moreover, Epo can decrease the production of pro-inflammatory cytokines and of the associated apoptotic injury, as it happens in adult stroke and neonatal models of hypoxia-ischemia treated with rEPO [32, 33]. Higher circulating Epo concentrations might produce more benefits than lower concentrations [34]. In various experimental models, Epo demonstrates a neuroprotective effect particularly after neuronal damage related to ischemia-reperfusion events [35]. Early treatment after HI with a high dose of Epo (5,000 U/kg) reduces tissue loss, preserving brain volume [36], and enhances neurogenesis, probably through a shift from astrocytic to neuronal cell fate [37]. A therapeutic strategy with lower multiple Epo doses, such as 1,000 U/kg, did not result in significant neuroprotection from early neuronal damage, even when combined with deferoxamine, an iron chelator which has been shown to decrease OS [38].

A recent randomized prospective study reported that repeated, low doses (300 or 500 U/kg every day) of Epo were safe and resulted in improved neurological outcome for patients with moderate neonatal HIE at 18 months of age [39]. A high survival rate with no or minor cerebral sequelae was observed in adult patients treated with hypothermia and early high doses of Epo therapy (40 000 IU) for the first 48 h, soon after resuscitation from cardiac arrest, in a small pilot study [40]. However, safety concerns appeared from some adult Epo trials reporting adverse effects related to vascular thrombosis, intracerebral hemorrhage, and brain edema [41].

Inhibitors of NOS

Nitric oxide (NO) is a free radical that is formed in high concentrations during and after hypoxia-ischemia. Three enzymes can catalyze the formation of NO: neuronal NOS (nNOS), inducible NOS (iNOS), and the endothelial NOS (eNOS) [42]. NO can react with superoxide to form peroxynitrite, which can cause nitration of proteins, predominantly on tyrosine residues [43].

Iminobiotin inhibits both the neuronal and inducible isoforms of nitric oxide synthase [44]. Otherwise, in vivo, it provides neuroprotection probably hindering apoptotic pathways. Nijboer et al demonstrated that treatment with 2-iminobiotin provided gender-specific long- and short-term neuroprotection in female newborn rats with hypoxia-ischemia via inhibition of the cytochrome c-caspase 3 neuronal death pathway [45].

Allopurinol

Allopurinol and its metabolite oxypurinol are inhibitors of xanthine oxidase, the enzyme involved in superoxide production especially during reperfusion damage (Fig. 22.3). Allopurinol has also additional effects, directly scavenging the toxic hydroxyl free radical and mainly chelating the non-bound protein iron (NBPI), particularly at high doses [46]. Allopurinol is converted into oxypurinol, which crosses the blood-brain barrier more easily. In asphyxiated infants who received allopurinol, NO serum level decreased significantly from 24 h to 72–96 h after birth [47]. The treated newborns had also better neurodevelopmental outcomes. Long-term neuroprotective effects of allopurinol administered after moderate perinatal asphyxia were also demonstrated in a secondary analysis of two randomized controlled trials [48]. Allopurinol may also be promising when administrated immediately before delivery in suspected fetal asphyxia. In a recent pilot study, Torrence et al. administrated to 53

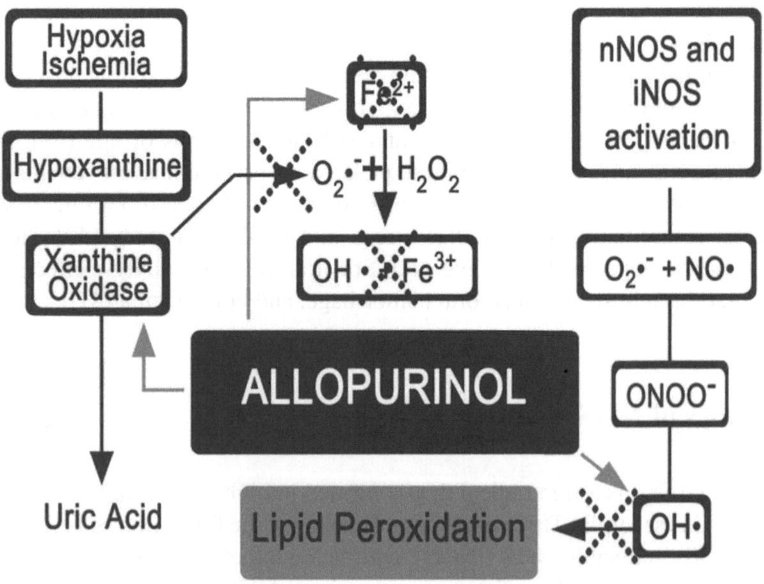

Fig. 22.3 Antioxidant effects of allopurinol

pregnant women in labor (54 fetuses with a gestational age >36 weeks and signs of fetal hypoxia) 500 mg of allopurinol or placebo intravenously. They found a reducing effect on biomarkers of neuronal damage and NPBI [49]. It is possible that allopurinol has no positive effect when administered late and at low doses [50]. Much larger trials are needed, to assess allopurinol as an adjunct to therapeutic hypothermia in infants with moderate and severe encephalopathy and to exclude important effects on mortality and adverse long-term neurodevelopmental outcomes [51].

Albumin

Albumin is the main protein of plasma, representing about 60 % of all plasma proteins. Marzocchi et al. demonstrated albumin carbonylation in newborns with higher NBPI levels and poor neurodevelopmental outcome [52]. Since NBPI may produce hydroxyl radicals through the Fenton reaction, the major target of OS induced by NBPI is its carrier, albumin. As albumin is a major extracellular antioxidant, its susceptibility to oxidation can be expected to decrease plasma antioxidant defenses and increase the likelihood of tissue damage due to OS in the newborn. Nitrated albumin was found significantly increased in patients who developed moderate or severe encephalopathy compared to those who had a normal neurological evolution or developed mild encephalopathy [53]. There is evidence that albumin significantly enhances neurological function and may decrease brain edema and infarction if administered 4 h after ischemia occurrence in adult rats [54]. In clinical trials, it is observed that administration of albumin may cause side effects on the lungs [55]. To reduce this side effect, albumin in low doses may be administered in association with docosanoids. Docosanoids are derivates of docosahexaenoic acid (DHA), which is a major product of the oxidative lipid degradation of the membrane after cerebral ischemia; its bioproducts can inhibit the infiltration of leukocytes and reduce expression of NF-κB. DHA treatment before the insult confers neuroprotection in a rat model of cerebral hypoxia-ischemia [56]. DHA improves function without affecting the brain volume loss in a rat model of hypoxia-ischemia and inflammation-induced perinatal brain injury [57].

Deferoxamine

Deferoxamine is a chelating agent and its target is the formation of hydroxyl radicals from free iron during reperfusion. Deferoxamine can cross the blood-brain barrier and chelate NBPI, thus reducing the severity of brain injury and improving cerebral metabolism in animal models of hypoxia-ischemia [58]. Deferoxamine treatment in lambs limited the hypoxia-ischemia induced NBPI increase in plasma and in cortical tissue [7, 8]. Negative effects on hemodynamics when administrated at high doses in preterm baboons have been observed [59].

Magnesium Sulfate

Magnesium is essential for key cellular processes, like glycolysis, oxidative phosphorylation, protein synthesis, and DNA and RNA aggregation [60]. Moreover, magnesium can influence mechanisms implicated in cell death due to the production of pro-inflammatory cytokines during the inflammatory response ([61]), and through the noncompetitive voltage-dependent inhibition of the NMDA-type glutamate receptor, it can reduce calcium entry in the cells leading to the prevention of the excitotoxic calcium-induced injury [62]. It also has anticonvulsant properties and hemodynamic effects by increasing cerebral blood flow [63]. Recent evidences demonstrate that fetal exposure to magnesium sulfate in women at risk for preterm delivery significantly reduces the risk of cerebral palsy without increasing the risk of death [64–66].

In adult rats with cerebral ischemia, magnesium sulfate was demonstrated to be more effective, increasing neuronal survival rate, than either treatment used alone [67, 68]. Magnesium has important side effects: it can provoke hemodynamic instability, hypotension, bradycardia, and delayed intraventricular conduction, including complete atrioventricular block [60]. It appears that magnesium sulfate is ineffective in delaying birth or preventing preterm birth when used as a tocolytic [69]. Furthermore, high cumulative doses of magnesium sulfate may be associated with increased infant mortality [70]. The evidence to date confirms the efficacy of magnesium sulfate therapy for women with eclampsia and preeclampsia and for reduction of cerebral palsy in the setting of threatened preterm delivery. However, magnesium sulfate should not be used as a tocolytic for preterm labor.

N-Acetylcysteine

N-Acetylcysteine (NAC) is a scavenger of oxygen radicals and restores intracellular glutathione levels, attenuating reperfusion injury and decreasing inflammation and nitric oxide (NO) production in adult models of stroke [71]. It has low toxicity and it is able to cross the placenta and the blood-brain barrier. In an animal model of hypoxia-ischemia, NAC treatment combined with systemic hypothermia prevented brain tissue loss, with increased myelin expression and improved the short-term functional outcomes of labyrinthine and cerebellar integration [72]. Consistently with this, Cakir et al. reported that after spinal cord ischemia, NAC and hypothermia alone were associated with limited protective effects, whereas the combination of NAC and hypothermia resulted in the highly significant recovery of the spinal cord function [73]. NAC may also have anti-inflammatory effects, by reducing intracerebral levels of tumor necrosis factor-α, interleukin-1β, and inducible NO synthase, when administered in pregnant female rats 2 h before the administration of endotoxin lipopolysaccharide [74]. In a randomized clinical trial on preterm newborns, therapy with NAC by continuous infusion during the first 6 days after birth, resulted in no changes in bronchopulmonary dysplasia incidence (the primary outcome), nor in incidence of periventricular leukomalacia or intraventricular hemorrhage, in the treated babies ([75]).

Melatonin

Melatonin is the main product of the pineal gland, with high antioxidant and anti-inflammatory properties [76, 77], and is synthesized starting from tryptophan [78]. When synthesized, it is quickly released in all biological fluids such as bile, cerebrospinal fluid [79], blood, saliva [80], semen [81], and amniotic fluid [82], and it functions as time-giver in the regulation of the circadian rhythm [83]. This rhythm in mammals is generated by an endogenous circadian master clock located in the suprachiasmatic nucleus (SCN) of the hypothalamus. It participates in several neuroendocrine and physiological processes and is considered also a tissue factor, a paracoid, an autocoid, an antioxidant, and sometimes a hormone depending on its physiological actions. Among melatonin effects in humans, it plays a role in sleep regulation, in the sexual maturation, and in thermoregulation [84]. Indeed, melatonin has a broad antioxidant spectrum, a direct or indirect effect, and an anti-inflammatory property [85, 86]. It is particularly interesting for its ability to cross all physiological barriers [87] and to be widely distributed in tissues, cells, and subcellular compartments. Working as a direct antioxidant, it is able to scavenge dangerous FR, such as OH•, O_2–, H_2O_2 and ONOO–, and as an indirect one, it induces the production of antioxidant enzymes, including glutathione peroxidase, glutathione reductase, glucose-6-phosphate dehydrogenase, and superoxide dismutase and increases the efficiency of mitochondrial electron transport [20, 88, 89].

Particularly its ability to scavenge the dangerous "·OH" is much higher than other antioxidants including mannitol, glutathione, and vitamin E [90, 91]. Moreover, melatonin has no pro-oxidant effects unlike many other antioxidants and does not interfere with the thrombolytic and neuroprotective actions of other drugs [92, 93]. In addition, melatonin can inhibit the expression of adhesion molecules therefore curbing PMN infiltration and tricking the cascade of inflammation [76]. Moreover, melatonin may inhibit the NF-κB expression, which is a nuclear transcription factor involved in inflammation development by interacting with its receptor MT1. Blocking NF-κB melatonin acts as an anti-inflammatory molecule through the inhibition of the biochemical events following NF-κB activation, such as the NO and prostaglandins production by iNOS and COX-2 [94]. Although many studies for supplementation of melatonin in pregnancy are from animals, recent evidences (e.g., that melatonin crosses the placenta) have suggested that supplements with melatonin might be a good way of prevention of preeclampsia in humans [95]. Furthermore, there is evidence that fetal growth restriction is associated with significant OS often resulting in small-for-gestational-age neonates [96, 97]. Melatonin is able to increase glutathione peroxidase activity in human chorion [98] and is able to improve placental efficiency, restore birth weight, and increase the expression of placental manganese superoxide dismutase and catalase in undernourished pregnancy in rats [99].

Melatonin could be a useful drug in preterm delivery, a condition highly susceptible to OS injury due to the need of oxygen use for neonatal resuscitation and to the immaturity of the antioxidant systems. Unfortunately, despite its anti-inflammatory and antioxidant functions, melatonin is not currently available for neonatal therapeutic trials, but when it was used as a compassionate drug in neonatal sepsis, bronchopulmonary dysplasia, and neonatal asphyxia, preliminary findings supported the possibility for its wider evaluation in perinatal medicine [100–102]. Other studies found also that melatonin administration after HI in immature rats has an excellent and long-lasting benefit on ischemic outcomes and could represent a potentially safe approach to perinatal brain damage in humans [103]. We demonstrated melatonin neuroprotective effects when administrated before the neonatal hypoxic-ischemic insult in an animal model with a reduction of oxidative damage [104] and a long-term neuroprotective effect. In addition, postischemia intraperitoneal administration of melatonin significantly protected the brain from injury and reduced infarct volume, mainly in the hippocampus and cerebral cortex [103]. Melatonin treatment also improved functional recovery into adulthood [103].

Based on all these data, melatonin is a well-documented multifunctional molecule that may be a useful therapeutic agent for the treatment of neonatal hypoxic-ischemic encephalopathy. Melatonin is safe, nontoxic, and available in pure form for human use. These results of animal experimental models and human case reports provide fundamental information on the need and potential usefulness of clinical trials to evaluate melatonin as a neuroprotective drug.

Lutein

Lutein belongs to the family of carotenoids. Studies conducted both in vitro and in vivo have identified several properties of lutein, showing a defensive action that occurs through the neutralization of FR and singlet oxygen. Studies report that lutein is able to reduce the risk of developing some ocular diseases or slow down their progression [105]. Lutein may play a role in tissue defense through functional mechanisms using the phenomenon of deactivation (quenching) of singlet oxygen and of reactive oxygen species [106]. This action gives the molecule different activities: antioxidant, anti-inflammatory, and anti-apoptotic properties [107]. Lutein and its isomer zeaxanthin in the macular pigment may play an important role in protecting the eyes of the newborn from the damage of light, thanks to their ability to absorb blue light and their antioxidant action [108]. Our group demonstrated that lutein administration in newborns increased the levels of biological antioxidant potential (BAP), decreasing OS marker levels in healthy term newborns, suggesting potential for its testing in clinical trials to protect newborns from perinatal OS [109]. There was also a difference between breastfed and formula-fed infants. Breastfed infants had higher mean serum lutein concentrations than infants who consumed formula unfortified with lutein. These data suggested that approximately four times more lutein is needed in infant formula than in human milk to achieve similar serum lutein concentrations among breastfed and formula-fed infants [110]. Manzoni et al. found no treatment-related adverse effect in 229 preterm infants supplemented with lutein. They found no significant differences in the threshold of ROP between treated versus not treated infants [111]. The same occurred for NEC and BPD. Interestingly, they found that the progression rate from early ROP stages to threshold ROP was decreased by 50 % [111], showing how lutein/zeaxanthin supplementation in preterm infants is well tolerated and can interfere with ROP progression. Rubin et al. assessed lutein safety, and they demonstrated that supplemented infants had lower plasma C-reactive protein and that plasma lutein levels correlated with the full field electroretinogram-saturated response amplitude in rod photoreceptors in a cohort of 203 preterm newborns. Finally, the supplemented group also showed greater rod photoreceptor sensitivity [112]. All these data suggest that lutein may be a promising drug against oxidative injury.

Conclusion

The relationship between FR production and perinatal FR diseases is complex. Clearly, FR damage results from many pathogenic influences. Hypoxia, ischemia-reperfusion, neutrophil and macrophage activation, Fenton reaction, endothelial cell xanthine oxidase, phospholipid metabolism, nitric oxide, mitochondrial oxidative metabolism impairment, and proteolytic pathways are all implicated. FR production by such mechanisms contributes to the pathogenesis of perinatal diseases, but each is only one of many factors responsible for disease occurrence. Each step in the oxidative injury cascade has become a potential target for therapeutic intervention. To date, studies have not adequately addressed the consequences of altering the oxidative or immune balance when OS is suspected. The administration of antioxidants for suspected OS is still not accepted for clinical use because treatments would be usually started after resuscitation of an asphyxiated newborn. The challenge for the future is the early identification of high-risk babies to target an antioxidant therapy preventing OS injury. Furthermore, it would be important to better clarify the drug intervention strategy (with antioxidants, anti-inflammatory, and anti-apoptotic properties) and to verify its impact on infants' health.

References

1. Halliwell B. Free radicals, antioxidants and human disease: curiosity, cause, or consequence? Lancet. 1994;344:721–4.
2. Thibeault D. The precarious antioxidant defenses of the preterm infant. Am J Perinatol. 2000;17:167–81.
3. Kowaltowski AJ, Vercesi AE. Mitochondrial damage induced by conditions of oxidative stress. Free Rad Biol Med. 1999;26:463–71.

4. McCord JM. Oxygen-derived free radicals in postischemic tissue injury. New Engl J Med. 1985;312:159–63.
5. Evans PJ, Evans R, Kovar IZ, Holton AF, Halliwell B. Bleomycin-detectable iron in the plasma of premature and full-term neonates. FEBS Lett. 1992;303:210–2.
6. Aruoma OI. Characterization of drugs as antioxidant prophylactic. Free Rad Biol Med. 1996;20:675–705.
7. Shadid M, Buonocore G, Groenendaal F, et al. Effect of deferoxamine and allopurinol on non protein bound iron concentrations in plasma and cortical brain tissue of newborn lambs following hypoxia ischemia. Neurosci Lett. 1998;22:5–8.
8. Shadid M, Moison R, Steendijk P, Hiltermann L, Berger HM, van Bel F. The effect of antioxidative combination therapy on post hypoxic-ischemic perfusion, metabolism, and electrical activity of the newborn brain. Pediatr Res. 1998;44(1):119–24.
9. Livrea MA, Tesoriere L, Bongiorno A, Pintaudi AM, Ciaccio M, Riccio A. Contribution of vitamin A to the oxidation resistance of human low density lipoproteins. Free Radic Biol Med. 1995;18(3):401–9.
10. Packer L, Weber SU, Rimbach G. Molecular aspects of alpha-tocotrienol antioxidant action and cell signalling. J Nutr. 2001;131:369S–73.
11. Shin SM, Razdan B, Mishra OP, Johnson L, Delivoria-Papadopoulos M. Protective effect of alpha-tocopherol on brain cell membrane function during cerebral cortical hypoxia in newborn piglets. Brain Res. 1994;653(1–2):45–50.
12. Osakada F, Hashino A, Kume T, Katsuki H, Kaneko S, Akaike A. Neuroprotective effects of alpha-tocopherol on oxidative stress in rat striatal cultures. Eur J Pharmacol. 2003;465:15–22.
13. Iwasa H, Aono T, Fukuzawa K. Protective effect of vitamin E on fetal distress induced by ischemia of the uteroplacental system in pregnant rats. Free Radic Biol Med. 1990;8(4): 393–400.
14. Raju TN, Langenberg P, Bhutani V, Quinn GE. Vitamin E prophylaxis to reduce retinopathy of prematurity: a reappraisal of published trials. J Pediatr. 1997;131(6):844–50.
15. Shichiri M, Yoshida Y, Ishida N, Hagihara Y, Iwahashi H, Tamai H, Niki E. α-Tocopherol suppresses lipid peroxidation and behavioral and cognitive impairments in the Ts65Dn mouse model of Down syndrome. Free Radic Biol Med. 2011;50(12):1801–11.
16. van Tits LJ, Demacker PN, de Graaf J, Hak-Lemmers HL, Stalenhoef AF. alpha-tocopherol supplementation decreases production of superoxide and cytokines by leukocytes ex vivo in both normolipidemic and hypertriglyceridemic individuals. Am J Clin Nutr. 2000;71:458–64.
17. Darlow BA, Graham PJ. Vitamin A supplementation to prevent mortality and short- and long-term morbidity in very low birthweight infants. Cochrane Database Syst Rev. 2011;10:CD000501.
18. Kirkwood B, Humphrey J, Moulton L, Martines J. Neonatal vitamin A supplementation and infant survival. Lancet. 2010;376(9753):1643–4.
19. West Jr KP, Christian P, Labrique AB, Rashid M, Shamim AA, Klemm RD, et al. Effects of vitamin A or beta carotene supplementation on pregnancy-related mortality and infant mortality in rural Bangladesh: a cluster randomized trial. JAMA. 2011;305(19):1986–95.
20. Chen YC, Tain YL, Sheen JM, Huang LT. Melatonin utility in neonates and children. J Formos Med Assoc. 2012;111(2):57–66.
21. Beyer RE. The role of ascorbate in antioxidant protection of biomembranes: interaction with vitamin E and coenzyme Q. J Bioenerg Biomembr. 1994;26:349–58.
22. Nakai A, Shibazaki Y, Taniuchi Y, Oya A, Asakura H, Koshino T, Araki T. Vitamins ameliorate secondary mitochondrial failure in neonatal rat brain. Pediatr Neurol. 2002;27:30–5.
23. Bass WT, Malati N, Castle MC, White LE. Evidence for the safety of ascorbic acid administration to the premature infant. Am J Perinatol. 1998;15(2):133–40.
24. Tan S, Parks DA. Preserving brain function during neonatal asphyxia. Clin Perinatol. 1999;26(3):733–47.
25. Hisanaga K, Sagar SM, Sharp FR. Ascorbate neurotoxicity in cortical cell culture. Ann Neurol. 1992;31(5):562–5.

26. Then SM, Mazlan M, Mat Top G, Wan Ngah WZ. Is vitamin E toxic to neuron cells? Cell Mol Neurobiol. 2009;29:485–96.
27. Richter HG, Camm EJ, Modi BN, Naeem F, Cross CM, Cindrova-Davies T, et al. Ascorbate prevents placental oxidative stress and enhances birth weight in hypoxic pregnancy in rats. J Physiol. 2012;590:1377–87.
28. Chappell LC, Seed PT, Kelly FJ, Briley A, Hunt BJ, Charnock-Jones DS, et al. Vitamin C and E supplementation in women at risk of preeclampsia is associated with changes in indices of oxidative stress and placental function. Am J Obstet Gynecol. 2002;187:777–84.
29. Roberts JM, Myatt L, Spong CY, et al. Vitamins C and E to prevent complications of pregnancy-associated hypertension. N Engl J Med. 2010;362:1282–91.
30. Conde-Agudelo A, Romero R, Kusanovic JP, Hassan SS. Supplementation with vitamins C and E during pregnancy for the prevention of preeclampsia and other adverse maternal and perinatal outcomes: a systematic review and metaanalysis. Am J Obstet Gynecol. 2011; 204:503.
31. Solaroglu I, Solaroglu A, Kaptanoglu E, Dede S, Haberal A, Beskonakli E, Kilinc K. Erythropoietin prevents ischemia-reperfusion from inducing oxidative damage in fetal rat brain. Childs Nerv Syst. 2003;19(1):19–22.
32. Kumral A, Baskin H, Gokmen N, Yilmaz O, Genc K, Genc S, et al. Selective inhibition of nitric oxide in hypoxic-ischemic brain model in newborn rats: is it an explanation for the protective role of erythropoietin? Biol Neonate. 2004;85(1):51–4.
33. Sun Y, Calvert JW, Zhang JH. Neonatal hypoxia/ischemia is associated with decreased inflammatory mediators after erythropoietin administration. Stroke. 2005;36(8):1672–8.
34. Bierer R, Peceny MC, Hartenberger CH, Ohls RK. Erythropoietin concentrations and neurodevelopmental outcome in preterm infants. Pediatrics. 2006;118(3):e635–40.
35. Maiese K, Li F, Chong ZZ. New avenues of explorations for erythropoietin. J Am Med Assoc. 2005;293:90–5.
36. Kellert BA, McPherson RJ, Juul SE. A comparison of high-dose recombinant erythropoietin treatment regimens in brain-injured neonatal rats. Pediatr Res. 2007;61:451–5.
37. Gonzalez FF, McQuillen P, Mu D, et al. Erythropoietin enhances long-term neuroprotection and neurogenesis in neonatal stroke. Dev Neurosci. 2007;29:321–30.
38. van der Kooij MA, Groenendaal F, Kavelaars A, Heijnen CJ, van Bel F. Combination of deferoxamine and erythropoietin: therapy for hypoxia–ischemia-induced brain injury in the neonatal rat? Neurosci Lett. 2009;451:109–13.
39. Zhu C, Kang W, Xy F, et al. Erythropoietin improved neurological outcomes in newborns with hypoxic–ischemic encephalopathy. Pediatrics. 2009;124:e218–26.
40. Cariou A, Claessens Y-E, Pène F, et al. Early high dose erythropoietin therapy and hypothermia after out-of-hospital cardiac arrest: a matched control study. Resuscitation. 2008;76: 397–404.
41. Ehrenreich H, Weissenborn K, Prange H, et al. For the EPO Stroke Trial Group. Recombinant human erythropoietin in the treatment of acute ischemic stroke. Stroke. 2009;40(12): e647–56.
42. Samdani AF, Dawson TM, Dawson VL. Nitric oxide synthase in models of focal ischemia. Stroke. 1997;28(6):1283–8.
43. Beckman JS, Koppenol WH. Nitric oxide, superoxide, and peroxynitrite: the good, the bad, and ugly. Am J Physiol. 1996;271:C1424–37.
44. Sup SJ, Green BG, Grant SK. 2-Iminobiotin is an inhibitor of nitric oxide synthases. Biochem Biophys Res Commun. 1994;204(2):962–8.
45. Nijboer CH, Heijnen CJ, Groenendaal F, May MJ, van Bel F, Kavelaars A. Strong neuroprotection by inhibition of NF-kappaB after neonatal hypoxia-ischemia involves apoptotic mechanisms but is independent of cytokines. Stroke. 2008;39(7):2129–37.
46. Marro PJ, Mishra OP, Delivoria-Papadopoulos M. Effect of allopurinol on brain adenosine levels during hypoxia in newborn piglets. Brain Res. 2006;1073–1074:444–50.

47. Gunes T, Ozturk MA, Koklu E, Kose K, Gunes I. Effect of allopurinol supplementation on nitric oxide levels in asphyxiated newborns. Pediatr Neurol. 2007;36(1):17–24.
48. Kaandorp JJ, van Bel F, Veen S, Derks JB, Groenendaal F, Rijken M, Roze E, Venema MM, Rademaker CM, Bos AF, Benders MJ. Long-term neuroprotective effects of allopurinol after moderate perinatal asphyxia: follow-up of two randomised controlled trials. Arch Dis Child Fetal Neonatal Ed. 2012;97:162–6.
49. Torrance HL, Benders MJ, Derks JB, Rademaker CM, Bos AF, Van Den Berg P, Longini M, Buonocore G, et al. Maternal allopurinol during fetal hypoxia lowers cord blood levels of the brain injury marker S-100B. Pediatrics. 2009;124(1):350–7.
50. Kelen D, Robertson NJ. Experimental treatments for hypoxic ischaemic encephalopathy. Early Hum Dev. 2010;86(6):369–77.
51. Chaudhari T, McGuire W. Allopurinol for preventing mortality and morbidity in newborn infants with hypoxic-ischaemic encephalopathy. Cochrane Database Syst Rev. 2012. doi:10.1002/14651858.CD006817.
52. Marzocchi B, Perrone S, Paffetti P, Magi B, Bini L, Tani C, Longini M, Buonocore G. Nonprotein-bound iron and plasma protein oxidative stress at birth. Pediatr Res. 2005;58(6): 1295–9.
53. Wayenberg JL, Ransy V, Vermeylen D, Damis E, Bottari SP. Nitrated plasma albumin as a marker of nitrative stress and neonatal encephalopathy in perinatal asphyxia. Free Radic Biol Med. 2009;47(7):975–82.
54. Liu Y, Belayev L, Zhao W, Busto R, Belayev A, Ginsberg MD. Neuroprotective effect of treatment with human albumin in permanent focal cerebral ischemia: histopathology and cortical perfusion studies. Eur J Pharmacol. 2001;428(2):193–201.
55. Ginsberg MD, Hill MD, Palesch YY, Ryckborst KJ, Tamariz D. The ALIAS Pilot Trial: a dose-escalation and safety study of albumin therapy for acute ischemic stroke-I: physiological responses and safety results. Stroke. 2006;37(8):2100–6.
56. Berman DR, Mozurkewich E, Liu Y, Barks J. Docosahexaenoic acid pretreatment confers neuroprotection in a rat model of perinatal cerebral hypoxia-ischemia. Am J Obstet Gynecol. 2009;200:305.e1-6.
57. Berman DR, Liu YQ, Barks J, Mozurkewich E. Docosahexaenoic acid confers neuroprotection in a rat model of perinatal hypoxia-ischemia potentiated by Escherichia coli lipopolysaccharide-induced systemic inflammation. Am J Obstet Gynecol. 2010;202(5):469.e1-6.
58. Peeters-Scholte C, Braun K, Koster J, Kops N, Blomgren K, Buonocore G, et al. Effects of allopurinol and deferoxamine on reperfusion brain injury in newborn piglets after neonatal hypoxia-ischemia. Pediatr Res. 2003;54(4):516–22.
59. deLemos RA, Roberts RJ, Coalson JJ, deLemos JA, Null Jr DM, Gerstmann DR. Toxic effects associated with the administration of deferoxamine in the premature baboon with hyaline membrane disease. Am J Dis Child. 1990;144:915–9.
60. Marret S, Doyle LW, Crowther CA, Middleton P. Antenatal magnesium sulphate neuroprotection in the preterm infant. Semin Fetal Neonatal Med. 2007;12(4):311–7.
61. Shogi T, Miyamoto A, Ishiguro S, Nishio A. Enhanced release of IL-1beta and TNF-alpha following endotoxin challenge from rat alveolar macrophages cultured in low-mg(2+) medium. Magnes Res. 2003;16(2):111–9.
62. Ovbiagele B, Kidwell CS, Starkman S, Saver JL. Potential role of neuroprotective agents in the treatment of patients with acute ischemic stroke. Curr Treat Options Cardiovasc Med. 2003;6:441–9.
63. Kim CR, Oh W, Stonestreet BS. Magnesium is a cerebrovasodilator in newborn piglets. Am J Physiol. 1997;272(1 Pt 2):H511–6.
64. Costantine MM, Weiner SJ, Eunice Kennedy Shriver National Institute of Child Health and Human Development Maternal-Fetal Medicine Units Network. Effects of antenatal exposure to magnesium sulfate on neuroprotection and mortality in preterm infants: a meta-analysis. Obstet Gynecol. 2009;114(2 Pt 1):354–64.

65. Doyle LW, Crowther CA, Middleton P, Marret S, Rouse D. Magnesium sulphate for women at risk of preterm birth for neuroprotection of the fetus. Cochrane Database Syst Rev. 2009;1:CD004661.
66. Conde-Agudelo A, Romero R. Antenatal magnesium sulfate for the prevention of cerebral palsy in preterm infants less than 34 weeks' gestation: a systematic review and metaanalysis. Am J Obstet Gynecol. 2009;200(6):595–609.
67. Zhu H, Meloni BP, Bojarski C, Knuckey MW, Knuckey NW. Post-ischemic modest hypothermia (35 degrees C) combined with intravenous magnesium is more effective at reducing CA1 neuronal death than either treatment used alone following global cerebral ischemia in rats. Exp Neurol. 2005;193(2):361–8.
68. Zhu H, Meloni BP, Moore SR, Majda BT, Knuckey NW. Intravenous administration of magnesium is only neuroprotective following transient global ischemia when present with post-ischemic mild hypothermia. Brain Res. 2004;1014(1–2):53–60.
69. Ramsey PS, Rouse DJ. Magnesium sulfate as a tocolytic agent. Semin Perinatol. 2001; 25(4):236–47.
70. Azria E, Tsatsaris V, Goffinet F, Kayem G, Mignon A, Cabrol D. Magnesium sulfate in obstetrics: current data. J Gynecol Obstet Biol Reprod. 2004;33(6 Pt 1):510–7.
71. Khan M, Sekhon B, Jatana M, et al. Administration of N-acetylcysteine after focal cerebral ischemia protects brain and reduces inflammation in a rat model of experimental stroke. J Neurosci Res. 2004;76:519–27.
72. Jatana M, Singh I, Singh AK, Jenkins D. Combination of systemic hypothermia and N-acetylcysteine attenuates hypoxic–ischemic brain injury in neonatal rats. Pediatr Res. 2006;59:684–9.
73. Cakir O, Erdem K, Oruc A, Kilinc N, Eren N. Neuroprotective effect of N-acetylcysteine and hypothermia on the spinal cord ischemia–reperfusion injury. Cardiovasc Surg. 2003;11:375–9.
74. Paintlia MK, Paintlia AS, Barbosa E, Singh I, Singh AK. N-Acetylcysteine prevents endotoxin-induced degeneration of oligodendrocyte progenitors and hypomyelination in developing rat brain. J Neurosci Res. 2004;78:347–61.
75. Ahola T, Lapatto R, Raivio KO, et al. N-Acetylcysteine does not prevent bronchopulmonary dysplasia in immature infants: a randomized controlled trial. J Pediatr. 2003;143:713–9.
76. Welin AK, Svedin P, Lapatto R, Sultan B, Hagberg H, Gressens P, et al. Melatonin reduces inflammation and cell death in white matter in the mid-gestation fetal sheep following umbilical cord occlusion. Pediatr Res. 2007;61(2):153–8.
77. Tütüncüler F, Eskiocak S, Başaran UN, Ekuklu G, Ayvaz S, Vatansever U. The protective role of melatonin in experimental hypoxic brain damage. Pediatr Int. 2005;47(4):434–9.
78. Vijayalaxmi Jr TCR, Reiter RJ, Herman TS. Melatonin: from basic research to cancer treatment clinics. J Clin Oncol. 2002;20(10):2575–601.
79. Rousseau A, Petrén S, Plannthin J, Eklundh T, Nordin C. Serum and cerebrospinal fluid concentrations of melatonin: a pilot study in healthy male volunteers. J Neural Transm. 1999;106(9–10):883–8.
80. Vakkuri O, Leppäluoto J, Kauppila A. Oral administration and distribution of melatonin in human serum, saliva and urine. Life Sci. 1985;37(5):489–95.
81. Bornman MS, Schulenburg GW, Reif S, Oosthuizen JM, De Wet EH, Luus HG. Seminal plasma melatonin and semen parameters. S Afr Med J. 1992;81(9):485–6.
82. Kivelä A, Kauppila A, Leppäluoto J, Vakkuri O. Serum and amniotic fluid melatonin during human labor. J Clin Endocrinol Metab. 1989;69(5):1065–8.
83. Agez L, Laurent V, Guerrero HY, Pévet P, Masson-Pévet M, Gauer F. Endogenous melatonin provides an effective circadian message to both the suprachiasmatic nuclei and the pars tuberalis of the rat. J Pineal Res. 2009;46(1):95–105.
84. Shanahan TL, Czeisler CA. Physiological effects of light on the human circadian pacemaker. Semin Perinatol. 2000;24(4):299–320.
85. Paradies G, Petrosillo G, Paradies V, Reiter RJ, Ruggiero FM. Melatonin, cardiolipin and mitochondrial bioenergetics in health and disease. J Pineal Res. 2010;48(4):297–310.

86. Tan DX, Manchester LC, Terron MP, Flores LJ, Tamura H, Reiter RJ. Melatonin as a naturally occurring co-substrate of quinone reductase-2, the putative MT3 melatonin membrane receptor: hypothesis and significance. J Pineal Res. 2007;43(4):317–20.
87. Ceraulo L, Ferrugia M, Tesoriere L, Segreto S, Livrea MA, Turco Liveri V. Interactions of melatonin with membrane models: portioning of melatonin in AOT and lecithin reversed micelles. J Pineal Res. 1999;26(2):108–12.
88. Reiter RJ, Tan DX, Burkhardt S. Reactive oxygen and nitrogen species and cellular and organismal decline: amelioration with melatonin. Mech Ageing Dev. 2002;123(8):1007–19.
89. Acuña-Castroviejo D, Martín M, Macías M, Escames G, León J, Khaldy H, et al. Melatonin, mitochondria, and cellular bioenergetics. J Pineal Res. 2001;30:65–74.
90. Tamura H, Takasaki A, Taketani T, Tanabe M, Kizuka F, Lee L, et al. Melatonin as a free radical scavenger in the ovarian follicle. Endocr J. 2013;60(1):1–13.
91. Reiter RJ, Tan DX, Manchester LC, Qi W. Biochemical reactivity of melatonin with reactive oxygen and nitrogen species: a review of the evidence. Cell Biochem Biophys. 2001;34(2):237–56.
92. Mishra OP, Delivoria-Papadopoulos M. Lipid peroxidation in developing fetal guinea pig brain during normoxia and hypoxia. Brain Res Dev Brain Res. 1989;45(1):129–35.
93. Esposito E, Cuzzocrea S. Antiinflammatory activity of melatonin in central nervous system. Curr Neuropharmacol. 2010;8(3):228–42.
94. Carrillo-Vico A, García-Mauriño S, Calvo JR, Guerrero JM. Melatonin counteracts the inhibitory effect of PGE2 on IL-2 production in human lymphocytes via its mt1 membrane receptor. FASEB J. 2003;17(6):755–77.
95. Briceño-Pérez C, Briceño-Sanabria L, Vigil-De Gracia P. Prediction and prevention of preeclampsia. Hypertens Pregnancy. 2009;28(2):138–55.
96. Gupta P, Narang M, Banerjee BD, Basu S. Oxidative stress in term small for gestational age neonates born to undernourished mothers: a case control study. BMC Pediatr. 2004;4:14.
97. Longini M, Perrone S, Kenanidis A, Vezzosi P, Petraglia F, Buonocore G, et al. Isoprostanes in amniotic fluid: a predictive marker for fetal growth restriction in pregnancy. Free Radic Biol Med. 2005;38(11):1537–41.
98. Okatani Y, Wakatsuki A, Shinohara K, Kaneda C, Fukaya T. Melatonin stimulates glutathione peroxidase activity in human chorion. J Pineal Res. 2001;30(4):199–205.
99. Richter HG, Hansell JA, Raut S, Giussani DA. Melatonin improves placental efficiency and birth weight and increases the placental expression of antioxidant enzymes in undernourished pregnancy. J Pineal Res. 2009;46(4):357–64.
100. Gitto E, Reiter RJ. Early indicators of chronic lung disease in preterm infants with respiratory distress syndrome and their inhibition by melatonin. J Pineal Res. 2004;36:250e5.
101. Gitto E, Reiter RJ, Cordaro SP, La Rosa M, Chiurazzi P, Trimarchi G, et al. Oxidative and inflammatory parameters in respiratory distress syndrome of preterm newborns: beneficial effects of melatonin. Am J Perinatol. 2004;21:209e16.
102. Gitto E, Reiter RJ, Sabatino G, Buonocore G, Romeo C, Gitto P, et al. Correlation among cytokines, bronchopulmonary dysplasia and modality of ventilation in preterm newborns: improvement with melatonin treatment. J Pineal Res. 2005;39:287e93.
103. Carloni S, Perrone S, Buonocore G, Longini M, Proietti F, Balduini W. Melatonin protects from the long-term consequences of a neonatal hypoxic-ischemic brain injury in rats. J Pineal Res. 2008;44(2):157–64.
104. Buonocore G, Perrone S, Bracci R. Free radicals and brain damage in the newborn. Biol Neonate. 2001;79(3–4):180–6.
105. Kijlstra A, Tian Y, Kelly ER, Berendschot TT. Lutein: more than just a filter for blue light. Prog Retin Eye Res. 2012;31(4):303–15.
106. Vijayapadma V, Ramyaa P, Pavithra D, Krishnaswamy R. Protective effect of lutein against benzo(a)pyrene-induced oxidative stress in human erythrocytes. Toxicol Ind Health. 2014;30(3):284–93.
107. Li SY, Fung FK, Fu ZJ, Wong D, Chan HH, Lo AC. Anti-inflammatory effects of lutein in retinal ischemic/hypoxic injury: in vivo and in vitro studies. Invest Ophthalmol Vis Sci. 2012;53(10):5976–84.

108. Moukarzel AA, Bejjani RA, Fares FN. Xanthophylls and eye health of infants and adults. J Med Liban. 2009;57:261–7.
109. Perrone S, Longini M, Marzocchi B, Picardi A, Bellieni CV, Proietti F, et al. Effects of lutein on oxidative stress in the term newborn: a pilot study. Neonatology. 2010;97(1):36–40.
110. Bettler J, Zimmer JP, Neuringer M, DeRusso PA. Serum lutein concentrations in healthy term infants fed human milk or infant formula with lutein. Eur J Nutr. 2010;49(1):45–51.
111. Manzoni P, Guardione R, Bonetti P, Priolo C, Maestri A, Mansoldo C, et al. Lutein and zeaxanthin supplementation in preterm very low-birth-weight neonates in neonatal intensive care units: a multicenter randomized controlled trial. Am J Perinatol. 2013;30(1):25–32.
112. Rubin LP, Chan GM, Barrett-Reis BM, Fulton AB, Hansen RM, Ashmeade TL, et al. Effect of carotenoid supplementation on plasma carotenoids, inflammation and visual development in preterm infants. J Perinatol. 2012;32(6):418–24.

Index

P.A. Dennery et al. (eds.), *Perinatal and Prenatal Disorders*,
Oxidative Stress in Applied Basic Research and Clinical Practice,
DOI 10.1007/978-1-4939-1405-0

MIX
Papier aus verantwortungsvollen Quellen
Paper from responsible sources
FSC® C105338

If you have any concerns about our products,
you can contact us on
ProductSafety@springernature.com

In case Publisher is established outside the EU,
the EU authorized representative is:
Springer Nature Customer Service Center GmbH
Europaplatz 3, 69115 Heidelberg, Germany

Printed by Libri Plureos GmbH
in Hamburg, Germany